Lecture Notes in Civil Engineering

Volume 535

Lecture Notes in Civil Engineering (LNCE) publishes the latest developments in Civil Engineering—quickly, informally and in top quality. Though original research reported in proceedings and post-proceedings represents the core of LNCE, edited volumes of exceptionally high quality and interest may also be considered for publication. Volumes published in LNCE embrace all aspects and subfields of, as well as new challenges in, Civil Engineering. Topics in the series include:

- Construction and Structural Mechanics
- Building Materials
- Concrete, Steel and Timber Structures
- Geotechnical Engineering
- Earthquake Engineering
- Coastal Engineering
- Ocean and Offshore Engineering; Ships and Floating Structures
- Hydraulics, Hydrology and Water Resources Engineering
- Environmental Engineering and Sustainability
- Structural Health and Monitoring
- Surveying and Geographical Information Systems
- Indoor Environments
- Transportation and Traffic
- Risk Analysis
- Safety and Security

To submit a proposal or request further information, please contact the appropriate Springer Editor:

- Pierpaolo Riva at pierpaolo.riva@springer.com (Europe and Americas);
- Swati Meherishi at swati.meherishi@springer.com (Asia—except China, Australia, and New Zealand);
- Wayne Hu at wayne.hu@springer.com (China).

All books in the series now indexed by Scopus and EI Compendex database!

Dayong Li · Yu Zhang

Editors

Advances in Frontier Research on Engineering Structures II

 Springer

Editors
Dayong Li
China University of Petroleum
Qingdao, China

Yu Zhang
China University of Petroleum
Qingdao, China

ISSN 2366-2557   ISSN 2366-2565 (electronic)
Lecture Notes in Civil Engineering
ISBN 978-981-97-6237-8   ISBN 978-981-97-6238-5 (eBook)
https://doi.org/10.1007/978-981-97-6238-5

This work was supported by Guangzhou KEO Info Technology Co., Ltd.

This Springer imprint is published by the registered company Springer Nature Singapore Pte Ltd.
The registered company address is: 152 Beach Road, #21-01/04 Gateway East, Singapore 189721, Singapore

If disposing of this product, please recycle the paper.

Preface

Titled with *Advances in Frontier Research on Engineering Structures II* and available on the latest research progress of structural engineering, this book gathers a bunch of papers. It consists of the following six parts: Engineering Mechanics and Construction Material Simulation, High Performance Concrete and Engineering Component Research, Structural Seismic Resistance and Energy Consumption Analysis, Structural Dynamics and Building Bearing Capacity Analysis, 3D Printed Concrete and Visualization Monitoring, and Engineering Design and Structural Simulation.

The manuscripts featured in this book have been carefully selected from a diverse range of sources. They include contributions from 2023 9th International Conference on Architectural, Civil and Hydraulic Engineering (ICACHE 2023), research institutions, and invited experts in related fields. Each manuscript has undergone a rigorous peer-review process to ensure the highest quality of content and to guarantee the relevance and significance of the research presented. The works in this book can promote the development of civil and structural engineering and resource sharing with flexibility and high efficiency, thereby promoting scientific information interchange among scholars from top universities, research centers, and high-tech enterprises working all over the world.

Featuring the most cutting-edge research directions and achievements related to civil and structural engineering, this book provides the most comprehensive research in the field of structural engineering and a more comprehensive understanding of the latest results of cross-research in this field. Meanwhile, it can also help researchers and engineers understand the research frontier, as well as discover the solutions to engineering problems.

We would like to acknowledge all the authors for their contributions and the expert reviewers for their time to review the manuscripts rigorously. We are also thankful to all the members and advisors of this book.

This book is believed to be beneficial to develop relevant subjects, and we do hope that readers can learn desired knowledge or experience and that it will serve as a reference for researchers and practitioners in academia and industry related to civil and structural engineering. Meanwhile, we welcome any advice or suggestion on this book.

Qingdao, China   Dayong Li
December 2023   Yu Zhang

Contents

High Performance Concrete and Engineering Component Research

Engineering Mechanics and Construction Material Simulation

Effect of Water-Binder Ratio, Alkali Mass Fraction and Lightweight Aggregate Size on Mechanical Properties of Alkali-Activated Concrete

Peng Deng, Zhiwei Niu, Kun Yang, Yan Liu, and Xianglong Zhang

Abstract Alkali-Activated Ceramsite Concrete (AACC), a novel green building material, is important for the reutilization of industrial waste. Although there has been considerable progress in research on AACC, further investigation is still needed on ceramsite size, early strength, etc. In this paper, 11 groups containing 198 standard cubic specimens were fabricated to investigate the effects of water-binder ratio, alkali mass fraction and ceramsite diameter on compressive strength and splitting tensile strength at different ages. And the slump constant of the mixtures was tested for different water-binder ratios and alkali mass fractions during fabrication. The results showed that the slump and strength increased with the increase of water-binder ratio. As the alkali mass fraction increased the slump decreased and then increased but the strength increased and then decreased. Therefore the proper value of alkali mass fraction was about 6%. When the diameter of ceramsite increased, the compressive strength first increased and then decreased. Thus the 5–15 mm diameter was the optimum. Splitting tensile strength always decreased with the increase of diameter of ceramsite.

Keywords Alkali-activated concrete · Ceramsites diameter · Alkali mass fraction · Concrete ages

P. Deng · Y. Liu (✉)
Shandong Provincial Key Laboratory of Civil Engineering Disaster Prevention and Mitigation, Shandong University of Science and Technology, Qingdao 266590, Shandong, China
e-mail: ly966@sina.com

P. Deng
e-mail: dengpeng1226@sdust.edu.cn

P. Deng · Z. Niu · K. Yang · Y. Liu · X. Zhang
College of Civil Engineering and Architecture, Shandong University of Science and Technology, Qingdao 266590, Shandong, China

© The Author(s) 2025
D. Li and Y. Zhang (eds.), *Advances in Frontier Research on Engineering Structures II*, Lecture Notes in Civil Engineering 535, https://doi.org/10.1007/978-981-97-6238-5_1

1 Introduction

Concrete obtained by replacing cement with slag and fly ash, and replacing coarse aggregate with ceramsites is a new eco-friendly building material. The material has gained the attention of researchers for its light weight, high strength, excellent thermal insulation and cost-effectiveness.

Slag and fly ash are common solid wastes discharged from industrial production and their reutilization in construction materials is economical and environmentally friendly. He [1] and Song [2] investigated the effect of fly ash and slag on the properties of cementitious composite binder and found that slag and fly ash could improve the degree of compaction of the structure. The addition of alkali activator resulted in high early strength and good durability. The effect of three different types of alkali activator (i.e., NaOH, Na_2SO_4 and Na_2SiO_3) on the mechanical properties of the binder was investigated by Qing [3], Shi et al. [4] the results showed that Na_2SiO_3 was the most effective in activation, followed by Na_2SO_4, and the least by NaOH. The experimental results of Farhan et al. [5] showed that the strength of alkali-activated fly ash or slag concrete was similar to that of Portland cement concrete at normal grades and greater than that at high strength grades. Lloyd [6], Ferdous [7] etc. investigated the effect of water-binder ratio on the compressive strength and workability of alkali- activated fly ash based polymer concrete, and found that the strength decreased and workability increased with its growth. In the study of alkali- activated lightweight aggregate concrete, some scholars use ceramsites to partially or completely replace natural crushed aggregate. Zheng et al. [8] conducted compressive tests on 252 AACC specimens to study the effect of water-cement ratio and modulus of activator on compressive strength. The test results showed that the compressive strength decreases with the increase of water-cement ratio and modulus of activator. Jiao [9] conducted basic mechanical tests on 408 AACC specimens. The test results indicate that the peak compressive strain of AACC is lower than that of ordinary concrete and the modulus of elasticity is higher than that. The test results of Cheng [10] revealed that the strength, cracking resistance and tensile creep of AACC decrease with the increase of ceramsites content.

Most of the existing studies only pay attention to the strength of AACC after 28 days of curing age and ignore its early high strength, which is important for shortening construction period, especially for pavement rehabilitation and so on. In addition, a large number of studies have focused on the type and replacement rate of lightweight aggregate, without considering the effect of coarse aggregate size. In this paper, 198 AACC specimens were fabricated to investigate the effect of water-binder ratio, alkali mass fraction and ceramsites size on the properties of AACC at different ages.

2 Test Overview

2.1 Raw Materials and Test Groups

The chemical compositions of slag and fly ash are listed in Table 1. River sand, whose diameter is medium-coarse grade, was used as fine aggregate. Coarse aggregate was made of shale clay ceramsites with compressive strength of 3.3 MPa. Water glass was used as an alkali activator with the original solution having sodium silicate concentration of 37.22% and modulus of 3.4. Sodium hydroxide solution was used to reduce the modulus of water glass. The content of slag, fly ash as well as aggregates is listed in Table 2.

In order to study the effect of water-binder ratio, alkali mass fraction, and ceramsites diameter on compressive and splitting tensile strength at different ages, a total of 198 specimens in 11 groups as shown in Table 3 were designed.

Table 1 Chemical composition of slag (%)

	CaO	SiO_2	Al_2O_3	MgO	MnO	Fe_2O_3	FeO	CaS	TiO
Slag	34.8	30.10	20.50	7.20	4.40	2.40	–	1.60	–
Fly ash	3.32	45.12	35.2	1.70	–	6.76	5.44	–	2.46

Table 2 Content of slag, fly ash and aggregates (kg/m^3)

Slag	Fly ash	Ceramsite	Sand
336.0	114.0	468.0	753.6

Table 3 Test groups

No.	Sodium silicate/ $kg\ m^{-3}$	Sodium hydroxide/ $kg\ m^{-3}$	Water/ $kg\ m^{-3}$	Distribution of ceramsite particle size/%		
				5–10 mm	10–15 mm	15–20 mm
W40A4D15	105.15	52.64	104.16	40	60	–
W40A6D10	157.72	79.00	60.22	100	–	–
W40A6D15	157.72	79.00	60.22	40	60	–
W40A6D20	157.72	79.00	60.22	20	40	40
W40A8D15	210.30	105.32	16.28	40	60	–
W43A4D15	105.15	52.94	120.05	40	60	–
W43A6D15	157.72	79.00	76.86	40	60	–
W43A8D15	210.30	105.32	33.68	40	60	–
W46A4D15	105.15	52.94	158.94	40	60	–
W46A6D15	157.72	79.00	93.51	40	60	–
W46A8D15	210.30	105.32	51.08	40	60	–

2.2 Test Procedure

The slump of concrete with different water-binder ratio and alkali mass fraction, ranging from 5 to 15 mm in diameter, was measured during fabrication. After the specimens had been prepared, they were divided into three groups and cured for 3, 7 and 28 days respectively for compressive and splitting tensile strength testing according to the standard [11]. The failure process of the specimen was observed during loading and the ultimate load was recorded.

3 Experimental Results and Analysis

The results of the compressive and splitting tensile strength tests were averaged over three times and are listed in the Table 4. It can be seen that AACC, whose compressive strength at 3d and 7d reached 76.4% and 86.4% of the compressive strength at 28d, respectively, had a high early strength. Meanwhile, its splitting tensile strength at 3d and 7d reached 69.2% and 83.5% of that at 28d, respectively.

Here, W40A4D15 means that the water-binder ratio is 0.40, the alkali mass fraction is 4%, and the ceramsites diameter is 15 mm.

Table 4 Strength test results (MPa)

No.	Compressive strength			Splitting tensile strength		
	3d	7d	28d	3d	7d	28d
W40A4D15	24.62	26.34	31.32	2.24	2.63	3.08
W40A6D10	23.80	25.37	31.00	2.63	3.34	3.98
W40A6D15	28.20	32.15	35.34	2.55	3.21	3.53
W40A6D20	20.60	22.83	28.37	1.87	2.28	2.93
W40A8D15	21.26	25.84	30.61	2.39	2.58	3.02
W43A4D15	22.51	25.07	29.62	2.14	2.50	2.89
W43A6D15	28.08	30.81	32.82	2.33	2.90	3.26
W43A8D15	20.03	25.33	29.24	2.12	2.53	2.85
W46A4D15	19.33	21.14	29.58	1.93	2.11	2.77
W46A6D15	26.85	29.76	31.93	1.82	2.32	3.15
W46A8D15	23.34	27.83	28.55	1.75	2.28	2.89

Fig. 1 Slump test results

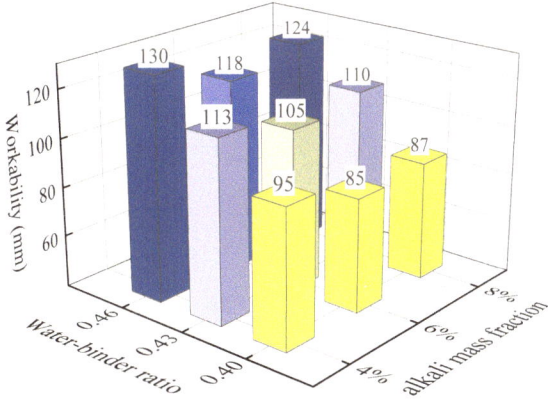

3.1 The Effect of Water-Binder Ratio and Alkali Concentration on Workability

The slump was similarly taken as an average of three results as shown in the Fig. 1. It is observed that as the water-binder ratio increased, the slump increased. This phenomenon may be attributed to the fact that the viscosity of slurry and alkali concentration decrease with the increase of free water content, which result in a lower rate of hydration. Thus the internal fine aggregate are not fully bonded to the hydration products, which leads to an increase in the factor of porosity. Therefore, the workability improves.

The slump first decreased and then increased with the increase of alkali mass fraction. This indicates that the activity of slag and fly ash was improved as the alkali content ranged from 4 to 6%. An alkali mass fraction of 6% is about the appropriate amount to optimize the activity of both. With the increase in alkali mass fraction, the activity of slag and fly ash decreased, and the hydration products are reduced and destroyed by the high concentration of alkali mass fraction, resulting in the lower bond performance of concrete.

3.2 The Effect of Water-Binder Ratio and Alkali Concentration on Concrete Strength

The comparison of compressive strength and splitting tensile strength of AACC after 28d is shown in Figs. 2 and 3. As the water-binder ratio increased, the number of pores occupied by free water within the concrete increased. In addition, the pores between aggregate and slurry mentioned in the slump analysis above remained after concrete setting. Therefore, these aspects could be used to explain the decrease of AACC strength.

Fig. 2 Compressive strength

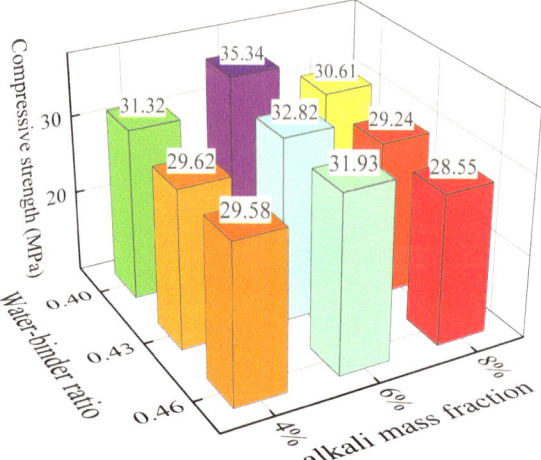

Fig. 3 Splitting tensile strength

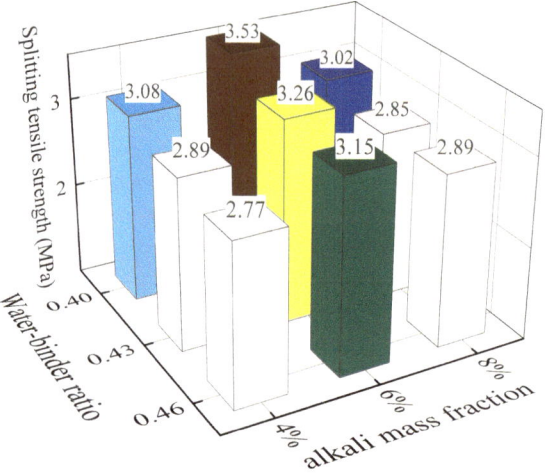

The strength of AACC increased and then decreased as the alkali mass fraction increased. This also indicates that the optimum alkali mass fraction is around 6% and the enhancement in slag-fly ash activity from 4 to 6% strengthen the obtained AACC. Continued increase in alkali mass fraction beyond what is optimum enhanced the alkalinity of the mixture and the internal hydration products were corroded by the excess alkalinity, which caused micro-cracks. Thus, the AACC strength decreased.

3.3   The Effect of Ceramsite Aggregate Diameter on Concrete Strength

The test results show that ceramsite with diameter of 5–15 mm resulted in higher compressive strength of AACC. During the forming process of AACC, large size aggregates in contact with each other were more likely to create pores inside the concrete, which were filled by fine aggregates and binder based on slag and fly ash. When the pore was filled, the degree of compaction was enhanced and the compressive strength of AACC was increased. However, when the ceramsites diameter was too large, the pore was also larger and the mass of binder was limited. Thus the pore filled by small and medium size aggregates. The mutual contact between the aggregates generated smaller pores inside the concrete, which resulted in reduced strength. When the aggregate size was small, the gaps between the coarse aggregate were small and their interaction depended mainly on the bonding of binder. And the strength of the binder was higher than that of large ceramsites. Thus the strength of AACC with small particle size ceramsite was higher than that with large particle size ceramsite (Fig. 4).

The analysis of the splitting tensile strength results shows that the smaller the aggregate size, the higher the AACC splitting tensile strength. This is due to the fact that the smaller diameter ceramsites were more uniformly distributed inside the concrete, which made the whole concrete more compact. Moreover, considering the size effect, smaller ceramsite was less likely to break. Therefore, AACC with small diameter ceramsite possessed higher splitting tensile strength.

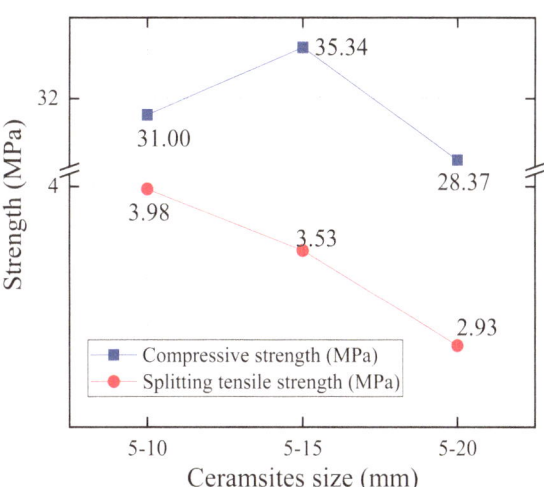

Fig. 4   Effect of ceramsite diameter on concrete strength

4 Conclusions

The early strength of AACC is high. The compressive strength and splitting tensile strengths at 3d can reach 76.4% and 69.2% of that at 28d, respectively. And it can reach 86.4% and 83.5% at 7d.

The slump, compressive and splitting tensile strength of AACC decrease when the water-binder ratio increases. With the increase of alkali mass fraction, the slump first decreases and then increases, and the strength first increases and then decreases. At 6% alkali mass fraction, the slump is minimum and the strength of AACC is maximum.

The compressive strength of AACC first increases and then decreases with the increase of the ceramsites diameter. Higher compressive strength of AACC can be obtained when the diameter is 5–15 mm. Splitting tensile strength decreases with the increase of ceramsite diameter.

Future studies could further refine the particle size range to obtain more precise optimal ceramsites size and further investigate AACC from a microscopic point of view.

References

1. He X.Y., Zhang C., Su Y., Wang Y.B., Yang J., & Chen W., et al. (2019) Properties of cement pastes containing high volume granulated blast furnace slag (GBFS). *Concrete*, (9), 5.
2. Song, B. X., & Liu, S. H., et al. (2022). Compressive strength, water and chloride transport properties of early CO2-cured Portland cement-fly ash-slag ternary mortars. *Cement and Concrete Composites*, *134*, 104786.
3. Su, Q., & Tao, Z. (2020). Experiment research on the mechanical performance of alkali-activated slag cementitious material. *Materials Science and Engineering*, *768*, 8981–8992.
4. Shi, C. (1996). Strength, pore structure and permeability of alkali-activated slag mortars. *Cement and Concrete Research*, *12*, 1789–1800.
5. Farhan, N. A., Sheikh, M. N., & Hadi, M. N. S. (2019). Investigation of engineering properties of normal and high strength fly ash based geopolymer and alkali-activated slag concrete compared to ordinary Portland cement concrete. *Construction and Building Materials*, *196*, 26–42.
6. Lloyd, N. A., & Rangan, B. V. (2010). Geopolymer concrete with fly ash. *Australian Journal of Structural Engineering*, *21*(3), 1482–1493.
7. Ferdous, M. W., Kayali, O., & Khennane, A. (2013). A detailed procedure of mix design for fly ash basedgeopolymer concrete. In *4th Asia-Pacific conference on FRP in structures* (Vol. 12, pp. 11–13). Melbourne.
8. Zheng, W. Z., Hang, W. X., & Jiao, Z. Z., et al. (2017). Experiment research on basic performance of alkali-activated slag ceramsite concrete. *Journal of Beijing University Technology*, *43*(8), 1182–1189.
9. Jiao, Z. Z. (2019). *Study on basic mechanical properties of alkali-activated slag cementitious material block masonry*. Harbin Institute of Technology.
10. Cheng, K. J. (2016). *Study on autogenous shrinkage crack resistance of alkali-activated slag ceramsite*. Fuzhou: Fuzhou University.
11. Ministry of Housing and Urban-Rural Development of the People's Republic of China. (2011). Standard for test methods of concrete physical and mechanical properties: GB/T 50081-2019. China Architecture and Building Press.

Recent Advance of Vibration Control Techniques in Structures

Junming Yuan

Abstract The demand of seismic resistance in structures and constructions is increasing, the Vibration control techniques are gaining more attention. Many researchers have investigated different vibration control systems to reduce the response of structures in earthquakes, including passive, active, semi-active, and hybrid vibration control systems. This paper represents the historical background and development of each technique is represented. Moreover, the advantage and limits of these devices are argued. Finally, the applications and the future prospects of these techniques are discussed.

Keywords Vibration control · Active control · Passive control · Semi-active control · Hybrid control

1 Introduction

Construction industries have rapid development in recent two decades in China. The number of high-rise buildings and skyscrapers is increasing, which requires the improvement and revolution of dynamic-resistance techniques. When an earthquake occurs, the crust will vibrate and release energy rapidly. When the seismic wave caused by the earthquake reaches the surface of the ground. The energy will result in the vibration of buildings and impact the safety of structures. The occurrence of earthquakes normally has a huge impact on the stability of buildings and constructions. In 2008, Wenchuan earthquake which was estimated a magnitude of 8.0 resulted in the death of 0.83 million. The seismic performance of structures has got more attention, and vibration control is becoming a key role to reduce the damage from the earthquake. Passive control does not require additional energy, it makes the kinetic energy of the seismic waves can be converted to another form of energy that can be absorbed safely by the building, like heat [1]. The earliest application of passive vibration control is the viscoelastic materials. In earlier 1969, viscoelastic dampers

J. Yuan (✉)
Chang'an Dublin International College, Chang'an University, Xi'an 710000, China
e-mail: 1165692244@qq.com

D. Li and Y. Zhang (eds.), *Advances in Frontier Research on Engineering Structures II*,
Lecture Notes in Civil Engineering 535, https://doi.org/10.1007/978-981-97-6238-5_2

(VD) were investigated by Mahmoodi. 10,000 dampers were set up in the original World Trade Center in New York. Now, VD is the most popular and cheapest method to protect the building from the damage of an earthquake. Unlike passive control, the operation of active vibration control devices needs external energy [2]. They can make passive control automatic and adjust the system in real-time. However, this method normally has high costs and low reliability. In active vibration control approaches Active mass driver (AMD) is the most popular device to control the vibration of the structures. AMD was proposed to reduce the vibration of high buildings in 1980 to improve the performance of passive-tuned mass dampers (TMD) [3]. The passive control techniques can resist the vibration without massive energy while the active control devices can improve the performance of passive control systems by the external power. However, a novel technology was investigated. The semi-active system has been proven in reducing building vibration in 1996 [4]. Compared to the traditional active control system, it needs less energy inputted to reduce the vibration in the structure by using semi-active control technology [2]. Although the vibration control system is developing, only using one single system in structure has obvious limitations, like the low efficiency of passive control or the low reliability of active vibration control. So the combination of different control devices would be a promising future. The Hybrid control method will be argued, which is the combination of active control and passive control. In a hybrid system, two different control devices can mutually reinforce to improve the stability and effectiveness of the system. For example, Passive control is used as a protective device for the structure under frequent external disturbances, and active control is used as the last line of defense against structural damage in case of external disturbances, but its design method is quite complex, it will get great advantages by investing and experiment for several times.

In this paper, several vibration control techniques will be introduced, including active control, passive control, semi-active control, and more complicated hybrid control. The advantages and drawbacks of these approaches are analyzed. Moreover, this essay will be based on these control techniques and compare some specific devices and systems, finding the promising development of vibration control and figuring out the potential of smart vibration control.

2 Passive Vibration Control

Passive control refers to the addition of energy-dissipating devices or sub-structural systems to appropriate parts of the structure, or the structural treatment of certain components of the structure itself has changed the dynamic characteristics of the structural system. Passive control is control without external energy, and its control force is passively generated by the control device with the vibration and deformation of the structure. In this part, some classic essays will be argued, tuned mass damper, tuned Liquid Damper and viscoelastic dampers.

2.1 Tuned Mass Damper

The tuned mass damper (TMD) is a kind of classic passive vibration control device. It has a simple structure, low cost, and reliable operation, but it needs much space to be installed and has poor performance in intensive earthquakes [1]. The first time it was introduced by Frahm in 1911 [5]. Now, it has been used in high structures around the world, like John Hancock Tower in the USA and Taipei 101 Tower. As shown in Fig. 1, TMD uses a spring and a viscous damper to connect concrete or steel to the building. When the building vibrates, the energy which amassed in the deck or columns of buildings can be transmitted to the secondary mass (concrete and steel). To reduce the vibration effectively, the performance spring and mass ratio are designed to achieve maximum damping.

Now, the optimization of TMD is no longer a new topic. Abdullah et al. [6] indicated implementing a shared tuned mass damper (STMD) between two neighbor buildings can reduce the dynamic responses of the structure effectively. The relative displacement of these two buildings will increase due to the shared tuned mass damper, which means the two buildings hardly collide with each other. This technique is economical because only one TMD device was installed to control the vibration of two buildings. However, some studies argue that the STMD requires a large footprint in the buildings because of its large weight [6]. Some reaches were trying to find improved Alternative methods. A better implementation of dampers was argued by Kareem and Kline [7]. They investigated the workability of several TMD devices which were installed in a 186 m high regular roof, they found that the system will be more effective within a range of frequency when multiple smaller tuned mass dampers (MTMD) were set up rather than the implementation of a lager single damper.

Fu and Johnson indicated the position of multiple dampers can improve the efficiency of the system to resist vibration [8]. The authors studied the seismic resistance performance of distributed mass damper systems (DTMD), each floor has different TMD devices which shows engineers will find it difficult to generate a DTMD system.

Fig. 1 The structure of tuned mass damper

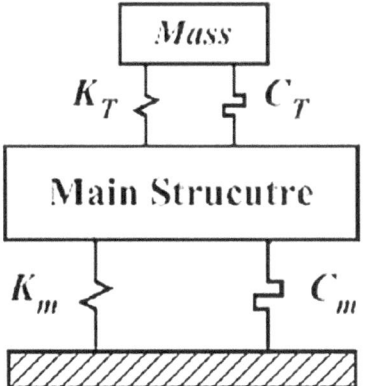

F. Yang et al. find a novel DTMD technique has an excellent performance in increasing vibration and has higher calculation efficiency in designing the system.

2.2 Tuned Liquid Damper (TLD) and Tuned Liquid Column Damper (TLCD)

TLD and TLCD are considered a kind of TMD. In TLD or TLCD, the secondary mass is replaced by the liquid, like water. As the structure is shown in Fig. 2, part (a) indicates those liquid normally is placed in horizontal pipes with an orifice or tanks to counteract the vibration with the shaking of the liquid. Part (b) shows some details in the container.

The first time TLD was introduced by Helmut F. Bauer in 1986 [9]. The author mentioned a novel liquid damping device which was considered the earliest TLD device. Some modification and improvement have been investigated in recent years, Fujino et al. found the beating and the modulated amplitude when the TLC control the structures, which means the damping of the traditional TLD is not sufficient [10]. The author changes the bottom of the tank and makes it rough. At the same time, the other tank was closed. However, these two improvements cannot increase the damping. To improve the damping effectively, a study found that flow-damping devices (FDD) can be added to the TLD [11]. FDDs are designed as floating objects or screens, which can increase the shake of the liquid in the tank of the TLDs to improve the efficiency of energy dissipating. After the problem caused by insufficient damping was solved, the studies began to find the high performance of TLD in different situations. Sarkar and Gudmestad improved the TLD system and got a pendulum-tuned liquid damper [12]. This technique uses a U-shape pendulum which doesn't add to the conventional TLD to make the system workable for the tower structure like chimneys and wind turbines.

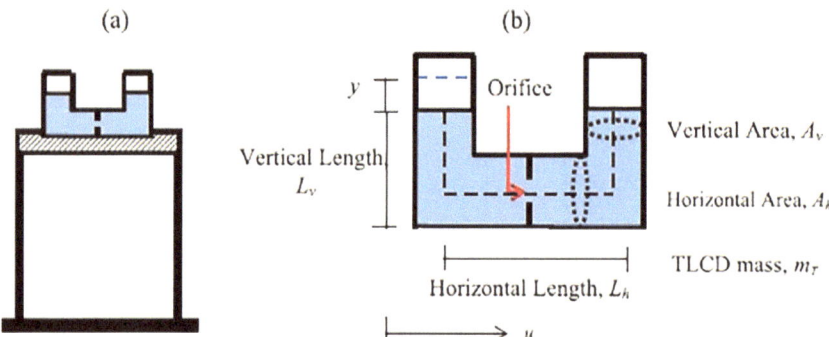

Fig. 2 The structure of the tuned liquid column damper

2.3 Viscoelastic Dampers (VD)

VDs are one of the most popular seismic-resistance devices installed in structures to control vibration. The shear deformation of the viscoelastic solid layers which are caught in two plates made from steels will dissipate the energy of vibration [13]. The components of VD system are shown in Fig. 3 [14].

VDS has been proven can reduce the vibration caused the earthquake because it can contribute to improving both the stiffness and damping of structures.

Zhang M. and Pang H. investigated the seismic resistance of VDs, by testing the performance of a 10-story building with reinforced concrete under different earthquake waves. The authors found the VDs can effectively reduce the maximum drift angle of the buildings [15].

The performance of VDs is controlled by the property of the viscoelastic materials. In engineering, the viscoelastic material normally is rubber because of its unique properties. After the deformation of rubber, it can return to its original shape, the energy is dissipated during this process. So, the damping of the viscoelastic materials is very essential. The high-damping VDs have high efficiency to consume energy. To improve the damping, a method for materials was found, super fine black carbon was added to generate rubber compounds. At the same time, an elastomer compound consisting of synthetic rubber and black carbon was introduced to achieve the high damping of the VDs. Iaboviste et al. indicated that the excellent performance of VDs is related to the environmental temperature and excitation frequencies [16].

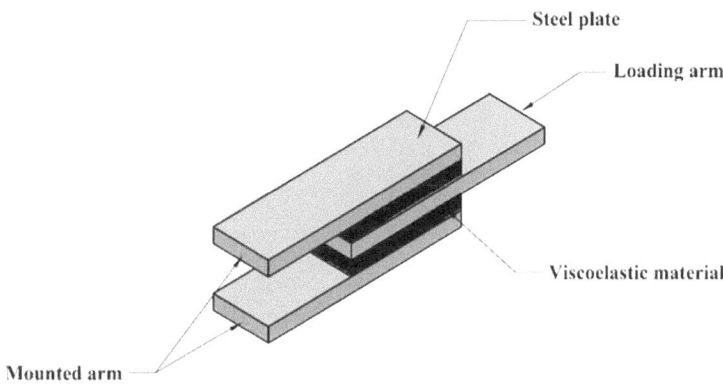

Fig. 3 The structure of VD system

3 Active Vibration Control

Active vibration control is to effectively control the system through the measured vibration information. Through the analysis of the actual operation, the corresponding control scheme is put forward. The system exerts a specific effect on the controlled object by manipulating the actuator so as to restrain or eliminate the oscillation. Due to its high effectiveness, strong adaptability and other potential advantages. The active vibration control techniques is getting more and more attention. However, it shows its low reliability and high cost because of its complex structure and the optimal location of the sensors. The general structure is shown in Fig. 4.

3.1 Active Mass Damper (AMD)

Active mass damper can collect feedback from the response of the essential location in the structure, then the information can be analyzed by the computer to make sure it is appropriate. The controller will accept the information and drive the mass to impose the inertial force to control the vibration. But the system always demands excessive energy. To overcome this problem, Yen-Po Wang et al. proposed a high-performance active mass driver (HP-AMD). The system combines a hydraulic actuator with a mechanical pully system to control the vibration with less energy. Except for the combination of new devices, the efficiency of the system can be improved by using new algorithms. Amini et al. found a new approach to reduce the response of structures in the earthquake efficiently by using particle, discrete wavelet transforms (DWT) and LQR algorithm to find the most efficient way to control the force offered by AMD [17].

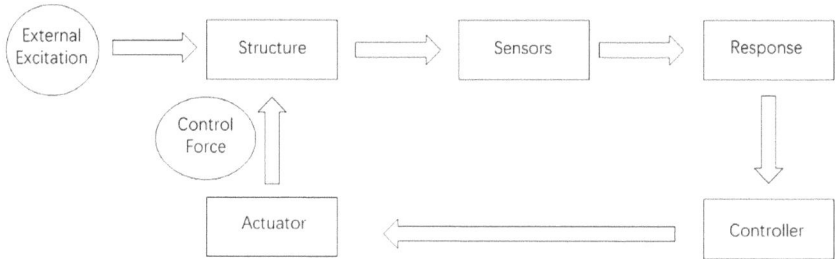

Fig. 4 Active vibration control system

3.2 Active Tendon (AT)

A large number of experimental studies on the project have been carried out by using this method, and good results have been obtained. This method was invented by Eugène Freyssinet in the 1960s. The device combined the pre-stressed tendon and electrohydraulic servomechanism to convert the force to control the vibration. After AT was introduced, many researchers were studying the improvement and optimal of AT system. Some methods are found to assess the system. Aldemir et al. investigated the performance of AT system in a building with multiple floors. A performance index was used to minimize mechanical in the buildings, which can control the vibration effectively [18]. At the same time, some more effective approaches are proposed to improve the efficiency of the system. Cazzulani et al. used the fiber sensors on a cantilever system to control the vibration actively, and a modified LQC algorithm was used to drive the system [19]. A function can control the gain matrix and measure the velocity of positions in the structures.

3.3 Active Bracing System (ABS)

In an active bracing system, active control devices are installed between floors or between the ground and the first story to control the response of buildings. LQR theory was normally used to control the ABS system, there are three algorithms, static output feedback LQR, and their effectiveness was proved by Loh et al. [20]. The author tested the control algorithm in a full-scale building and found ABS system can reduce the response of structure by up to 50%. ABS is not can be controlled by the LQR algorithm only, Lu introduced a discrete-time modal control scheme that can control the response of structures efficiently [21].

4 Semi-Active Control

Semi-active vibration reduction is to effectively adjust and control its dynamic characteristics according to the specific requirements of the project. This design has the characteristics of good energy saving effect and good energy saving effect. The basic method to suppress the structural vibration is installing a semi-active device that can adaptively adjust the flexibility and damping [3]. Commonly used semi-main Dynamic control systems include Active variable stiffness systems (AVS) and magnetorheological (MR) dampers.

4.1 Active Variable Stiffness System (AVS)

According to a certain control law, the system can dynamically switch the rigidity of the controlled components between each stiffness in each test stage. In order to make the system leave the resonance point as far as possible in each sampling period, in order to achieve the effect of vibration reduction. Variable stiffness devices generally consist of a support and an electro-hydraulic servo system disposed between the support and the controlled structure. In 1998, semi-active variable flow dampers were installed in the walls of the Shizuoka city building in Japan, each damper had a valve to control the flow of liquid. This system can provide a maximum damping force of 1000kN which can reduce the vibration effectively.

4.2 Magnetorheological (MR) Damper

Under the action of magnetic fields, magnetorheological materials will change from liquid to semi-solid with controllable yield strength in a very short time, and they can be used to make semi-active control devices with fast response. However, reducing the vibration by utilizing the property of the material is not efficient. Some studies indicated that efficiency can be improved by using advanced algorithms. Laflamme and others apply MMR damper to the damping effect of a 39-story building and apply it to practical engineering [22]. Mexican hat wavelet is used as the regulator of neural networks. The response of buildings in an earthquake can be reduced effectively. Although the MR damper should use an algorithm to control, like active control techniques. The cost of an MR damper is quite low. Tse et al. proposed the rare implementation of active control devices in the buildings is caused by the high long-term maintenance cost, then they evaluated the MR damper system in the structures which is 240 m tall and with 60 floors [23]. A simple LQR algorithm is used to optimize the system. The result showed the cost of semi-active only occupies 2% of the total budget.

5 Hybrid Vibration Control

Hybrid control is a combination of active and passive control. Hybrid control helps alleviate the constraints and limitations of a single control method and can take advantage of both active and passive control. The first building in the world to install a hybrid mass damper (HMD) control system was the 7-story building of the Shimizu Corporation Technical Research Institute in Tokyo, Japan (1992) control the translational and rotational movements of the structure. The work theory of a hybrid system is shown in Fig. 5

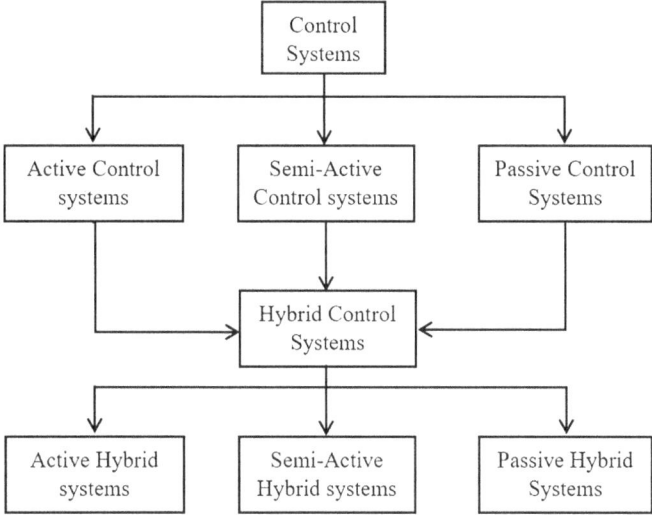

Fig. 5 Hybrid vibration control system

Many reaches have investigated the combination of active devices and passive systems. In 2005, Park et al. indicated a combination of a passive BI system and an active actuator [24]. The hybrid system was tested to reduce the vibration caused by the earthquake in a cable-stayed bridge. The authors found that the hybrid systems perform better than the passive control system and they were marginally superior to the active control system. At the same time, the system consists of semi-active control devices and a passive control system showed its high efficiency to mitigate the earthquake responses of structures. A sliding isolator and an MR damper were combined by Sahasrabudhe and Nagarajaiah to minimize the vibration of an earthquake-excited bridge [25]. The authors used a bridge model with 1:20 scaled under a shake-table test and several earthquake motions, the data of the test were analyzed. The result of the study showed that the MR damper have higher efficiency to mitigate the bearing displacement compare to a high damping or a low-damping passive system. In general, the hybrid system normally performs better than the traditional active or passive system.

6 Discussion and Conclusion

In conclusion, the design, development, and verification of the new seismic system are discussed in depth, in order to provide reference for the overall seismic design of all kinds of buildings in our country. Ultimately, the findings demonstrate significant advancements in seismic engineering. Passive control systems that rely on specific dynamic movements for tuning are widely accepted today, yet their effectiveness is constrained by bandwidth limitations. In recent years, scholars at home and abroad have conducted extensive research on it. In order to improve the performance of power systems, a variety of methods have been proposed to improve the performance of power systems. However, the financial investment associated with implementing an active control system would be one of the most evident drawbacks of utilizing active control techniques. Furthermore, there have been advancements in hybrid designs that incorporate various technologies to merge the properties of damping materials and structural systems. It is crucial to investigate methods for further decreasing energy usage and overall expenses, enabling buildings to streamline the project while maintaining excellent performance. It is vital to consider the impact of diverse external factors on the vibration of structures, enhance the robustness of vibration management, and ensure the safety of architectural constructions. It is crucial to try to integrate the technological advancements from various fields. Including new materials, sensors, calculations, and information technology.

References

1. Almajhali, K. Y. M. (2023). Review on passive energy dissipation devices and techniques of installation for high rise building structures. *Structures*, *51*, 1019–1029. ISSN 2352-0124.
2. Djerouni, S., Elias, S., Abdeddaim, M., & Domenico, D. D. (2023). Effectiveness of optimal shared multiple tuned mass damper inerters for pounding mitigation of adjacent buildings. *Practice Periodical on Structural Design and Construction*, 28(1). https://doi.org/10.1061/(ASCE)SC. 1943–5576.0000732.
3. Chang, J. C., & Soong, T. T. (1980). Structural control using active tuned mass dampers. *Journal of the Engineering Mechanics Division, 106*, 1091–1098. https://doi.org/10.1061/JMCEA3.0002652
4. Nonami, K., & Mizuno, T. (1996). Active structural damping of beam structures. In Proceedings 3rd international conference on motion and vibration control. J-STAGE.
5. Frahm, H. (1911). Device for damping vibrations of bodies, US Patent 989 958, April 18.
6. Kapasakalis, K., Mantakas, A., Kalderon, M., Antoniou, M., Sapountzakis, E. J. (2023). Performance evaluation of distributed extended KDamper devices for seismic protection of mid-rise building structures. *Journal of Earthquake Engineering, 0*(0), 1–26.
7. Elias, S., & Matsagar, V. (2019). Seismic vulnerability of a non-linear building with distributed multiple tuned vibration absorbers. *Structure and Infrastructure Engineering, 15*(8), 1103–1118.
8. Fu, T. S., & Johnson, E. A. (2011). Distributed mass damper system for integrating structural and environmental control in buildings. *Journal of Engineering Mechanics, 137*(3), 205–213.
9. Bauer, H. F. (1984). Oscillations of immersible liquids in a rectangular container: A new damper for excited structures. *Journal of Sound and Vibration, 93*, 117–133.
10. Fujino, Y., Pacheco, B. M., Chaiseri, P., & Sun, L. M. (1988). Parametric study on tuned liquid damper (TLD) using circular containers by free oscillation experiment. *Struct Eng/Earthq Eng, 5*, 177–187.
11. Konar, T., & Ghosh, A. D. (2021). Flow damping devices in tuned liquid damper for structural vibration control: A review. *Archives of Computational Methods Engineering, 28*, 2195–2207. https://doi.org/10.1007/s11831-020-09450-0
12. Sarkar, A., & Gudmestad, O. T. (2013). Pendulum type liquid columns damper (PLCD) for controlling vibrations of a structure–theoretical and experimental study. *Engineering Structures, 49*, 221–233.

13. Soong, T. T., & Spencer, B. F. (2002). Supplemental energy dissipation: State-of-the-art and state-of-the-practice. *Engineering Structures, 24*(3), 243–259.
14. Alhasan, A. A., Vafaei, M., & C Alih, S. (2022). Viscoelastic dampers for protection of structures against seismic actions. Innovative Infrastructure Solutions, 7, 309.
15. Zhang, M., & Pang, H. (2020). Analysis of damping performance of frame structure with viscoelastic dampers. *China Journal of Applied Mechanics, 37*(1), 418–426.
16. Jaboviste, K., Sadoulet-reboul, E., Peyret, N., Arnould, C., Collard, E., & Chevallier, G. (2019). On the compromise between performance and robustness for viscoelastic damped structures. *Mechanical Systems and Signal Processing, 119*, 65–80.
17. Amini, F., Hazaveh, N. K., & Rad, A. A. (2013). Wavelet PSO-based LQR algorithm for optimal structural control using active tuned mass dampers. *Computer-Aided Civil and Infrastructure Engineering, 28*(7), 542–557.
18. Aldemir, U., Yanik, A., & Bakioglu, M. (2012). Control of structural response under earthquake excitation. *Comput-Aided Civ Infrastruct Eng, 27*(8), 620–638.
19. Cazzulani, G., Cinquemani, S., Comolli, L. (2012, April 26). Enhancing active vibration control performances in a smart structure by using fiber Bragg gratings sensors. In *Sensors and smart structures technologies for civil, mechanical, and aerospace systems.*
20. Loh, C. H., Lin, P. Y., & Chung, N. H. (1999). Experimental verification of building control using active bracing system. *Earthquake Engineering and Structural Dynamics, 28*, 1099–1119.
21. Lu, L.-Y. (2001). Discrete-time modal control for seismic structures with active bracing system. *Journal of Intelligent Material Systems and Structures, 12*, 369–381.
22. Laflamme, S., Slotine, J. J. E., & Connor, J. J. (2011). Wavelet network for semi-active control. *Journal of Engineering Mechanics, 137*(7), 462–474.
23. Tse, K., Kwok, K., & Tamura, Y. (2012). Performance and cost evaluation of a smart tuned mass damper for suppressing wind-induced lateral-torsional motion of tall structures. *Journal of the Structural Engineering. American Society of Civil Engineers, 138*(4), 514–525.
24. Park, K., Jung, H., & Lee, I. (2003). Hybrid control strategy for seismic protection of a benchmark cable-stayed bridge. *Engineering Structures, 25*(4), 405–417.
25. Sahasrabudhe, S. S., & Nagarajaiah, S. (2005). Semi-active control of sliding isolated bridges using MR dampers: An experimental and numerical study. *Earthquake Engineering and Structural Dynamics, 34*(8), 965–983.

Experimental and Analyses on the Deterioration of Mechanical Properties of Transport Bridges Based on Hot and Humid Marine Environment

Yunkai Chen, Peng Deng, and Yao Wang

Abstract The mechanical properties of in-service concrete bridges in the sea are subject to accelerated deterioration due to the harsh environment in which they are located, which poses a serious challenge to bridge transportation work. Therefore, it is necessary to analyse and study the mechanical property evolution of lightweight aggregate concrete in the marine environment and transport engineering design. In order to study the mechanical property degradation law of nano-modified ceramic concrete under the action of dry and wet cycles and temperature, tests were conducted on nano-modified ceramic concrete specimens at different temperatures (30 °C, 40 °C, 50 °C) and different numbers of dry and wet cycles (15, 45, 75), respectively. The test results showed that with the increase of the number of cycles, the change of compressive strength showed a typical three-phase law, while the split tensile strength showed the law of deterioration, strengthening, and then deterioration. With the increase of temperature, the compressive strength increases and then decreases, while the splitting tensile strength decreases continuously, and the decrease is greater. Under the combined effect of the two factors, three damage modes of compressive and split tensile were analysed. Finally, the uniaxial compression constitutive model and the formula for the attenuation change of mechanical properties of nano-modified ceramic concrete were fitted based on the test data under dry and wet cycles and temperature.

Keywords Dry–wet cycle · Ceramsite concrete · Mechanical properties · Constitutive model · Intensity attenuation

Y. Chen (✉) · P. Deng · Y. Wang
Shandong University of Science and Technology, Qingdao 266590, Shandong, China
e-mail: 1752030194@qq.com

P. Deng
e-mail: dengpeng1226@sdust.edu.cn

Y. Wang
e-mail: wangyao0911@sdust.edu.cn

D. Li and Y. Zhang (eds.), *Advances in Frontier Research on Engineering Structures II*,
Lecture Notes in Civil Engineering 535, https://doi.org/10.1007/978-981-97-6238-5_3

1 Introduction

The serious structural deterioration and failure of marine concrete in dry and wet cyclic environments is one of the most unfavourable factors affecting its durability [1], which hinders the development of marine engineering. As a man-made lightweight aggregate material, the durability of concrete can be improved by using it as an aggregate [2], but the microstructure of concrete prepared from ceramic particles has more defects. Nanomaterials possess special properties such as small size effect and surface effect [3], which can be incorporated into concrete to improve the microstructure and durability of concrete. In order to further improve the mechanical properties of vitrified concrete, this paper selects nanomaterials to be incorporated into vitrified concrete so that it can better serve the complex environment.

Many scholars at home and abroad have carried out a lot of research on the durability of marine concrete. Ganjian et al. [4] experimental study shows that with the increase of the number of dry and wet cycles, the durability of salt-etched concrete shows a deterioration law that first rises and then accelerates the decline; Tan Jiong [5] experimental study in the dry and wet cycles, the larger the temperature will produce a tensile stress cracking the concrete surface, and ultimately measured a decline in the compressive strength of 10%; Plateau [6] experimentally investigated that the free deformation of concrete with different water-cement ratios under the action of wet and dry cycles is different, and the longer the cycle of wet and dry cycles, the longer the free deformation. Meanwhile, the research on the modification of concrete by nanomaterials is more mature, Ye et al. [7] and Erhan et al. [8] explored the effect of nano-SiO_2 on the durability of concrete, and tested that nano-SiO_2 promotes the hydration reaction, which in turn improves its durability; Li Guohua et al. [9] concluded that the nano-SiO_2 greatly improves the performance of concrete in the later stage, the reason It is because it stimulates the material activity. Most of the studies have discussed the durability of nanoconcrete under the effect of a single factor, and it is inaccurate to analyse it only unilaterally in practical marine engineering. Therefore, the combined effect of dry and wet cycles and temperature is more in line with the actual working conditions, and little research has been reported in this area.

Therefore, the group carried out an experimental study of nano-SiO_2-modified ceramic concrete in a wet and dry cycling environment [10], controlling different seawater solution temperatures (30, 40 and 50 °C), and different numbers of wet and dry cycling (15, 45, and 75 times), and the concentration of the seawater solution was kept unchanged, to explore the effects of temperature and the number of wet and dry cycling on the mechanical properties of nano-materials-modified concrete, and to investigate the mechanical properties of nano-materials-modified concrete under the effect of temperature and the number of wet and dry cycling. Based on the fitting of explicitly wet and hot marine environment ceramic concrete under the action of the number of dry and wet cycles, temperature control parameters, the nano-modified ceramic concrete stress–strain ontological relationship model was established, and

Table 1 Mix proportion

Strength	Water cement ratio	Cement	SiO$_2$ mixing amount	Water	Sand ratio (%)
LC30	0.35	336.14	6.86	120	40
LC35	0.29	406.7	8.3	120	40

at the same time, the concrete compressive strength and split tensile strength attenuation formulae were established, to provide a reference for its application in marine engineering.

2 Materials and Methods

2.1 Mixing Ratios

According to JGJ/T12-2019 [11] and GB/T50082-2009 [12], the proportion of ceramic concrete was designed, and the mechanical properties of ceramic concrete were tested to obtain the ceramic concrete with different strength grades (LC30 and LC35), and the preferred The best mix ratio of nano-SiO$_2$ modified vitrified concrete. In order to reduce the error in the test, four groups of water-cement ratios (0.38, 0.35, 0.32 and 0.29) were designed for the test, and the compressive strength test was carried out by using 100 mm × 100 mm × 100 mm standard cube specimens after 28d standard curing. The water-cement ratios of 0.35 and 0.29 were finally selected as the ratios for the final test through the measured data, and the ratios are shown in Table 1.

2.2 Test Methods

Firstly, the water-absorbed ceramic granules were put into the mixing tank, river sand was placed, and finally silicate cement was placed, in order to prevent the aggregates from sinking, partially watered mixing was carried out first, so that the cement was slightly bonded to the aggregates, and then water and water reducer were added for wet mixing, and SiO$_2$ nanomaterials were uniformly sprinkled on the mix in the process of mixing. The mixed nano-SiO$_2$ ceramic concrete was poured into the test moulds of 100 mm × 100 mm × 100 mm and 150 mm × 150 mm × 300 mm specimens, vibrated densely and smoothed, and the specimens were put into a curing room with relatively smooth temperature and humidity, the temperature was about 20 °C, the relative humidity was more than 95%, and the age of the curing period was 28 d. The test was carried out in a curing room with a temperature of about 20 °C, relative humidity of more than 95%, and the age of the curing period was 28d.

Comprehensive consideration of test factors and according to the specification to consider different temperatures (30, 40 and 50 °C), different wet and dry cycle times (15 times, 45 times, 75 times) design wet and dry cycle test. Before the start of the dry and wet cycle, the specimen should be dried first. The total time for specimens to be immersed, put into solution, and discharged from solution during the wet/dry cycle is 16 h. The drying temperature is about 80 °C, the drying time is 11 h, the cooling time is 1 h, and the wetting time is one day, and the total time of a wet/dry cycle process is 24 h. The choice of this wet/dry cycle system is more in line with the service environment of the offshore concrete, and the data obtained will also be more representative. In order to facilitate the later experimental research, the naming rules are as follows, LC30-30-15, representing the strength of the ceramic concrete-infiltration solution temperature-the number of wet and dry cycles.

3 Results and Discussion

3.1 Test Data

The test was carried out using 100 mm × 100 mm × 100 mm specimen blocks, which were immersed in artificial seawater solution at 30, 40 and 50 °C, and cycled wet and dry for 15, 45 and 75 times respectively, and the specimen blocks did not corrode. In accordance with the requirements of the first chapter of the dry and wet cycle to complete the test under the planned conditions, the size conversion factor of 0.95. Processing the data to obtain the LC30, LC35 compressive strength of 31.9 MPa and 38.2 MPa, respectively, and the rest of the numbered strengths are shown in Table 2:

3.2 Analysis of Damage Modes Under the Combined Effect of Temperature and Dry–Wet Cycles

When concrete specimens are tested for compressive strength, bond damage between aggregate and cement, splitting damage of aggregate and tensile damage of cement usually occurs, see Figs. 1 and 2. The first type of damage is manifested by the appearance of vertical cracks first, which extend and penetrate due to the stress concentration, and the degree of damage becomes more serious with the increase in the number of wet and dry cycles. The second type of damage is that transverse and longitudinal cracks are produced in the concrete under pressure, and the transverse crack in the middle of the specimen reaches the ultimate tensile strain of the concrete and extends through. The third type of damage is the initial loading, only a few parts of the tensile stress is less than the bond strength of the place, then cracking will occur, cracking through the weak parts of the weak, and ultimately gradually to the surrounding expansion and brittle damage.

Table 2 Compressive strength of nano modified ceramsite concrete

Specimen number	Sand rate (%)	Water-to-cement ratio	Gelling material	Strength/Mpa
LC30-30-15	40	0.35	343	33.5
LC30-40-15	40	0.35	343	37.4
LC30-50-15	40	0.35	343	35.2
LC35-30-15	40	0.29	415	39.3
LC35-40-15	40	0.29	415	43.2
LC35-50-15	40	0.29	415	41.6
LC30-30-45	40	0.35	343	31.0
LC30-40-45	40	0.35	343	36.0
LC30-50-45	40	0.35	343	33.0
LC35-30-45	40	0.29	415	37.1
LC35-40-45	40	0.29	415	42.6
LC35-50-45	40	0.29	415	39.5
LC30-30-75	40	0.35	343	27.6
LC30-40-75	40	0.35	343	31.5
LC30-50-75	40	0.35	343	28.2
LC35-30-75	40	0.29	415	31.9
LC35-40-75	40	0.29	415	39.5
LC35-50-75	40	0.29	415	34.4

Fig. 1 Breakdown of bond between aggregate and cement

Fig. 2 Tensile damage of cement

The splitting damage modes of concrete are generally centre cracking damage, local crushing damage and secondary cracking damage. The splitting tensile strength damage mode under dry and wet cycles is centre cracking damage, and at the initial stage, due to the low number of cycles, the concrete surface only appears to spall on the concrete corners and edges. In the middle stage, the peak of concrete splitting tensile strength appeared, the surface was more intact, and the macroscopic increase in splitting tensile strength was also evidenced, see Fig. 3a. In the later stage, under 75 cycles, the concrete splitting tensile strength showed a decreasing trend, and the appearance of spalling on the concrete surface and the exposure of aggregates with fine cracks developing parallel to the prismatic edges, see Fig. 3b.

3.3 Effect of Temperature and Dry–Wet Cycles on the Compressive Strength of Nano-material Vitrified Concrete

The change rule of compressive strength of nano-material concrete under the combined effect of temperature and wet/dry cycles is shown in Fig. 4. At the same temperature, with the increase of the number of wet and dry cycles, the compressive strength shows the trend of increasing first and then accelerating the deterioration. Taking LC30-30 working condition as an example, the compressive strength of LC30-30-15 reached 33.5Mpa, which was 1.6Mpa higher than that of the specimen without wet and dry cycles, and the strength was increased by 4%; under the LC30-30-45 working condition, compared with the 15 times of wet and dry cycles, the compressive strength was decreased by 2.5Mpa, and the loss rate of compressive strength was 7%.

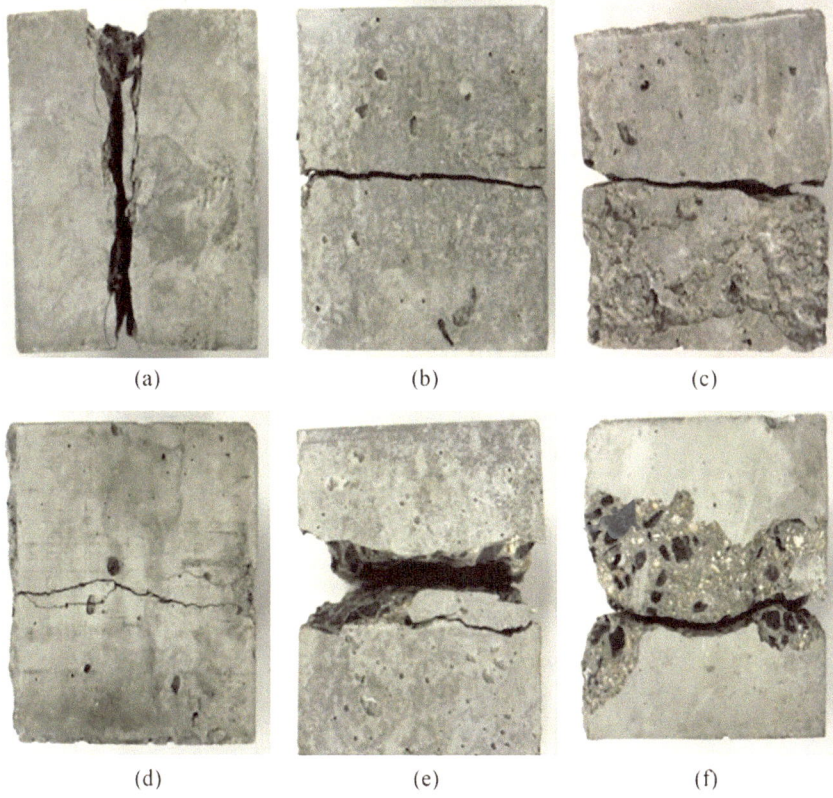

(a) (b) (c)

(d) (e) (f)

Fig. 3 Specimen failure mode

Fig. 4 Effect of different number of wet and dry cycles and temperature on compressive strength of concrete

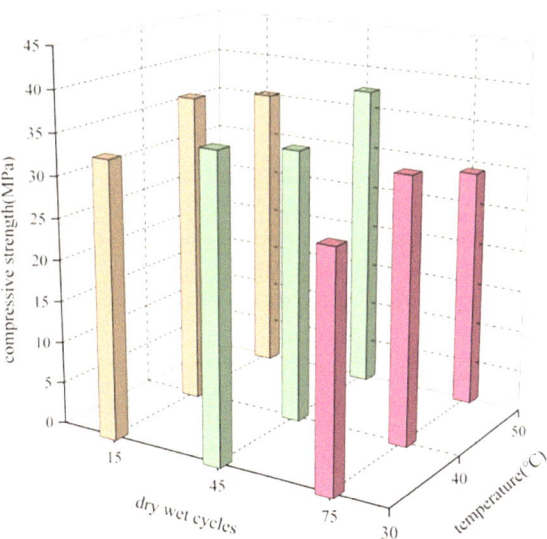

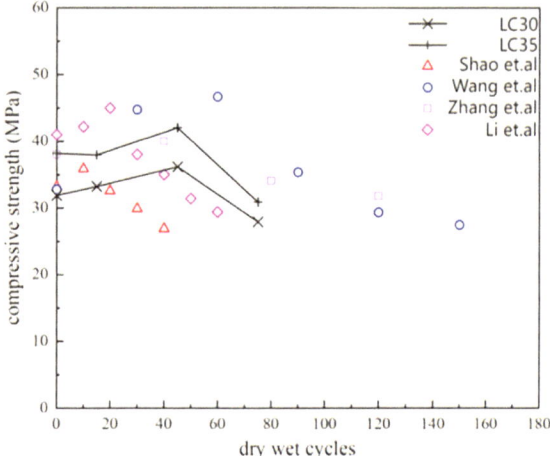

Fig. 5 The compressive strength are compared with the literature

Under the same number of wet and dry cycles, with the increase of temperature, its strength also shows a trend of first increase and then decrease. Taking LC35-30 working condition as an example, LC35-30-15 enhanced the strength of the specimen by 1.1 Mpa than the specimen without wet and dry cycles, with a strength enhancement of 2.8%, and the compressive strength of LC35-30-45 was reduced by 2.2 Mpa compared with 15 cycles, with a loss of strength of less than 1%, whereas LC35-30-75 had a larger change in the strength compared with 45 cycles, which was reduced by 5.2 Mpa, with a strength Loss of strength is 14%.

In order to better carry out the analysis of the mechanism of the effect of the number of dry and wet cycles, the compressive strength values in the test were statistically compared and analysed with the relevant data [13–16], as shown in Fig. 5.

As can be clearly seen from Fig. 6, the change in concrete compressive strength exhibits a similar pattern, but the decline rate of the test data in this paper is significantly more rapid than that of the other tests. This is due to the fact that the other tests only considered the effect of unilateral factor changes in dry and wet cycles on the compressive strength of concrete, ignoring the role that other influencing factors may play, such as the coupling effect of temperature. This can be proved from two aspects, on the one hand, existing studies have directly confirmed that the effect of temperature on the compressive strength of concrete is more significant, which is also reflected in the test data of this paper.

On the other hand, the addition of nanomaterials can effectively increase the compactness of concrete and improve the mechanical properties of concrete, which has been confirmed in related studies. In this paper, nanomaterials are also added, but the compressive strength decline rate of the test data in this paper is even faster than that of the other test data without nanomaterials, which can directly prove that the combined effect of dry and wet cycles and temperature will promote each other and accelerate the deterioration of concrete, so it is imperfect to consider only from the one-sided influencing factors.

Fig. 6 Stress strain curve under LC35-strength

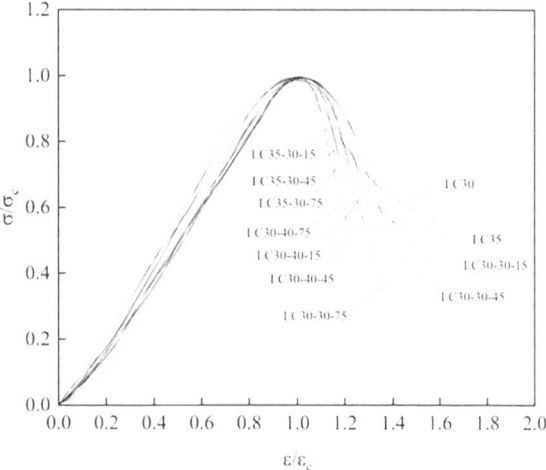

3.4 Stress–Strain Constitutive Equations for Nano-modified Ceramic Concrete

The stress–strain curve of axial compression of concrete is a comprehensive representation of mechanical properties. In order to obtain the stress–strain curve of modified ceramic concrete under dry and wet cycling conditions, this paper pastes strain gauges in the middle of the side of the prismatic concrete specimen, the strain gauges are aligned and centred, and the strain gauges are pasted on both sides in order to exclude the influence of bias, connected to the static collector, and finally, the power supply lead links the static strain testing system to the computer for data acquisition.

The stress–strain curve of vitrified concrete under axial compressive loading can be divided into four stages: elastic, elasto-plastic, plastic and destructive under dry–wet cycling in a hot and humid marine environment. In this chapter, the uniaxial compressive stress–strain full curve test is carried out on SiO_2-modified ceramic concrete prismatic cylinders. The uniaxial compressive stress–strain full curve process, as the most basic kind of ontological relationship of concrete, contains important mechanical parameters, modulus of elasticity, peak stress, and peak strain. It is an important mechanical basis for the finite element analysis of concrete.

From the figure, it can be seen that the overall modulus of elasticity under LC35 strength is smaller than under LC30 strength, indicating that the water-cement ratio is different, the contact reaction area of the amount of cement per unit volume is also different, the smaller the water-cement ratio, the greater the efficacy of the cement and the reaction, the internal such as the pore space and the micro-cracks between the concrete will be smaller, concrete is dense, the better the overall ductility, and the smaller the modulus of elasticity. The modulus of elasticity gradually decreases with increasing temperature, and the decrease decreases with decreasing water-cement ratio. As the concrete is a slow heat conduction speed object, temperature change process inside and outside always exists in the surface of a certain temperature difference, and the internal material of the components of the contraction and expansion deformation is not consistent, will produce fine cracks, due to the role of dry and wet cycle, the internal generation of substances to help fill the fine cracks, ultimately leading to the modulus of elasticity to show a certain rule of change. In the same strength and the same temperature, with the increase in the number of wet and dry cycles, the modulus of elasticity shows a gradual decline in the trend, the smaller the water-cement ratio, the smaller the magnitude of the decline. It can be seen that the wet and dry cycle on the deformation capacity of concrete has a promotional effect, with the cycle, the cracks and pores within the concrete is filled, the crystalline material on the pore wall of the expansion of more than the tensile strength of the concrete material, the pore will be expanded or even the formation of through the interconnection of the cracks. As the wet-dry cycle is still in progress, the newly generated hydration crystallisation products will continue to fill these micropores and penetrating cracks, thus showing a decrease in the modulus of elasticity on a macroscopic scale.

Based on the measured stress–strain curves of the tests, the curves with clearly visible data in the descending section of each group of specimens were selected for the dimensionless analysis, and the horizontal coordinate was expressed by $\varepsilon/\varepsilon_c$, and the vertical coordinate was expressed by σ/σ_c (where ε_c and σ_c are the peak strain and peak stress, respectively). Eleven sets of specimen data were selected for analysis, taking the values corresponding to LC30, LC30-30-15, LC30-30-45, LC30-30-65, LC30-40-15, LC30-40-45, LC30-40-65, LC35, LC35-30-15, LC35-30-45, and LC35-30-65 respectively. The dimensionless stress–strain curves are shown in Fig. 6. As can be seen in Fig. 6: the rising section of vitrified concrete with different temperatures and different numbers of dry and wet cycles does not change significantly before the peak stress is reached, and the curves of each rising section basically overlap, but in the descending section after the peak stress is reached, the curves are more discrete and show an increasingly steeper tendency with the increase of the temperature and the number of dry and wet cycles. The overall shape of the curves was basically similar to the change process of concrete, and both experienced the development from elasticity, elasto-plasticity, descending section to residual section. The descending segment after the peak point of vitrified concrete after the combined effect of temperature and dry and wet cycles is obviously steeper than that of natural vitrified concrete, thus indicating the existence of larger damage defects.

Based on the dimensionless stress–strain curve in Fig. 6, it can be seen that the full curve is similar to that of ordinary concrete in overall shape, so it can be fitted in the form of uniaxial compression eigenstructure equation of ordinary concrete (Eq. (1)), and the results of the fitting are shown in Fig. 7.

$$\begin{cases} y = ax + (3 - 2a) + (a - 2)x^3 \\ y = \dfrac{x}{b(x - 1)^2 + x} \end{cases} \tag{1}$$

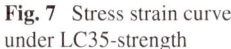

Fig. 7 Stress strain curve under LC35-strength

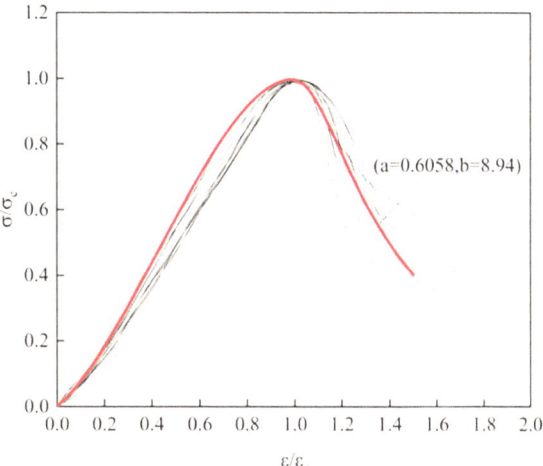

where: parameters a and b are the parameters of the equations controlling the ascending and descending segments, respectively. $x = \varepsilon/\varepsilon_c$ and $y = \sigma/\sigma_c$.

As can be seen in Fig. 7, after statistical regression and data fitting, when a = 0.6058, the fitted curve of the ascending section and the test curve are basically close to each other, and when b = 8.94, the fitted curve of the descending section basically overlaps with the mean value of the test curve, so it is suggested that the expression of Eq. 1 can be adopted for the calculation of vitrified concrete under the combined effect of temperature and dry and wet cycles, and the control parameters of its equations can be taken as follows: a = 0.6058 and b = 8.94.

3.5 Equation for the Change of Compressive Strength Attenuation of Nano-modified Ceramic Concrete Under the Combined Effect of Temperature and Dry–Wet Cycle

The existing equations only consider the effect of unilateral factors on the compressive strength of concrete, which is often inaccurate for the prediction of strength decay in complex marine environments. Therefore, in this paper, based on experimental data and existing equations, the test takes into account the effects of temperature, the number of dry and wet cycles and the compressive strength of nano-modified ceramic concrete without the action of temperature and dry and wet cycles, and establishes an equation for the change of the strong compressive strength attenuation of ceramic concrete under the hot and humid oceanic environment based on the analytical fitting method such as the least-squares method:

$$\alpha = f_{cu}^{T,C,cu} = (1.141T^{0.538} - 0.11T - 3.233) \times (1.397C^{0.318} - 0.063C + 6.771)$$

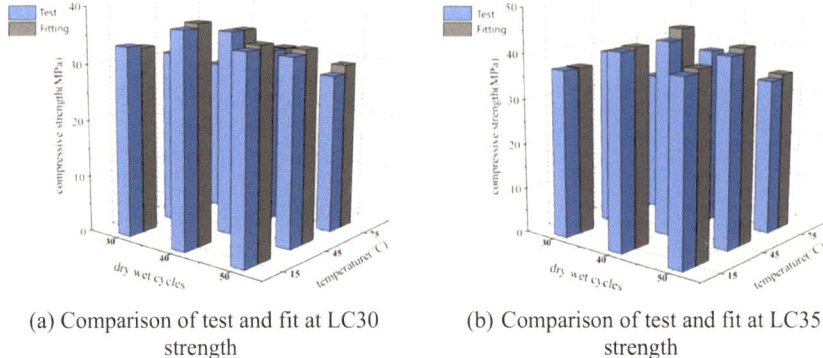

(a) Comparison of test and fit at LC30   (b) Comparison of test and fit at LC35
strength   strength

Fig. 8 Comparison result

$$\times (0.193f_{cu} + 0.09) \tag{2}$$

where $f_{cu}^{T,C,cu}$ is the compressive strength of vitrified concrete, f_{cu} is the compressive strength of vitrified concrete without wet and dry cycle test, T is the temperature, and C is the number of wet and dry cycles.

The fitting formula fitting correlation coefficient R^2 is 0.96, according to the above fitting formula obtained results and test data comparison shown in Fig. 8, through the comparison found that the calculated value and the fitted value error within 5%, indicating that the formula fitting degree is high.

4   Conclusions

(1) The vast majority of the damage of nano-modified ceramic concrete is through aggregate damage, and a few specimens have the phenomenon of sectional Sensei separation.

(2) Under the joint action of dry and wet cycle and temperature, the increase of cycle number and temperature makes the compressive strength of ceramic concrete show the law of first strengthening and then accelerated deterioration. The initial strength increase of ceramic concrete with larger water-cement ratio is larger in dry and wet cycles.

(3) Under the joint action of dry and wet cycles and temperature, the increase of cycle number and temperature makes the splitting tensile strength of ceramic concrete show the change rule of enhancement first and then deterioration. The experimental data of compressive strength and splitting tensile strength of nano-modified ceramic concrete under different cycle times and different temperatures are compared with those shown in the literature, and the change rules coincide with each other.

(4) Based on the stress–strain curve under the combined effect of temperature and dry–wet cycle, the stress–strain constitutive relationship of vitrified concrete is derived.

(5) The use of Origin nonlinear fitting were derived to derive the attenuation law pattern of the strength coupling of nano-modified ceramic concrete with different number of dry and wet cycles, different temperatures and different uncirculated, respectively, and the fitting results have small error.

References

1. Cody, R. D., Cody, A. M., Spry, P. G., & Lee, H. (2001). Reduction of concrete deterioration by ettringite using crystal growth inhibition techniques. *Department of Geological Atmospheric Sciences, Iowa State University*, 111–116.
2. He, K. C., Guo, R. X., & Ma, Q. M. (2016). Experimental research on high temperature resistance of modified lightweight concrete after exposure to elevated temperatures. *Advances in Materials Science and Engineering, 206*(16), 1–6.
3. Li, Z. D., & Meng, D. (2020). Analysis on macroscopic property and microcoscopic control mechanism of nano-SiO$_2$ modified concrete. *Bulletin of the Chinese Ceramic Society, 39*(07), 2145–2153. (in Chinese).
4. Ganjian, E., Canpdat, F., Claisse, P., et al. (2015). Special issue on sustainable construction materials. *Journal of Materials in Civil Engineering, 23*(7), B2015001.
5. Tan, J. (2020). Study on durability test of foam concrete subjected to drying-wetting cycles. *Journal of Transport Science and Engineering, 36*(03), 14–18. (in Chinese).
6. Gao, Y., Zhang, J., & Sun, W. (2012). Measurement of moisture and deformation of concrete under dry wet cycle. *Journal of Tsinghua University (Sciences of Technology), 52*(2), 144–149. (in Chinese).
7. Ye, Q., Zhang, Z. N., & Kong, D. Y., et al. (2003). Comparison of properties of high strength concrete mixed with nano SiO$_2$ and silica fume. *Journal of Building Materials*, (04), 381–385. (in Chinese).
8. Erhan, G., Mehmet, G., & Oday, A. A. (2016). Effect of nano silica on the workability of self-compacting concretes having untreated and surface treated lightweight aggregates. *Construction and Building Materials, 115*, 371–380.
9. Li, G. H., & Gao, B. (2007). Effect of nano-SiO2 on salt crystallization cycle performance of concrete. *Journal of Southwest Jiaotong University, 42*(1), 70–74. (in Chinese).
10. Wang, Y. (2021). *Experimental study on mechanical properties and durability of modified ceramsite concrete in hot and humid marine environmen.* Shandong University of science and technology. (in Chinese).
11. Lightweight aggregate concrete application technical standards. China Construction Industry Press (2021).
12. Test methods for long-term performance and durability of normal concrete. China Construction Industry Press (2009).
13. Shao, H. J., Li, Z. L., & Xiao, S. P., et al. (2021). Mechanical properties and microstructure of concrete under drying-wetting cycles. *Bulletin of the Chinese Ceramic Society, 40*(09), 2949–2955. (in Chinese).
14. Wang, K. (2020). *Experimental study on deterioration of mechanical properties of concrete under sulfate attack in dry wet alternating environment.* Zhengzhou University. (in Chinese).
15. Zhang, T. H. (2019). *Study on the deterioration law of concrete performance under the coupling of sulfate attack and dry wet cycle.* Xijing University. (in Chinese).

16. Li, S. C., Zhang, F., & Zhu, J. P., et al. (2009). Test on deterioration of mechanical properties of concrete under seawater erosion environment. *Journal of Highway and Transportation
Research and Development, 26*(12), 35–38. (in Chinese).

Enhancing the Performance of Oil Well Cement Strength with Graphene Oxide

Zihe Li, Jiangtao Xu, Xia Miao, Wei Cao, Jinqi Zhang, Duyou Lu, and Xiao Yao

Abstract This study aims to use graphene oxide to improve the strength properties of oil well cement. The fluidity, setting time, strength properties, and hydration products of oil well cement paste with different GO dosages (0–0.05 wt%) were investigated. The fluidity of cement paste decreased linearly with increasing GO dosage was indicated. The addition of GO not only promoted cement hydration but also reduced the R-index of portlandite, which densified the microstructure of hardened paste and thus enhanced the strength properties. By using a 0.02% dosage of graphene oxide, the compressive strength of oil well cement pastes increased by 28.9% and 11.1% at 1 day and 28 days, respectively. Meanwhile, the flexural strength of cement paste increased by 72.5 and 34.1% compared to control. The enhancement of strength properties reduces the brittleness of oil well cement pastes, and improve wellbore integrity and cementing quality.

Keywords Oil well cement · Graphene oxide · Strength · Hydration products

1 Introduction

Cementing is the essential part of the process of well construction. It's targets to isolate oil, gas, and liquid formations and guarantee effective oil production, and seal the annular area between the casing and the rock formation. Oil well cement, widely used for well cementing, exhibits microscopic defects in its hardened paste, including poor crack resistance, a high brittleness coefficient, and a tendency to develop micro-cracks. Cracks in the cement sheath will cause tensile failure and deteriorated sealing performance in extreme environments like hydraulic fracturing, high temperature differences, and pressure squeeze. This compromises the wellbore's

Z. Li · J. Xu (✉) · X. Miao
Sinopec Key Laboratory of Cementing and Completion, Beijing 102206, China
e-mail: xujiangtao18@163.com

Z. Li · J. Xu · W. Cao · J. Zhang · D. Lu · X. Yao
College of Materials Science and Engineering, Nanjing Tech University, Nanjing 210009, Jiangsu Province, China

© The Author(s) 2025
D. Li and Y. Zhang (eds.), *Advances in Frontier Research on Engineering Structures II*,
Lecture Notes in Civil Engineering 535, https://doi.org/10.1007/978-981-97-6238-5_4

integrity and cementing quality, endangering the safety and successful operation of oil and gas wells [1]. Various reinforcing additives are used to improve the embrittlement of oil well cements, including fibrous, latex and powdered rubber. However, these compounds also compromise the strength properties, compatibility, and thermal stability of oil well cement [2, 3]. In order to achieve high-quality cementation, it is essential to investigate superior reinforcing materials to increase the strength and fragility of oil well cement paste.

As a two-dimensional nanomaterial, graphene oxide (GO) exhibits unique characteristics., such as high strength, resilience, and extensive surface area. Numerous experimental findings [4, 5] indicate that incorporating GO into cementitious materials may enhance the strength of hardened paste. Wang [6] reported that 27.1% compressive strength and 16.4% of flexural strength were improved with the addition of 0.08 wt% GO in cement paste. Similar improvement in strength properties was also observed by Lv [7], who discovered that GO could promote the formation of regular hydrates and that it could compact the microstructure of the hardened cement paste. Pan [8] found GO may serve as a barrier to crack propagation because of its special 2D geometry and high aspect ratio, thereby enhancing cement properties. Because of its abundant carboxylic groups, GO is prone to agglomerate in cement paste, which can lead to unstable dispersion [9, 10]. The application of GO in the cement matrix is hampered by these agglomerates. According to several studies, polycarboxylate superplasticizer [11, 12] can enhance the GO's dispersion in cement paste.

Currently, little attention was paid to oil well cement and most research on the use of GO in cement composites is concentrated on regular Portland cement. Although the physical composition of oil well cements is similar to that of the ordinary Portland cement. Limited by stringent service requirements, the service environment of oil well cement is different from those of OPC. The oil well cement is exposed to underground environment with high temperatures. It is well known that the hydration products of cement change with temperature increased. Therefore, GO may have different effects in oil well cement with the ordinary Portland cement, the feasibility and mechanical properties of GO reinforced oil well cement need to be studied.

In this study, the optimal proportion of PCE for dispersing GO was determined by visual inspection and UV–vis. Subsequently, the flowability, setting time, strength properties, and hydration products of oil well cement paste with various GO dosages were investigated after curing at 50 °C. Hydration products were investigated using XRD analysis.

Table 1 Chemical composition of oil well cement/wt%

SiO$_2$	CaO	TiO$_2$	Fe$_2$O$_3$	Al$_2$O$_3$	SO$_3$	K$_2$O	MgO	Na$_2$O	LOI
21.74	63.33	0.22	5.28	3.71	1.80	0.52	1.70	0.14	1.56

Table 2 Mineral composition of oil well cement/wt%

C$_3$S	C$_2$S	C$_3$A	C$_4$AF	CaSO$_4$	LOI
49.76	24.78	0.93	16.03	4.36	4.14

2 Experimental

2.1 Materials

The oil well cement was purchased from China Gezhouba Group Corporation. The chemical composition and mineral composition of oil well cement was shown in Tables 1 and 2 respectively. Aqueous dispersions of GO (4g/L) used was purchased from Suset Piezotronics Ltd. Polycarboxylate superplasticizer (PCE) was produced from Sobute New Materials Co., Ltd and calcium hydroxide (CH) was produced by Shanghai Lingfeng Chemical Reagent Co., Ltd, both of them were used to study the dispersibility of GO.

The X-ray Photoelectron Spectroscopy (XPS, Thermo Scientific K-Alpha) was used to test the oxygen content of GO, the results were shown in Fig. 1a, b. In Fig. 1a, it is clear that the oxygen content of GO is 35.27%, originating from the hydroxyl, carboxyl, epoxy, and carbonyl groups resulting in the oxidation of graphite flakes. The C1s pattern of GO was presented in Fig. 1b, revealing the proportion of C=O/C–O (286.6 eV), C=C/C–C (285.1 eV), and O–C=O (288.5 eV) are 30.4%, 50.5%, and 19.1%, respectively. Additionally, the Atomic Force Microscope (AFM, Bruker Dimension Icon) was used to test the size of GO, and the results were shown in Fig. 1c, d, indicating that the GO nanosheets exhibit sizes ranging from 0.5 to 2 μm, and the height of graphene oxide is about 1 nm.

2.2 Testing Method

2.2.1 The Dispersion of GO in CH

A saturated solution of calcium hydroxide, simulating the high alkalinity and Ca^{2+} content found in cement paste environments, was prepared to investigate the dispersibility of GO. The mass ratio of polycarboxylate superplasticizer (PCE) to GO was selected as GO-PCE1, GO-PCE2, GO-PCE4, and GO-PCE6 (the mass ratio of PCE to GO is represented by the number following PCE). Visual inspection and

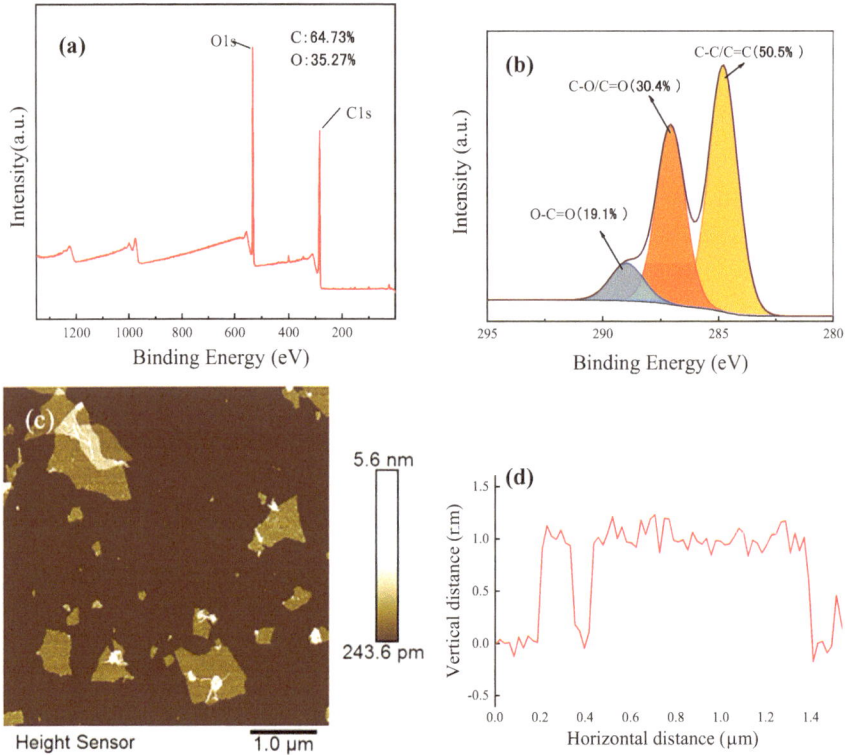

Fig. 1 XPS spectrum of graphene oxide (**a**) high-resolution C1s XPS spectra for graphene oxide (**b**) AFM micrograph of graphene oxide (**c**) height of graphene oxide (**d**)

UV spectrophotometer (Inesa-N4S) at a wavelength of 210 nm were used to assess the dispersibility of GO.

2.2.2 Preparation of Oil Well Cement Paste

GO was added at a weight percentage of 0–0.05 in relation to oil well cement while the dispersant was administered at twice the weight of GO, based on the GO dispersion results. A constant water/cement ratio of 0.44 was used to produce the cement paste. Table 3 shows a detailed experimental design. The fresh pastes were mixed and then poured into Ø25.4 × 25.4 mm molds for compressive strength and (20 mm × 20 mm × 80 mm) molds for flexural strength, and then cured in water at 50 °C.

Table 3 Formulation design for cement pastes

Sample	Cement (%)	GO (%)	PCE (%)
Control	100	0	0
GO-1	100	0.01	0.02
GO-2	100	0.02	0.04
GO-3	100	0.03	0.06
GO-4	100	0.04	0.08
GO-5	100	0.05	0.10

2.2.3 Properties of Cement Paste

Cement paste was tested according to GB/T 1346-2011 procedures for flowability and setting time. The compressive strengths of hardened paste were determined utilizing the MTS CDT1305-2 compressive strength tester with a loading rate of 1 mm/min. Likewise, the flexural strength of hardened cement paste was assessed via the MTS CDT1305-2 concrete three-point flexural strength tester with a loading rate of 10 N/s.

2.2.4 Hydration Products

To identify mineral composition and determine hydration products, X-ray diffraction (XRD) was used. The XRD (Model Rigaku-25001) radiation source is Cu Kα ($\lambda =$ 1.54059 Å), the measurement range of 2θ was 5–70° with the scanning speed is 5 °/min.

3 Results and Discussion

3.1 Characterize of Dispersive Properties of GO in CH

As an important factor influence on the improvement of the strength development, the dispersion of GO in the cement composites is a direct evaluation mothed. A simulation experiment was carried out by dispersing GO in a saturated calcium hydroxide (CH) solution to determine the state of optimum dispersion of GO in cement paste. The dispersion performance of different mass ratios of GO to PCE in CH was visually observed for up to 4 h, as depicted in Fig. 2. It is clear that the control without dispersant exhibited severe agglomeration of GO. However, the incorporation of PCE considerably enhanced the dispersion of GO within the CH solution. With the extension of time, groups 1:4 and 1:6 gradually began to appear GO agglomerates and settled, which investigated that excessive PCE is detrimental to GO

Fig. 2 States of
sedimentation of the
GO-PCEn with various time
intervals

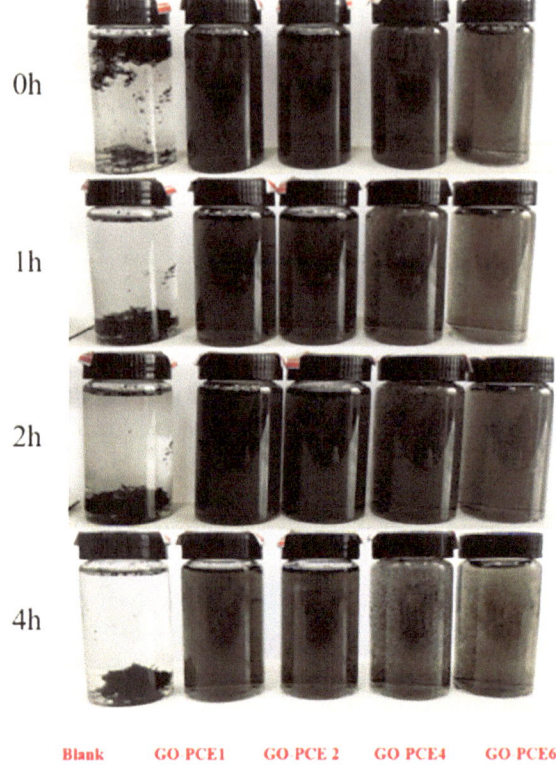

dispersion. Based on visual observation, the GO-PCE1 and GO-PCE2 samples, with mass ratios of 1:1 and 1:2 respectively, demonstrated the most effective dispersion.

To characterize the dispersion of GO, UV–Vis spectroscopy is a widely used technique. The degree of dispersion of GO in a CH solution is directly correlated to higher absorbance in the UV–Vis test. Figure 3 shows the absorbance of the samples after dispersing 4 h. Absorbance values of PCE containing samples were significantly higher than PCE free control. It is evident that PCE plays a significant role in the dispersion of GO in the CH solution. This phenomenon is due to PCE's electrostatic repulsive and steric hindering properties. The absorbance values of the samples demonstrate an initial increase followed by a subsequent decrease with an increasing mass ratio of GO to PCE. Excessive PCE has the potential to form micelles [10, 13], which is consistent with visual observations. The highest absorbance value is observed in the GO-PCE2 sample, suggesting that the optimal ratio is 1:2. Therefore, a mass ratio of 1:2 was selected for the subsequent experiments.

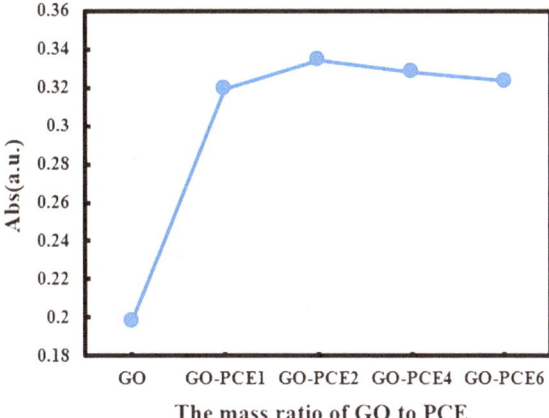

Fig. 3 Absorbance of GO-PCE with various GO and PCE mass ratios in CH at 4 h

3.2 Fluidity and Setting Time of Oil Well Cement Paste

The fluidity of cement paste with various GO dosages was investigated in Fig. 4, which investigated that the fluidity of oil well cement paste exhibited a linear decrease with increasing GO dosages. This finding is consistent with similar results by previous reports, the GO has trapped a significant amount of water and exerted a detrimental impact on the workability of cement paste used in oil well. Compared to control, the samples containing 0.05% GO exhibited a fluidity of 140 mm, which decreased by 30%. This indicate that GO may deteriorate the pumping performance of oil well cement paste.

Figure 5 presents the setting time results for oil well cement paste. It was observed that the final setting time of cement paste decreased gradually with increasing GO dosages, while the initial setting time showed an increase followed by a decrease.

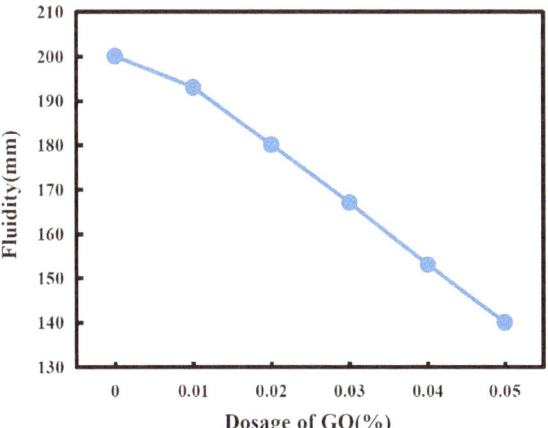

Fig. 4 Fluidity of oil well cement paste with GO

Fig. 5 Setting time of oil well cement paste with GO

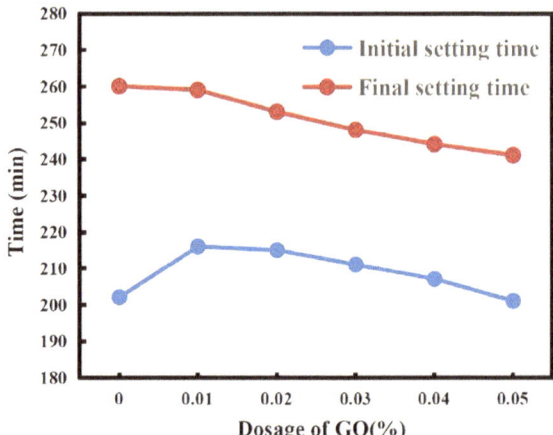

The addition of graphene oxide led to minor alterations in the setting time of cement pastes, with the degree of variation ranging between 0.5% and 4%.

3.3 Compressive and Flexural Strength of Cement Paste

The strength performance of oil well cement paste with different GO dosages at 50 °C was illustrated in Fig. 6. With the increasing of curing age, both the compressive and flexural strength of cement paste increased. Compared to the control, the inclusion of GO in cement paste led to a significant boost in both compressive and flexural strengths, particularly during the early age of curing. This enhancement demonstrates a pattern of initial increase followed by subsequent decrease as the GO dosages increase. However, the compressive and flexural strength trends of cement paste exhibit dissimilarity. As illustrated in Fig. 6a, compared to the control, the compressive strength of the oil well cement paste with GO dosage of 0.02% is enhanced by 28.8% and 11.1% at 1 and 28 days, respectively. On the contrary, an increased quantity of GO at 0.05wt% results in a reduction of 3.7% and 13.0% at 1 and 28 days, respectively.

This may be explained by the higher dosage of GO in the cement paste aggregate, which created a barrier and prevented the development of hydration products, thereby reduced the degree of cement hydration. This finding agrees with existing literature [14, 15], which supported the notion that an appropriate addition of GO improves strength, while excessive admixture adversely affects strength. The flexural strength of the oil well cement paste was shown in Fig. 6b, indicating a more pronounced improvement in flexural strength compared to compressive strength in samples with GO additions. Specifically, the flexural strength increases by 72.5% and 34.1% at 1 and 28 days, respectively, with a 0.02% GO dosage. Interestingly, even with the addition of 0.05wt% GO, a significant increase of 33.1% at 1 day and 11.3% at

Fig. 6 The compressive
(**a**) and flexural strength
(**b**) of oil well cement paste
with various GO dosages

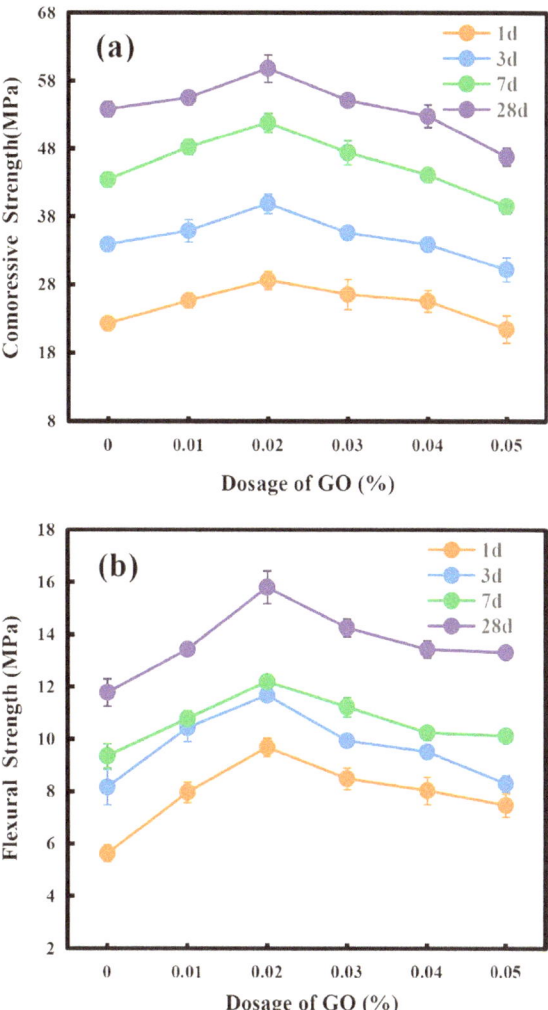

28 days is observed. This confirmed the substantial enhancement of oil well cement, particularly in terms of flexural strength. Additionally, the fact that flexural strength improved more than compressive strength in GO-modified samples suggests that GO additions lower the brittleness of cement paste and improve its toughness.

3.4 XRD Analysis

The XRD patterns of cement paste with different GO additions after curing in 50°C was shown in Fig. 7, which investigated that the main phases present in all samples

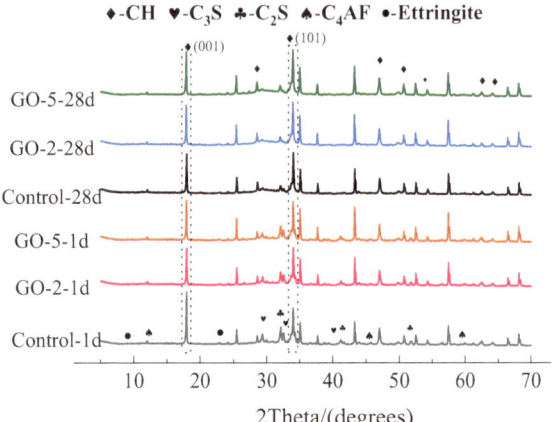

Fig. 7 XRD of Control, GO-2, GO-5 at 1 day and 28 days

were C_3S, C_2S, C_4AF, portlandite, and ettringite. After 1 day of curing, the intensities of the diffraction peaks corresponding to C_3S and C_2S in GO-2 and GO-5 samples were lower than those of the control, indicating that GO additions could promote the reaction of the silicate phase and accelerate the early hydration of cement. After curing in 28 days, this phenomenon will be weakened because each sample exhibited similar hydration products.

Furthermore, it is noteworthy that the inclusion of GO caused significant alterations in the diffraction peaks of portlandite, particularly in early stage. When curing time are 1 and 28 days, the samples containing GO showed higher diffraction peaks in the portlandite (101) plane and lower peaks in the (001) plane, when compared to the control. In accordance with literature, the R-index of portlandite plays an important role in the strength development of cementitious materials containing nanoparticles, mechanical properties increase as the R-index of the portlandite decreases [16, 17]. The R-index of portlandite can be calculated according to Eq. (1).

$$R = 1.35 \times Intensity\ (001)/Intensity\ (101) \tag{1}$$

Table 4 shows the R-index of samples. It is obvious that the R-index of samples with GO were lower than that of the control, with the minimum values in GO-2 sample. The decreased R-index could possibly be linked to the interaction between the oxygenated functional groups of GO with the cement pore solution containing Ca^{2+}, which controls the crystal morphology of portlandite [9, 12]. Overall, GO additions may accelerate the hydration of oil well cement paste and improve the microstructure of oil cement paste, thereby contributing to the strength development.

Table 4 The R-index of GO-0, GO-2, GO-5 at 1 day and 28 days

Age (days)	Sample	R
1	Control	1.89
	GO-2	1.34
	GO-5	1.38
28	Control	1.30
	GO-2	1.26
	GO-5	1.30

4 Conclusions

The study examines the impact of GO on the oil well cement's efficiency at 50 °C. The following conclusions can be drawn based on the experimental findings:

1. The fluidity of oil well cement paste exhibited a linear decrease with increasing GO dosages, and the final setting time decreased with increasing GO dosage, while the initial setting time of cement paste increased and then decreased with increasing GO dosages.
2. The strengths performance of cement paste increased with the dosages of GO, especially in early curing stage. However, excessive dosages of GO showed a negative impact on compressive strength. When the GO dosage is 0.02%, the compressive strength of oil well cement paste increased by 28.8% and 11.1% at 1 and 28 days respectively, and the flexural strength increased by 72.5% and 34.1% after 1 and 28 days respectively.
3. GO promoted the reaction of the silicate phase and modified the crystal orientation of portlandite, resulting in a more compact and dense paste. These may be responsible for the enhancement of strengths in oil well cement hardened paste with GO addition.
4. The use of GO can enhance both the toughness and brittleness of oil well cement paste addressing the microscopic defects and enhancing its performance as a cement sheath for well construction.

Overall, the results suggest that by enhancing the strength properties and toughness of the cement paste, GO is a potential nano-strengthening additive for the cement used in oil well. The enhancement can lead to improved wellbore integrity and cementing quality.

Acknowledgements This research was funded by the Sinopec Well Cementing and Completion Key Laboratory Open Fund (21-GWJ-KF-1). The work was supported by the National Natural Science Foundation for Young Fund of China (52102025) and the Priority Academic Program Development (PAPD) of Jiangsu Higher Education Institutions.

References

1. Ulm, F. J., & James, S. (2011). The scratch test for strength and fracture toughness determination of oil well cements cured at high temperature and pressure. *Cement and Concrete Research, 41*(9), 942–946.
2. Zhang, C., Cai, J., Xu, H., Cheng, X., & Guo, X. (2020). Mechanical properties and mechanism of wollastonite fibers reinforced oil well cement. *Construction and Building Materials, 260*, 120461.
3. Li, M., Yang, Y., & Guo, X. (2015). Mechanical properties of carbon fiber reinforced oil well cement composites. *AMCS, 32*(3), 782–788.
4. Wang, X., & Zhong, J. (2023). Revisiting the strengthening mechanisms of graphene oxide reinforced cement: Effects of dispersion states. *Cement and Concrete Research, 170*, 107189.
5. Chintalapudi, K., & Pannem, R. M. R. (2020). The effects of Graphene Oxide addition on hydration process, crystal shapes, and microstructural transformation of Ordinary Portland Cement. *Journal of Building Engineering, 32*, 101551.
6. Wang, L., Zhang, S., Zheng, D., Yang, H., Cui, H., Tang, W., & Li, D. (2017). Effect of Graphene Oxide (GO) on the morphology and microstructure of cement hydration products. *Nanomaterials, 7*(12), 429.
7. Lv, S., Hu, H., Zhang, J., Luo, X., Lei, Y., & Sun, L. (2017). Fabrication of GO/cement composites by incorporation of few-layered GO nanosheets and characterization of their crystal/chemical structure and properties. *Nanomaterials, 7*(12), 457.
8. Pan, Z., He, L., Qiu, L., Korayem, A. H., Li, G., Zhu, J., Collins, F., Li, D., Duan, W., & Wang, M. (2015). Mechanical properties and microstructure of a graphene oxide–cement composite. *Cement and Concrete Composites, 58*, 140–147.
9. Wang, Q., Wang, J., Lv, C., Cui, X., Li, S., & Wang, X. (2016). Rheological behavior of fresh cement pastes with a graphene oxide additive. *New Carbon Materials, 31*(6), 574–584.
10. Ghazizadeh, S., Duffour, P., Skipper, N. T., Billing, M., & Bai, Y. (2017). An investigation into the colloidal stability of graphene oxide nano-layers in alite paste. *Cement and Concrete Research, 99*, 116–128.
11. Plank, J., Sakai, E., Miao, C. W., Yu, C., & Hong, J. X. (2015). Chemical admixtures—Chemistry, applications and their impact on concrete microstructure and durability. *Cement and Concrete Research, 78*, 81–99.
12. Wang, Q., Zhan, D., Qi, G., Wang, Y., & Zheng, H. (2020). Impact of the microstructure of polycarboxylate superplasticizers on the dispersion of graphene. *New Carbon Materials, 35*(5), 547–558.
13. Zhao, L., Zhu, S., Wu, H., Zhang, X., Tao, Q., Song, L., Somh, Y., & Guo, X. (2020). Deep research about the mechanisms of graphene oxide (GO) aggregation in alkaline cement pore solution. *Construction and Building Materials, 247*, 118446.
14. Kang, D., Seo, K. S., Lee, H., & Chung, W. (2017). Experimental study on mechanical strength of GO-cement composites. *Construction and Building Materials, 131*, 303–308.
15. Qureshi, T. S., & Panesar, D. K. (2019). Impact of graphene oxide and highly reduced graphene oxide on cement based composites. *Construction and Building Materials, 206*, 71–83.
16. De Matos, P. R., Andrade Neto, J. S., & Campos, C. E. M. (2021). Is the R index accurate to assess the preferred orientation of portlandite in cement pastes? *Construction and Building Materials, 292*, 123471.
17. Liu, J., Fu, J., Yang, Y., & Gu, C. (2019). Study on dispersion, mechanical and microstructure properties of cement paste incorporating graphene sheets. *Construction and Building Materials, 199*, 1–11.

Intelligent Analysis Research on Reservoir Dam Structure Settlement Prediction in Coal Mining Subsidence Area

Qiang Chang, Hao Yan, Daiyao Zhao, and Ning Zhang

Abstract When constructing a dam in a mined-out area, it is necessary to handle the mined-out area effectively. Analyzing the distribution of mined-out areas is essential to ensure the safety of dam structures. In response to this issue, a machine learning-based method for analyzing the settlement of reservoir dam structures in coal mining subsidence areas is proposed. The method includes the following steps: first, analyze the geological conditions of the mined-out area and calculate the conditions for each depth of the mined-out area. Secondly, analyze the deformation mechanism of the mined-out area and calculate the impact of the mined-out area at representative depths. Next, use machine learning methods to regressively fit the deformation displacement and obtain the deformation function. Finally, validate the predicted depth based on on-site monitoring data and drilling information. The study concludes that deeper mined-out areas lead to larger total displacement, and as the depth increases, the displacement caused by self-weight stress gradually increases. Additionally, the displacement response of the dam load decreases with the increase in depth. The predicted depths align well with the actual depths. The proposed method is reasonable and feasible, providing a basis for reinforcement schemes in mined-out areas.

Keywords Mined-out area · Roof Strata · Gaussian process regression · Deformation · Monitoring

1 Introduction

During the process of coal mining, a large number of goafs are generated. If these goafs are not properly managed, they may have adverse effects on the surrounding environment and human safety. Goafs formed during coal mining will eventually collapse after a certain period of time, and comprehensive management of these

Q. Chang (✉) · H. Yan · D. Zhao · N. Zhang
Powerchina Guiyang Engineering Corporation Limited, Guiyang 550081, China
e-mail: 276455902@qq.com

© The Author(s) 2025
D. Li and Y. Zhang (eds.), *Advances in Frontier Research on Engineering Structures II*,
Lecture Notes in Civil Engineering 535, https://doi.org/10.1007/978-981-97-6238-5_5

collapsed areas is widely regarded as a global challenge [1, 2]. Machine learning technology for the treatment of reservoir dam structures in mined-out areas refers to the research on the structural management of reservoir dam bodies in mined-out areas using machine learning methods. The structural integrity and stability of the dam body need to be ensured. Therefore, employing machine learning techniques for the treatment of reservoir dam structures in mined-out areas can effectively improve the treatment results.

In the research on comprehensive management of mined-out areas, many scholars have conducted relevant studies. For example, based on reflection wave exploration data from actual engineering projects, Qi [3] used numerical simulation methods based on finite difference method to establish geological models of various types of collapse areas, and conducted forward numerical simulation on these geological models. Zhou [4] proposed measures for mined-out area treatment including boundary detection, grouting filling, no-sealing edge grouting method, and grouting effect detection and acceptance, addressing risks such as subsidence and collapse during interval shield construction in coal mining areas. The results of later shield construction and subway operation showed that the reinforced mined-out areas using this technology were safe and compact. In the current popular field of artificial intelligence, many scholars combine machine learning with mined-out area treatment. For instance, Song [5] developed an intelligent evaluation method for gas drainage based on Long Short Term Memory (LSTM) networks, forming an intelligent evaluation method for gas drainage in mined-out areas. Li [6] and others established a prediction model for mined-out area stability time based on support vector machine through the prediction of stability time test samples in mined-out areas.

In the research on the treatment of reservoir dam structures in mined-out areas, machine learning can utilize abundant data resources and powerful computing capabilities to conduct complex analysis and modeling [7, 8]. This paper proposes using machine learning to identify the influence of different factors on the stability of dam structures based on historical and experimental data, and predicts the response and performance of structures under different working conditions.

2   Project Overview

The ravine in question is a continuous channel approximately 2 km long, oriented nearly north–south (SN direction). It is situated beneath the watershed of high mountains, with sloping hilly terrain on both sides, symmetrically distributed along the banks. The upper mountain ranges generally run in a northeast direction, forming a string of bead-shaped, chessboard-like circular hills and hillocks with interlacing ravines in the lower sloping hilly area. It is located in the pass region of the towering mountain range on the left bank. The overall longitudinal slope of the ravine bottom is gentle, with gradients less than 5°. The terrain on both sides of the dam site is steep, resembling a "V"-shaped ravine. The upstream and downstream sections have relatively open terrain, especially the downstream section. The elevation of the ravine

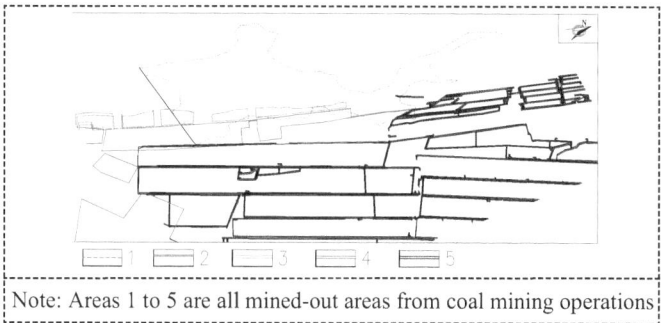

Note: Areas 1 to 5 are all mined-out areas from coal mining operations

Fig. 1 Schematic diagram of Li Zi Ya reservoir and surrounding coal mine distribution

bottom ranges from 873 to 910 m above sea level, while the hilltop elevations on both banks are generally between 970 and 990 m. The hilltop elevations of the upper mountain ranges on the left bank range from 1280 to 1400 m, and on the right bank, they range from 1100 to 1200 m.

This area near the reservoir and its surroundings have a history of intense coal mining activity, with several coal mines distributed in the region (see Fig. 1). Mining activities near the reservoir have been ongoing for a long time, resulting in extensive mined-out areas. At present, all coal mines have been closed down.

This reservoir is located at the southern boundary of the coal mining field, adjacent to mining faces 16,032 (2007, 2008, 2009 comprehensive mining), 16,011 (2004, 2005 comprehensive mining), and the old mining areas of 1992 and 1993. There are +680 south main haulage roadways connecting the new and old mining faces, forming a grid-like layout. There are 3 m wide protective coal pillars between adjacent working faces. After projecting the mining faces onto the surface, the left bank area of the reservoir is partially within the mining face 16,032, with a depth of 171.3–197.9 m in the goaf area. The mining area extends northeast along the direction of the strata, gradually moving away from the reservoir. Except for the old mining areas, which were manually mined, other mining faces utilized mechanical comprehensive mining methods. The roof management method involves natural subsidence, and the immediate roof rock is carbonaceous mudstone. These mining faces are located near the inclined core, mostly on the northwestern wing of the syncline, with a few on the southeastern wing. The coal seam dip angle ranges from 3 to 35°. According to relevant engineering data, the mining height near the reservoir area is 2.02 m.

3 Predicting Subsidence in Mined-Out Areas Based on Machine Learning

3.1 Gaussian Process Machine Learning

Gaussian Process Regression (GPR) is a machine learning method based on Bayesian theory [9, 10]. Bayesian regression methods effectively avoid overfitting issues seen in other machine learning methods. By defining a distribution over functions, each possible function is assigned a prior probability, with functions having higher likelihoods given higher prior probabilities.

In the process of establishing a Gaussian Process Regression model, a key factor determining the accuracy of the regression model is the choice of the covariance function, also known as the kernel function. However, not all covariance functions are effective; their suitability depends on whether they satisfy certain properties of the covariance matrix, such as symmetry and positive definiteness. Some commonly used covariance functions include:

(1) The Squared Exponential Covariance Function (Squared Exponential Covariance Function)

$$k_{se}(x_p, x_q) = \sigma_f^2 \exp(-\frac{1}{2l^2}\|x_p - x_q\|^2) + \sigma_n^2 \delta_{pq} \qquad (1)$$

In the equation, l, σ_f and σ_n are hyperparameters. l represents the isotropic length scale, with a number of scales equal to the dimensions of the variables. When the exponent approaches 1, it indicates a higher correlation between two data points; as it approaches 0, it signifies a lower correlation between the two data points. σ_f represents the degree of local correlation, and σ_n represents the standard deviation of the noise.

(2) The Matérn Covariance Function (Matern Covariance Function).

When the parameter $v = 3/2$ in the Matérn covariance function, the expression is as follows:

$$k_{v=3/2}(x_j, x^*) = \sigma_f^2 \left(1 + \frac{\sqrt{3\|x_j - x^*\|^2}}{l}\right) \exp\left(-\frac{\sqrt{3\|x_j - x^*\|^2}}{l}\right) + \sigma_n^2 \delta \qquad (2)$$

When the parameter v = 5/2 in the Matérn covariance function, the expression is as follows:"

$$k_{v=5/2}(x_j, x^*) = \sigma_f^2 \left(1 + \frac{\sqrt{5\|x_j - x^*\|^2}}{l} + \frac{5\|x_j - x^*\|^2}{3l^2}\right) \exp\left(-\frac{\sqrt{5\|x_j - x^*\|^2}}{l}\right)$$
$$+ \sigma_n^2 \delta \qquad (3)$$

In the equation l, σ_f and σ_n are hyperparameters, with the same meanings as in the squared exponential covariance function.

(3) The Linear Covariance Function (Linear Covariance Function, abbreviated as LIN)

$$k_{\text{LIN}}(x_j, x^*) = \frac{x_j x^*}{l^2} + \sigma_n^2 \delta \tag{4}$$

Similarly, we can combine the above covariance functions to create new functions that satisfy the requirements of symmetry and positive definiteness. This process can be guided by the construction of kernel functions in support vector machines.

3.2 Calculation of Differential Settlements in Mined-Out Areas with Different Dam Bodies

To evaluate the feasibility of constructing a reservoir in the karst geological conditions near the mined-out areas, and to study the reservoir's conditions in the complex geological environment affected by both karst and mined-out areas, mathematical methods will be employed to calculate the geological model.

After the coal seams underground are extracted, the roof rock in the mined-out area undergoes complex movements and deformations. The immediate roof rock layer in the mined-out area moves downward and bends under the influence of its own weight and the overlying rock layers. Scholars from both domestic and international research have conducted extensive on-site and theoretical studies on the mechanisms of movement and deformation in coal mining rock layers. These studies roughly classify the overlying rock layers into three different mining impact zones (collapse zone, fault zone, and bending zone) based on their degree of destruction.

When the additional stress depth of the dam foundation reaches the contact point of the water-conducting fault zone, it will cause the mined-out area to "activate," leading to further deformation in the mined-out area. According to the "Code for Rock and Soil Engineering in Mined-Out Areas of Coal Mines" the following formulas are derived:

(1) When $H \leq k\,(H_{li} + D_Z)$, the dam load will cause the mined-out area to "activate," and the foundation will be in an unstable state.
(2) When $H \geq k\,(H_{li} + D_Z)$, the dam load will not cause the mined-out area to "activate," and the foundation will be in a stable state.

Where, H is the depth of mining in the mined-out area, in meters; H_{li} is the height of the water-conducting fault zone, in meters; D_Z is the depth affected by the dam load, in meters; k is the safety factor, taken as 1.5.

The simulation involves establishing a 2D finite element computational model to analyze the settlement in the backfilled area of a coal mine in both longitudinal and transverse directions. Assuming a horizontal coal mining area with a height of 2 m, different burial depths are considered along the cross-sections at the bottom of the dam. The horizontal distance between the intersection point of the dam's top (bottom) and the slope surface (additional stress boundary at the foundation) extended at a 45° angle to the vertical stress direction is taken as the width of the mined-out area at a certain depth.

The stress field is modeled using a gravitational field, with the top boundary representing the actual ground elevation. The ratio of effective stress inside and outside the vertical surface is set to 0.5, and the density of the overlying layer is 27 kN/m³.

The material behavior is described using the Mohr–Coulomb constitutive model, considering the rock mass as an isotropic elastic body. Geotechnical parameters are shown in Table 1.

Please establish calculation models for the longitudinal and transverse sections as shown in the Fig. 2.

H1 (20 m), H2 (40 m), H3 (58 m), H4 (73 m), H5 (98 m), H6 (137 M), H5 (196 M) were used as the roof depth of the goaf for simulation calculation, and the calculation results were as follows: constrained left and right boundary X direction displacement, bottom boundary Y direction displacement (Fig. 3).

The vertical displacement diagram (Fig. 3) shows that the displacement is mainly concentrated at the riverbed bottom and the middle of the mined-out area. The

Table 1 Calculation parameters

Material	Density	Elastic modulus (Mpa)	Tensile strength (Mpa)	Poisson's ratio	Cohesion (Kpa)	Internal friction angle (°)
p3c	2680	6.73E + 04	8.00E + 00	0.21	1.96E + 03	40
p3l	2660	1.94E + 04	4.45E + 00	0.24	1.59E + 03	30
p2m	2660	1.94E + 04	8.00E + 00	0.21	1.59E + 03	40

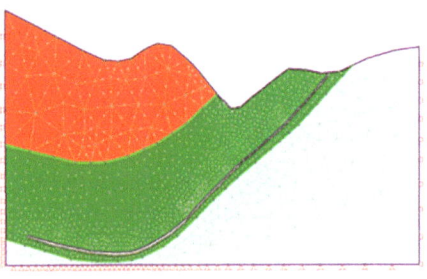 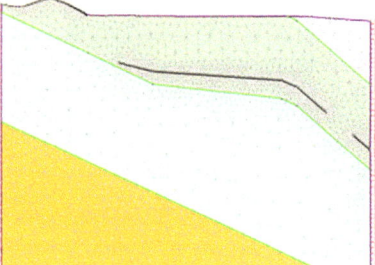

Fig. 2 Structural model diagram

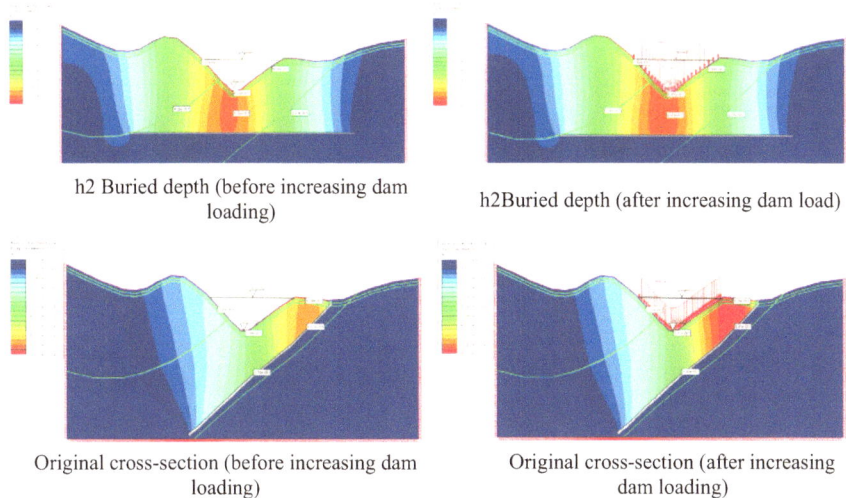

h2 Buried depth (before increasing dam loading)

h2Buried depth (after increasing dam load)

Original cross-section (before increasing dam loading)

Original cross-section (after increasing dam loading)

Fig. 3 Cross-sectional displacement diagram

maximum displacement occurs at a depth of h5, approximately 1.04 m. According to the displacement diagrams at different depths, it can be observed that as the depth increases, the displacement values due to self-weight gradually increase. Simultaneously, the response of the mined-out area to the dam load decreases with increasing depth.

The simulation calculations were conducted for different depths of the mined-out area roof: h1 (50 m), h2 (62 m), h3 (75 m), h4 (82 m), h5 (98 m), and h6 (106 m). The results include constrained displacements in the X-direction at the left and right boundaries and displacements in the Y-direction at the bottom boundary (Fig. 4).

The displacement diagrams in both longitudinal and transverse sections indicate that the total displacement is greatest at a depth of h1 in the longitudinal section, measuring approximately 1.34 m. At the same depth (h1) in the longitudinal section, the displacement due to the load is about 0.60 m. The depth-displacement curve shows that as the depth increases, the displacement caused by self-weight gradually increases. Simultaneously, the displacement response of the mined-out area to the dam load decreases with increasing depth. In the transverse section displacement diagram of the original shape of the mined-out area, significant displacement is observed on the right bank of the dam.

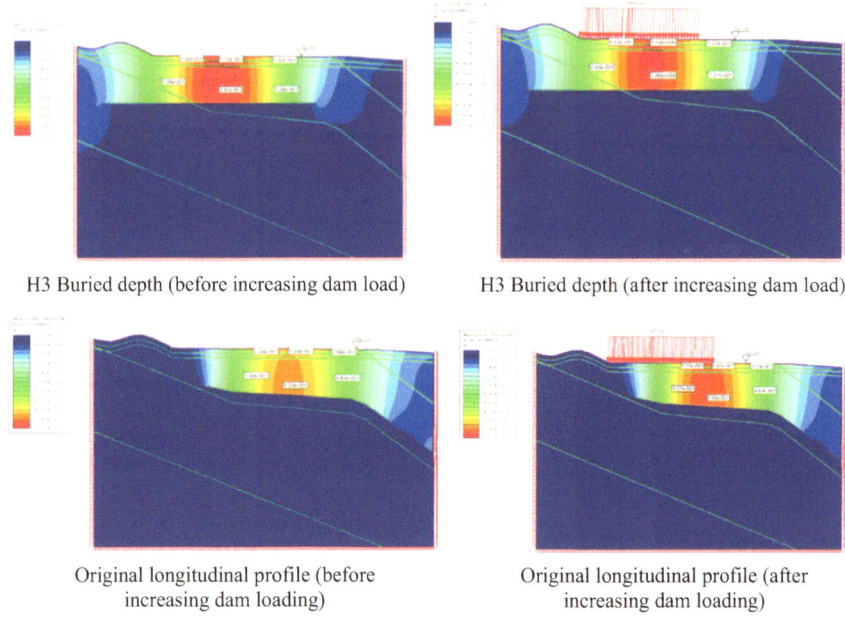

H3 Buried depth (before increasing dam load)   H3 Buried depth (after increasing dam load)

Original longitudinal profile (before   Original longitudinal profile (after
increasing dam loading)   increasing dam loading)

Fig. 4 Longitudinal displacement diagram (only shown in the picture)

3.3 Prediction of Settlements in Different Dams Based on Gaussian Process in Mined-Out Areas

The approach for this Gaussian process machine learning involved fitting and learning from 8 sample points of transverse deformation displacement obtained through calculations and 7 sample points of longitudinal deformation. The Gaussian process regression fitting curve was obtained, as shown in Fig. 5. Furthermore, based on the measured deformation, the height of the mined-out area was analyzed inversely. According to the detection data (0.654, 0.620), the height of the mined-out area (h) was determined to be 81 m.

To validate the validity of this machine learning process, a drilling experiment was conducted in the area. According to the drilling results, the depth of the mined-out area was approximately 85 m, which was in close agreement with the predicted result. The fitting results of this study are reasonable and reliable.

According to Tables 2 and 3, the results are consistent with the predictions and align with the actual situation. The depth-displacement curve indicates that as the depth increases, the displacement caused by self-weight stress gradually increases. Simultaneously, the displacement response of the mined-out area to the dam load decreases with increasing depth. The original shape displacement predicted using the Gaussian process is essentially consistent with the calculated original displacement.

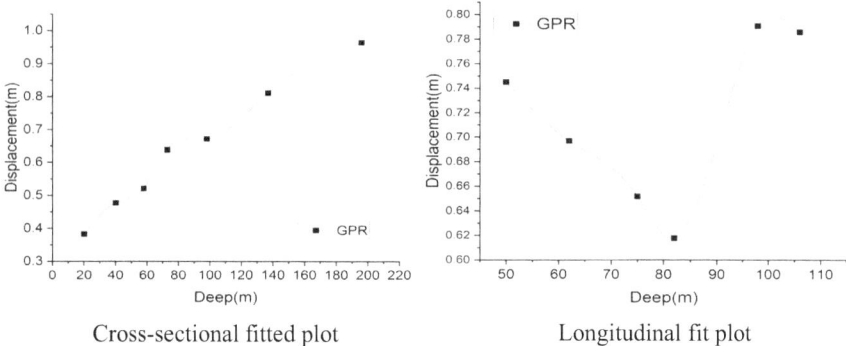

Cross-sectional fitted plot   Longitudinal fit plot

Fig. 5 GPR curve prediction plot

Table 2 Cross-section displacement statistical prediction table

Buried deep	Deep (m)	Before damming (m)	After damming (m)	Difference (m)
H1	20	0.383	0.787	0.404
H2	40	0.477	0.791	0.314
H3	58	0.521	0.822	0.301
H4	73	0.638	0.845	0.207
H5	98	0.671	0.82	0.149
H6	137	0.811	0.91	0.099
H7	196	0.964	1.04	0.076
Original		0.617	0.891	0.274

Table 3 Longitudinal section displacement statistical prediction table

Buried deep	Deep (m)	Before damming (m)	After damming (m)	Difference (m)
H1	50	0.745	1.34	0.595
H2	62	0.697	1.19	0.493
H3	75	0.652	1.03	0.351
H4	82	0.618	0.895	0.277
H5	98	0.791	1.06	0.269
H6	106	0.786	1.03	0.244
Original		0.523	0.665	0.142

4   In Conclusion

The mined-out area has a significant adverse impact on the stability of the surrounding environment and the construction of new dams. In order to verify the depth and layout of the mined-out area, a method based on Gaussian process learning for mined-out area depth prediction and risk analysis was proposed in this study. Through this method, the following conclusions can be drawn:

The deeper the mined-out area, the larger the corresponding total displacement deformation. As the depth increases, the displacement caused by self-weight stress also gradually increases. Simultaneously, the displacement response of the mined-out area to the dam load decreases with increasing depth.

The depth obtained through inverse analysis based on Gaussian process fitting is consistent with the on-site monitoring and drilling results. Gaussian process, as a small sample machine learning method, demonstrates good rationality and has certain research value.

References

1. Li, W. (2022). Research on railway route selection and prevention measures in deep buried karst mined-out areas. *Journal of Railway Engineering, 39*(12), 17–23.
2. Li, T., Li, J., Zhang, Q., et al. (2021). Dual Application of Transient Electromagnetic Method in Bauxite Exploration and Mined-Out Area Detection[J]. Resource Information and Engineering, 2021, 36(5):
3. Qi Di. Analysis and Application of Reflection Wave Group Characteristics in Coal Subsidence Area[D]. Jilin University, 2022.
4. Xin, Z. (2022). Key Technologies for Coal Mined-Out Area Treatment in Metro Construction Sections[J]. *Shanxi Architecture, 48*(22), 81–84.
5. Song, S. (2019). Research on intelligent evaluation method of gas drainage in mined-out areas based on deep learning. Xi'an University of Science and Technology, 2019.
6. Guo, S., Guo, G., Li, H., Cui, H. (2020). Stability evaluation of mining sites based on dimensionality reduction fuzzy C-means clustering algorithm. *Coal Science and Technology, 48*(10), 6.
7. Naranjo-Pérez, J., Infantes, M., Jiménez-Alonso, J. F., et al. (2020). A collaborative machine learning-optimization algorithm to improve the finite element model updating of civil engineering structures. *Engineering Structures, 225*, 111327.
8. Zhou, K., Xie, D. L., Xu, K., et al. (2023). A machine learning-based stochastic subspace approach for operational modal analysis of civil structures. *Journal of Building Engineering, 76*, 107187.
9. Su, G. S. (2009). Impingement pressure hazard prediction based on Gaussian process machine learning. *Journal of Liaoning Technical University: Natural Science Edition, 5*, 762–765.
10. Zhang, Y. (2013). *Gaussian process model and dynamic intelligent feedback analysis for predicting and identifying nonlinear behavior of underground engineering rock mass.* Guangxi University.

Research on Cumulative Damage of Reinforced Concrete Short Columns

Hongyu Zhou, Qi Tang, Xiaohua Zhao, and Yanan Liu

Abstract Reinforced concrete columns are the main load-bearing elements in many structures and are also the elements that bear seismic effects when earthquakes occur. In high-rise buildings, bridges, and other building structures column members also gradually accumulate damage due to fatigue cyclic loading and gradually reduce their load-bearing performance. The design and fabrication of reinforced concrete column members, the application of equal amplitude cyclic loading to them, so that they enter different damage stages, the study of their fatigue damage after the fatigue damage of the load-bearing capacity, deformation capacity, and damage morphology and other damage, to reveal the fatigue damage on the shear performance of the members of the law of influence. The monotonic pushover and low circumferential fatigue proposed static tests of reinforced concrete columns were carried out to study the fatigue shear damage characteristics, damage mechanism, and mechanical properties of the member such as the bearing capacity after the damage. The test process is divided into two stages. In the first stage, a certain number of equal-amplitude cyclic loads are applied to the reinforced concrete column members to bring them to different degrees of damage, and the cumulative damage development law is studied. In the second stage, the members after the accumulated damage were subjected to damage tests under monotonic loading to study their residual bearing capacity and dimensional effects in the damaged state.

Keywords Reinforced concrete short column · Cumulative damage · Low cycle fatigue · Shear failure

H. Zhou (✉) · Q. Tang
Faculty of Architecture, Civil and Transportation Engineering, Beijing University of Technology, Beijing, China
e-mail: ZHYktztgyx@163.com

X. Zhao
China Academy of Building Research, Beijing, China

Y. Liu
Beijing General Municipal Engineering Design and Research Institute Co. Ltd., Beijing, China

© The Author(s) 2025
D. Li and Y. Zhang (eds.), *Advances in Frontier Research on Engineering Structures II*, Lecture Notes in Civil Engineering 535, https://doi.org/10.1007/978-981-97-6238-5_6

1   Introduction

The damage process in which the load acts repeatedly at the same amplitude or variable amplitude for a long time is called fatigue [1]. Structural members will produce irreversible damage in some parts, and at the same time, with the cyclic crack initiation or further development of the load, the components completely fail.

Frangois [2] conducted a three-bending point test and found that increasing the stress ratio is beneficial to improving fatigue life. El-Ragaby [3] conducted failure tests on 5 full-scale concrete bridge decks under variable amplitude fatigue loads. Under the peak fatigue load level, the effect of the top reinforcement ratio is significant, and the bond strength between steel and concrete is the key factor affecting the fatigue life of the structure. Katakalos and Papakonstantinou [4] found through experiments that the maximum deflection increases with the number of cycles. Zhu et al. [5] found that the load range has an important effect on the fatigue performance of concrete beams filled with fiber-reinforced polymers and that the higher the intensity of the cyclic loading, the shorter the fatigue life. Malek et al. [6] studied the damage development, material degradation, residual bearing capacity, and plastic damage assessment of unconfined and glass fiber-reinforced polymer-confined concrete cylinders under low-cycle fatigue loading, and found that under the same stress/strength ratio, the constrained and unconstrained number of failure cycles for confined concrete is almost the same, but the magnitude of stress for confined concrete is much higher. Zhu [7] studied the low-cycle fatigue cumulative damage and horizontal bearing capacity of reinforced concrete piers damaged by bending and shearing. As the value increases, the strength and stiffness of the specimen degrade seriously when it is close to failure, and the residual deformation of the specimen increases with the increase in the number of loading cycles. Yuan et al. [8] considered the real response of bridge piers under the joint action of corrosion damage and real multidimensional earthquake action, established two numerical models of corroded bridge piers, and verified them with the experimental results in the literature. It turns out that the trajectory of the biaxial loading path and the degree of corrosion has a significant effect on the peak force, deformation capacity, and dissipated energy of the piers. Xiao [9] and Peng et al. [10] divided the concrete damage evolution curve obtained from the test into three stages, namely, the initial growth stage, the stable growth stage, and the rapid development stage.

Previous studies on the cumulative damage of reinforced concrete members mainly focused on the performance of beam members, and explored the influence of cyclic loading times on the mechanical properties of different stirrup forms, different stressed tendons, and different cross-section forms. When studying the degradation of mechanical properties caused by cumulative damage of reinforced concrete short columns, they also focus on exploring the changes in macro indicators such as component strength and stiffness, without quantitative analysis of test results and data, and deriving their calculation formulas. Therefore, in the course of the experiment, this paper not only observes the occurrence of macroscopic phenomena, but also pays more attention to the processing of experimental data, and analyzes and summarizes its laws from a mathematical point of view.

2 Test Piece Production and Test Plan

This test studies the mechanical properties of reinforced concrete columns in compression-shear failure form. The design axial compression ratio is 0.6, the design shear-span ratio is 2, the cross-sectional size of the specimen is b × h × H0 = 300 × 300 × 546 mm, and the space height required for the fixture is 248 mm, the total height of the column. The design strength grade C30 is selected for the concrete, HRB400 is selected for the longitudinal reinforcement, and HPB300 is selected for the stirrup. The specific dimensions and reinforcement conditions are shown in Fig. 1.

After the design of the specimen is completed, the design drawings and parameters are handed over to the component factory for fabrication of the specimen. And after going through the steps of reinforcement cage tying, mold support, concrete casting, mold dismantling, and component maintenance, the specimen is transported to the structural laboratory hall of the Beijing Institute of Technology for further maintenance, and the test is conducted only after 28 days of maintenance.

Before testing the members, the mechanical properties of the steel and concrete materials were measured using mechanical testing machines, and the material test data of the steel bars are shown in Table 1. For the determination of the basic properties of the concrete materials, test blocks were made using the same materials as the reinforced concrete members were cast, and a total of 24 test blocks were made and measured after being maintained under the same maintenance conditions as the reinforced concrete test columns and the individual data were obtained. After processing, the axial compressive strength of concrete cubes was 27.66 MPa and the splitting tensile strength was 2.25 MPa.

The test process is divided into two stages. In the first stage, a certain number of equal-amplitude cyclic loads are applied to the reinforced concrete column members, so that they enter different damage states, and the cumulative damage development law is studied. In the second stage, the cumulatively damaged components are subjected to failure tests under monotonic loading to study their residual bearing capacity and size effects in the damaged state.

The test in this paper was conducted in the Key Laboratory of Urban Safety and Disaster Engineering of Beijing Institute of Technology, and the specimen was loaded by the 20 KN proposed static testing machine, and the test was completed by applying axial and horizontal loads and restraint conditions to the specimen according to the set monotonic loading regime. As the column root boundary restraint conditions are simulated through the reinforced concrete base, the base needs to be fixed to the ground by I-beam and bolts to prevent the base from shifting and turning during the test. The ground level is also checked and adjusted to prevent damage to the column base caused by stress concentration.

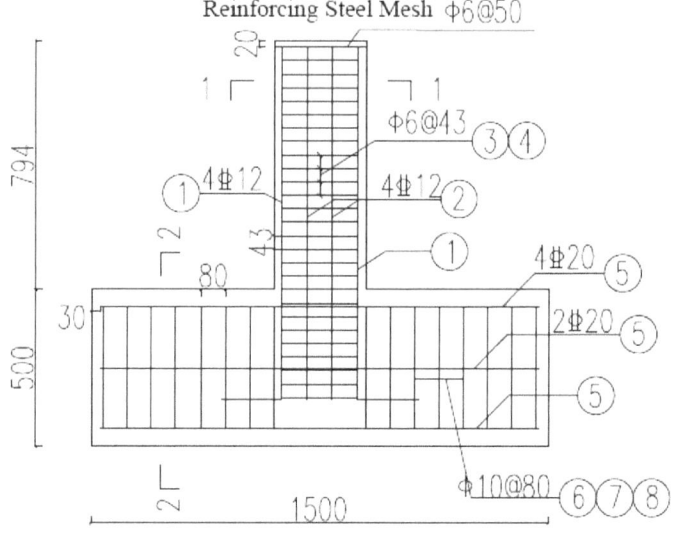

Column dimensional drawing

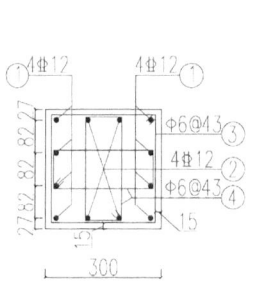

1-1 Section reinforcement diagram

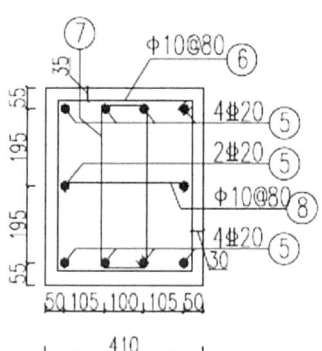

2-2 Section reinforcement diagram

Fig. 1 Specimen size and reinforcement diagram

Table 1 Mechanical properties of reinforcing steel materials

Serial number	Diameter of steel bar (mm)	Reinforcement grade	Yield strength (MPa)	Tensile strength (MPa)
1	6	HPB300	486.67	669.42
2	10	HPB300	453.33	593.43
3	12	HRB400	471.67	705.35
4	20	HRB400	433.67	695.91

The axial load is applied first, and then the horizontal load is applied after the test is stabilized, the specimen is preloaded before the formal loading, and the preload load value is set at 76 KN. For the monotonic test loading, the first half of the load control loading is selected, and the loading speed is 5 KN/s; after loading to 342 KN, the displacement control loading is switched, and the loading speed is 0.02 mm/s until the specimen is damaged. For the damage test choose an equal amplitude cyclic loading system, to carry out the cyclic loading test process first applied thrust (i.e., to the W side loading for forward loading), followed by the application of tension (i.e., to the E side loading for reverse loading); forward loading once, reverse loading once for the completion of a cycle of loading, cycle 30 times after the end of the cumulative damage loading, start monotonic push overloading, until the specimen damage. Cyclic loading with displacement control loading, specimen YJ-2 ~ 4 displacement rates of 0.4%, 0.7%, 0.8%, respectively, calculated column top displacement of 2.18, 3.82, and 4.37 mm.

3   Test Results

Before you begin to format your paper, first write and save the content as a separate text file. Keep your text and graphic files separate until after the text has been formatted and styled. Do not use hard tabs, and limit the use of hard returns to only one return at the end of a paragraph. Do not add any kind of pagination anywhere in the paper. Do not number text heads-the template will do that for you.

Finally, complete content and organizational editing before formatting. Please take note of the following items when proofreading spelling and grammar:

3.1   *Experimental Phenomenons*

During the loading process, the cracks of the undamaged components increased and developed with the increase of the horizontal displacement, and the failure of the specimen finally showed a 45° failure from the tension surface to the compression surface, a large amount of concrete fell off, and the steel bars on the failure surface were embossed and separated from the concrete and hoops Rib constraints protrude. During the application of cyclic loading, the cracks on the loading surface and the side of the observation surface close to the loading surface alternately initiate and develop cracks. After ten cycles of loading, the crack development is basically stable and does not change. Formation of horizontal and vertical cracks. During the failure test of the damaged component, the vertical cracks near the column feet continued to sprout and develop, and the failure was sudden. The angle between the failure surface and the horizontal plane was 45°. The stirrup deformed and the column completely lost its bearing capacity. The final damage of the components is slightly different. In terms of crack development, deformation, and other indicators, it is generally from a non-damaged state to a damaged state, and each index gradually increases with the increase of the damage degree.

3.2 Analysis of Test Results

The cumulative damage test results are organized and analyzed, and the displacement-load curves of the cumulative damage process of specimens YJ-2, YJ-3, and YJ-4 are shown in Fig. 2.

In the process of repeated loading, the cracks near the foot of the column of specimen YJ-2 were most obvious in the first cycle, and the cracks continued to develop in the subsequent loading, and the cracks changed less after the fifth cycle of loading, and the cracks were basically stable and no longer changed after the tenth cycle of loading. At the end of the cyclic loading, there were a small number of through cracks and a large number of microcracks not exceeding 100 mm, and the crack width was not large. The damage process curve of specimen YJ-2 was more rounded and fuller and was date shaped. The reason was that the maximum strain value of the reinforcement was reached in the first cycle of the repeated loading process, and the strain of the reinforcement decreased slowly in the subsequent cycles of loading due to the cracking of the concrete and the decrease of the stiffness of the member, and the preset displacement loading was carried out again. Therefore, during the repeated loading of member YJ-2, because the tensile and compressive strains of the reinforcement are small, there is no obvious bond damage between the reinforcement and concrete, and the cracks will not sprout along the longitudinal reinforcement, and the concrete protection layer on the outside of the column structure will not be damaged, so there is no obvious "pinch shrinkage" in the hysteresis curve.

Specimen YJ-3 and YJ-4 repeated the loading process in the first cycle that appears through cracks, the concrete surface slightly spalling, producing a little vertical

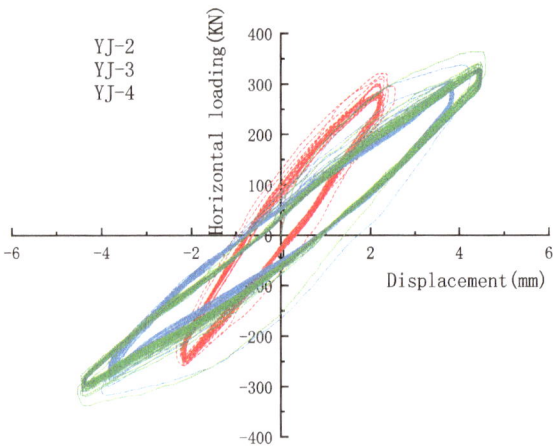

Fig. 2 Comparison of displacement-load curves in the damage process of specimens

micro-crack; in the subsequent loading horizontal crack width increased significantly, and continued to produce vertical cracks. After the fifth cycle of loading, the concrete at the bottom of the column became blocky and spalled off, and some oblique cracks were produced at the bottom of the column. Thereafter, horizontal and vertical cracks continued to develop, and by the tenth cycle of loading the crack development was basically stable and no longer developed. At the end of the final cycle of loading, there were vertical micro-cracks near each horizontal crack and oblique crack, and the concrete at the foot of the column was spalling, but the crack width of the oblique section of specimen YJ-4 was slightly larger than that of YJ-3. Specimen YJ-3 and YJ-4 cumulative damage process displacement-load curve is shuttle-shaped, and specimen YJ-4 curve is more narrow, the reason is that at the beginning of the cyclic loading, the specimen bearing reinforcement is constantly subjected to the alternating effect of approximately equal to the yield load tension and pressure, the specimen damage is constantly superimposed, and at the same time, under the action of alternating tension and compression stress, the bond between the reinforcement and concrete gradually failed, concrete damage is obvious, resulting in bond slip. Therefore, the vertical microcracks of specimens YJ-3 and YJ-4 developed significantly, and the hysteresis curve at the beginning of the damage process cycle had an obvious "pinching phenomenon", and because the loading displacement of specimen YJ-4 was larger, the damage caused was more and the hysteresis curve was more elongated. Jian [11] pointed out that the displacement of the member is composed of the horizontal displacement formed by the bond slip, shear, and bending deformation of the reinforcement and concrete. At the same time, it is due to the formation and development of split crack bonds that the protective layer of concrete is no longer restrained, and with the alternation of tension and compression, the cumulative concrete damage accumulates internally, showing a gradual increase in the crack width of the specimen, which is the reason for the spalling and peeling of the concrete at the foot of the columns of specimens YJ-3 and YJ-4. Meanwhile, according to the research related to fatigue of related concrete, it can be concluded that under the alternating action of tensile and compressive stresses, the internal microcracks of concrete gradually develop, which will lead to the gradual decrease of compressive elastic modulus of concrete. Therefore, as the number of cyclic loading increases, the horizontal load of the member gradually decreases when the load amplitude is achieved.

The magnitude of the horizontal load when the specimen reaches the loading amplitude in each cyclic loading process is not equal but has obvious degradation in the first 5 cycles. In particular, the degradation speed of the horizontal load is inconsistent between the various hysteresis loops, and the degradation rate of the horizontal load is the largest from the first cycle to the second cycle, and then gradually decreases. In the subsequent cycles, the total amount of horizontal force degradation is less than that of the first 5 cycles, which shows that the damage inside the component gradually accumulates and develops, and gradually converges to a stable state.

3.3 Energy Consumption Analysis

Combined with the experimental phenomena, it can be concluded that the energy dissipation in the first loading cycle is the largest, and the area of the hysteresis loop becomes smaller as the number of loadings increases, that is, the energy consumption of the damage volume gradually decreases; at the same time, the area of the hysteresis loop decreases gradually during the first few cycles. The hysteresis loop area of the first cycle and the second cycle of the specimens YJ-3 and YJ-4 is nearly twice, which is significantly reduced.

The energy dissipation of each loading cycle of specimens YJ-2, YJ-3, and YJ-4 is normalized to compare the energy dissipation capacity changes of specimens during cyclic loading, as shown in Fig. 3. The difference in energy dissipation between the first and second cyclic loading cycles is the largest, and the energy dissipation of specimens YJ-2, YJ-3, and YJ-4 in the second cyclic loading cycle is higher than that in the first cyclic loading cycle. The dissipated energy is reduced by 14, 33, and 40%. According to the research of Rodrigues [12], the damage caused in the first cyclic loading cycle leads to the reduction of the stiffness and strength of the specimen, which reduces the energy dissipation capacity of the specimen in the subsequent cycle. At the same time, it can be found that the hysteresis loop area ratio of specimens YJ-3 and YJ-4 is smaller than that of specimen YJ-2, so it can be considered that the larger displacement loading in the initial state will reduce the energy dissipation capacity of the specimen.

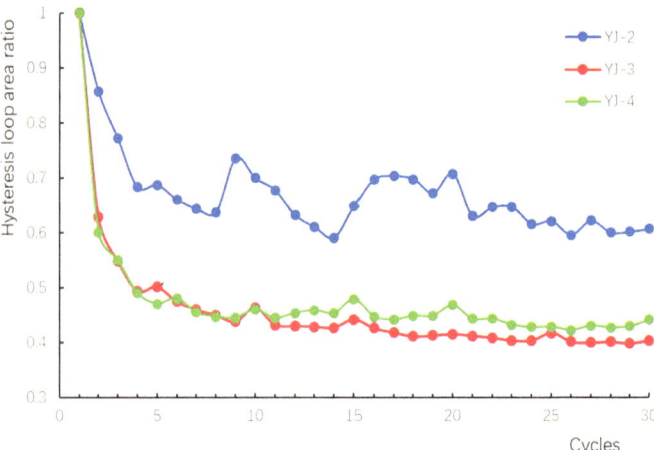

Fig. 3 Hysteresis loop area ratio of specimen Example of a figure caption

Table 2 Main test result parameters

Specimen No.	Yield displacement (mm)	Ultimate load (kN)	Ultimate displacement (mm)	Ductility factor	Yield load (Py/kN)
YJ-1	2.82	321	6.40	2.27	321
YJ-2	2.66	320	8.56	3.31	312
YJ-3	3.63	275	7.48	1.97	285
YJ-4	4.38	318	8.49	1.94	330

3.4 Ductility Evaluation

The ductility of a member is evaluated by the ductility coefficient, which is calculated as follows:

$$Ductility\ coefficient = \frac{ultimate\ displacement}{yield\ displacement} \tag{1}$$

Compared with specimen YJ-1, the ductility coefficient of specimen YJ-2 increases, and the ductility is slightly improved; while the ductility coefficients of specimens YJ-3 and YJ-4 decrease, and the ductility coefficient decreases as the loading displacement rate increases and the ductility becomes worse. It means that the cyclic loading of horizontal displacement within the yield displacement can improve the ductility of the member, while the cyclic loading of the displacement exceeding the yield displacement of the undamaged state will reduce the ductility of the member, and the larger the loading amplitude is, the smaller the ductility is. The specific calculation results are shown in Table 2.

4 Conclusions

From the test results, it can be seen that the reinforced concrete short column under compression and lateral horizontal force composite stress state, the member damage suddenly, the bearing capacity drop suddenly, belong to the brittle damage, in line with the design expectations. At the same time, in the cyclic loading process, with the growth of the number of loading hysteresis loop area becomes smaller, that is, the damage volume energy consumption is gradually reduced, the specimen in 5 cycles of loading after the damage volume energy consumption reduction slowed down, the specimen damage volume energy consumption in 10 cycles of loading after the damage tends to stabilize. The energy dissipation of the specimen in the first loading cycle is the largest, with the growth of the number of loading hysteresis loop area becomes smaller, that is, the damage volume energy dissipation gradually reduced. The larger the loading displacement during cyclic loading, the lower the energy dissipation capacity of the specimen.

The final damage of the member is slightly different, from the crack development, deformation, and other indicators, roughly from the state of no damage to the state of damage with the increase in the degree of damage and each indicator gradually increased. The reinforced concrete short column under compression and lateral horizontal force composite stress state, the member damage suddenly, bearing capacity drop suddenly, belonging to the brittle damage specimen in the first loading cycle of the maximum energy dissipation, with the number of loading growth hysteresis loop area becomes smaller, that is, the damage volume energy dissipation gradually reduced. The larger the loading displacement in the cyclic loading process, the lower the energy dissipation capacity of the specimen.

Acknowledgements Supported by the National Natural Science Foundation of China under the project "Study on the Evolution of Cumulative Damage of Reinforced Concrete Structures and its Interaction Mechanism of Dimensional Effects (51978012)".

References

1. Nemes, J. A., & Spéciel, E. (1996). Use of a rate-dependent continuum damage model to describe strain-softening in laminated composites. *Computers & Structures, 58*(6), 1083–1092. https://doi.org/10.1016/0045-7949(95)00229-4
2. Frangois, B. A. A. A. (1992). Damage of concrete in fatigue. *Journal of Engineering Mechanics.* https://kns.cnki.net/kcms/detail/detail.aspx?FileName=SJCE4C404C95723FA35ABCB0A557B31A53D7&DbName=GARJ8099_1
3. El-Ragaby, A., El-Salakawy, E., & Benmokrane, B. (2007). Fatigue life evaluation of concrete bridge deck slabs reinforced with glass FRP composite bars. *Journal of Composites for Construction, 11*(3), 258–268. https://doi.org/10.1061/(ASCE)1090-0268(2007)11:3(258)
4. Katakalos, K., & Papakonstantinou, C. G. (2009). Fatigue of reinforced concrete beams strengthened with steel-reinforced inorganic polymers. *Journal of Composites for Construction, 13*(2), 103–112. https://doi.org/10.1061/(ASCE)1090-0268(2009)13:2(103)
5. Zhu, Z., Ahmad, I., & Mirmiran, A. (2009). Fatigue modeling of concrete-filled fiber-reinforced polymer tubes. *Journal of Composites for Construction, 13*(6), 582–590. https://doi.org/10.1061/(ASCE)1090-0268(2009)13:6(582)
6. Malek, A., Scott, A., Pampanin, S., Gregory MacRae, M. A., & Marx, S. (2018). Residual capacity and permeability-based damage assessment of concrete under low-cycle fatigue. *Journal of Materials in Civil Engineering, 30*(6). https://doi.org/10.1061/(ASCE)MT.1943-5533.0002248
7. Zhu, J. (2020). *Study on low-cycle fatigue cumulative damage and lateral capacity of reinforced concrete pier columns subjected to flexural-shear failure* [Ph.D. Thesis, Dalian University of Technology]. https://kns.cnki.net/kcms/detail/detail.aspx?FileName=1020093907.nh&DbName=CDFD2021
8. Yuan, W., Li, X., Pang, X., Tian, C., Li, Z., Zhou, P., & Wang, Y. (2023). Cyclic behavior of rectangular bridge piers subjected to the coupling effects of chloride corrosion and bidirectional loading. *Buildings (Basel), 13*(2), 425. https://doi.org/10.3390/buildings13020425
9. Xiao, W. (2018). *Experimental study on shear fatigue properties of reinforced concrete beams* [Master's thesis, Zhejiang University]. Available from https://kns.cnki.net/kcms/detail/detail.aspx?FileName=1018244311.nh&DbName=CMFD2018

10. Peng, L. M., Liu, N., & Shi, C. H. (2016). Experimental study on dynamic amplitudes for cumulate damage characteristics of tunnel invert concrete. *Journal of Railway Science and Engineering*, *13*(06), 1091–1099. https://kns.cnki.net/kcms/detail/detail.aspx?FileName=CST D201606013&DbName=CJFQ2016

11. Yi, W., Zhou, Y., Hwang, H., Cheng, Z., & Hu, X. (2018). Cyclic loading test for circular reinforced concrete columns subjected to near-fault ground motion. *Soil Dynamics and Earthquake Engineering, 1984*(112), 8–17. https://doi.org/10.1016/j.soildyn.2018.04.026

12. Rodrigues, H., Varum, H., Arede, A., & Costa, A. (2012). A comparative analysis of energy dissipation and equivalent viscous damping of RC columns subjected to uniaxial and biaxial loading. *Engineering Structures, 35*, 149–164. https://doi.org/10.1016/j.engstruct.2011.11.01

Numerical Simulation of the Effect of Joint Parameters on the Mechanical Behaviors of the Fractured Sandstone

Yue Zhang, Miaoran Du, and Peijie Yin

Abstract Joint parameters affect the mechanical properties of sandstone significantly. This paper investigates the influence of fracture on the mechanical properties of roof plate bedded sandstone in Yuanjue Cave by using the Discrete Fracture Network (DFN) model. The effects of joint density and dip on the mechanical behavior of bedded sandstone under triaxial compression are first examined. And, the fractal dimension is used to quantify the relationship between joint density and strength parameters. The results show that: (1) The compressive strength and elastic modulus of bedded sandstone decrease with increasing joint density. The compressive strength shows a 'U' distribution with increasing joint dip from 0° to 90°, while the elastic modulus increases gradually. (2) The bedded sandstone shows a very high anisotropic characteristic, and the weak plane of the layers influences the strength and deformation damage of the bedded structure. (3) The fractal dimension has a significant negative correlation with rock strength and elastic modulus, with correlation coefficients of 0.834 and 0.965.

Keywords Discrete fracture network · Numerical simulation · Joint parameters · Mechanical behavior · Fractal dimension

1 Introduction

Bedded sandstone found in Yuanjue Cave is a typical heterogeneous rock with different scales and shapes of fractures, such as bedding, joints, and cracks. The characteristics of these joints significantly affect the mechanical properties of this sandstone, altering its strength, stiffness, deformation, and failure behavior. A large number of studies have been carried out to investigate the mechanical behavior of joint characteristics through indoor experiments [1–3], whereas laboratory tests mainly focused on the mechanical characteristics of samples with fewer and more regular fractures. It was unable to accurately reproduce the bedded sandstone with severe

Y. Zhang · M. Du · P. Yin (✉)
School of Highway, Chang'an University, Xi'an 710061, China
e-mail: peijie.yin@chd.edu.cn

© The Author(s) 2025   77
D. Li and Y. Zhang (eds.), *Advances in Frontier Research on Engineering Structures II*,
Lecture Notes in Civil Engineering 535, https://doi.org/10.1007/978-981-97-6238-5_7

weathering and many irregular fractures [4], leading to inaccurate analysis of the relationship between rock fracture characteristics and mechanical properties. With the development of computer technology, numerical simulation has become a powerful tool for studying fractured rock masses. Feng et al. [5] developed an elasto-plastic cellular automaton (EPCA) and its associated code to reproduce uniaxial compression tests, advancing the application of numerical simulations in the field of rock mechanics characteristics. Bahaaddini et al. [6] studied the effects of joint orientation and spacing on the mechanical characteristics of rock using particle flow software PFC3D and concluded that joint orientation and spacing had significant effects on the mechanical characteristics of rock. Wang et al. [7] utilized modeling techniques based on particle mechanics to simulate the mechanical behavior of jointed rock masses and examined the influence of normal stress, friction coefficient, and joint cohesion strength on the mechanical behavior of jointed rocks. Hu et al. [8] developed a numerical model of equivalent crystalline rock with fracture network based on a particle flow software platform and elucidated the effects of discrete fracture network on rock strength and deformation properties from a mesoscopic perspective.

The aforementioned studies have displayed how joint parameters affect rock mechanical properties. However, most rock models established by scholars are relatively uncomplicated, as they do not consider the interplay between the bedding structure and fracture features of the rock. Moreover, it is challenging to quantitatively analyze the influence of various parameters on rock mechanical properties during the analysis of experimental results, which inevitably limits the interpretation of the results.

This paper presents a bedded sandstone model with a fracture network based on field statistics of the occurrence, trace length, and distribution of bedded sandstone fractures in Yuanjue Cave, using discrete element numerical simulation software (3DEC). The study investigates the impact of fracture characteristics on various mechanical parameters of the rock mass including the triaxial compressive strength, elastic modulus, cohesion, internal friction angle, and others through triaxial compression test discrete element simulation.

2 Models and Methods

2.1 Establishing of Numerical Model

Joint Parameters Statistics

Yuanjue Cave is located in Ziyang City, Sichuan Province. The roof sandstone is cut by fractures, joints, and other structural planes, and deteriorates continuously under the influence of external forces, eventually forming blocks. Using the line method, a total of 166 fractures on the roof of Yuanjue Cave were collected for trace length, dip direction, dip angle, position, and other information, which are shown in Fig. 1.

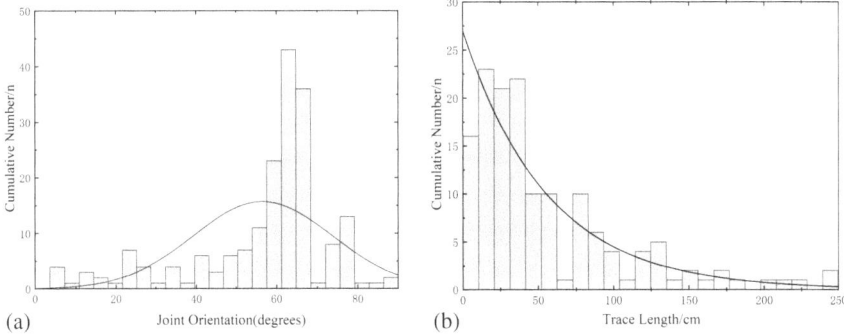

Fig. 1 Statistical distribution of fracture angle and length

In the DFN module of 3DEC, the traces, attitude, and position of the joints are determined by the corresponding probability functions. The main distribution functions include uniform distribution, power-law distribution, Gaussian distribution, and user-defined distribution. As shown in Fig. 1a, the angle of the joints follows a normal distribution, and its probability density function shown in Eq. 1.

$$P(x_1) = \frac{e^{\frac{-(x_1-\mu_1)^2}{2\sigma_1^2}}}{\sigma_1\sqrt{2\pi}} \tag{1}$$

where represents the dip of the joints; and represents the mean and standard deviation of the Gaussian distribution. After calculation, and are taken as 57.5° and 86.9°.

As shown in Fig. 1b, trace lengths in nature typically conform to a sub-distribution, and its probability density function shown in Eqs. 2 and 3.

$$f(x) = x^{-\alpha-1} \tag{2}$$

$$F(x) = \int_{-\infty}^{\infty} x^{-\alpha-1}dx = x^{-\alpha} + C \tag{3}$$

where α represents power-law distribution index; C is a constant.

Test Methods

According to the statistical data in the previous text, discrete element models with fixed interlayer spacing and bedding angles of 0, 30, 60, and 90 were established in 3DEC software with dimensions of $40 \times 40 \times 100$ mm, which is shown in Fig. 2.

Table 1 presents the mechanical properties of the model, which were derived from pertinent literature on Yuanjue cave bedded sandstone. The Mohr–Coulomb criterion was utilized for the calculation model. The model underwent loading, beginning with a confining pressure that brought it up to the hydrostatic pressure level. Then, the

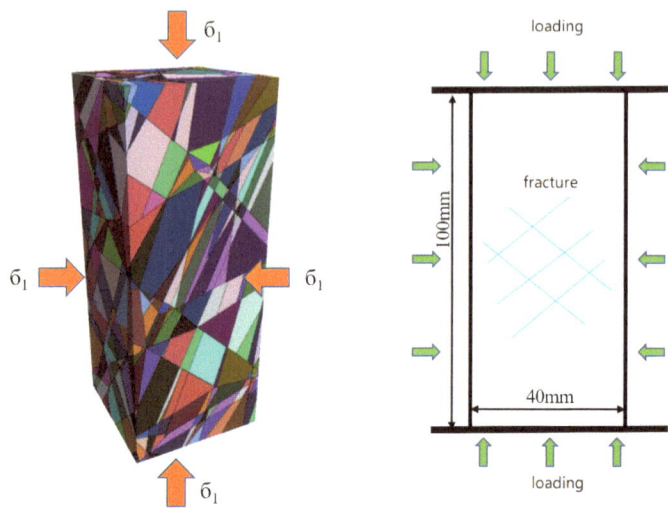

Fig. 2 Numerical model schematic

Table 1 Mechanical parameters of rock block and structural plane

Parameter	Bulk modulus (GPa)	Shear modulus (GPa)	Tensile strength (MPa)	Friction angle (°)	Cohesive (MPa)	Tangential stiffness (GPa)	Normal stiffness (GPa)
Rock	1.74	0.348	1.62	45	2.8	/	/
Plane of weakness	0.4	0.08	0.8	32	1.2	/	/
Joints	/	/	/	23	0.2	14	14

confining pressure remained fixed so that the body stress in the model could initialize and a state of equilibrium could be resolved. Subsequently, displacement-controlled loading was implemented at a velocity of 0.005 mm/s until an instance of failure within the model manifest. Stress–strain details throughout the course of the event were documented and a stress–strain curve was produced.

3   Results and Discussions

3.1   *Effect of Joint Density on Mechanical Behavior*

The geometric parameters of the fracture, such as dip angle and size, were kept constant, where the joint dip angle was 45°, and the dip angle variance was 0; the joint trace length l = 5 m, and the trace length variance was 0; the joint density of the sample was set to 10, 20, 30, 40, and 50 m-2 respectively, and the model was subjected to a triaxial compression test under the confining pressure of 5Mpa. The stress–strain curves of specimens different bedding angles under the triaxial compression as shown in Fig. 3 and the variation of compressive strength and elastic modulus with the number of joints is shown in Fig. 4.

The stress–strain curve shape is basically consistent under different fracture density conditions and different layer angle conditions. As the fracture density increases, the elastic stage of the curve gradually lengthens, the stress–strain curve gradually becomes slower, the corresponding elastic modulus gradually increases, and the peak stress at failure increases.

The compressive strength of specimens with different bedding angles exhibits significant anisotropy. When the bedding angle (α) is 60°, the compressive strength of specimens with the same joint density decreases significantly. This is primarily due to the fact that α is close to the maximum friction angle (45° + $\varphi/2$ = 67°), causing local shear slip failure between the bedding planes and becoming the main factor controlling specimen failure, resulting in changes in the curve. When α is 90°, the curve becomes steeper and both the elastic modulus and compressive strength reach their maximum values. At this stage, fractures form along the direction of axial

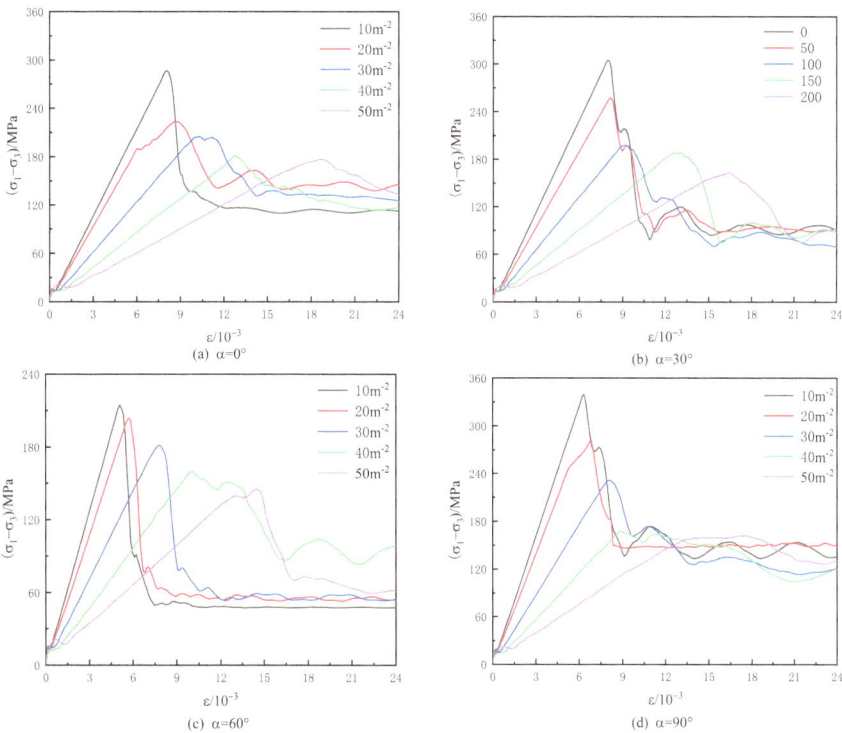

Fig. 3 Stress-strain curves of specimens with different number of joints

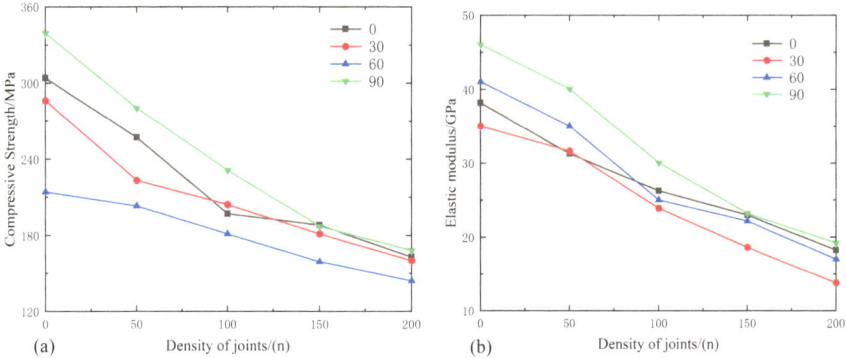

Fig. 4 Variation of compressive strength and elastic modulus with number of joints

stress. Even though these fractures can weaken the integrity of the specimen, the load is mainly supported by the mass of the model with minimal effect from the joints on its load-bearing capacity, leading to higher values.

Under different bedding angles, the joint density and model mass strength are negatively linearly correlated. As the joint density increases, the compressive strength of the model mass gradually decreases. In the case of specimen $\alpha = 0°$, when the joint density is 10 m^{-2}, compressive strength of the model is 304.2 MPa; when the joint density increases to 20 m^{-2}, the peak strength drops to 257.8 MPa, a decrease of 15.25%; when it increases from 20 to 30 m^{-2}, the compressive strength becomes 197.3 MPa, a decrease of 23.46%; when it increases from 30 to 40 m^{-2}, the compressive strength becomes 181.1Mpa, a decrease of 8.21%; when it increases from 40 to 50 m^{-2}, the compressive strength becomes 168.6Mpa, a decrease of 6.91%; it can be seen that the effect of joint density on model peak strength has stage characteristics, before 30 m^{-2}, the rising rate gradually increases, and after that the rising rate gradually slows down, and there is a marginal effect of density on compressive strength.

Under different bedding angles, the joint density also shows a negative linear correlation with the elastic modulus of the rock mass. As the joint density increases, the elastic modulus of the model decreases accordingly. However, it is worth noting that when the joint density is 10 m^{-2}, the maximum value of the elastic modulus of the model is 46.45 GPa and the minimum value is 38.15 GPa, a difference of 17.6%. When the joint density is 20, 30, 40, and 50, the difference between the maximum and minimum values of the elastic modulus is 15.2%, 13.5%, 18.3%, and 17.4%, respectively. The main reason is that as the joint density increases, the degree of fragmentation of the model also increases accordingly, which is the main factor affecting the elastic modulus of the model, so the elastic modulus of the rock mass under the same density does not differ much.

3.2 Effect of Joint Orientation on Mechanical Behavior

With all other parameters controlled, the density is set at 30 m^{-2}; the trace length l is 5 m with a variance of 0. The joint dip angles (θ) are respectively 0°, 15°, 30°, 45°, 60°, 75°, and 90°. Under a confining pressure of 5 MPa, triaxial compression tests are conducted on the rock samples. The stress–strain curves of specimens with different bedding angles under the triaxial compression as shown in Fig. 5 and the variation of compressive strength and elastic modulus with joint orientation is shown in Fig. 6.

Under different bedding angles, the stress–strain curve shapes of the model with different joint dip angles during the loading process are basically consistent, all going through the compaction stage, elastic stage, yield stage, and failure stage. As the dip angle changes from 0° to 90°, the elastic stage of the curve gradually shortens, the peak position first decreases and then increases, corresponding to a gradual increase in the elastic modulus, and the peak stress at the time of rock sample failure shows a trend of first decreasing and then increasing. Under the same bedding angle conditions,

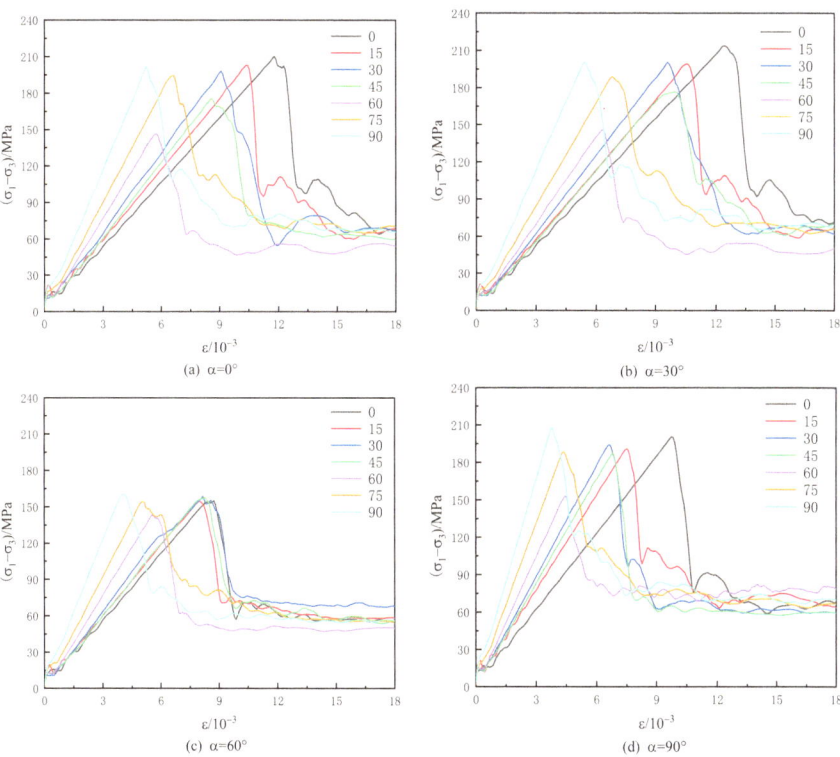

Fig. 5 Stress-strain curves of specimens with different joint orientation

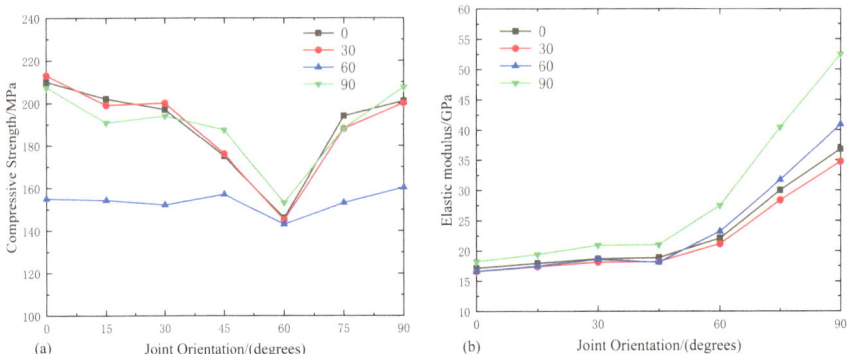

Fig. 6 Variation of compressive strength and elastic modulus with joint orientation

the minimum compressive strength of the sample occurs at $\beta = 60°$, which is close to the maximum friction angle $(45° + \varphi/2 = 67°)$ in Mohr–Coulomb criterion.

The samples show obvious anisotropy. Compared with other bedding angles, when the bedding angle $\alpha = 60°$, the compressive strength of the sample shows a large decrease. The reason is still that the weak plane between the beddings has undergone local shear slip failure, which has become the main factor controlling sample failure. This once again shows that the weak surface of bedding is the fundamental reason for causing anisotropy in strength and deformation failure characteristics of bedded sandstone.

As the joint dip angle increases, the peak strength of the model shows a U-shaped curve, indicating its impact on compressive strength. The compressive strength is relatively large at $0°$ and $90°$, and smaller around $60°$. The compressive strength shows a slow trend of decreasing and increasing from $0°$ to $30°$ and $75°$ to $90°$, and a faster trend from $30°$ to $60°$ and $60°$ to $75°$.

As the bedding angle increases, the elastic modulus of the model gradually increases. When $\beta = 0°$, the matrix and bedding weak plane are perpendicular to the loading direction, with stiffness mainly controlled by the bedding weak plane, resulting in a smaller elastic modulus. When $\beta = 90°$, they are parallel to the loading direction, with stiffness mainly controlled by the matrix, resulting in a larger elastic modulus.

3.3 Correlation Study Between Mechanical Behavior and Fractal Dimension

The preceding analysis demonstrates that the variation in the mechanical characteristics of bedded sandstone under different dip angles is primarily related to the maximum friction angle. Subsequently, we will conduct a quantitative examination of how different joint densities affects the mechanical characteristics under the same joint dip angle. Bagde et al. [9] proposed that there is a significant negative correlation between fractal dimension and rock strength. As the fractal dimension increases, the strength of the specimen decreases. The fractal dimension index is used to quantitatively describe the relationship between fracture characteristics and the strength and deformation characteristics of bedded sandstone which is presented in Eq. 4.

$$\lg(N_i) = \lg(V_0) - D\lg(\delta_i) \tag{4}$$

where N_i represents the number of cubes of size δ_i required to cover a 3D fracture body; V_0 is a constant; D is the fractal dimension.

By using the cubic coverage function in the 3DEC software, the fractal dimension of the 3D fracture space distribution can be calculated through Eq. 4. The advantage of this method is that the obtained fractal dimension can comprehensively reflect the complexity, roughness, orientation, and opening degree of 3D fractures. The data obtained were collated into Table 2.

In summary, the functional expression of the fractal dimension (D) of the specimen with the triaxial compressive strength (σ) and the elastic modulus (G) can be obtained as shown in Eqs. 5 and 6

$$\sigma = -93.75D + 499.86 \tag{5}$$

$$G = -18.67D + 7.61 \tag{6}$$

A curve function was used for regression analysis of dimension and strength in Table 2. The linear fitting relationship expression is shown in Fig. 7, with determination coefficients R^2 of 0.834 and 0.965, indicating a strong negative correlation between fractal dimension and compressive strength, elastic modulus parameters. As the fractal dimension increases, the complexity of particle pore distribution increases, resulting in a decrease in rock triaxial compressive strength, elastic modulus.

Table 2 Experimental main indicator parameters

Number of joints (n)	Fractal dimension (D)	Compressive strength (MPa)	Elastic modulus (GPa)
0	2.126	304.34	38.151
50	2.350	257.266	31.25
100	2.570	197.154	26.252
150	2.866	181.679	22.97
200	3.141	168.454	18.242

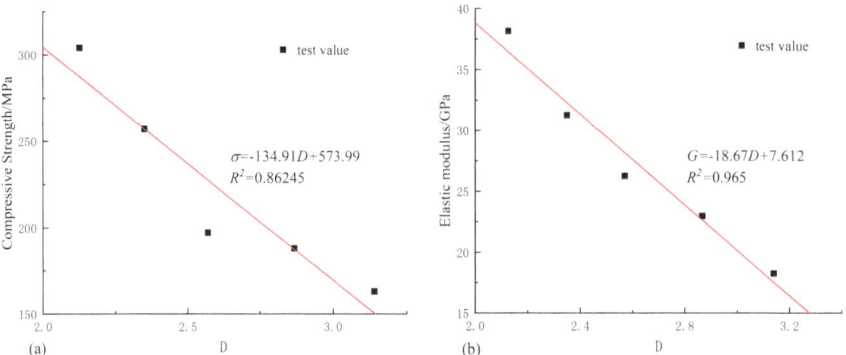

Fig. 7 Correlations among compressive Strength, elastic modulus and fractal dimension

4 Conclusion

This paper establishes a 3D discrete element model of bedded sandstone with weak interlayers. Based on this, the relationship between the mechanical characteristics of rock mass and joints of different dip angle and densities under different bedding angles is further explored. The following conclusions can be drawn.

(1) As the joint density increases, the triaxial compressive strength and elastic modulus of rock mass with different angle bedding gradually decrease; as the joint dip angle increases from 0° to 90°, the compressive strength of rock mass with different angle bedding shows a U-shaped curve change of first decreasing and then increasing, while the elastic modulus gradually increases.

(2) The weak bedding plane is the weak link of the bedded sandstone and the fundamental cause of the anisotropy of the strength and deformation failure characteristics of the bedded sandstone. In the design and deformation stability analysis of engineering slopes and tunnels related to bedded sandstone, the influence of bedding angle on the mechanical properties and failure modes of rock mass cannot be ignored.

(3) The fractal dimension of the 3D fracture network within the rock mass exhibits a significantly negative correlation with its triaxial compressive strength and elastic modulus, with correlation coefficients of 0.862 and 0.965, respectively. This implies that as the fractal dimension increases, both the triaxial compressive strength and the elastic modulus tend to decrease.

This study has certain limitations: The model formulated in this paper is established on a constant confining pressure. As the influence of confining pressure on anisotropy is considerable, it is uncertain whether the model can be broadened to encompass rock masses with distinct confining pressures. It is imperative to continuously refine and enhance this model.

References

1. Tavallali, A., & Vervoort, A. (2010). Failure of layered sandstone under Brazilian test conditions: Effect of micro-scale parameters on macro-scale behaviour. *Rock Mechanics and Rock Engineering, 43*(5), 641–653.
2. Yang, S., Yin, P., Li, B., & Yang, D. (2020). Behavior of transversely isotropic shale observed in triaxial tests and Brazilian disc tests. *International Journal of Rock Mechanics and Mining Sciences, 133*, 104435.
3. Wang, P., Liu, Z., Cai, M., & Labuz, J. F. (2022). Shear behavior of synthetic rough jointed rock mass with 3D-printed jointing. *Arabian Journal of Geosciences, 15*(5), 1–13.
4. Li, X., Liu, J., Gong, W., Xu, Y., & Bowa, V. M. (2022). A discrete fracture network based modeling scheme for analyzing the stability of highly fractured rock slope. *Computers and Geotechnics, 141*, 104558.
5. Feng, X., Pan, P., & Zhou, H. (2006). Simulation of the rock microfracturing process under uniaxial compression using an elasto-plastic cellular automaton. *International Journal of Rock Mechanics and Mining Sciences, 43*(7), 1091–1108.
6. Bahaaddini, M., Hagan, P., Mitra, R., & Hebblewhite, B. K. (2016). Numerical study of the mechanical behavior of nonpersistent jointed rock masses. *International Journal of Geomechanics, 16*(1), 04015035.
7. Wang, P., Cai, M., Ren, F., Li, C., & Yang, T. (2017). A Digital image-based discrete fracture network model and its numerical investigation of direct shear tests. *Rock Mechanics and Rock Engineering, 50*(7), 1801–1816.
8. Hu, X., Xie, N., Zhu, Q., Chen, L., & Li, P. (2020). Modeling damage evolution in heterogeneous granite using digital image-based grain-based model. *Rock Mechanics and Rock Engineering, 53*(11), 4925–4945.
9. Bagde, M., Raina, A., Chakraborty, A., & Jethwa, J. (2002). Rock mass characterization by fractal dimension. *Engineering Geology, 63*(1), 141–155.

Research on the Conversion Relationship of Nuclear-TNT Blast Loading Based on a Protective Door Test Under TNT Blast Loading

Shangwei Dong, Zhimin Tian, Ce Tian, and Jiuyi Li

Abstract To solve the technical problem of converting the nuclear blast loading to equivalent chemical blast loading, based on a certain protective door test under TNT blast loading, taking the same maximum dynamic response of the protective door as the equivalent principle, numerical simulation method, impulse equivalence method, and *Duhamel* integration method are carried out to obtain the peak overpressure of nuclear blast under different TNT blast, conversion function of nuclear-TNT blast loading is fitted as a consequence. The fitting results show that all three methods are reasonable and effective. In comparison, when the stand-off distance z is greater than or equal to 1.71 m/kg$^{1/3}$, the *Duhamel* integration method is more suitable, while when z is less than 1.71 m/kg$^{1/3}$, the impulse equivalence method is.

Keywords Protective door tests · TNT blast loading · Nuclear blast loading · Conversion relationship

1 Introduction

As the international situation is becoming more and more complex and severe, the nuclear threat to civil defense engineering is becoming more and more serious, it is necessary to find out the law of destruction of nuclear blast waves for the civil defense facilities and equipment and do targeted protection in this way. However, with the signing of the International Convention on the Comprehensive Nuclear Test Ban Treaty, no country is allowed to carry out nuclear blast tests in any form and at any location. Against this background, equivalent conversion of nuclear blast using the relevant theory of chemical blast has become a technical challenge.

Protective doors are important protective equipment for the mouth of civil defense engineering, which also bears the brunt of the threat of nuclear blast shock waves.

S. Dong · Z. Tian · C. Tian (✉) · J. Li
Institute of Defence Engineering, Academy of Military Sciences, People's Liberation Army, Beijing 100850, China
e-mail: wwwtiance@126.com

D. Li and Y. Zhang (eds.), *Advances in Frontier Research on Engineering Structures II*, Lecture Notes in Civil Engineering 535, https://doi.org/10.1007/978-981-97-6238-5_8

Therefore, the equivalent method can be carried out using the blast shock wave acting on the protective door as the object of this paper. In recent years, experts and scholars at home and abroad have done a series of studies on the blast-resistant performance of various types of protective doors. Chen et al. [1] found that the strain-rate effects and nonlinear contacts between the door leaf and the doorframe steel plate have considerable influence on the responses of the arched reinforced concrete blast door through full-scale in-situ tests and three-dimensional finite element analysis. Li et al. [2] established the numerical model of interlayered high-damping rubber blast door based on test results, and the antiknock performance of it under thermobaric shock wave was analyzed. Veeredhi et al. [3] used computational methods to study the extent of damage that is likely to be caused by an air blast load of TNT on a stiffened steel plate (door structure). What's more, Chen et al. [4], Guo et al. [5] and Ganorkar et al. [6] also used numerical simulation methods to design the blast doors of new structures under chemical blast loading. Meng et al. [7] proposed a type of airtight blast door using topology and shape optimization. All the research above is based on chemical blast loading, there are fewer studies considering nuclear blast loading as a comparison. As an exploration, Li et al. [8] presented a nonlinear transient dynamic analysis of the two-leaf vaulted steel-coated concrete protective door under nuclear blast wave loadings. However, research on protective doors under chemical blast loading and nuclear blast loading is relatively independent with little connection.

In this paper, a TNT field blast test of a composite flat-type protective door is selected, taking the same maximum dynamic response of the protective door as the equivalent principle, numerical simulation method, impulse equivalence method, and Duhamel integration method are used respectively to obtain the peak overpressure of nuclear blast loading corresponding to different standoff distances of TNT blast loading. With these results, the conversion function of the nuclear-TNT blast loading is fitted.

2   TNT Blast Tests on the Protective Door

Meng et al. [9] have conducted a TNT blast test on a protective door at a site in Xuyi, Jiangsu Province. This type of protective door is prepared with SMC (sheet molding compound) composite material, with a flat plate type as the main structure, and stiffness enhancement is set by the form of multi-stage reinforcement. The exterior dimensions are 1200×2200 mm with a thickness of 85 mm, of which the main plate is 8 mm thick.

The protective door and basic information about the test are shown in Fig. 1:

Three displacement gauges are located on the back side of the door while two overpressure gauges are located on the front, as shown in Fig. 1b and c. The protective door is four-sided-supported by a doorframe, and the TNT charge is hung above the door, as shown in Fig. 1d. There are two blast conditions carried out, one is that the weight of the TNT charge $w = 0.2$ kg, the distance between the TNT charge and the protective door $s = 1.0$ m, and the other is that $w = 0.5$ kg, $s = 1.2$ m. Define the

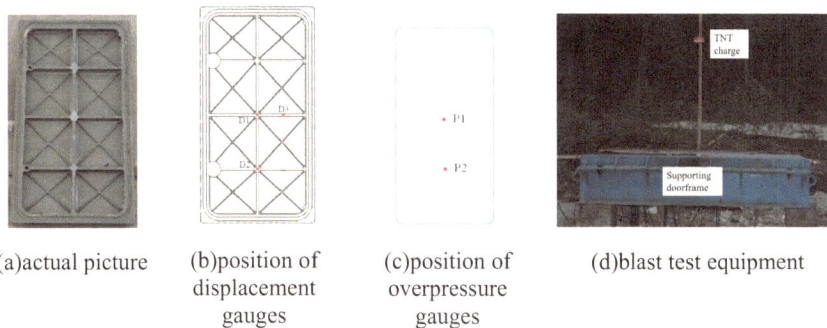

(a)actual picture   (b)position of   (c)position of   (d)blast test equipment
                     displacement   overpressure
                     gauges   gauges

Fig. 1 Protective door and test situation

standoff distance $z = s/w^{1/3}$, so for condition 1, $z = 1.71$ m/kg$^{1/3}$; for condition 2, $z = 1.51$ m/kg$^{1/3}$.

3 Equivalence Methods of Nuclear-TNT Blast Loading

3.1 *Numerical Simulation Method*

Using dynamic finite element analysis software ANSYS/LS-DYNA, the protective door and its doorframe mentioned in the test are modeled. Based on the symmetry, a 1/4 model is built, as shown in Fig. 2.

Fig. 2 Finite element model of the protective door

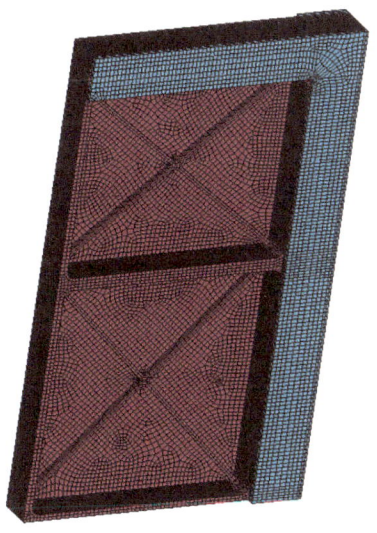

The material properties are defined as follows: SMC is a kind of linear elastic material, which can be simulated by MAT_ELASTIC with density $\rho = 1850$ kg/m³, Young's modulus $E = 7.74$ GPa, Poisson's ratio $\lambda = 0.3$, tensile strength $T = 63$ MPa, failure strain $\varepsilon = 0.0123$. The doorframe can be simulated using an intrinsic plastic strengthening model (Mat_Plastic_Kinematic) with yield strength $f_y = 400$ MPa, density $\rho = 7850$ kg/m³, and Young's modulus $E = 206$ GPa.

TNT blast loading can act on the surface of the protective door through the *Conwep* method, inputting the keywords LOAD_BLAST_ENHANCED [10]. The simulation scene of the protective door under the weight of the TNT charge $w = 0.2$ kg is shown in Fig. 3.

Reliability Verification of the Model Two working conditions, which are the same as the test, are calculated and the numerical simulation results are compared with the test results as follows:

The results of the overpressure data comparison are shown in Fig. 4.

Fig. 3 Simulation scene of the protective door under TNT blast loading

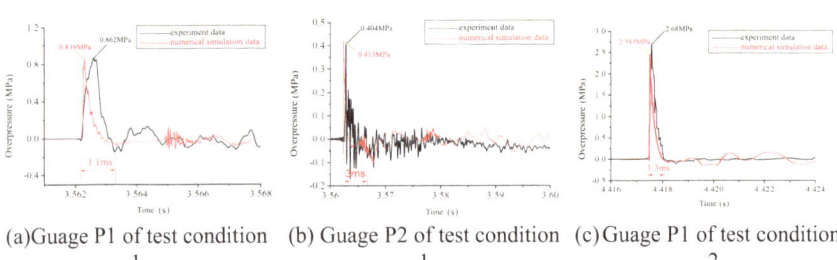

(a)Guage P1 of test condition 1   (b) Guage P2 of test condition 1   (c)Guage P1 of test condition 2

Fig. 4 Comparison of overpressure time-history curves

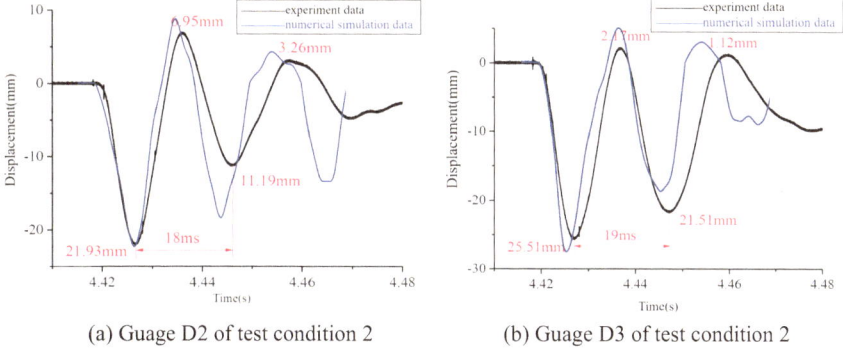

(a) Guage D2 of test condition 2   (b) Guage D3 of test condition 2

Fig. 5 Comparison of displacement time-history curves

From the above three sets of overpressure data, it can be seen that the peak overpressure and positive pressure times of the test and numerical simulation are close, which indicates that the blast loading applied by the *Conwep* method is in line with the actual situation.

The results of the displacement data comparison are shown in Fig. 5.

From the above two sets of displacement data, it can be seen that the displacement change of gauge D3 is basically consistent. The positive and negative peaks of gauge D2 obtained by test and numerical simulation are close, but after reaching the positive peak, the rate of decay of the test displacement is greater than that of the numerical simulation, in addition, the period of the displacement change of the test is in the range of 18–19 ms, while that of numerical simulation is in the range of 15–16 ms, although the difference is not large, may be due to the effect of the structural damping in the test. As for the different decay rates after the peak of the displacement curves, the post-peak decay does not have much effect on the subsequent equivalent calculations since the maximum dynamic response is mainly considered in this paper. In summary, the numerical simulation method used in this paper is reasonable and effective.

Supplementary Calculation Conditions Considering the small number of test conditions, the numerical simulation method is used, and several more sets of calculation conditions are performed. The standoff distances of TNT blasts z are selected by 1.89 m/kg$^{1/3}$, 1.79 m/kg$^{1/3}$, 1.62 m/kg$^{1/3}$, and 1.36 m/kg$^{1/3}$, respectively.

The time-history curve of gauge P1 of one of the calculation conditions ($w = 0.4$ kg, $s = 1.2$ m, $z = 1.62$ m/kg$^{1/3}$) is shown in Fig. 6. To facilitate the analyses in subsequent sections, the curve is fitted by an exponential function.

The comparison results of time-history curves of the three displacement gauges D1, D2, and D3 under this calculation condition are shown in Fig. 7, which can be seen that the displacement changing trends of the three gauges are almost the same, and the gauge with the highest displacement amplitude among the three is D1, which indicates that the position in the protective door where the dynamic response is greatest is the center of it.

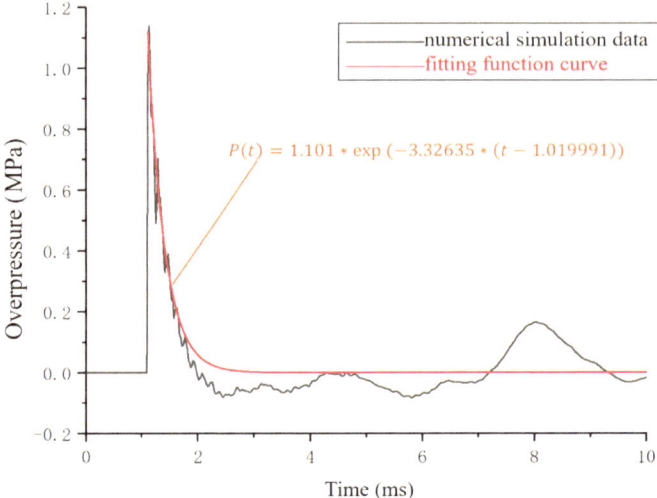

Fig. 6 Time-history curve of gauge P1 under calculation condition $z = 1.62 \text{ m/kg}^{1/3}$

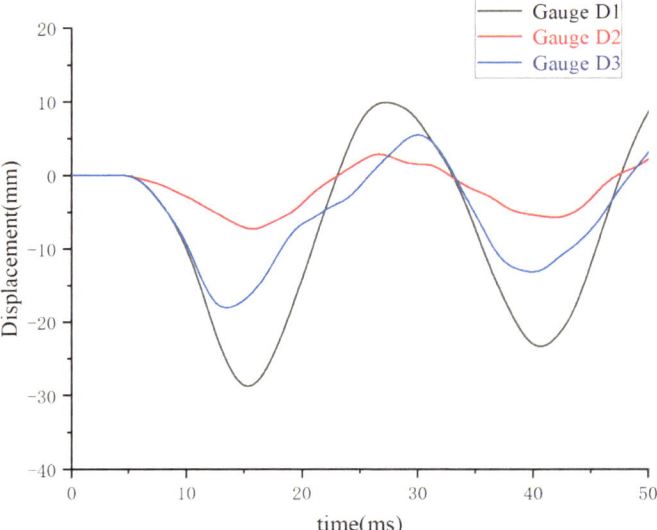

Fig. 7 Comparison of time-history curves of three displacement gauges under calculation condition $z == 1.62 \text{ m/kg}^{1/3}$

The calculation results of supplementary conditions, together with the existing conditions of the tests, are summarised in Table 1.

Equivalent Nuclear Blast Loading The equivalence of nuclear-TNT blast is carried out on the principle that the maximum dynamic response is the same, i.e. so that the

Table 1 Calculation results of TNT blast conditions

Condition No	w/kg	s/m	z(m/kg$^{1/3}$)	Fitting function $P(t)$ of overpressure time-history curves	Peak displacement on D1/mm
1	0.5	1.5	1.89	$0.617 * \exp(-5.51044 * (t - 1.42992))$	19.6
2	0.3	1.2	1.79	$0.786 * \exp(-4.01025 * (t - 1.09991))$	22.6
3	0.2	1.0	1.71	$0.839 * \exp(-4.81129 * (t - 0.92))$	23.7
4	0.4	1.2	1.62	$1.101 * \exp(-3.32635 * (t - 1.019991))$	28.7
5	0.5	1.2	1.51	$1.862 * \exp(-4.0344 * (t - 0.9499507))$	39.3
6	0.4	1.0	1.36	$2.08 * \exp(-4.78544 * (t - 0.729968))$	41.5

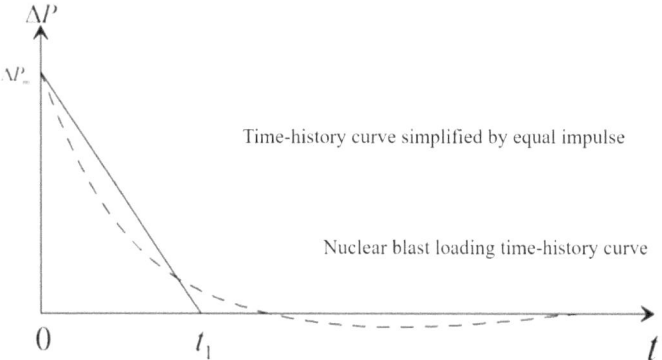

Fig. 8 A typical nuclear blast loading curve

peak displacement at the center point of the protective door (gauge D1) under the action of a TNT blast loading and a nuclear blast loading are the same. A typical nuclear blast loading curve is shown in Fig. 8.

In Fig. 8, ΔP_m is the peak overpressure of a nuclear blast loading, t_1 is the positive overpressure duration of the nuclear blast wave. According to the People's Air Defense Basement Design Code[11], if the design is guided by commonly used level IV of human defense, the positive overpressure duration simplified by equal impulse can be taken as $t_1 = 0.38$ s. The nuclear blast loading shown in Fig. 8 can be expressed by Eq. 1:

$$\Delta P(t) = \Delta P_m \left(1 - \frac{t}{t_1} \right) (0 < t < t_1) \qquad (1)$$

The nuclear blast loading expressed in Eq. (1) is uniformly applied in the form of the load curve keyword (DEFINE_CURVE) on the surface of the protective door, and the solution of the dynamic response of the door is carried out. Peak displacement at gauge D1 of each TNT blast calculation condition can correspond to a nuclear

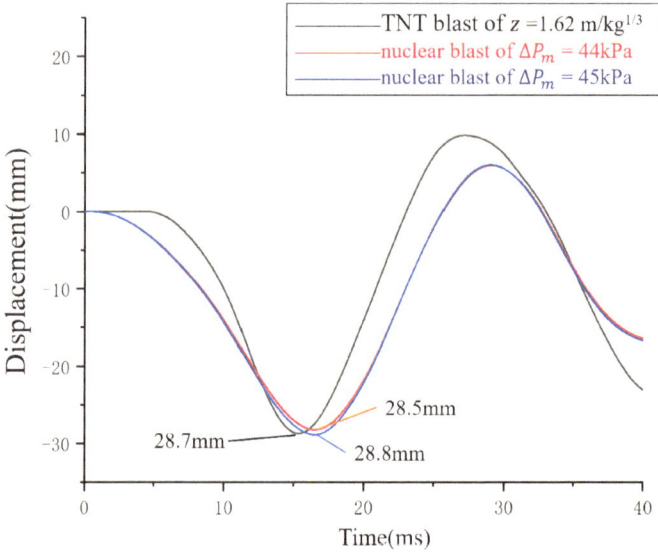

Fig. 9 Comparison of time-history curves of gauge D1 under the TNT blast loading of $z = 1.62$ m/kg$^{1/3}$ and different nuclear blast loadings

Table 2 Nuclear-TNT blast loading Correspondence from numerical simulation method

Condition No.	$z\left(m/kg^{1/3}\right)$	Equivalent ΔP_m
1	1.89	18.5
2	1.79	24.5
3	1.71	28.5
4	1.62	44.5
5	1.51	50.5
6	1.36	70.5

blast loading of specific ΔP_m. For example, for TNT blast calculation condition 4 ($z = 1.62$ m/kg$^{1/3}$), the corresponding equivalent nuclear blast loading ΔP_m is between 44 and 45 kPa, as shown in Fig. 9. Take the middle value of them, i.e.44.5 kPa.

In the same way, the equivalent results of the nuclear blast loadings for the other five TNT blast conditions can be obtained, as shown in Table 2.

3.2 Impulse Equivalence Method

The idea behind the impulse equivalence method is that the impulse of the TNT and nuclear blast loadings acting on the protective door is equal in the time from the beginning of the dynamic response of the structure to the time when the maximum

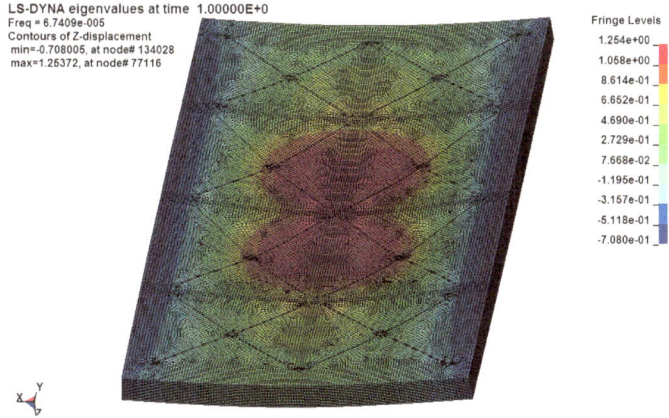

Fig. 10 Modal analysis results of the protective door

dynamic response is reached. As can be seen from the displacement time-history curves shown in Sect. 3.1, both in tests and numerical simulations, the maximum dynamic response of the protective door is reached within a very short time after the blast loading is applied. To quantitatively characterize this time, modal analysis is used to calculate the main vibration frequency of the door, i.e. the self-oscillation frequency $f = 67.41$ Hz, as shown in Fig. 10. Thus, the self-oscillation period $T = 14.83$ ms. After comparison, the time for the door to reach maximum dynamic response is approximately $T/2$, i.e. 7.417 ms.

Integrate the overpressure time-history curve, taking the time when the peak overpressure arrives as the initial moment of integration, we obtain:

$$i_{T/2} = \int_0^{7.417} P(t)dt \tag{2}$$

In Eq. (2), for TNT blast loading, $P(t)$ is expressed in Table 1; for nuclear blast loading, $P(t)$ is expressed by Eq. (1).

Take the TNT calculation condition 4 ($z = 1.62$ m/kg$^{1/3}$) as an example, $i_{T/2} = 0.31237$ MPa.ms.

For the nuclear blast loading shown in Eq. (1), $i_{T/2}$ increases linearly with ΔP_m. If $\Delta P_m = 1$ kPa, then $i_{T/2}$ will be 0.00734 MPa.ms. Thus, a TNT blast loading with $z = 1.62$ m/kg$^{1/3}$ produces the same maximum dynamic response impulse as a nuclear blast loading with $\Delta P_m = 42.6$ kPa.

As a result, an equivalent correspondence between TNT blast loading and nuclear blast loading can be obtained, see Table 3:

Table 3 Nuclear-TNT blast loading Correspondence from impulse equivalence method

Condition No.	$z(\mathrm{m/kg^{1/3}})$	Equivalent ΔP_m
1	1.89	15.3
2	1.79	20.8
3	1.71	26.7
4	1.62	42.6
5	1.51	50.4
6	1.36	68.5

3.3   Duhamel Integration Method

Because the material of the protective door is linearly elastic, and it can be seen from the TNT blast tests and the numerical simulation results that in all conditions, the protective door is basically manifested as elastic vibration. Therefore, an equivalent single-degree-of-freedom linear elastic system can be used for dynamic response analysis in this paper.

According to *Duhamel* integration[12],

$$y(t) = \frac{1}{m\omega_D} \int_0^t P(\tau) e^{-\xi\omega(t-\tau)} \sin[\omega_D(t-\tau)] d\tau \tag{3}$$

where, $y(t)$ is the dynamic displacement of a single-degree-of-freedom structural system over time, m is the equivalent mass of the structural system, which can be taken as unit mass 1.0, ω is the self-oscillation circular frequency of the structural system, $\omega = 2\pi f$, f is the self-oscillation frequency of the protective door, taken as 67.41 Hz, ω_D is the self-oscillating circular frequency of the protective door with damping, $\omega_D = \omega\sqrt{1-\xi^2}$, ξ is the damping ratio, which can be taken as 0.05 in this paper. $P(\tau)$ is the time history of overpressure, the same meaning as that in Sect. 3.2.

Substituting the overpressure time-history curves of every TNT blast condition and the nuclear blast loading with unit overpressure ($\Delta P_m = 1$ kPa) into *Duhamel* integration, we obtain the time-history curves of the dynamic displacement of the door caused by each loading. The comparison of dynamic displacement time history between the TNT blast loading of $z = 1.62$ m/kg$^{1/3}$ and the nuclear blast loading of $\Delta P_m = 1$ kPa is shown in Fig. 11.

It can be seen from Fig. 11. that the trends shown by the two integral curves are basically the same, with the dynamic displacement rising sharply to reach the negative peak at the beginning of the loading, and then showing regular changes of damped vibration. The maximum dynamic displacement caused by the TNT blast loading of $z = 1.62$ m/kg$^{1/3}$ is 0.01376, while that of the nuclear blast loading of $\Delta P_m = 1$ kPa is 0.00036. The value of the former is 38.2 times higher than that of the latter.

From Eq. (1) and *Duhamel* integration Eq. (3), it can be obviously seen that varying the ΔP_m of the nuclear blast loading results in a linear change in the dynamic

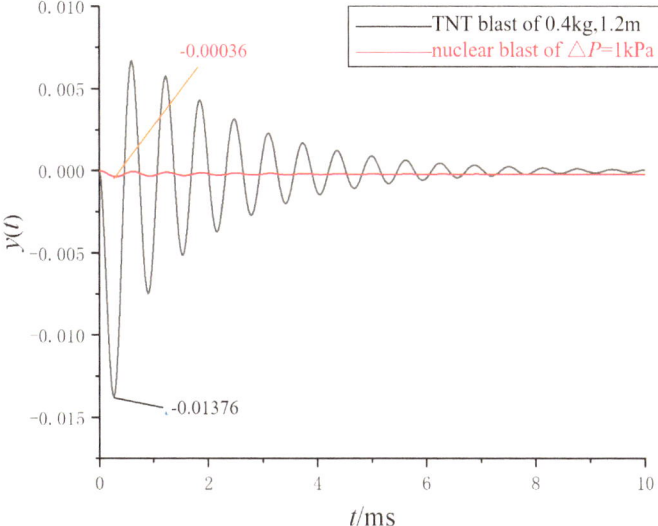

Fig. 11 The comparison of dynamic displacement time history between the TNT blast loading of $z = 1.62$ m/kg$^{1/3}$ and the nuclear blast loading of $\Delta P_m = 1$ kPa

Condition No.	z(m/kg$^{1/3}$)	Equivalent ΔP_m
1	1.89	19.2
2	1.79	23.5
3	1.71	27.7
4	1.62	38.2
5	1.51	44.2
6	1.36	62.9

Table 4 Nuclear-TNT blast loading Correspondence from the *Duhamel* integration method

displacement, while the shape of the integral curve remains unchanged. In this way, it can be approximated that the maximum dynamic response caused by the TNT blast loading of $z = 1.62$ m/kg$^{1/3}$ is numerically equal to that of the nuclear blast loading with $\Delta P_m = 38.2$ kPa, i.e. the equivalence process is completed. The TNT blast loadings of other conditions are treated in this same manner. Nuclear-TNT blast loading correspondence is shown in Table 4.

4 Analysis of the Conversion Results

The data from Tables 2, 3, and 4 are summarised in Fig. 12:

It can be seen from Fig. 12. that the curves obtained by the three methods of conversion are basically the same, indicating that the three conversion methods are

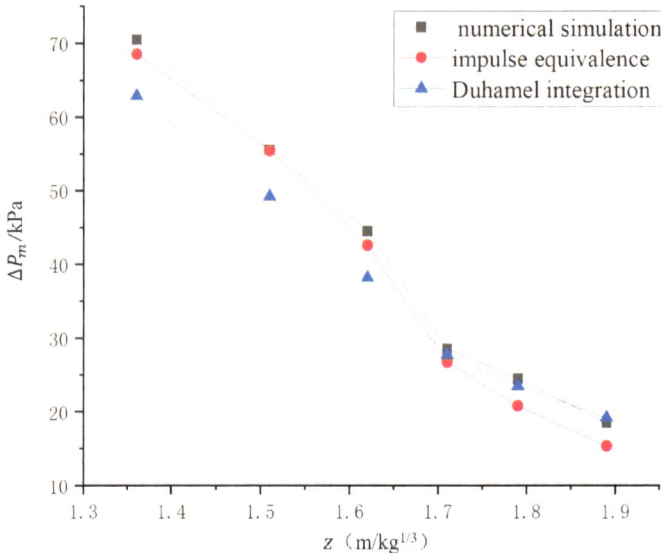

Fig. 12 The comparison of conversion results using three methods

reasonable. When z is greater than or equal to 1.71 m/kg$^{1/3}$, results from the *Duhamel* integral method are closer to the results obtained by the simulation, while the conversion results from the impulse equivalence method are relatively small. When z is less than 1.71 m/kg$^{1/3}$, the conversion results obtained by the impulse equivalence method are gradually close to the simulation results, while the conversion results obtained by the *Duhamel* integration method are gradually far away from the simulation results as z decreases.

Numerical simulation results can more realistically restore the nuclear blast loading on the protective door, and the conversion results of a relatively high degree of accuracy, but the amount of calculation is much more than the other two methods. Thus, it's more efficient to use the impulse equivalence method and the *Duhamel* integration method. More detailedly, when z is greater than or equal to 1.71 m/kg$^{1/3}$, the *Duhamel* integration method is suitable for fitting the conversion function, while when z is less than 1.71 m/kg$^{1/3}$, the impulse equivalence method is more suitable.

Function-fitting results are as follows:

$$\Delta P_m = z^{-5.63484} \quad z < 1.71 \text{ m/kg}^{1/3}$$

$$\Delta P_m = z^{-5.70164} \quad z \geq 1.71 \text{ m/kg}^{1/3} \tag{4}$$

5 Conclusions

The main results and conclusions obtained in this paper through the TNT blast load test acting on the protective door are as follows:

(a) The nuclear-TNT blast loading correspondences are obtained by three different equivalence conversion methods: simulation method, impulse equivalence method, and *Duhamel* integration method. It is proven that all these three methods are reasonable.

(b) When z is greater than or equal to 1.71 m/kg$^{1/3}$, the *Duhamel* integration method is more suitable, while when z is less than 1.71 m/kg$^{1/3}$, the impulse equivalence method is. Conversion functions are fitted using these two methods on different ranges.

This paper establishes the nuclear-TNT blast loading converting method by a certain composite protective door with a flat plate type, but the method is also suitable for other protective structures that can be simplified to a liner elastic SDOF system.

References

1. Chen, L., Fang, Q., Zhang, Y.-D., Zhang, Y., & Fan, J.-Y. (2010). Numerical and Experimental Investigations on the blast-resistant properties of arched RC blast doors. *Int. J. Prot. Struct., 1*, 425–441.
2. Li, X.; Miao, C.; Wang, Q.; Geng, Z. Antiknock performance of interlayered high-damping-rubber blast door under thermobaric shock wave. Shock. Vib. 2016, 2420893.
3. Veeredhi, L.S.B.; Ramana Rao, N.V. Studies on the impact of explosion on blast resistant stiffened door structures. J. Inst. Eng. Ser. A 2015, 96, 11–20.
4. Chen, W., & Hao, H. (2012). Numerical study of a new multi-arch double-layered blast-resistance door panel. *International Journal of Impact Engineering, 43*, 16–28.
5. Guo, D., Li, Z., Wang, Q., & Hou, X. (2013). Research on blast-resistant properties of concrete-filled steel and steel tube blast doors. *Protection Engineering, 35*, 38–43. (In Chinese).
6. Ganorkar, K., Goel, M. D., & Chakraborty, T. (2023). Numerical analysis of double-leaf composite stiffened door subjected to blast loading. *Journal of Performance of Constructed Facilities, 37*, 04022067.
7. Meng, Y., Li, B., & Wang, Y. (2016). Structure design of new airtight blast door based on topology and shape optimization method. *Geotechnical and Geological Engineering, 34*, 703–711.
8. Li, E., Sheng, X., & Wang, J. (2011). Dynamic analysis of vaulted protective doors under nuclear and conventional blast loadings. *Protection Engineering, 33*, 16–21. (In Chinese).
9. Meng, F., Zhang, B., Zhao, Z., Xu, Y., Fan, H., & Jin, F. (2016). A novel all-composite blast-resistant door structure with hierarchical stiffeners. *Composite Structures, 148*, 113–126.
10. LSTC. (2017). *LS-DYNA keyword user's manual*. San Francisco, CA, USA.
11. GB 50038-2005. (2005). *Code for design of civil air defense basement*. Beijing (in Chinese).
12. Zhou, S., & Ward, H. (2016). *Structural dynamics*. Beijing Institute of Technology Press.

High Performance Concrete
and Engineering Component Research

Study on the Durability of Manufactured Sand Concrete with the Change of Mix Proportion Parameters

Liangyu Guo, Fushan Ma, and Binhui Zhao

Abstract Cement, aggregate, mineral admixture, concrete admixture, and mixing water were used as the main raw materials to prepare manufactured sand concrete. By adjusting the proportioning parameters of manufactured sand concrete, such as sand ratio and water-cement ratio, the influence of proportioning parameters on the durability of manufactured sand concrete was studied. Experimental results show that when the sand ratio is between 0.38 and 0.44, the mechanism of sand best workability and durability of the concrete strength, the best value for the mechanism of sand concrete, sand ratio range within the scope of the young stage of mechanism sand concrete stress is almost not affected by sand ratio increase, age stage mechanism of sand concrete stress along with the increase in sand ratio decreased significantly; The stress of natural medium sand concrete and manufactured sand concrete increases with the increase in the water-cement ratio, and the compressive strength of manufactured sand concrete is slightly higher than that of ordinary natural sand concrete. When the stone powder content is 10.5%, its filling performance is better, and the frost resistance of mechanized sand concrete can be improved by adjusting the stone powder content. When the powder volume is 158 dm^3, the manufactured sand concrete has good workability and the highest durability. The increase in the water-binder ratio will lead to the aggravation of carbonization depth, serious carbonization phenomenon, and reduced durability of machined sand concrete.

Keywords Mix proportion parameters · Manufactured sand concrete · Durability · Optimal sand yield · Stone powder content · Water-cement ratio

L. Guo
Qinghai Provincial High Grade Highway Construction Management Bureau, Xining, China

F. Ma (✉) · B. Zhao
CCCC First Highway Engineering Group Co. Ltd., Beijing, China
e-mail: 893238584@qq.com

© The Author(s) 2025
D. Li and Y. Zhang (eds.), *Advances in Frontier Research on Engineering Structures II*,
Lecture Notes in Civil Engineering 535, https://doi.org/10.1007/978-981-97-6238-5_9

1 Introduction

In recent years, with the rapid development of the economy, on the one hand, people's requirements for various buildings and infrastructure have been continuously improved. On the other hand, the state advocates energy conservation and environmental protection and pays more and more attention to environmental protection, which puts forward more requirements for the environmental protection performance of engineering materials.

The durability of manufactured sand concrete refers to the ability of manufactured sand concrete to maintain its performance under the action of various environmental factors. Under normal circumstances, various buildings and infrastructures built with manufactured sand concrete are permanent. Once an accident occurs, it will increase the difficulty of project reconstruction, cause major economic losses, and even cause casualties. Therefore, it is of great practical significance for engineering construction and infrastructure construction to study the durability of manufactured sand concrete [1–4].

At present, there is a lack of systematic research on the influencing factors of the configuration of manufactured sand concrete. With a more comprehensive understanding of manufactured sand and the more exquisite preparation technology of sand, people have broken through the singularity and limitations of the previous research on the durability of manufactured sand. It is no longer limited to the study of the durability of manufactured sand concrete with a single ratio parameter change, but gradually turns to the study of the influence of multiple mix proportion parameters on the durability of manufactured sand concrete. For example, Li proposed a study based on the effect of MB value on the early plastic cracking and shrinkage of high-strength manufactured sand concrete [5]. Song et al. [6] proposed a study based on the effect of lithology on the performance of manufactured sand concrete. Although these studies have made some contributions to the study of the durability of manufactured sand concrete, there are some problems in the ratio analysis, which has affected its applicability in practical engineering construction [7–11].

In order to solve the above problems, find out the most suitable mix proportion parameters, and provide a reliable basis and reference for the configuration of manufactured sand concrete in engineering construction, this paper studies the influence of the change of mix proportion parameters on the durability of manufactured sand concrete.

Table 1 Main raw material information

Raw material	Type	Manufacturer
Cement	P.O 42.5R	Beijing Taiding Chemical Materials Co., Ltd.
Fine aggregate	Manufactured sand	Hebei Zhuofi Mineral Products Co., Ltd.
Coarse aggregate	Limestone gravel	Hebei Jiegui Mineral Products Co., Ltd.
Mineral admixtures	200–325 coal fly ash	Lingshou County Baofeng Mica Processing Co., Ltd.
Concrete admixture	HB-32322 pumping agent	Langfang Hanbo Fireproof Material Co., Ltd.
Mixing water	Ordinary tap water	

Table 2 Performance of coarse aggregate

Varieties	Limestone gravel particle size/mm	Needle flake (%)	Tight packing density (g/cm^3)
A1	5–10	8–9%	1.8–1.9
A2	10–20	13–15%	1.9–2.0

2 Materials and Methods

2.1 Main Raw Materials

The raw materials used in this paper mainly include cement, aggregate, mineral admixture, concrete admixture and mixing water. Specific material information is shown in Table 1.

Among them, the coarse aggregate is matched with A1 and A2 limestone gravels with mud content less than 1%, and the matching ratio is 3:7. Performance indicators are shown in Table 2.

2.2 Test Method

In this paper, when studying the influence of the change of the mix proportion parameters on the durability of the manufactured sand concrete, the other parameters are unchanged, and the mix ratio performance of the manufactured sand concrete is tested according to the 'Specification for mix proportion design of ordinary concrete' (JGJ 55-2011) [12]. According to the code 'construction sand' (GB/T 14684-2022) [13], the composition of manufactured sand was tested. The compressive strength of manufactured sand concrete was measured according to the 'Standard for test methods of concrete physical and mechanical properties' (GB/T 50081-2019) [14].

3 Test Results and Analysis

3.1 Sand Rate

In the case of a certain amount of stone, the increase or decrease of sand content will affect the size of the sand rate. Because the main components of the aggregate in this experiment are manufactured sand and limestone gravel, the size of the sand rate will affect the slump and expansion of the manufactured sand concrete, which affects its durability in turn. Figure 1 shows the production and tests of the manufactured sand.

In order to make the manufactured sand concrete more durable, it is necessary to select the appropriate sand ratio to obtain the best workability and dense structure. In general, when the ratio of slump to expansion of manufactured sand concrete is 2:5, the durability of manufactured sand concrete is better. It can be seen that in this paper, when the sand ratio of manufactured sand concrete is 0.38–0.44, the ratio of the slump to the expansion of manufactured sand concrete is closer to 2:5, indicating that the workability of manufactured sand concrete is more reasonable and the durability is better. When the sand ratio is 0.44, the workability and durability of the manufactured sand concrete are the best.

3.2 Water Cement Ratio

The water-cement ratio law of concrete strength has been widely used in the study of the durability of manufactured sand concrete. The compressive strength of concrete can reflect the stress of concrete. Table 3 is the compressive strength of two kinds of concrete under different water-cement ratios.

It can be seen from Table 3 that in the two kinds of sand concrete, the relationship between the water-cement ratio and compressive strength is negatively correlated, and the compressive strength of manufactured sand concrete is slightly higher. The main reason is that the roughness and angle of the surface of the manufactured sand lead to high surface energy. In addition, the particle size is small, which can fill the gap of concrete, increase the density of concrete and improve the strength of concrete. At the same time, the surface of the manufactured sand concrete with calcium carbonate as the main component will undergo a weak chemical reaction to form a protective film, while the natural sand concrete with SiO_2 as the main component cannot produce the chemical reaction.

3.3 The Amount of Stone Powder

In manufactured sand concrete, stone powder can fill its gap, increase its compactness, and appropriately adjust the content of stone powder, which can improve its working

Fig. 1 Production and tests of the manufactured sand, **a** Production of manufactured sand, **b** Manufactured sand, **c** Test of manufactured sand concrete

(a) Production of manufactured sand

(b) Manufactured sand

(c) Test of manufactured sand concrete

Table 3 Compressive strength of different water-cement ratio

Water-cement ratio of sand concrete	Natural medium sand concrete strength (MPa)	Artificial sand concrete strength (MPa)
0.36	64	66
0.38	63	65
0.4	62	64
0.42	61	63
0.44	60	62
0.46	59	61
0.48	58	60
0.5	57	59
0.52	56	58
0.54	55	57

performance and durability, and better meet the construction requirements of efficient projects.

Figure 2 shows the laboratory tests on freeze–thaw of manufactured sand concrete, and Fig. 3 is the strength ratio of low and high-strength manufactured sand concrete with different stone powder content before and after freeze–thaw.

It can be seen from Fig. 3 that the stone powder content of low-strength manufactured sand concrete is negatively correlated with the strength ratio before and after freezing and thawing, while the strength ratio of high-strength manufactured sand concrete before and after freezing and thawing is positively correlated with the stone powder content. Therefore, in order to improve the durability of low-strength concrete with manufactured sand, it is necessary to reduce the stone powder content and improve its frost resistance. On the contrary, in manufactured sand high-strength concrete, in order to improve the durability of the manufactured sand concrete, it is necessary to increase the stone powder content to improve its frost resistance.

Fig. 2 Laboratory tests on freeze–thaw of manufactured sand concrete

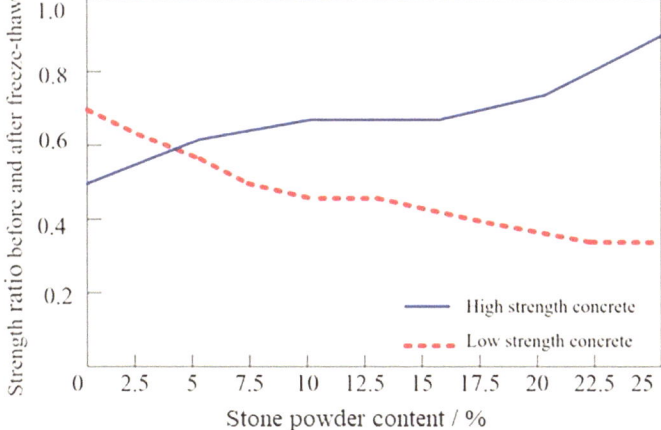

Fig. 3 Frost resistance of manufactured sand concrete with different stone powder content

3.4 Water-Binder Ratio

The durability of mechanical sand concrete is affected by its carbonation depth. The carbonation depth of mechanical sand concrete is related to the reason for the water gel. The reason is that the concrete cementitious material makes the concrete dense, and the carbonation rate of mechanical sand concrete is mainly related to its compactness.

Figure 4 shows the relationship between the water-binder ratio and carbonation depth of manufactured sand concrete with different carbonation ages. It can be seen that for manufactured sand concrete of different ages in Fig. 4, the water-binder ratio is positively correlated with the carbonization depth, and the carbonization age is positively correlated with the carbonization depth. It is proved that the increase in the water-binder ratio will lead to an increase in carbonation depth, serious carbonation and a decrease in the durability of manufactured sand concrete. The higher the carbonation age is, the more serious the carbonation of manufactured sand concrete is, and the worse the durability is.

Figure 5 is the carbonation depth of manufactured sand concrete with different water-binder ratios affected by different compressive strengths.

It can be seen from Fig. 5 that under a certain water-binder ratio, the relationship between compressive strength and carbonation depth of concrete is negatively correlated, and the lower the water-binder ratio, the lower the carbonation depth. This proves that under a certain water-binder ratio, the increase of compressive strength will increase the durability of manufactured sand concrete, and the larger the water-binder ratio is, the worse the resistance of manufactured sand concrete to carbonization is, and the lower the durability is.

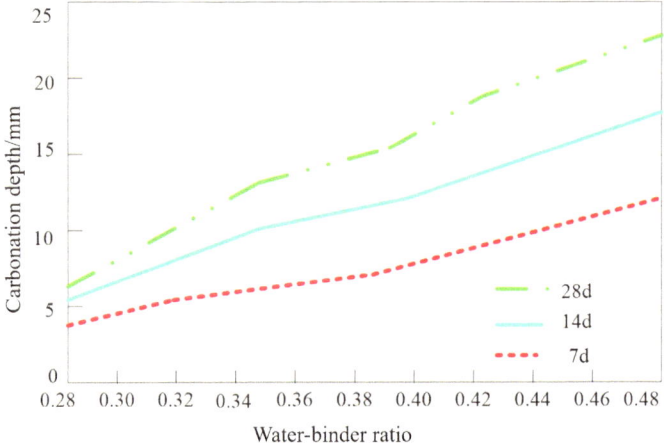

Fig. 4 Carbonization depth of concrete with different water-binder ratios

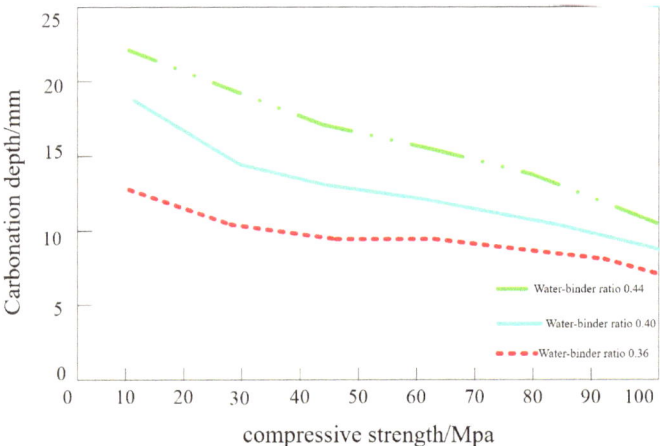

Fig. 5 Carbonation depth of different compressive strength

4 Conclusion

(1) In the case of a certain compressive strength, when the sand ratio is 0.44, the durability of artificial sand concrete is higher than that of natural sand concrete. In the manufactured sand concrete, when the sand ratio is between 0.38 and 0.44, the slump of the manufactured sand concrete is larger and the ratio of its expansion is closer to 2:5, and the durability of the manufactured sand concrete is better. When the sand ratio is 0.44, the durability of the manufactured sand concrete is the strongest.

(2) The stone powder content of low-strength manufactured sand concrete is negatively correlated with the strength ratio before and after freeze–thaw, while the strength ratio before and after freeze–thaw of high-strength manufactured sand concrete is positively correlated with the stone powder content. By adjusting the stone powder content in the manufactured sand concrete, its frost resistance can be adjusted, which in turn affects its durability.

(3) The increase of the water-binder ratio of manufactured sand concrete will aggravate the carbonation of manufactured sand concrete and reduce the durability of manufactured sand concrete. Under a certain water-binder ratio, the compressive strength of manufactured sand concrete is positively correlated with its durability.

(4) The content of cementitious material in manufactured sand concrete has a certain correlation with durability. When the quality of the cementitious material reaches 355 kg/m^3, the manufactured sand has high durability, which can be used as a reference and basis for manufactured sand concrete.

Acknowledgements This work was financially supported by the Science and Technology Plan Project of Qinghai Province (No. 2021-SF-140), the project name is 'Key Technology Research on the Application of Manufactured Sand in Ordinary Strength Grade Concrete in Alpine Areas'.

References

1. Guo, R. P., Zhang, Y. Z., Xia, J. L., et al. (2020). Study on the influence of manufactured sand and stone powder content on the durability of sleeper concrete. *Concrete, 8*, 118–121.
2. Lai, Y. C., Liu, D. W., Huang, L. J., et al. (2019). Durability evaluation of manufactured sand concrete structures based on cloud model. *Silicate Bulletin, 10*, 275–281.
3. Xu, X., Xia, J. L., Zhou, Y. X., et al. (2019). Preparation and performance study of C55 concrete for manufactured sand T beam. *Concrete, 04*, 139–142.
4. Jin, Y. Q., Liu, W., & Yang, X. T. (2020). Research on grain shape characteristics of fine manufactured sand based on sphericity ratio. *Railway Building, 60*(02), 148–151.
5. Li, P. S. (2018). Effect of MB value of manufactured sand on early plastic cracking and dry shrinkage of high-strength concrete. *Road Traffic Technology (Applied Technology Edition), 01*, 52–54.
6. Song, S. M., Cheng, C., & Yang, N. (2018). Research on the influence of mechanism sandstone on the properties of mortar and concrete. *Concrete, 11*, 27–30.
7. Li, J. (2020). Research on the working performance and strength influencing factors of manufactured sand concrete. *New Building Materials, 47*(05), 26–28, 32.
8. Chen, Y., Yang, L. H., & Zheng, Y. B., et al. (2020). Effect of sand-binder ratio on flexural and compressive properties of hybrid fiber reinforced manufactured sand mortar. *Concrete, 11*, 101–105+109.
9. Deng, C., Yan, J. J., & Ye, X. S. (2018). Study on the effect of manufactured sand content on the mechanical properties and volume stability of concrete. *New Building Materials, 45*(4), 42–46.
10. Wang, Z., Han, Z. L., Li, H. J., et al. (2019). Study on the performance of manufactured sand concrete for prestressed structures in railway engineering. *Railway Architecture, 59*(04), 16–21.

11. Liu, D., Wu, H. L., Lan, C., et al. (2020). Study on the influence of manufactured sand powder content on the performance of airport pavement concrete. *Construction Technology, 02*, 104–107.
12. GJ 55-2011. (2011). *Specification for mix proportion design of ordinary concrete*. China Architecture & Building Press.
13. GB/T 14684-2022. (2022). *Sand for construction*. China Standards Publishing House.
14. GB/T 50081-2019. (2019). *Standard for test methods of concrete physical and mechanical properties*. China Architecture & Building Press.

Study on Performance of Reactive Diluted Cold Patch Asphalt Fluid

Zhi Suo, Shijie Xu, Tao Hu, Yue Zhang, Yating Wang, and Sheng Yang

Abstract A reactive dilution type of cold patch asphalt liquid was developed in this study using acrylate epoxidized soybean oil (AESO). The AESO was employed as a reactive diluent to replace a portion of conventional solvents. The curing of the cold patch asphalt liquid was facilitated through photoinitiation to enhance the road performance of the cold patch asphalt mixture. An orthogonal experimental design method was used to optimize the proportions of various components, followed by laboratory experiments to investigate the performance of the developed reactive dilution type of cold patch asphalt liquid. The research results indicated that AESO could replace a portion of solvents and significantly reduce the viscosity of the asphalt. The best improvement in the rheological properties of the cold patch asphalt was achieved when the dosage of AESO was 4%. Further increasing the dosage of AESO did not significantly enhance the performance improvement and instead increased viscosity, which was unfavorable for construction. SEM observations revealed the formation of a uniform network structure in the asphalt after curing with AESO, which can enhance the road performance of the cold patch asphalt mixture. The road performance of the reactive dilution type of cold patch asphalt mixture was also preliminarily evaluated through Marshall stability tests.

Keywords Epoxy soybean oil acrylate (AESO) · Cold patch asphalt · Cold patch asphalt mixture · Strength

Z. Suo · S. Xu (✉) · T. Hu · Y. Zhang · Y. Wang · S. Yang
Beijing University of Civil Engineering and Architecture, Beijing, China
e-mail: xshijie136@163.com

Z. Suo
e-mail: suozhi@bucea.edu.cn

T. Hu
e-mail: mmiwm123@163.com

S. Yang
e-mail: h13253815123t@163.com

© The Author(s) 2025
D. Li and Y. Zhang (eds.), *Advances in Frontier Research on Engineering Structures II*,
Lecture Notes in Civil Engineering 535, https://doi.org/10.1007/978-981-97-6238-5_10

115

1 Introduction

The asphalt pavement is subjected to various forms of distress due to factors such as vehicle loads and environmental conditions during its service life. One of the most common distresses is potholes, which can generate impact loads 1.5–2.5 times greater than normal traffic loads [1], resulting in significant damage to the asphalt pavement. Pothole repairs typically involve the use of hot-mix asphalt and cold-patch asphalt mixtures [2]. Although hot-mix asphalt technology is well-established, it has drawbacks such as high energy consumption and susceptibility to temperature-related construction issues.

Cold patch asphalt mixtures offer advantages such as on-demand availability, minimal construction waste, and reduced carbon emissions, making them widely used for pothole repairs. However, they often suffer from performance limitations. Traditional solvent-based cold patch asphalt mixtures exhibit low early strength, poor low-temperature performance, and inadequate durability. Moreover, solvent usage typically accounts for 20–30% of the asphalt content [3], which can have environmental implications due to solvent volatilization. These factors have restricted the widespread adoption and application of solvent-based cold patch asphalt mixtures. Therefore, the development of a low-solvent, high-performance cold patch asphalt emulsion is of significant importance for road maintenance purposes.

Currently, plant oil derivatives are widely used in road engineering [4–6]. AESO is synthesized from soybean oil through epoxidation and acrylic acid esterification. It possesses excellent workability, low raw material cost, and easy availability, making it extensively employed in various fields such as UV-curable coatings, plastics, and rubber [7]. The photopolymerization mechanism of epoxy soybean oil acrylate involves the addition reaction between the acrylate monomers and epoxy groups under light irradiation, leading to the formation of a crosslinked structure and producing stable cured products [8, 9].

Considerable research has been conducted on the curing effect of photoinitiators on AESO and the modification of asphalt by AESO. For example, Migle et al. [10] investigated the influence of photoinitiators on the elastic modulus and tensile strength of AESO at different temperatures, demonstrating that AESO can undergo a curing reaction at room temperature. Lee et al. [11] used AESO to prepare photopolymerizable crack repair materials, which showed excellent adhesion and chemical resistance and could be used for repairing cracks in chemical containers and transportation pipelines. Chen et al. [12] studied the effect of AESO on the rheological properties of asphalt and found that AESO not only improves the fatigue and low-temperature performance of asphalt mixtures but also reduces the aging level of asphalt mixtures. Chen et al. [13] utilized epoxy soybean oil, photoinitiators, and other additives to develop recycled asphalt rejuvenators, which have been applied in the Chongqing region and exhibit excellent thermal stability.

In this study, AESO was introduced as a modifier into cold patch asphalt emulsion. The feasibility of using AESO in cold patch asphalt mixtures was investigated by studying the viscosity, curing time, rheological properties, microstructure, and Marshall stability of the cold patch asphalt emulsion.

2 Materials and Experimental

2.1 Raw Materials

Asphalt

Panjin 70# base asphalt, which is provided by Beijing Municipal Road and Bridge Building Materials Group Co., Ltd. The performance parameters of the asphalt were tested according to the Standard Test Methods of Bitumen and Bituminous Mixtures for Highway Engineering (JTG E20-2011). The test results are shown in Table 1.

Diluent

Diesel has a higher flash point and moderate volatility. Considering factors such as solubility, safety, and volatility rate, 0# diesel fuel was chosen as the asphalt diluent. The 0# diesel was purchased from a local gas station.

AESO

AESO was purchased from Shandong Yaotong Industrial Company. It is a light-yellow liquid with a viscosity of 1.5–2.5 Pa·s.

Photoinitiators

The photoinitiator is a key material that promotes the curing of epoxy soybean oil acrylate, determining the rate and quality of the curing reaction. The photoinitiator used in this study was purchased from Nanjing Jiazhong Chemical Technology Co., Ltd. It is a yellow powder with a melting point of 127 °C to 133 °C.

Aggregate and Gradation

This study used limestone from the Hebei region as aggregates and mineral powder. The performance tests were conducted in accordance with the relevant specifications of the Test Methods of Aggregate for Highway Engineering (JTG E42-2005), and the test results met the requirements. The Marshall stability test was performed using the LB-13 gradation for the cold patch asphalt mixture.

Table 1 Panjin 70# asphalt parameters

Test item	Technique requirements	Test result	Test method
Penetration (25 °C)/0.1 mm	60–80	64.1	T 0604
Ductility (15 °C)/cm	>150	≥ 100	T 0605
Softening point/ °C	≥ 43	47.6	T 0606

Table 2 Orthogonal test protocol design

Test No	AESO	Diesel	Photoinitiators	viscosity	Curing time
1	2	10	3.0	5.66	220
2	4	12	3.5	2.56	150
3	6	14	4.0	2.03	130
4	2	12	4.0	2.21	120
5	4	14	3.0	1.59	200
6	6	10	3.5	3.84	160
7	2	14	3.5	1.39	150
8	4	10	4.0	4.81	130
9	6	12	3.0	2.89	230

Orthogonal Experimental Design

The orthogonal experimental design method was used to determine the optimal ratio of components for the cold patch asphalt mixture. Based on the literature review and recommended dosages from manufacturers, with asphalt (A) as the base at 100, and diluent, epoxy soybean oil acrylate, and photoinitiator as factors B, C, and D, respectively, three levels were set. The initial ratio was set as 70# asphalt: diesel fuel: AESO: photoinitiator = 100: 12: 4: 3.5. The amounts of each component were adjusted based on this initial ratio. The L_9 (3^3) orthogonal experimental design table was used to arrange the experiments. The design of the orthogonal experimental plan is shown in Table 2.

2.2 Experimental Design

Brookfield Viscosity Test

The viscosity of the cold patch asphalt is a critical indicator of its workability and ease of construction. The variations in the dosage of each component have a significant impact on the viscosity of the asphalt. Therefore, in the process of determining the component ratio, viscosity should be the main controlling parameter. The viscosity of the cold patch mixture at 60 °C is tested according to the Standard Test Methods of Bitumen and Bituminous Mixtures for Highway Engineering (JTG E20-2011) T0625. A No. 27 rotor is used with a rotational speed of 50 revolutions per minute. Based on the test results of viscosity and curing time, the component ratios for several cold patch asphalt mixtures are determined.

Dynamic Shear Rheometer (DSR) Test

This study utilized an MCR102 Dynamic Shear Rheometer (DSR) to conduct temperature sweep tests on the cold patch asphalt. The objective was to analyze the variations of complex modulus (G*), phase angle (δ), and rutting factor (G*/sinδ) of the cold patch asphalt with temperature. The temperature range of the test was from 34 °C to 76 °C, with an incremental gradient of 6 °C. The applied strain was 12%, and the frequency was set at 10 rad/s. The diameter of the parallel plates was 25 mm, with a loading gap of 1 mm.

Bending Beam Rheometer (BBR) Test

The Cannon Instrument Company Bending Beam Rheometer (BBR) was used to test the low-temperature creep behavior of the cold patch asphalt. The test aimed to obtain the creep stiffness modulus S(t) and the rate of change of stiffness modulus m at low temperatures. The test temperatures were set at -6, -12, -18, and -24 °C. Specimens were prepared in aluminum molds with dimensions of 127 mm (length), 6.35 mm (thickness), and 12.7 mm (width). A constant load was applied, and the deflection of the beam's center was continuously measured within 240 s.

Scanning Electron Microscopy (SEM)

Scanning Electron Microscopy was used to observe the curing behavior of the cold patch asphalt before and after solidification. The samples were coated with a gold–palladium alloy and observed under the SEM. The magnification used was 1000 times, allowing for a detailed examination of the asphalt paste's curing process.

Fourier Transform Infrared (FTIR) Test

The Nicolet IS20 Fourier Transform Infrared (FTIR) spectrometer, equipped with an attenuated total reflectance (ATR) accessory, was used to investigate the changes in functional groups of the cold patch asphalt before and after curing. The scanning range was from 500 to 4000 cm^{-1}, with a resolution of 4 cm^{-1}. This FTIR analysis allowed for the examination of the variations in functional groups present in the cold patch asphalt, providing insights into its chemical structure and composition before and after the curing process.

Marshall Stability

The strength development of the reaction-dilution type cold patch asphalt mixture developed in this study consists of two parts: the curing of epoxy soybean oil acrylate and the complete evaporation of the diluent to form the final strength. Before the cold patch material achieves its full strength, it needs to have a certain level of strength to resist the effect of traffic loads. In this study, the initial strength of the reaction-dilution type cold patch asphalt mixture was evaluated following the test method outlined in Standard Test Methods of Bitumen and Bituminous Mixtures for Highway Engineering (JTG E20-2011). Marshall compaction was performed on both sides of the sample for 75 blows, and the Marshall Stability was immediately tested after demolding. The Technical Specifications for Construction of Highway Asphalt Pavements (JTG F40-2004) provide clear instructions for the test method of the forming strength of cold patch materials. It states that after compacting the sample on both sides using a Marshall compactor for 50 blows, the sample should be placed vertically on its side in a 110 °C oven for 24 h, followed by an additional

25 blows of Marshall compaction. Afterward, the sample should be stored vertically at room temperature for 24 h before conducting the Marshall stability test.

3 Results and Discussion

3.1 *Experimental Results of Physical Performance*

Brookfield viscosity test and curing time observations were conducted on 9 groups of cold patch asphalt mixtures designed using an orthogonal experimental design. The experimental results are shown in Figs. 1 and 2.

From the experimental results of Brookfield viscosity shown in Fig. 1, it can be observed that increasing the dosage of AESO significantly reduces the viscosity of the asphalt. AESO can act as a diluent to reduce the viscosity of the asphalt, but its effect as a diluent alone is limited. It is still necessary to add a diluent to further reduce the viscosity of the asphalt. Extensive research represented by SHAR has shown that a viscosity of less than 3 Pa·s is more favorable for achieving a balance between construction, workability, and performance [14]. Therefore, from the perspective of construction and workability, it is not advisable to use compositions 1#, 6#, 8#, and 9# as components in the cold patch mixture.

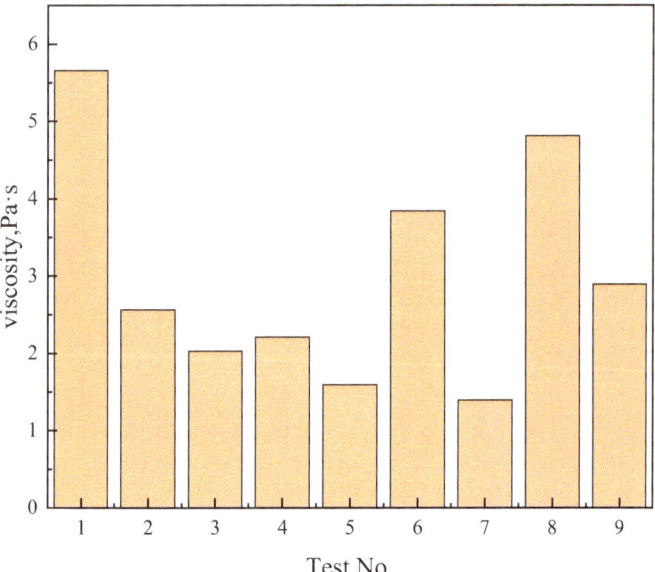

Fig. 1 Cold patch asphalt viscosity

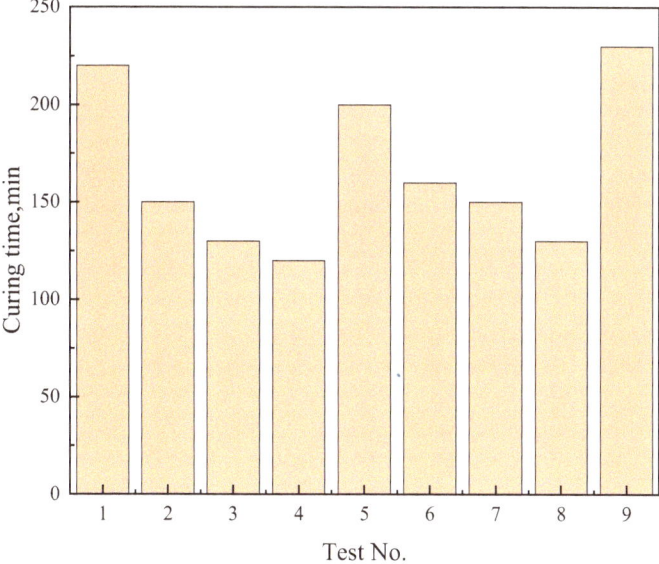

Fig. 2 Curing time of cold patch asphalt

The curing condition of the asphalt was judged by observation and tactile sensation. As shown in Fig. 2, the curing rate of the asphalt increases with the increase in the dosage of the photoinitiator, while the variations in the dosages of other components do not significantly affect the curing time of the asphalt. Overall, within 4 h, all 9 groups of asphalt mixtures achieved curing. A 4% dosage of the photoinitiator can achieve curing of the reactive diluent asphalt in approximately 2 h, which is much faster than the reactive diluent asphalt system with a 3% dosage. Research has shown that a 4% dosage of the photoinitiator can achieve a good balance between the curing rate and the economic performance of AESO [15]. Therefore, in the subsequent research of this paper, a 4% dosage of the photoinitiator is chosen.

3.2 Rheological Performance Analysis

Based on the viscosity test results, three groups of mixtures with a 12% content of 0# diesel were selected for DSR and BBR tests to study the rheological performance of cold patch asphalt with different AESO dosages.

The temperature sweep test results of AESO-modified asphalt are shown in Fig. 3a and b. As the temperature increases, the complex modulus G* and the rutting factor G*/sinδ of the asphalt rapidly decrease in the initial stage and then the curves become smoother. The phase angle δ initially increases linearly and then increases slowly. This is because as the temperature rises, the asphalt transitions from an elastic state to a viscous state. When the temperature reaches the termination temperature of the experiment, the asphalt's elastic properties almost disappear, and the viscous properties become dominant. As shown in Fig. 3c, when the temperature exceeds 64 °C, the variation of G*/sinδ for the matrix asphalt and the modified asphalt with the three AESO dosages is relatively small. In addition, at the same temperature, the G*/sinδ values of the modified asphalt with the three AESO dosages are all higher than that of the 70# matrix asphalt, indicating that the addition of AESO can improve the high-temperature performance of the asphalt. With an increase in the AESO content, the rutting factor curves of the 4% and 6% dosages overlap, indicating that a higher AESO dosage does not have a significant effect on improving the high-temperature performance of the cold patch asphalt mixture and may even increase viscosity, which is not conducive to construction.

From Fig. 4a, the influence of AESO on the stiffness of the binder can be observed. At −24 °C, the stiffness of the matrix asphalt could not be measured. The 2 and 4% dosages of AESO have similar effects in reducing the stiffness of the asphalt, and only the 6% dosage of AESO-modified asphalt meets the specification requirements, with a decrease in stiffness of 30 MPa. At a testing temperature of −18 °C, the matrix asphalt still does not meet the specification requirements. At this temperature, the three dosages have similar effects in reducing the stiffness of the asphalt, with the 4 and 6% dosages slightly better than the 2% dosage. At −12 and −6 °C, there is no significant difference in the stiffness effects of the three dosages, and only a 2% dosage of AESO is sufficient to significantly improve the stiffness of the asphalt.

For the m value, it can be observed from Fig. 4b that as the AESO concentration level increases, the m value increases. With the decrease in temperature, the m value decreases. At −24 °C, the 6% dosage of AESO meets the standard requirements. It can be concluded that AESO enhances the flexibility of the asphalt binder and reduces the cracking resistance of the asphalt mixture at low temperatures. For the four test temperatures, it was found that the m value of the asphalt significantly increases with the increase in the AESO concentration level. These results indicate that AESO helps with stress relaxation, thereby improving the low-temperature crack resistance of the binder.

3.3   Microscopic Results Analysis

SEM Results Analysis

Through scanning electron microscopy (SEM), the changes in cold patch asphalt liquid before and after curing can be observed. The specimens were magnified 1000 times on the surface, and from Fig. 5a, it can be clearly seen that there were no significant changes in the surface before curing. Figure 5b shows the sample after half an hour, where partial curing products can be observed in the cold patch liquid. Figure 5c displays the fully cured image, showing the formation of a crosslinked network that is uniformly dispersed. The crosslinked network is a mixture formed

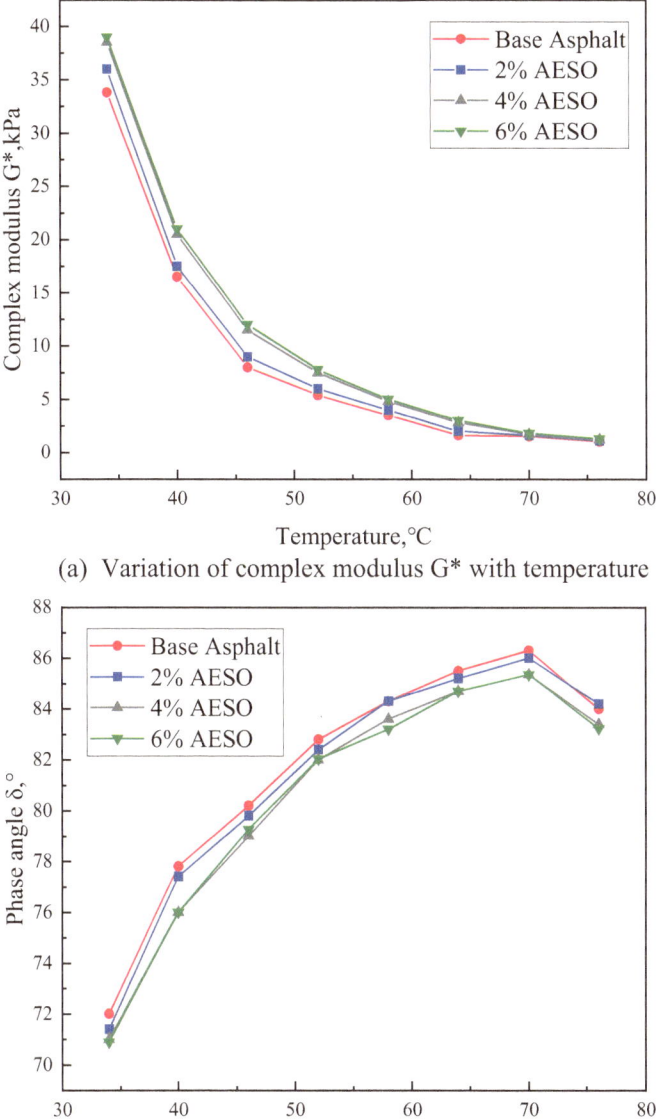

(a) Variation of complex modulus G* with temperature

(b) Variation of phase angle with temperature

Fig. 3 DSR test results

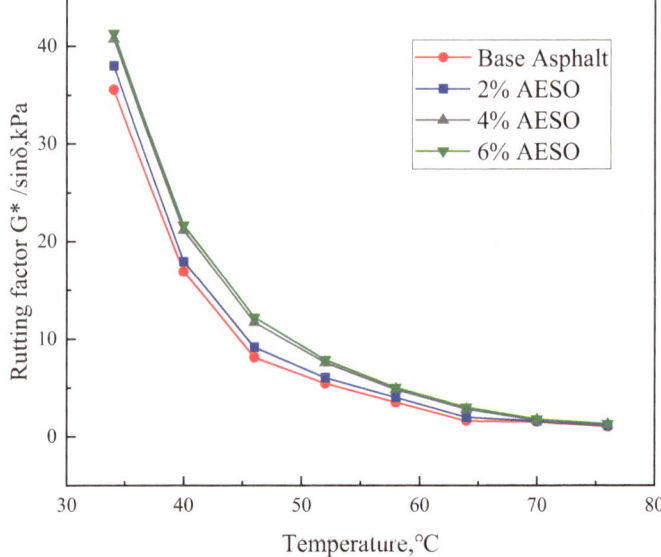

(c) Variation of rutting factor G*/sinδ with temperature

Fig. 3 (continued)

by the curing of AESO with the asphalt, and it is interlaced within the asphalt, which can enhance the overall strength of the asphalt mixture.

FTIR Results Analysis

Using the same formulation of reactive diluent asphalt as in the previous section, observations were made before and after curing with 4% AESO content, as shown in Fig. 6. By comparing the positions of the absorption peaks in the infrared spectra of the reactive diluent cold patch asphalt before and after curing, it can be observed that the absorption peaks overlap in the range of 4000–600 cm^{-1}. However, there are noticeable differences at two specific positions: the absorption peak at 1636.31 cm^{-1}, attributed to the stretching vibration of carbon–carbon double bonds in the acrylic ester group, disappears after asphalt curing, indicating the completion of the polymerization reaction of the functional group; the absorption peak at 984.20 cm^{-1}, attributed to the bending and wagging vibrations of the carbon-hydrogen bonds in the acrylic ester group, also disappears after asphalt curing, indicating the completion of the polymerization reaction of the functional group.

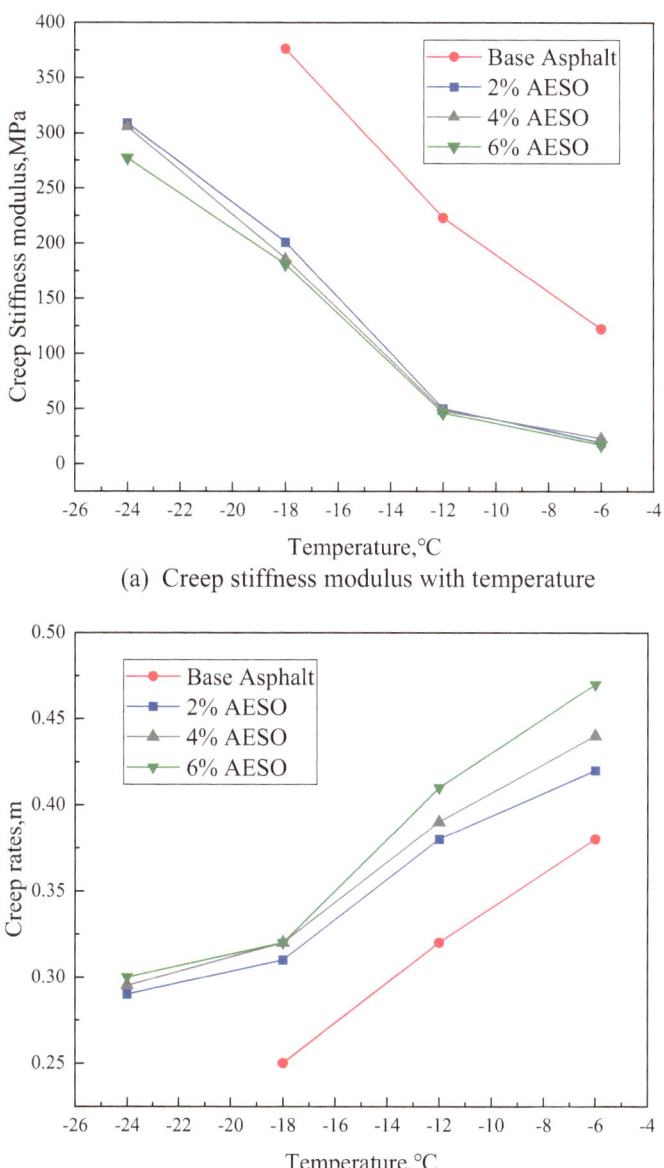

(a) Creep stiffness modulus with temperature

(b) Creep rates with temperature

Fig. 4 BBR test results

Fig. 5 Microphotographs
SEM: **a** before curing;
b curing in progress; **c** after
curing

(a)

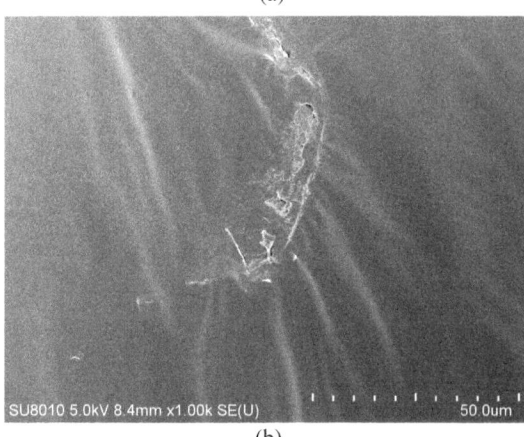

(b)

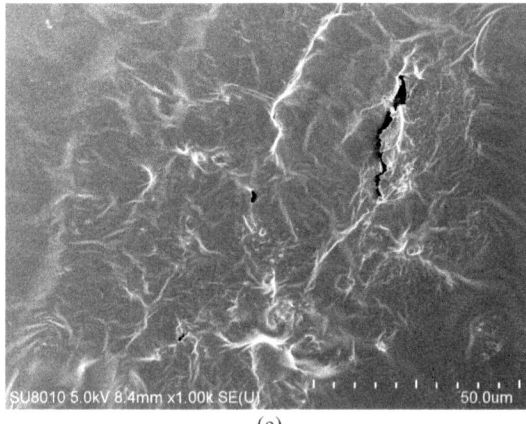

(c)

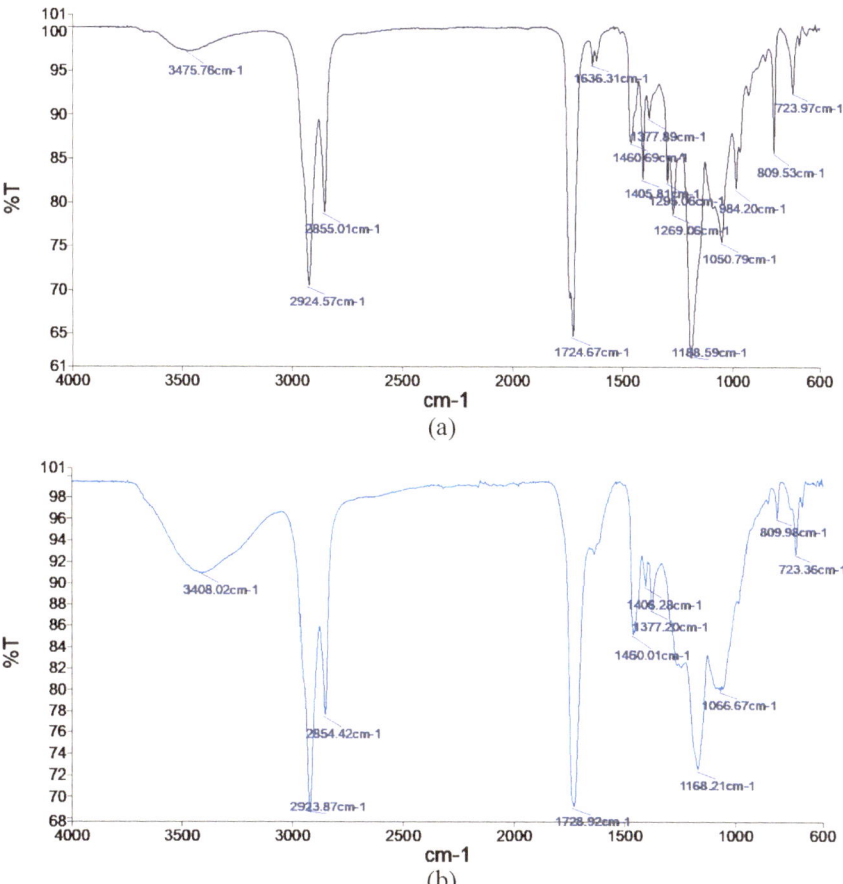

Fig. 6 Infrared spectrum image: **a** before curing; **b** after curing

3.4 Strength Analysis

Table 3 shows the initial and compaction strengths of cold patch asphalt mixtures. Currently, there is still a significant difference in the standards for the initial strength of cold patch asphalt mixtures in the academic community. For example, Japanese scholars have pointed out that the initial strength of cold patch materials only needs to meet 0.5–0.8 kN, while the research team at Tongji University has set higher requirements for cold patch asphalt mixtures, stipulating that the initial strength should be greater than 1.5 kN [16].

It can be observed that the initial stability of AESO cold patch asphalt mixtures is much higher than the requirements for the initial strength of cold patch asphalt mixtures proposed by the aforementioned scholars. This is because the reduction in solvent dosage significantly improves the initial stability of the cold patch asphalt

Table 3 Initial strength and forming strength

Marshall stability	1	2	3	Mean value
Initial strength (kN)	5.83	5.62	5.82	5.79
Forming strength (kN)	8.23	8.25	8.12	8.17

mixture. As the solvent volatilizes, the asphalt properties are restored, and at the same time, AESO gradually cures, leading to a continuous increase in the strength of the asphalt mixture. The compaction strength shows a significant improvement compared to the initial strength, far exceeding the requirement of stability greater than 3 kN in 2004. It also approaches the requirement for high-grade hot mix asphalt mixtures with stability greater than 8 kN.

4   Conclusion

The performance of reaction-diluted cold patch asphalt emulsion was studied using viscosity tests, DSR tests, BBR tests, SEM analysis, FTIR spectroscopy, and Marshall stability tests to evaluate the strength of the mixture. Based on the experimental results and analysis, the main conclusions are as follows:

1. With an increase in AESO content, the viscosity of the asphalt significantly decreases, indicating that AESO can replace a portion of diesel as a reactive diluent. When the AESO content reaches 6%, the performance improvement of the cold patch asphalt emulsion is not significant compared to 4%. The formulation of the reaction-diluted cold patch asphalt emulsion is determined as follows: 70# Asphalt: Diesel: AESO: Photoinitiator = 100:12:4:4.
2. SEM observations revealed the formation of a network structure in the asphalt after AESO curing. This network structure improves the performance of the asphalt and enhances the strength of the cold patch asphalt mixture. FTIR results indicate that AESO curing involves a polymerization reaction.
3. By serving as a reactive diluent and reducing the diesel content, AESO significantly enhances the initial strength of the cold patch asphalt mixture, reaching 5.79 kN. As AESO cures and forms a network structure in the asphalt, and as the diluent evaporates, the strength of the cold patch asphalt mixture continues to increase, with a compacted strength of up to 8.17 kN.
4. Further investigations are needed to verify the high-temperature performance, low-temperature performance, and water stability of the reaction-diluted cold patch asphalt mixture. A comprehensive evaluation of its road performance is necessary. Additionally, research on gradation is needed to improve the composition of the reaction-diluted cold patch asphalt emulsion.

Acknowledgements The research is supported by the National Key R&D Program of China (2022YFC3803405 and 2021YFB2601204), the National Natural Science Foundation of China

(Grant No. 52078024), the R&D Program of Beijing Municipal Education Commission (KM202110016011), BUCEA Post Graduate Innovation Project (PG2023045). The authors gratefully acknowledged their financial support.

References

1. Wang, R. (2017). *Research on the rapid repairing technology of asphalt pavement pothole.* Changan University.
2. Huang, H., Ren, J., & Li, M. (2020). Development and evaluation of solvent-based cold patching asphalt mixture based on multiscale. *Advances in Materials Science and Engineering, 2020,* 1–16.
3. Xia, D. (2015). *Pothole repair experimental study for cold patch asphalt.* Northeast Forestry University.
4. Suo, Z., Yan, Q., & Ji, J. (2021). The aging behavior of reclaimed asphalt mixture with vegetable oil rejuvenators. *Construction and Building Materials, 299,* 123811.
5. Li, N. (2015). *Study on micro-macro properties and regeneration mechanism of vegetable oil reclaimed asphalt.* Zhejiang Normal University.
6. Chen, H. (2021). *Research on performance of waste vegetable oil preformed foamed asphalt cold recycled mixture.* Beijing University of Civil Engineering and Architecture.
7. Ying, Q., Tang, H. & Luo, D. (2021) Curing and application of acrylated epoxy soybean oil. *Thermosetting Resin, 1,* 50–54+60.
8. Yang, Z., Li, P., & Chu, Z. (2022). Research progress of vegetable oil-based UV curable materials. *Journal of South China Agricultural University, 43*(1), 1–12.
9. Yan, P. (2014). *Preparation and modification of UV-cured acrylated epoxidized soybean oill.* Hubei University.
10. Lebedevaite, M., & Ostrauskaite, J. (2021). Influence of photoinitiator and temperature on photocross-linking kinetics of acrylated epoxidized soybean oil and properties of the resulting polymers. *Industrial Crops and Products, 161,* 113210.
11. Lee, T., Park, Y., & Lee, S. (2019). A crack repair patch based on acrylated epoxidized soybean oil. *Applied Surface Science, 476,* 276–282.
12. Chen, C., Podolsky, J., & Williams, R. (2020). Rheological properties and effects of aging on acrylated epoxidised soybean oil monomer-modified asphalt binder. *Road Materials and Pavement Design, 21*(2), 347–373.
13. Chen, X. (2022). *A kind of recycling agent for thermal recycling of waste asphalt and its preparation method.*
14. Zheng, M., Jin, J., & Liu, X. (2022). Development and performance evaluation of solvent cold patching asphalt. *Bulletin of the Chinese Ceramic Society, 41*(1), 342–353.
15. Li, P. (2016). *Environment friendly UV-cured coatings based on itaconic acid and soybean oil.* North University of China.
16. Yang, S. (2022). *Study on performance of reactive dilution cold patching asphalt mixture.* Beijing University of Civil Engineering and Architecture.

Experimental Study on the Behaviour and Deformation Prediction of Underwater Deep Cement Mixing (DCM) Foundation

Ruiyi Huang, Ran Tao, Zhongxin Cao, and Yun Peng

Abstract Based on the marine project of Hong Kong Integrated Waste Management Facilities Phase 1, this paper presents the static loading test of the underwater DCM foundation, and compares the instrumentation monitoring results with the predicted results by finite element analysis. In conclusion, the predicted total DCM compression and settlement is significantly higher than the actual site-measured records. Therefore, the adopted design parameters for the DCM are appropriate and conservatively under-estimate the field strength, which satisfies the requirement as stipulated in the specification.

Keywords Deep foundation · Deep cement mixing · Static load test · Finite element analysis · Numerical simulation · Geological model

R. Huang
Department of Marine and Environmental Engineering, CCCC 2nd Harbor Engineering Company Ltd., National Enterprise Technology Center, Wuhan 430040, People's Republic of China
e-mail: huangruiyi@ccccltd.cn

R. Tao (✉)
Department of Science, Technology and Digitalization, China Communications Construction Company Ltd., Beijing 100088, People's Republic of China
e-mail: taoran1@ccccltd.cn

Z. Cao
Department of Engineering, Zhen Hua Engineering Company Ltd., Hong Kong 999077, People's Republic of China
e-mail: zxcao@chec.bj.cn

Y. Peng
State Key Laboratory of Coastal and Offshore Engineering, Dalian University of Technology, Dalian 116023, People's Republic of China
e-mail: yun_peng@dlut.edu.cn

© The Author(s) 2025
D. Li and Y. Zhang (eds.), *Advances in Frontier Research on Engineering Structures II*,
Lecture Notes in Civil Engineering 535, https://doi.org/10.1007/978-981-97-6238-5_11

1 Introduction

Deep cement mixing (DCM), as an environmental-friendly technique [1] for offshore ground improvement is recently adopted for underwater soft ground improvement projects in China more often [2]. This technique, without dredging, will cause limited disturbance to the marine ecological system [3]. However, systematic study on the compression behavior of underwater composite ground has seldom been conducted [4].

As for the marine project of Hong Kong Integrated Waste Management Facilities Phase 1, there were a total of 123 DCM clusters for static loading tests [5]. The static load test for DCM underground was carried out for design check for certification [6], according to the finite element analysis of the behavior and deformation prediction [7].

This paper aims to present the static loading test of DCM works and the instrumentation monitoring results with the loading of 14 layers of concrete blocks and unloading orderly. The findings of the results will be further presented and discussed in the following text.

2 Test Configuration and Instrumentation

2.1 Site Location

As shown in Fig. 1, the DCM static loading test is located in the reclamation area which is closely adjacent to the "Berth" DCM area, containing an 8 m-thick of the soft marine clay layer. It shall not take up the area with both permanent DCM work and future foundation works [8].

2.2 Static Loading Test Configuration

The vertical load of the DCM static loading test is formed by 14 layers of concrete blocks about 10.75 m high above seawater. The loading platform footprint is 15 × 15 m which is formed at the center area of DCM. Rock fill and Sand blankets are placed underneath the concrete blocks, which is in accordance with the permanent works design [9]. Figure 2 shows the configuration of the static loading test, whereas the site photo of the loading platform is shown in Fig. 3.

The setup of the Static Loading Test generally consists of three loading stages, which are Stage 1-design test load stage, Stage 2-additional load stage, and Stage 3-unloading stage. After the completion of applying Stage 1 loading, the load will be maintained for 7 days before the continuation of Stage 2 loading. During the loading stages, the concrete blocks were placed on the loading platform layer by layer from

Fig. 1 Site photo of static loading test

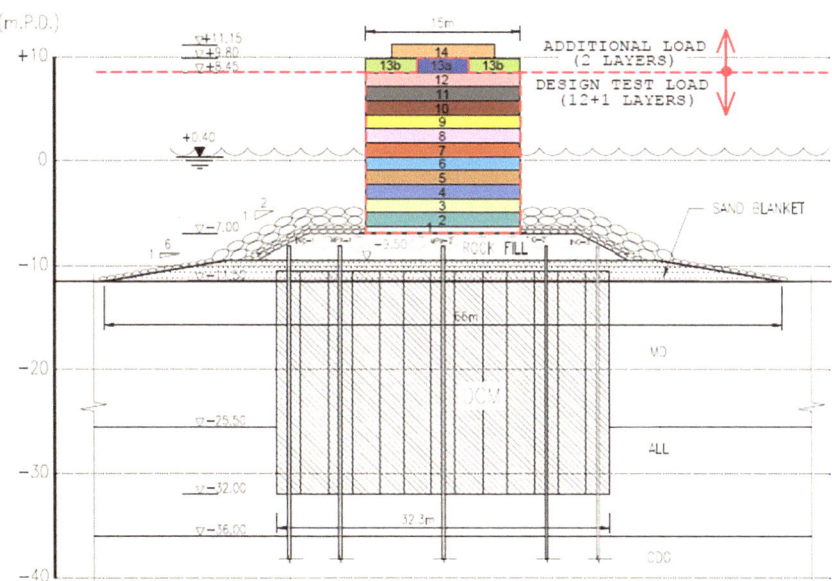

Fig. 2 Schematic configuration of static loading test

Fig. 3 Site photo of loading platform

−7 m to + 11.15 m and unloaded at descending level. Except for the first layer comprising thinner concrete slabs; the remaining layers were formed by standard concrete blocks.

2.3 Instrumentation Arrangement

There are a total of 6 Shape Array Vertical Inclinometers and 3 yield-point Multiple Rod Extensometers installed inside the DCM clusters for monitoring of DCM behavior during the static loading test period, as:

- Settlement behavior at pre-determined soil layers of DCM.
- Lateral deflection of DCM.

All instrumentation is routed (cable covered by steel tube) to the datalogger fixed on the prefabricated steel platform which is located about 5.8 m away from the static loading test area.

The instruments were installed in the orientation towards North and East direction, where Axis A (+ve) represents of North direction, Axis A (-ve) represents of South direction, Axis B (+ve) represents of East direction and Axis B (-ve) represents of West direction. To facilitate the analysis of results, the measured results from Axis A and Axis B of the instrument have been converted to align with the loading platform Axis X and Axis Y as shown in Fig. 4, for the orientation of the instruments.

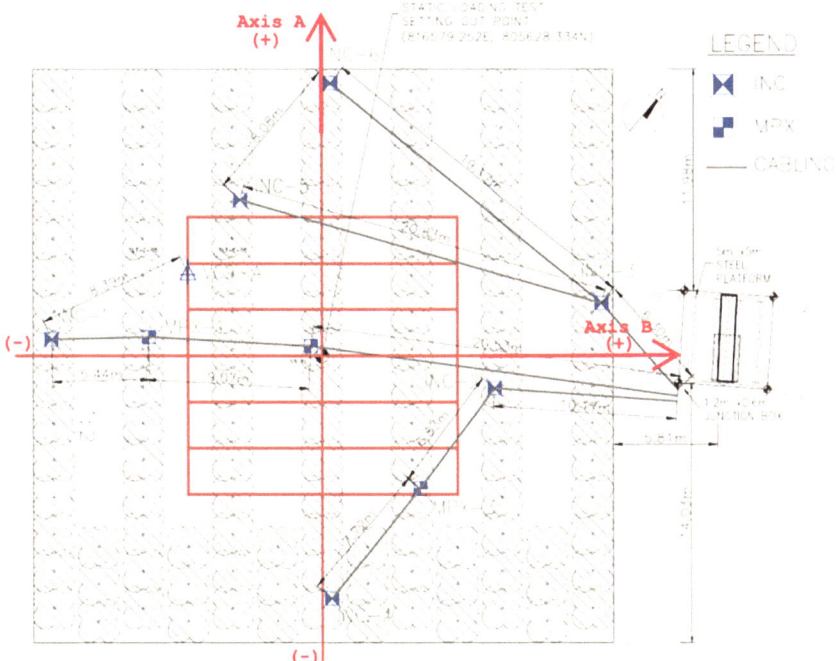

Fig. 4 General arrangement of instrumentation

3 Monitoring Data of Static Loading Test

3.1 Monitoring Frequency

The monitoring of the static loading test was commenced prior to the placement of concrete slabs to determine baseline reading. Readings were recorded by the datalogger at nominally 1 h intervals continuously after completion of concrete slab placing. The monitoring of the static loading test will continue until the complete removal of the last layer of concrete blocks.

4 Summary of Monitoring Results of Extensometers

The measured vertical movements of the three installed extensometers up to the completion of layer 14 are summarized as the maximum settlement during the loading period was recorded up to −9.28 mm.

In addition, the maximum residual settlement was recorded up to −8.4 mm.

Table 1 Descriptions of soils

Soil/Rock	Descriptions	Thickness (m)
Marine deposits (MD)	Very soft, grey, silty CLAY with occasional shell fragments	13.5
Alluvium (ALL)	Thick alluvium layer is anticipated. It consists of well-sorted to semi-sorted clays, silts, and occasionally gravels. Quartz fragments are present	11
CDG	Extremely weak, grey, spotted light brown, pink and white, completely decomposed granite	13

5   Summary of Monitoring Results of Inclinometers

The measured lateral deflections of the six installed inclinometers to the completion of layer 14 are reaching up to −4.26 mm.

6   Comparison of Measured Results with Predicted Results

6.1   Geological Model

The geological profile is adopted for the numerical analysis, with the descriptions of soils given in Table 1.

6.2   Geotechnical Parameters

The geotechnical parameters used for the numerical analysis incorporated a reduction factor of 1.2 with the shear strength parameters for the in-situ soils. The adopted value shown in Table 2 has been determined based on the co-relation with the UCS value as recommended.

Young's modulus of deep mixed ground for wet mixing is an important parameter that affects deformations of the DCM works [10].

6.3   Finite Element Analysis

Plaxis 3D 2017 was used to analyze the behavior and predict deformations of the DCM works for the static loading test. The adopted 3D Plaxis model is shown in Fig. 5, while the calculated principal effective stress vectors at the model boundaries

Table 2 Adopted geotechnical parameters

Soil type	Bulk density (kN/m³)	Drained parameters					Undrained parameters		
		SPT 'N' (−)	E' (MPa)	c' (kPa)	φ' (°)	ν' (−)	E_u (MPa)	C_u (kPa)	$ν_u$ (−)
Rock fill	18	N/A	20	0	40	0.3	N/A	N/A	N/A
Sand blanket (above DCM)	19	N/A	10	0	30	0.3	N/A	N/A	N/A
Marine (Clay/Silt)	16	N/A	D < 3 m, 2.4/1.15; D > 3 m, (0.6D + 1.8)/1.15	4	25	0.3	D < 3 m, 2.4; D > 3 m, 0.6D + 1.8	D < 3 m, 6; D > 3 m, 1.5D + 4.5	0.5
Alluvium (Sand)	20	z > -20mPD, 8; z ≤ -20mPD, -1.5z-22	z > −20 mPD, 8; z ≤ −20 mPD, −1.5z − 22	0	33	0.3	N/A	N/A	N/A
CDG	20	z > -67mPD, 33	z > -67mPD, 50	3	35	0.3	N/A	N/A	N/A
DCM Cluster	18	N/A	N/A	N/A	N/A	N/A	360 (note1)	551 (note2)	0.5
Concrete block	23	N/A	26.4×10^3	N/A	N/A		N/A	N/A	N/A

do not rotate, remaining vertically or horizontally close to the boundaries, indicating that the adopted model dimensions are sufficient.

The calculated DCM compression and settlement results are plotted against applied loading in Figs. 6 and 7, respectively, for assumed DCM Young's modulus value equals 360 MPa. It is shown that both the DCM compression and settlement increase almost linearly with increasing loading after 400 kPa which is larger than both the design loading of 242 kPa and the proposed test loading of 288.1 kPa. Even for 400 kPa, the calculated DCM compression is still in the order of about 30 mm, for the assumed Young's modulus value.

The calculated results of DCM compression and lateral movement corresponding to the applied Stage 1 and 2 loading indicate as within the specified limits shown in Table 3.

In addition, the Slope/W analysis with GeoStudio 2016 is for reference. The calculated factor of safety against overall slip failure is 8, which is larger than the minimum factor of safety for global stability, while for temporary conditions is 1.2.

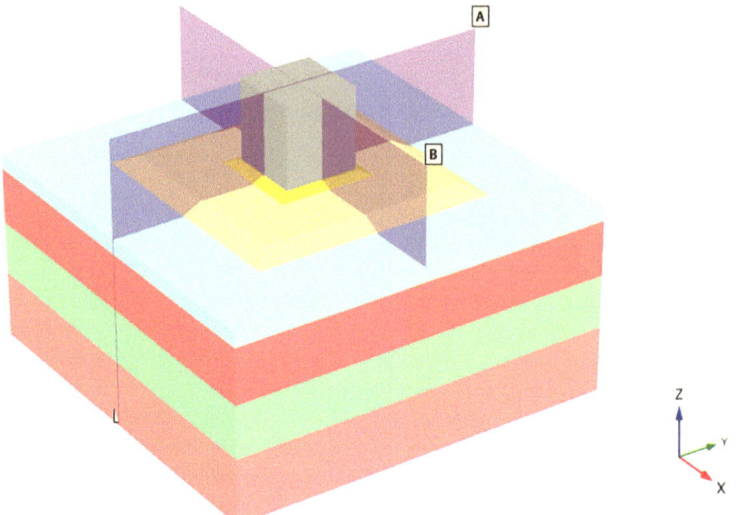

Fig. 5 Adopted 3D plaxis model

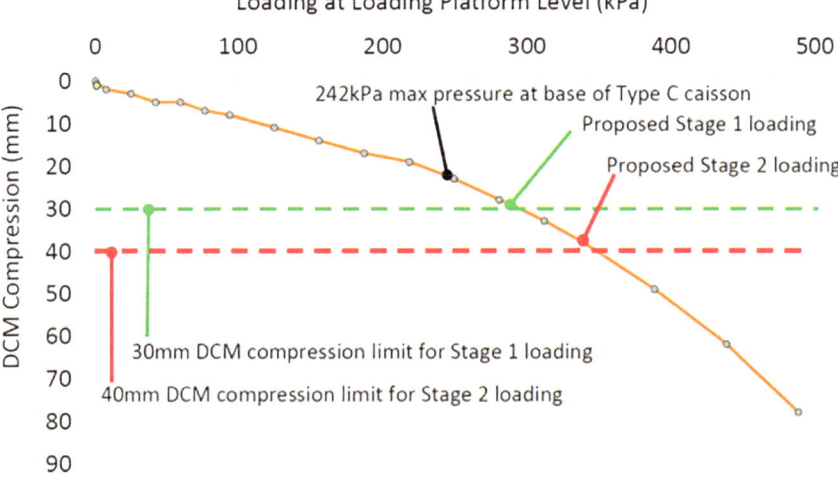

Fig. 6 Calculated DCM compression versus applied loading

6.4   Compression of DCM

The compression of the DCM-treated soil is a key performance indicator of the DCM installation. This can be deduced by the relative settlements at the top and bottom of the DCM zone.

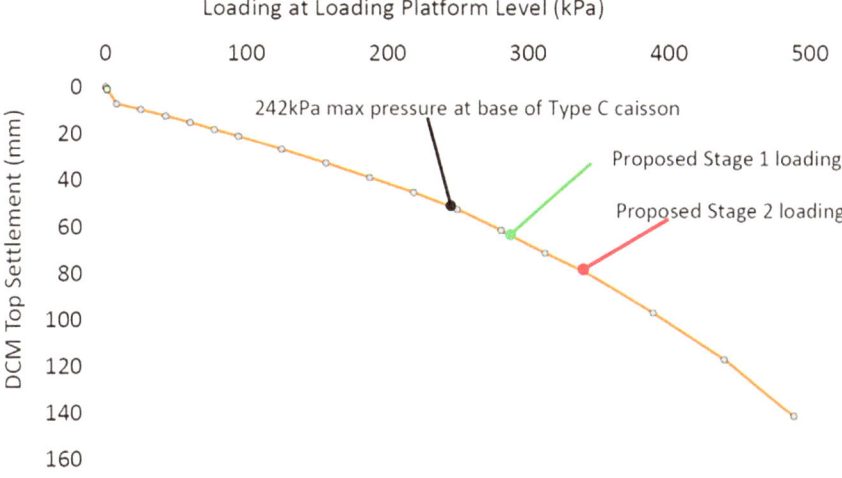

Fig. 7 Calculated settlement of DCM top versus applied loading

Table 3 Summary of calculated DCM vertical compression and lateral movements

Stage	DCM vertical movement		DCM vertical compression (mm)	DCM lateral movement (mm)	Conclusion
	At top (mm)	At base (mm)			
Load continues for 7 days after placing Stage 1 concrete blocks	64	35	29 < 30 mm	47 < 50 mm	Acceptable
Load continues for 3 days after placing Stage 2 concrete blocks	79	41	38 < 40 mm	58 < 60mm	Acceptable

The maximum cumulative settlement and compression of DCM during the loading period were identified to be −9.17 mm and −1.37 mm respectively as shown in Fig. 8. The corresponding maximum cumulative settlement and compression of DCM during the unloading period were identified to be −8.4 mm and −0.76 mm respectively as shown in Fig. 9. As the results show significantly variation value, the value of point which is comprehension to the location of the test loading area as located outside the loading area, is smaller than the value of points located within the loading area. The reason is due to that the points in the loading area absorbed more concentration load in the middle than the edge of the loading area.

The measured DCM compression result is plot together with predicted DCM compression results in Fig. 10. This demonstrates that the measured settlements are significantly lower than the prediction and the parameters adopted in the DCM design are appropriate but conservative.

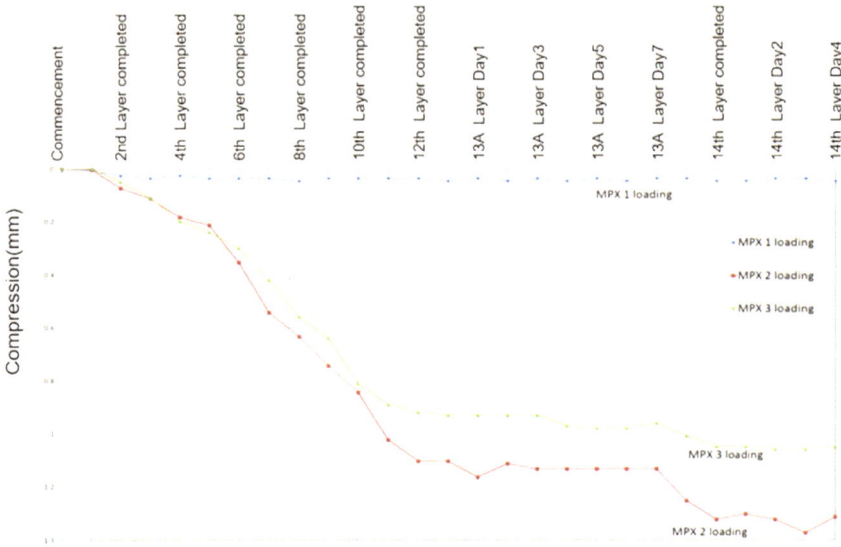

Fig. 8 Compression of DCM during the loading period

Fig. 9 Compression of DCM during unloading period

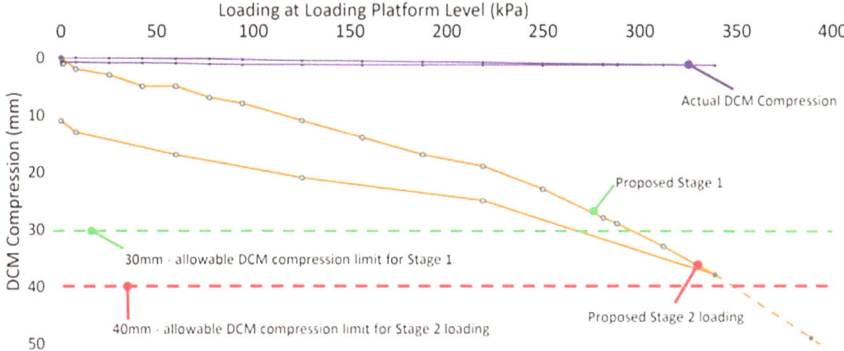

Fig. 10 Measured DCM compression versus predicted DCM compression

6.5 Lateral Movement of DCM

The measured deflections as recorded in inclinometers are generally small, with a maximum lateral deflection of + 10.54 mm.

The measured DCM lateral deflection is plotted together with predicted results in Fig. 11, where the legend of the graph of the measured DCM lateral movement is denoted by "INCxx Axis xx" and the predicted DCM lateral movement is denoted by "Predicted INCxx_Axis xx". From this figure, it is noted that the magnitude of the measured DCM lateral deflection is a lot lower than the predicted DCM movement.

6.6 The Behavior of DCM-Treated Soil

Prior to the static loading test, 5 DCM clusters were selected to carry out the elastic modulus test. To further analyze the elastic behavior of DCM, a stress–strain curve is plotted to study together with the elastic modulus results.

Similar to a typical elastic–plastic stress–strain curve, the strain drops are subject to stress relief. From Fig. 12, the maximum strength gain is 0.0642%, which is lower than the average strain at failure 0.785% (20% = 0.157%) recorded from the elastic modulus test. It is reasonable to believe that the DCM cluster is still within the elastic stage during both the loading and unloading periods of the static loading test, where the DCM also performs a high stiffness as the measure strains.

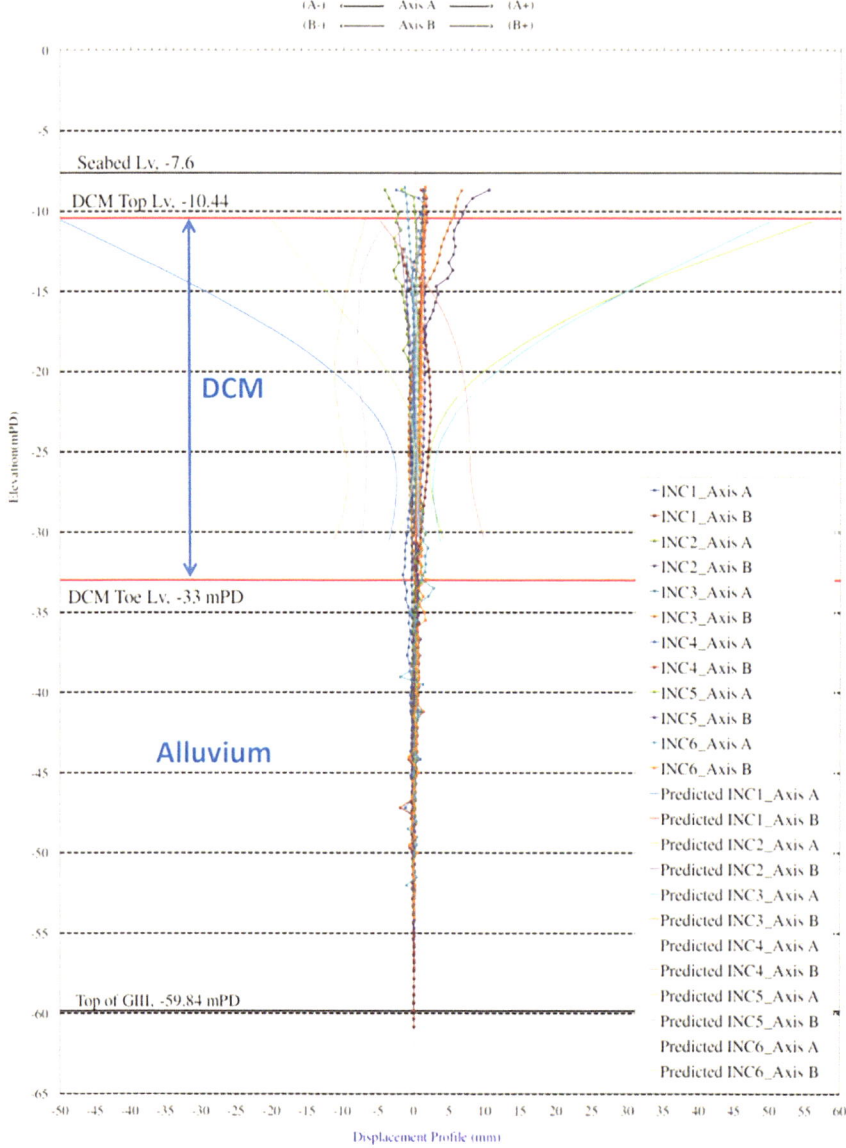

Fig. 11 Measured DCM lateral deflection versus predicted DCM lateral deflection (at the completion 14th layer)

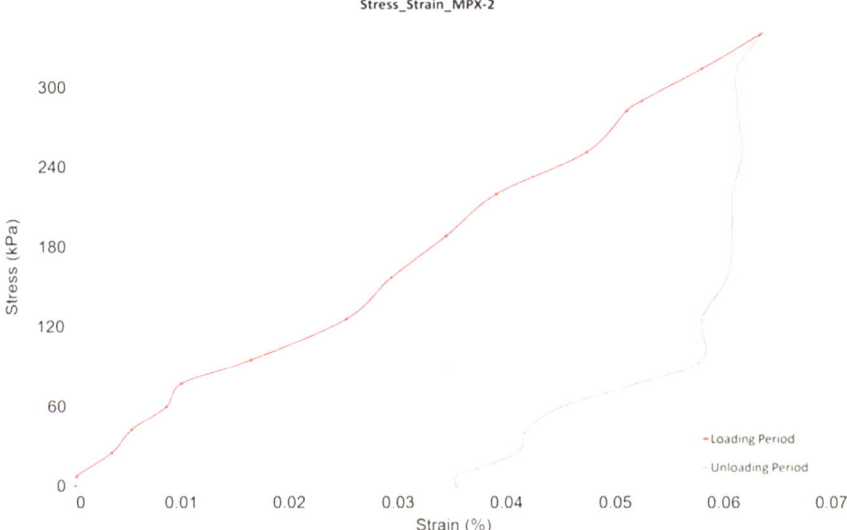

Fig. 12 Stress–Strain curve of DCM

7 Findings and Conclusions

This paper presents the static loading test details and the monitoring data. In conclusion, the predicted total compression varied from 2 to 38 mm and maximum lateral deflection reached 55.9 mm at stage 14, which are significantly higher than the actual site-measured records for compression varied from 0.0033 to 1.3733 mm, and 10.54 mm respectively. Based on the results of the DCM settlement analysis, it can be concluded that the adopted design parameters for the DCM are appropriate and conservatively under-estimate the field strength. Therefore, the predicted settlement of DCM underneath seawall/breakwater within a design life of 50 years will be within 500 mm, and satisfy the requirement as stipulated in the specification.

Acknowledgements The National Key Research and Development Program of China (2021YFB2600200, 2022YFB2603000)

References

1. Shiu Fung, H. (2020). *Public participation and EIA effectiveness: Empirical case studies in Hong Kong.*
2. Chung, P., Chu, F., Cheung, H., Yan, C., Cheung, C., & Wong, A. (2022). *Technical developments related to deep cement mixing method in Hong Kong* (pp. 249–259).

3. Yin, K., Zhang, L., Zou, H., Luo, H., & Lu, W. (2022). Key factors for deep cement mixing construction for undredged offshore land reclamation. *Journal of Geotechnical and Geoenvironmental Engineering, 148.*

4. Voottipruex, P., Bergado, D. T., Suksawat, T., Jamsawang, P., & Cheang, W. (2011). Behavior and simulation of deep cement mixing (DCM) and stiffened deep cement mixing (SDCM) piles under full scale loading. *Soils and Foundations, 51,* 307–320.

5. Cheung, H., Yan, C., Cheung, C., & Wong, A. (2022). *Deep cement mixing—The experience in Tung Chung East reclamation and challenges ahead* (pp. 348–360).

6. Verastegui, D., & Impe, W. F. (2006). Deep mixing in underwater conditions: A laboratory and field investigation. *Proceedings of the ICE—Ground Improvement, 10,* 15–22.

7. Wang, C., Xu, Y., & Dong, P. (2014). Plate load tests of composite foundation reinforced by concrete-cored DCM pile. *Geotechnical and Geological Engineering, 32.*

8. Liu, Z., Chen, P., Hu, L., Zhou, H., & Wang, X. (2019). Testing methods for underwater composite ground with deep cement mixing. *Port and Waterway, 552,* 155–162.

9. Lau, S., Yeung, K. M., Yu, K., & Bhanja, A. (2019). *DCM static loading test proposal method statement.*

10. Keaton, J. R. (2018). Young's modulus. In P. T. Bobrowsky & B. Marker (Eds.), *Encyclopedia of engineering geology. Encyclopedia of earth sciences series.* Springer.

Study on the Durability of Hydraulic Engineered Cementitious Composites

Yan Shi, Yupu Wang, Jiazheng Li, and Wenguang Jiang

Abstract To explore the durability of hydraulic engineered cementitious composites (HECC), this article uses hydraulic concrete permeability and frost resistance tests, evaluates them in combination with mechanical tests in high-temperature resistance and self-healing tests, and explores the influence of different maximum particle sizes of sand on permeability and frost resistance. In high-temperature resistance, the thermal stability of fibers and the influence of different curing temperatures on mechanical properties are explored, and the effects of different maximum particle sizes of sand, different water–cement ratios, and different ages on self-healing performance are explored. The results show that the finer the sand used in HECC is, the better its impermeability is and the permeability coefficient is in the order of 10^{-6} or 10^{-7}; the better the frost resistance is, the maximum frost resistance level can reach F300. As the curing temperature increases, the compressive and flexural strength first increase and then decrease, with a maximum increase of 43%. Compared to 0.25, a water adhesive ratio of 0.35 has better self-healing ability, with a strength ratio of 167% at 7 days and 72% at 28 days. The research results can provide a reference for further promoting the application of HECC in water conservancy engineering.

Keywords Hydraulic-engineered cementitious composites · Durability · Permeability · Frost resistance · High-temperature resistance · Self-healing

Y. Shi · Y. Wang (✉) · J. Li · W. Jiang
Changjiang River Scientific Research Institute of Changjiang Water Resources Commission, Wuhan 430010, China
e-mail: wangyupu29@126.com

National Center for Dam Safety Engineering Technology Research, Wuhan 430010, China

© The Author(s) 2025
D. Li and Y. Zhang (eds.), *Advances in Frontier Research on Engineering Structures II*,
Lecture Notes in Civil Engineering 535, https://doi.org/10.1007/978-981-97-6238-5_12

1 Introduction

Concrete is one of the most widely used materials in the field of construction engineering today, and its characteristics, such as simple structure, low manufacturing cost, and good plasticity, make it play an important role in the construction of infrastructure such as houses, bridges, roads, and water conservancy projects. As a building material, the service life of concrete is crucial. However, in the long-term use process, breakage, cracking, leakage, high temperature, cold and other issues tend to seriously affect the stability and safety of the structure, so the lack of durability of the problem for the quality of the project as well as the socioeconomic development of the community has brought a major challenge.

In the 1990s, Prof. Li of the University of Michigan and Prof. Leung of Massachusetts Institute of Technology proposed a new model to invent high-toughness cementitious composites (Engineered cementitious composites (ECC)), which have good durability with multi-cracking, strain hardening, and good self-healing ability [1]. To apply to water conservancy and hydropower projects, the Changjiang River Scientific Research Institute of Changjiang Water Resources Commission proposed the concept of HECC (Hydraulic Engineered Cementitious Composites, HECC) applicable to hydraulic construction [2], which was successfully applied for the first time in the localized part of the Batang hydropower station corridor. Compared with ECC materials, HECC can use a wider variety of raw materials, such as the use of engineering local sand and gravel to prepare the required fine aggregates, the maximum particle size can be increased to 1.25 mm, the 28 days compressive strength of C25–C40, 28 days modulus of elasticity is less than 20 GPa, and strain ranges from 1 to 3% [3].

To guarantee the long-term safe operation of water conservancy and hydropower projects, the durability performance of HECC is particularly important, but there are fewer related studies at present. This paper intends to carry out experimental research on the permeability, frost resistance, high-temperature resistance, and self-healing ability of HECC and analyze the influence of factors such as the maximum particle size of artificial sand, curing temperature, age, and other factors on its durability, enrich the relevant research results of HECC materials, and provide technical support for the popularization and application of HECC in water conservancy and hydropower projects.

2 Materials and Methods

2.1 Test Raw Materials

Cement is made of Esheng P.O42.5 ordinary Portland cement, and the test results of quality indexes are shown in Table 1, which show that the performance indexes of cement meet the relevant technical requirements of GB 175-2020 "Common Portland cement" for 42.5 ordinary Portland cement.

The quality index test results of the mineral admixture that adopts Jintang Class I fly ash are shown in Table 2. The results show that the quality index of fly ash meets the technical requirements of Class F Class I fly ash in DL/T 5055-2007 "Technical standard of fly ash concrete for hydraulic structures".

The test selected the short-cut PVA fiber products produced by Anhui Wanwei Company, and the quality indexes are shown in Table 3.

The sand used is artificial sand sampled from a hydropower project site, which is divided into 4 categories according to the maximum particle size after sieving,

Table 1 Basic properties of cement

Cement type	Specific surface area	Setting time		Compressive strength		Flexural strength	
	$(m^2 \cdot kg^{-1})$	Initial setting (min)	Final setting time (h:min)	MPa		MPa	
				3 days	28 days	3 days	28 days
Esheng42.5	334	181	4:19	27.5	47.4	6.3	8.6
GB175-2020	≥ 300	≥ 45 min	≤ 10 h	≥ 17.0	≥ 42.5	≥ 3.5	≥ 6.5

Table 2 Basic properties of fly ash

Type	Fineness (%)	Specific surface area $(m^2 \cdot kg^{-1})$	Moisture content (%)	The ratio of water requirements (%)	Compressive strength ratio (%)		Ignition loss (%)
					7 days	28 days	
Jintang	6.8	390	0.1	95	68	75	2.8
DL/T 5055-2007	≤ 12.0	–	≤ 1.0	≤ 95	–	–	≤ 5.0

Table 3 Quality indicators of fibers

Type	Diameter (μm)	Length (mm)	Density $(g \cdot cm^{-3})$	Breaking strength (MPa)	Elastic modulus (GPa)	Fracture elongation (%)
Wanwei	37	12	1.3	1800	34	6.6

Table 4 HECC basic mix ratio

Specimen number	The maximum particle size of artificial sand (mm)	Water–cement ratio	Water content per unit (kg·m^{-3})
D1	1.25	0.35	350
D2	0.63	0.35	355
D3	0.315	0.35	360
D4	0.16	0.35	365
D5	1.25	0.25	315
D6	0.63	0.25	320
D7	0.315	0.25	325
D8	0.16	0.25	330

and the maximum particle sizes are 1.25 mm, 0.63 mm, 0.315 mm, and 0.16 mm respectively.

The admixtures used in the test include a polycarboxylic acid high-performance water-reducing agent, cellulose thickening agent, defoamer, and so on.

2.2 HECC Mix Ratio

The water–cement ratio of the test was 0.25 and 0.35, and four kinds of artificial sand with different maximum particle sizes were used to prepare HECC, and the water consumption per unit was adjusted appropriately under the circumstance of controlling the comparable mobility (shown in Table 4). The amount of fly ash in each mixing ratio is 50%, fiber content is 26 kg·m^{-3} (2% volume dosage), water-reducing agent content is 0.6%, thickening agent content is 0.05% and defoamer content is 0.09%.

2.3 Test Methods

Compressive and Flexural Strengths

According to JC/T 2461-2018 "Standard test method for mechanical properties of ductile fiber reinforced cementitious composites", HECC specimens are molded and cured to the age of 7 and 28 days, and then the compressive test is conducted, with the specimen size of 100 mm × 100 mm × 100 mm. According to GB/T17671-2021 "Method of testing cement determination of strength (ISO Method)", the flexural strength test is conducted, with the specimen size of 40 mm × 40 mm × 160 mm.

Impermeability and Frost Resistance

The water–cement ratio of 0.35 and different maximum particle sizes of artificial sand are adopted, and the corresponding specimens were molded and tested for impermeability and frost resistance.

Before the test, some specimens were pre-compression to 40 and 70% of the maximum ultimate load to simulate different degrees of damage, and then carried out the impermeability test after unloading. The method of step-by-step pressurization is adopted to reach the maximum water pressure of 1.0 MPa, then, we lower the pressure after 24 h of stabilization and calculate the relative permeability coefficient or record the seepage pressure. The HECC specimens of 100 mm × 100 mm × 400 mm were molded, the specimens were standardly cured until 28 days of age, the rapid freeze–thaw test was carried out to determine the freezing-resistance grade, and the changes in the mechanical strength of the specimens before and after the freeze–thaw cycle are compared.

Thermal Stability and Crack Self-Healing Ability

The thermal stability test included the fiber itself and the HECC specimen. The PVA filament fibers were immersed in water at 100 °C for three days, taken out, and dried at 105 °C for 3 h. The appearance changes before and after heat exposure were observed, and the indexes of elongation at break, strength at break, and initial modulus were tested. At the same time, the test group with a water–cement ratio of 0.35 and a maximum grain size of sand of 1.25 mm was selected, and the demolded HECC specimens were placed in different temperatures for curing to compare the mechanical properties under different curing temperatures.

Based on the HECC compressive strength test at the age of 7 and 28 days, the specimens with cracks after the loading test were put into the curing room again for standard curing for 28 days, and then the second compressive strength test was carried out, and the self-healing ability of the HECC material was comprehensively evaluated by the strength ratio of the two tests.

3 Test Results and Analysis

3.1 Mechanical Properties of HECC

The durability of HECC is inseparable from the mechanical strength, and the mechanical properties of HECC specimens at the age of 7 and 28 days were tested first, and the results of the compressive and flexural strength tests under different water-cement ratios and the maximum particle size of artificial sand are shown in Fig. 1.

It can be seen that the compressive and flexural strengths of HECC are increased by decreasing water-cement ratios and prolonging the age of curing, and the compressive and flexural strengths of the HECC at the age of 7 and 28 d are in the range of

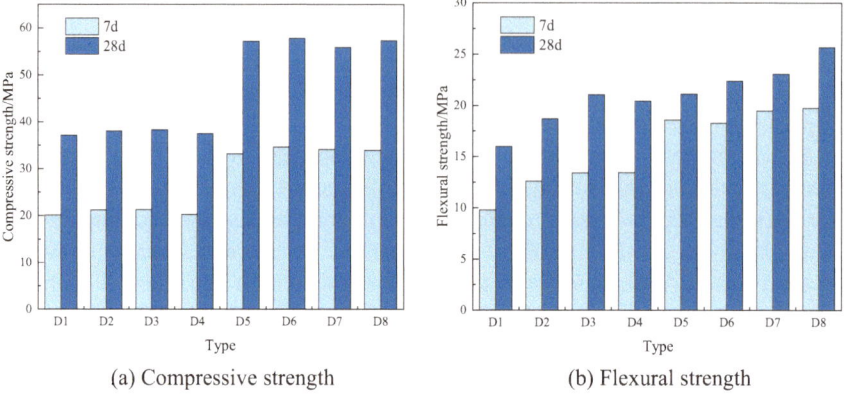

(a) Compressive strength   (b) Flexural strength

Fig. 1 HECC basic mechanical properties

20.1–33.9 and 37.1–57.8 MPa, and the flexural strength was 9.79–19.76 and 16–25.69 MPa at 7 and 28 days. At the same time, the maximum particle size of artificial sand decreased did not change the compressive strength of the HECC much, but it improved the flexural strength at all ages, which originated from the improvement of the interfacial microstructure in the interior of the HECC.

3.2 Impermeability of HECC

The impermeability of HECC has an important influence on the quality and safety of hydraulic buildings. Considering the possible crack state on-site, the relative permeability coefficient or permeable water pressure of HECC of different precompression specimens is texted, and the influence of the maximum particle size of artificial sand is considered to comprehensively evaluate the impermeability of HECC specimens under different working conditions, and the test results are shown in Table 5.

The test results show that the HECC specimens are impermeable with relative permeability coefficients of the order of 10^{-6} or 10^{-7} for the base specimen or the

Table 5 HECC impermeability test results

Type number	The maximum particle size of artificial sand (mm)	Relative permeability coefficient or seepage water pressure of different precompression specimens		
		0% (cm·h^{-1})	40% (cm·h^{-1})	70%
D1	1.25	1.6×10^{-6}	8.4×10^{-6}	0.6 MPa
D2	0.63	3.3×10^{-7}	6.3×10^{-6}	0.8 MPa
D3	0.315	3.0×10^{-7}	1.1×10^{-6}	5.4×10^{-4} cm·h^{-1}
D4	0.16	2.0×10^{-7}	9.8×10^{-7}	8.5×10^{-4} cm·h^{-1}

precompression 40% ultimate load. When there is no load, the permeability resistance of HECC is mainly due to the secondary hydration and volcanic ash effect of fly ash, which generates a low alkalinity hydrated calcium silicate gel that can effectively fill and block the capillary pores generated by the cement hydration in the previous period. So that the internal pore structure exists more as a large number of closed-type independent pore structures, which ultimately results in more difficulty for water molecules to penetrate the interior of the material under the action of water pressure [4].

Compared with the base specimen, after loading 40% of the ultimate load, the effect of the impermeability of the HECC specimen is not obvious, and after loading 70% of the ultimate load, the maximum grain size of sand 0.315 or 0.08 mm of the specimen is not permeable, and the relative permeability coefficient is in the order of 10^{-4}. But when the maximum grain size exceeds 0.63 mm, the specimen is permeable after subjecting to 0.6–0.8 MPa water pressure, which originates from the continuous increase of water pressure, and the formation of a certain number of connected pores and crack bands within the specimen, increasing the permeability of the HECC specimen. Due to the bridging effect of a large number of fibers of HECC, the cracks can be controlled within 60 μm, the permeability is proportional to the third power of the crack width, and there is almost no change in the permeability when the crack width is controlled within 100 μm [5], which makes the HECC still have a certain degree of resistance to the permeability when it is precompression with different degrees of loading. A comparison of the effect of precompression loading on the permeability of the specimen shows that the damaged cracks after HECC holding load are fine and non-connected, and can self-healing, repairing the tiny cracks as well as pores generated inside the specimen, thus avoiding the infiltration phenomenon and even maintaining the integrity of the specimen under the action of water pressure [6].

It has also been shown that HECC has high water content and fine pore structure, forming certain water transport channels inside the specimen. And when the specimen is loaded, the water will be squeezed out through or into the cracks and pores in the specimen. Due to its fine pore structure, the water transport needs to go through a long path and undergo many bends and rotations, and the trajectory of the water flow will be interfered with, which leads to a larger transient and steady-state resistance to flow, and the flow resistance, in this case, is larger. The flow resistance of HECC is large, and this process can cause a lot of friction, sticking, and other phenomena [7], which also makes HECC have a high impermeability.

3.3 Frost Resistance of HECC

Freeze resistance is an important basis for evaluating the durability of HECC materials, so the freeze–thaw resistance test of HECC materials was carried out. The results of the frost resistance test of HECC specimens at different sand maximum

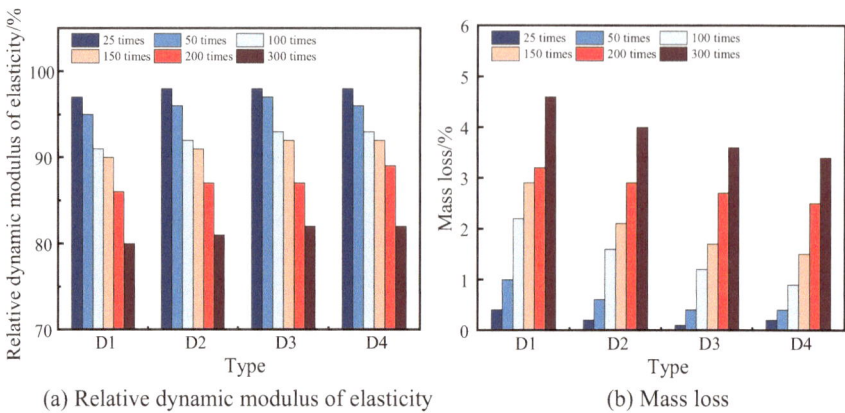

(a) Relative dynamic modulus of elasticity   (b) Mass loss

Fig. 2 Results of the HECC frost resistance test

grain sizes are shown in Fig. 2, including the relative dynamic elastic modulus and mass loss at different numbers of freeze–thaw cycles.

The evaluation criteria for the frost resistance grade are that the mass loss of the specimen is less than 5% and the relative dynamic modulus is more than 60% after being subjected to freeze–thaw cycles. Therefore, according to the test results, the frost resistance grade of HECC specimens in all test groups reached F300. Under the same number of freeze–thaw cycles, the finer the sand is, the lower the mass loss is and the higher the relative dynamic elastic modulus of the specimens is.

To compare the changes in the mechanical properties of the specimens before and after freezing and thawing, compressive and flexural strength tests were carried out, and the results are shown in Fig. 3. The results show that taking the test results of the initial state as a reference, the compressive strength of the ECC specimens decreased at a rate of 14.0–30.1% after 300 freezing and thawing cycles, while the flexural strength decreased more significantly, with a decrease rate of 70.6–75.3%, which indicates that the flexural strength is more sensitive to the internal cracks. This indicates that flexural strength is more sensitive to the presence of internal cracks and defects.

The main reason for the high frost resistance of HECC is that the internal pore volume allows the water in the capillary pores to have more space to release the pressure generated by volume expansion when freezing, thus reducing the damage to the material components. PVA fibers increase the ductility of HECC under tensile loading and slow down the damage to the HECC material caused by the stress generated by the water expansion [8]. In addition, due to the HECC species mixed with a large amount of fly ash, the secondary hydration reaction of fly ash has the advantage of late development, so that the structure is gradually dense, reducing the pore space inside the HECC and the freezing point of the pore water [9].

The finer the sand is, the number of particles and fibers between the contact points increase, providing more blocking force and tangential friction, which makes

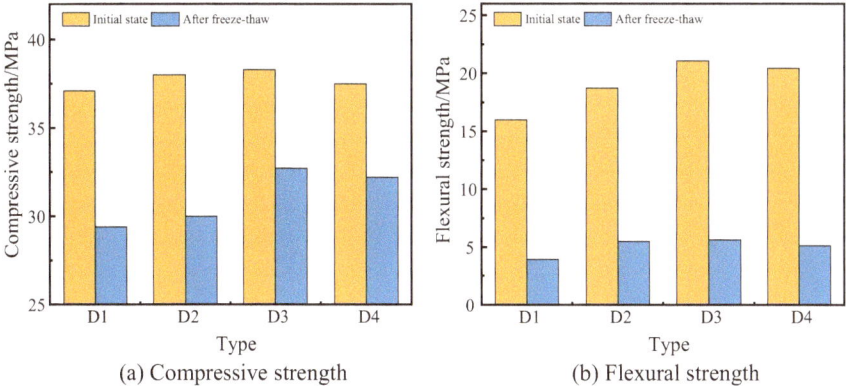

Fig. 3 Comparison of strength before and after 300 freeze–thaw cycles of HECC with different maximum particle sizes of sand

HECC have better toughness. And in the process of the freeze–thaw cycle, good toughness helps to offset cyclic stress and slows down the accumulation of damages, thus reducing the rate of quality loss and improving the resistance to freezing.

3.4 High-Temperature Resistance of HECC

Thermal Stability of Fibers

The fiber was heated to different temperatures to observe the appearance of the color changes after heating, see Fig. 4. It was found that the PVA fiber was put into the oven and gradually warmed up, and when heated up to 110 °C, the appearance of the color almost did not change; when heated up to 130 °C, the color changed slightly to light yellow, and the color changed from light yellow to dark yellow when heated up to 160 °C for 1 h. At 160 °C for 1 h, the color of PVA fiber changed from light yellow to dark yellow, but there was no dissolution or bonding phenomenon. The above results show that the PVA fiber used in HECC has good performance in high-temperature resistance and can meet the practical requirements.

The thermal stability of fibers in hot and humid environments was examined by accelerated weathering tests, the results of which are shown in Fig. 5 and Table 6. The results show that, after the environmental impacts of hot water immersion at 100 °C and drying and heating at 105 °C, the appearance of the filament fibers did not change much. Compared with the initial state, soaked and dried PVA fiber elongation at break has increased, the fracture strength is unchanged, and the initial modulus is slightly reduced. 105 °C oven for 3 days, it is found that the color is slightly yellowed, and the performance does not change significantly. It shows that the thermal stability of the fiber is good and can meet the technical requirements of HECC's long-term durability.

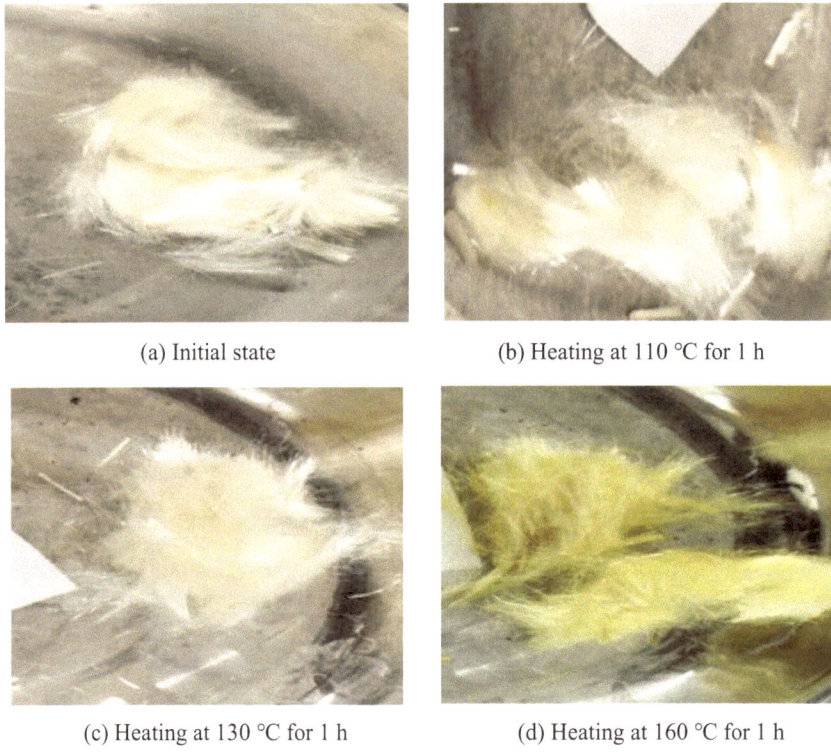

(a) Initial state   (b) Heating at 110 °C for 1 h

(c) Heating at 130 °C for 1 h   (d) Heating at 160 °C for 1 h

Fig. 4 Appearance and morphology of fibers heated to different temperatures for 1 h

(a) Normal fibers   (b) 3 days in 100 °C water,   (c) Place in a 105 °C
                         dried at 105 °C   oven for 3 days

Fig. 5 Changes in appearance and morphology of fibers under different conditions

Table 6 Test results of PVA fiber mechanical properties affected by temperature

Test metric	Measuring unit	Normal temperature state	3 days in 100°C water, dried at 105°C
Fracture elongation	%	6.56	10.41
Breaking strength	cN·dtex^{-1}	8.13	8.11
	MPa	1048.8	1046.2
Initial elastic modulus	cN·dtex^{-1}	228.9	207.47
	MPa	29.5	26.8

After 3 days in 100 °C water, the fiber structure relaxes, leading to the destruction of hydrogen bonds within the long-chain polymer, the breakage of the imide group and other bonding compounds, the formation of free radicals, and causing the cleavage of the main chain, resulting in a covered slip. The polymer molecular arrangement becomes loose, the molecular arrangement becomes loose, and the aggregates at all levels can participate more actively in the mechanical motion, which also contributes to the elongation [10]. As the structure becomes relaxed, the mechanical coupling of interactions within the fiber is weakened, and a slight decrease in the initial modulus of the fiber also occurs. At the same time, the fibers immersed in water become soft, which results in the formation of a stress-relieving zone on the surface of the fibers, in turn affecting the mechanical properties and certain structural characteristics [11].

Effect of Curing Temperature on Mechanical Properties of HECC

The D1 ratio is selected and five different curing temperature maintenance is molded in, the results of flexural strength and compressive strength shown in Fig. 6. It can be found in the compressive strength in the 7 days age, with the increase in maintenance temperature, the strength gradually increased in 80 °C to reach the maximum and then appeared to decline in 100 °C. At 28 d age, the law of change is still showing the law of the first rise and then decline in 60 °C to reach the maximum. The change rule of flexural strength is also similar, the maximum value of strength appeared at 80 °C and reached 11.18 and 11.69 MPa at the age of 7 days and 28 days, respectively.

It can be found that before 60 °C, the compressive and flexural strengths increased significantly with the increase of the curing temperature, and the maximum growth rate reached 63 and 43%, respectively. However, with the further increase in temperature, the strength showed a decrease of 5–15%. Meanwhile, the folding compression ratios at the age of 7 and 28 days under 20 °C–100 °C were 21–26% and 20–25%, respectively, and the folding compression ratios were maximum under the curing temperature of 20 °C. This is because, at higher temperatures, the initial rapid hydration causes more rapid precipitation of hydration products, the formation of more hydration products, and tighter pore structure, which improves the early strength of the material [12, 13], Hansen [14] proposed a new model that the curing temperature of 50 °C will be sacrificed to the physical bonding of water and increase the chemical bonding of water, the formation of hydration products will increase additional

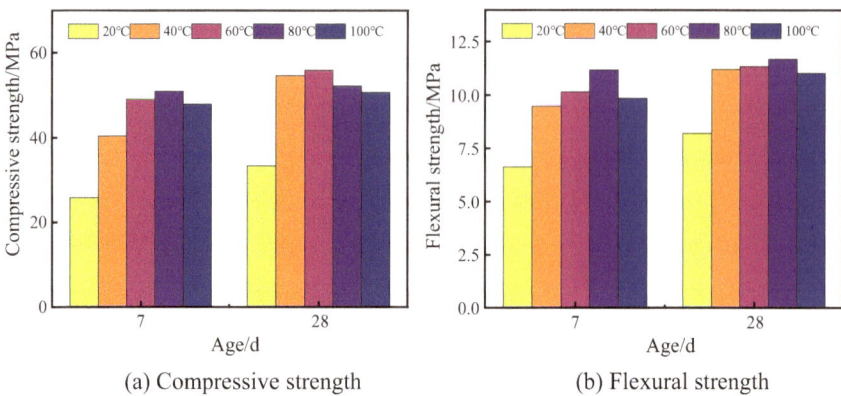

(a) Compressive strength   (b) Flexural strength

Fig. 6 Mechanical properties of HECC under different curing temperatures

strength. When certain mineral admixtures are added, the curing temperature increase will also increase the compressive and flexural strength to a certain extent [15], and the curing temperature increase will increase the rate of silicate polymerization, which will increase the density and hardness of the C–S–H gel.

However, as the temperature increases, the newly formed hydration product C–S–H gel has stabilized, and the external hydration product is denser and does not fill the capillary pores efficiently. To continue to warm up, a large number of hydration products generated in a short period do not have sufficient time to precipitate in an orderly manner, thus generating a disordered and porous structure, which increases the porosity and reduces the strength. It will also cause a certain degree of internal drying and water loss, which is not conducive to the hydration reaction of the cement so the strength of the specimen is reduced [16–18].

3.5 Self-Healing Ability of HECC

The results of the HECC specimen crack self-healing ability test based on the strength test are shown in Fig. 7.

After being cured again for 28 d, the ECC specimens can still withstand different degrees of pressure loading. 7 d old specimens after compressive strength test, cured again for 28 d, the strength ratio is 167–183% (water-cement ratio of 0.35), 110–124% (water-cement ratio of 0.25), which is higher than 100%, and the higher the water to glue ratio is, the higher the strength ratio is. 28 days old specimen compressive strength after the test was again maintained for 28 days. The strength ratios were 72–76% (water-cement ratio 0.35) and 60–72% (water-cement ratio 0.25), which were all lower than 100%, and the higher the water-cement ratio is, the higher the strength ratio is. But the relationship between the maximum particle size of sand and self-healing ability (strength ratio) is not obvious, and the two are not well correlated.

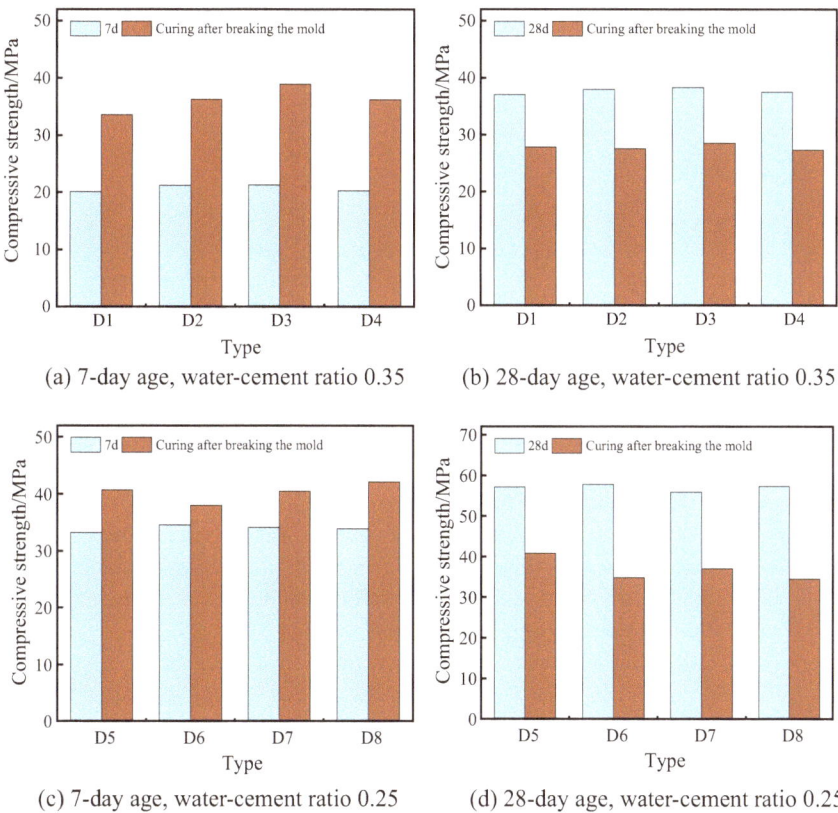

(a) 7-day age, water-cement ratio 0.35   (b) 28-day age, water-cement ratio 0.35

(c) 7-day age, water-cement ratio 0.25   (d) 28-day age, water-cement ratio 0.25

Fig. 7 Self-healing test results of HECC specimens with different sand particle sizes

The results show that HECC has a good self-healing ability, thanks to several factors: good crack control and multiple crack development; bridging of fibers reduces the cross-sectional area of the cracks; water retention in the turbulent zone on the lower side of the fibers; and the secondary hydration reaction of the mineral admixture fly ash [19, 20].

The higher the water-cement ratio is, the higher the self-healing ability observed in this test is, because of the water-cement ratio, the more water is supplied to the volcanic reaction of the fly ash, and the increase in the degree of reaction of the fly ash, i.e., the ability to repair the cracks is enhanced. The study also found the same law, using the water-cement ratio of 0.24, 0.3, 0.4 cement stone in the fly ash doping 20%, the compressive strength recovery rate increases with the increase of water–cement ratio, 101.59%, 102.02%, 102.88% respectively, [21]. The study found that the water-cement ratio should not be too small, otherwise, it will lead to dry and thick cement paste, which affects the fluidity and quality of concrete and makes its self-healing ability worse. It should not be too large, otherwise, the cohesion and water retention of the mix will be affected, and the number of cement paste per unit of the specific surface area is less, which makes the self-healing ability of the concrete in the later stage also decreased [22].

4 Conclusion

(1) The relative permeability coefficient of HECC specimens is in the order of 10^{-6} or 10^{-7}, and the finer the sand is, the better the impermeability is. After loading 40% of the ultimate load, it does not have too great an effect on the impermeability of the specimens; after loading 70% of the ultimate load, the specimens with the largest grain size of the sand of 0.315 and 0.16 mm are not permeable, and other specimens can withstand water pressure of 0.6–0.8 MPa.

(2) The frost resistance grade of HECC reaches F300, and with the decrease of the largest grain size of sand, the mass loss rate gradually decreases and the relative dynamic elastic modulus gradually increases under the same number of freezing and thawing cycles. Compared with the specimens that have not been subjected to freezing and thawing cycles, the reduction rate of compressive strength after 300 freezing and thawing cycles ranges from 14.0 to 30.1%, and the reduction rate of flexural strength ranges from 70.6 to 75.3%.

(3) HECC specimens with the increase of maintenance temperature, compressive strength, and flexural strength are showing a trend of increasing first and then decreasing, with the maximum increase in compressive strength of 63%, and flexural strength of 43%. But for fibers in the maximum of 160 °C oven for 1 h, 105 °C heating for 3 d, 100 °C hot water immersion, and other environments, the change in the performance of the changes is not obvious.

(4) The higher the water-cement ratio of HECC is, the higher the strength ratio is and the stronger the self-healing ability is. The strength ratio of specimens at the age of 7 days after re-conditioning can reach a maximum of 183%, and at the age of 28 days can reach 76%. But the relationship between the maximum particle size of sand and the self-healing ability (strength ratio) is not obvious, and the two are nearly not well correlated.

Acknowledgements This research was funded by the National Natural Science Foundation of China (Grant Nos.U2040222, 52179122), the Natural Science Foundation of Hubei Provincial (Grant No 2022CFD026), the Basic Scientific Research Service Funds of Central Level Public Welfare Research Institutes (CKSF2023304/CL).

References

1. Li, Q. H., & Xu, S. L. (2009). Performance and application of ultra-high toughness cementitious composite: A review. *Engineering Mechanics, 26*(S2), 23–67. https://kns.cnki.net/kcms/detail/detail.aspx?FileName=GCLX2009S2005&DbName=CJFQ2009

2. Li, J. Z. (2023). Material properties of Hydraulic Engineered Cementitious Composites (HECC) and their application in hydraulic structures. *Journal of Changjiang River Scientific Research Institute, 40*(02), 1–6+26. https://kns.cnki.net/kcms/detail/42.1171.TV.20220720. 0848.002.html

3. Li, Q. H., Gao, D., & Xu, S. L. (2012). Study on water permeability of UHTCC. *Journal of Hydraulic Engineering, 43*(S1), 76–84. https://doi.org/10.13243/j.cnki.slxb.2012.s1.017

4. Grassl, P., Wong, H. S., & Buenfeld, N. R. (2010). Influence of aggregate size and volume fraction on shrinkage induced micro-cracking of concrete and mortar. *Cement and Concrete Research, 40*(1), 85–93. https://doi.org/10.1016/j.cemconres.2009.09.012

5. Wang, K., Jansen, D. C., Shah, S. P., et al. (1997). Permeability study of cracked concrete. *Cement and Concrete Research, 27*(3), 381–393. https://doi.org/10.1016/S0008-8846(97)000 31-8

6. Reinhardt, H. W., & Jooss, M. (2003). Permeability and self-healing of cracked concrete as a function of temperature and crack width. *Cement and Concrete Research, 33*(7), 981–985. https://doi.org/10.1016/S0008-8846(02)01099-2

7. Rouchier, S., Janssen, H., Rode, C., et al. (2012). Characterization of fracture patterns and hygric properties for moisture flow modeling in cracked concrete. *Construction and Building Materials, 34*, 54–62. https://doi.org/10.1016/j.conbuildmat.2012.02.047

8. Sahmaran, M., Ozbay, E., & Yiicel, H. E. (2012). (2012) Frost resistance and microstructure of engineer-ed cementitious composites: Influence of fly ash and micro poly-vinyl-alcohol fiber. *Cement and Concrete Composites, 2*, 34. https://doi.org/10.1016/j.cemconcomp.2011.10.002

9. Cao, M. L., Xu, L., & Zhang, C. (2015). Review on micromechanical design, performance and the development tendency of engineered cementitious composite. *Journal of the Chinese Ceramic Society, 43*(05), 632–642. https://doi.org/10.14062/j.issn.0454-5648.2015.05.12

10. Zuo, B., Lu, X. L., & Zhang, R. P. et al. (2011). Advances in chain relaxation behaviors on the polymer surface. *Polymer Materials Science and Engineering, 27*(09), 183–186+190. https:// doi.org/10.16865/j.cnki.1000-7555.2011.09.048

11. Wang, J. C., Xiang, A. M., & Liu, X. J. (2008). Research on thermal-oxidative aging and stability of poly (vinyl alcohol). *China Plastics, 22*(09), 74–77. https://doi.org/10.19491/j.issn. 1001-9278.2008.09.016

12. Sajedi, F., & Razak, H. A. (2011). Effects of curing regimes and cement fineness on the compres-sive strength of ordinary Portland cement mortars. *Construction and Building Materials, 25*(4), 2036–2045. https://doi.org/10.1016/j.conbuildmat.2010.11.043

13. Al-gahtani, A. S. (2010). Effect of curing methods on the properties of plain and blended cement concretes. *Construction and Building Materials, 24*(3), 308–314. https://doi.org/10. 1016/j.conbuildmat.2009.08.036

14. Hansen, T. B. (2002). Temperature dependency of the hydration of dense cement paste systems containing micro silica and fly ash. *Nordic Concrete Research Publications, 27*, 27–34. https://www.Researchgate.net/publication/265811511_temperature_depenency_ of_the_hydration_of_dense_cement_paste_systems_containing_micro_silica_and_fly_ash

15. Aziz, M. A. A. E., Aleem, S. A. E., & Heikal, M. (2012). Physical-chemical and mechanical characteristics of pozzolanic cement pastes and mortars hydrated at different curing tempera-tures. *Construction and Building Materials, 26*(1), 310–316. https://doi.org/10.1016/j.conbui ldmat.2011.06.026

16. Escalante-garcia, J. I., & Sharp, J. H. (1998). Effect of temperature on the hydration of the main clinker phases in Portland cement: Part II, blended cement. *Cement and Concrete Research, 9*, 1259–1274. https://doi.org/10.1016/S0008-8846(98)00107-0

17. Martinez-ramirez, S., & Frias, M. (2009). The effect of curing temperature on white cement hydration. *Construction and Building Materials, 23*(3), 1344–1348. https://doi.org/10.1016/j. conbuildmat.2008.07.012

18. Luo, J., Cai, Z., & Yu, K., et al. (2019). Temperature impact on the micro-structures and mechanical properties of high-strength engineered cementitious composites. *Construction and Building Materials, 226*, 686–698. https://doi.org/10.1016/j.conbuildmat.2019.07.322

19. Kan, L. L., Wang, M. Z., & Shi, J. W., et al. (2015). Review on self-healing of engineered cementitious composites materials. Journal of Functional Materials, *46*(5), 6. https://kns.cnki.net/kcms/detail/50.1099.th.20150211.0935.001.html

20. Bao, W. B., Wang, D. X., & Wang, H. C. (2019). Experimental study on the self-healing performance of crack of green toughness cementitious composites. *Materials China*, *38*(04), 396–400. CNKI:SUN:XJKB.0.2019-04-010

21. Zhao, J., Yang, W., & Xu, K., et al. (2020). Effect of fly ash and slag on self-healing properties of oil well cement. *Modern Chemical Industry*, *40*(S1), 186–189+194. https://doi.org/10.16606/j.cnki.issn0253-4320.2020.S.040

22. Wu, S. Y., & Wang, L. C. (2018). A review of research progress on self-healing of concrete cracks. *China Concrete and Cement Products*, 271(11), 6–12. https://doi.org/10.19761/j.1000-4637.2018.11.002

Permeability and Compressive Strength of Pervious Cement Concrete with Small Size Aggregates

Bo Chen, Xiaochuang Zhang, Xiongfeng Wang, Feng Zhang, Lele Lv, Yin Bai, and Haoda Ma

Abstract To prepare high-performance pervious concrete with excellent mechanical properties and permeability, an orthogonal test method is used to investigate the effects of water-to-binder (W/B) ratio, grading of small size aggregates, and target continuous void content on the mechanical properties and permeability of pervious concrete. The results show that the W/B ratio and grading of small-size aggregates have negligible effects on compressive strength and permeability. The target continuous void content is the key parameter in determining the permeability of concrete. The compressive strength must be controlled within a certain range to ensure a certain range of permeability. Only one permeability parameter is suitable for pervious concrete mix design.

Keywords Pervious concrete · Permeability · Compressive strength · Small size aggregates

1 Introduction

Pervious concrete consists of hydraulic cementitious composites combined with an open-graded aggregate to produce a medium and low compressive strength with a typical continuous void content of 15–25%. For pervious concrete, mixed design parameters, such as water-to-binder ratio, aggregate size, and grading, have complex effects on the permeability of pervious concrete [1–3]. In the construction of a sponge city, to improve the porosity and water permeability of concrete, large coarse aggregates are typically used, and the strength of this concrete is often very low. To improve the strength of pervious concrete, some reinforcement measures are usually adopted, such as adding polymer, silica fume, fiber, and other reinforcement materials [4, 5].

B. Chen (✉) · X. Zhang · F. Zhang · L. Lv · Y. Bai · H. Ma
The National Key Laboratory of Water Disaster Prevention, Nanjing Hydraulic Research Institute, Nanjing 210029, China
e-mail: bchen@nhri.cn

X. Wang
POWERCHINA Zhongnan Engineering Corporation Ltd., Changsha 410014, China

© The Author(s) 2025
D. Li and Y. Zhang (eds.), *Advances in Frontier Research on Engineering Structures II*,
Lecture Notes in Civil Engineering 535, https://doi.org/10.1007/978-981-97-6238-5_13

Table 1 Physical properties of aggregates

Aggregate size (mm)	Apparent density (kg/m³)	Bulk density (kg/m³)	Void ratio (%)
2.36–4.75	2890	1680	41.9
4.75–9.5	2870	1680	41.5

The addition of sand also induces a significant increase in mechanical properties [6, 7]. It is difficult to balance permeability and mechanical properties in the preparation of permeable concrete as increases in one property typically induce decreases in the coinciding property.

With a low water-to-binder (W/B) ratio and small-size aggregates with optimized grading, pervious concrete with excellent performance can be prepared [8, 9]. In this paper, the orthogonal test method is used to investigate the effects of influencing factors, i.e., W/B ratio, aggregate grading (2.36–4.75 mm and 4.75–9.5 mm), target continuous void content, on the mechanical properties and permeability of pervious concrete, and to prepare high-performance pervious concrete with excellent mechanical properties and permeability.

2 Experiment

2.1 Materials

P·II 42.5R Portland cement produced by Nanjing Conch Cement Co., Ltd. was used. The physical and mechanical properties of this cement meet the requirements of Common Portland Cement (GB 175-2007).

Two kinds of aggregates with particle sizes of 2.36–4.75 mm and 4.75–9.5 mm were selected for the test. The basic properties of aggregates were tested according to Pebble and crushed stone for construction (GB/T14685-2011). The basic properties of the aggregates are shown in Table 1. Poly-carboxylic acid water reducer was used in the test, and the water reduction rate was 25%.

2.2 Pervious Concrete Mix Proportions

An L9(34) orthogonal table was used for the three factors, i.e., W/B ratio, aggregate grading, target continuous void content, and three-level orthogonal design, as in Table 2. Aggregate grading means the composition ratio of aggregate to particle size of 4.75–9.5 mm to 2.36–4.75 mm. The pervious concrete mix proportions were designed according to CJJ/T 135-2009 technical specifications for pervious cement concrete pavement. The mix proportions are in Table 3.

Table 2 Orthogonal table of pervious concrete test

Level	Factor A	Factor B	Factor C
	W/B ratio	Aggregate grading	Target continuous void content
Level 1	0.24	10 + 0	15%
Level 2	0.27	8 + 2	20%
Level 3	0.30	6 + 4	25%

Table 3 Pervious concrete mix proportions

Mix No.	W/B ratio	Aggregate grading	Target continuous void content (%)	Cement (kg/m^3)	Water (kg/m^3)	Aggregate (kg/m^3)	
						4.75–9.5 mm	2.36–4.75 mm
T1	0.24	10 + 0	15	505	121	1646	/
T2	0.24	8 + 2	20	415	100	1317	329
T3	0.24	6 + 4	25	326	78	988	659
T4	0.27	10 + 0	25	309	83	1646	/
T5	0.27	8 + 2	15	479	129	1317	329
T6	0.27	6 + 4	20	394	106	988	659
T7	0.30	10 + 0	20	375	113	1646	/
T8	0.30	8 + 2	25	294	88	1317	329
T9	0.30	6 + 4	15	456	137	988	659

2.3   Test Method

The pervious cement concrete permeability coefficient testing was conducted according to CJJ/T 135-2009. The continuous void content was conducted according to the ASTM C1754-2012 standard test method for density and void content of hardened pervious concrete. The compressive strength test was conducted according to the GB/T50081-2019 standard for test methods of concrete physical and mechanical properties.

3   Results and Discussion

The orthogonal test results of compressive strength, permeability coefficient, and continuous void content are in Table 4.

Table 4 Test results of pervious concrete

Mix No.	28 days Compressive strength/MPa	Permeability coefficient /(mm/s)	Continuous void content/%
T1	32.6	0.37	12.60
T2	29.8	1.98	18.78
T3	21.1	3.34	22.24
T4	18.5	4.22	22.73
T5	35.7	1.86	12.30
T6	30.8	2.33	19.88
T7	25.6	3.94	20.47
T8	19.4	5.79	24.24
T9	33.5	0.27	13.73

3.1   The Effect of Factors on Properties

By analyzing orthogonal test data, the average test results of compressive strength, permeability coefficient, and continuous void content under three influencing factors and three levels were obtained. The cumulative effect of these factors on the mechanical properties is in Fig. 1.

When the W/B ratio increases from 0.24 to 0.30, the average compressive strength is 27.8 MPa, 28.3 MPa, and 26.1 MPa, respectively, and the range is 2.1 MPa. The average permeability coefficient is 1.90 mm/s, 2.80 mm/s, and 3.33 mm/s, respectively, and the range is 1.43 mm/s. The continuous void content is 17.87%, 18.30%, and 19.48%, respectively, and the range is 1.61%. The W/B ratio has an insignificant effect on compressive strength and continuous void content but has a significant effect on the permeability coefficient.

With aggregate gradings of 10 + 0, 8 + 2, and 6 + 4, the average compressive strength is 25.6 MPa, 28.3 MPa, and 28.5 MPa, respectively, and the range is 2.9 MPa. The average permeability coefficient is 2.84 mm/s, 3.21 mm/s, and 1.98 mm/s, respectively, and the range is 1.23 mm/s. The continuous void contents are 18.60%, 18.44%, and 18.62%, respectively, and the range is 0.18%. The increase of the proportion of fine aggregate with a particle size of 2.36–4.75 mm has little effect on compressive strength and continuous void content but has a significant effect on the permeability coefficient.

When target continuous void content increases from 15 to 25%, the average compressive strength is 33.9 MPa, 28.7 MPa, and 19.7 MPa, respectively, and the range is 14.2 MPa. The average permeability coefficient is 0.83 mm/s, 2.75 mm/s, and 4.45 mm/s, respectively, and the range is 3.62 mm/s. The continuous void content is 12.88%, 19.71%, and 23.07%, respectively, and the range is 10.19%. With increasing target continuous void content, the compressive strength gradually decreases, and the permeability coefficient and continuous void content gradually

Fig. 1 Effects of factors on properties

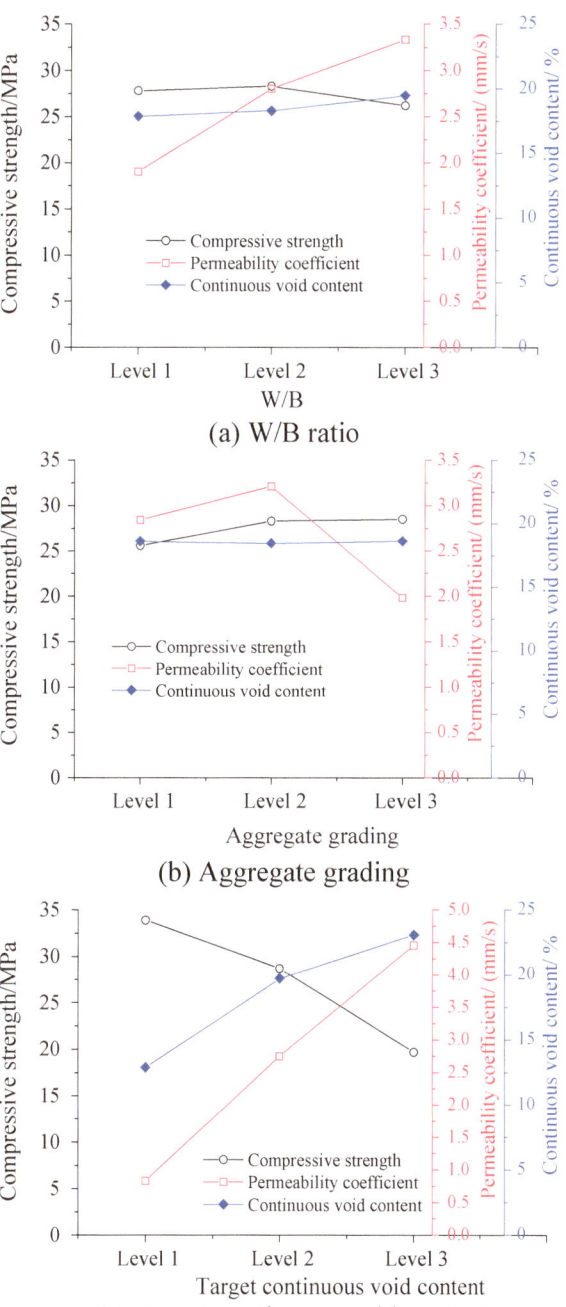

(a) W/B ratio

(b) Aggregate grading

(c) Target continuous void content

increase. The target continuous void content has a pronounced effect on compressive strength, continuous void content, and permeability coefficient [10].

W/B ratio and aggregate grading have little effect on compressive strength and continuous void content but have a significant effect on the permeability coefficient. The target continuous void content has a significant effect on the performance of previous concrete.

3.2   Relationship Between Performance Parameters

By analyzing the test results of pervious concrete, it can be found that there is a significant correlation between different performance parameters, as in Fig. 2. There is a negative linear correlation between continuous void content and permeability coefficient and compressive strength (see Fig. 2a, b). The higher the compressive strength of pervious concrete, the lower the concrete permeability. Therefore, to ensure the permeability of pervious concrete, the compressive strength of pervious concrete must be controlled within a certain range. From Fig. 2c, it can be found that the permeability coefficient has a positive linear correlation with continuous void content. The higher the permeability coefficient, the higher the continuous void content. Therefore, one permeability parameter can be selected as the control index of pervious concrete permeability, which is enough for concrete mix design.

4   Conclusions

W/B ratio and small-size aggregate grading have little effect on compressive strength and continuous void content but have a significant effect on the permeability coefficient. The target continuous void content has a significant effect on the performance of pervious concrete.

The compressive strength of pervious concrete must be controlled within a certain range to ensure appropriate permeability of pervious concrete.

One permeability parameter can be selected as the control index of pervious concrete permeability, which is enough for the concrete mix design.

Fig. 2 Relationship between compressive strength, permeability coefficient, and continuous void content

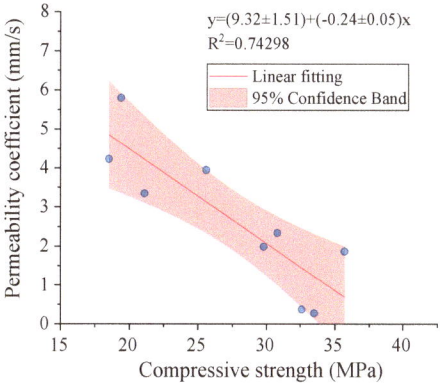

(a) Compressive strength vs. permeability coefficient

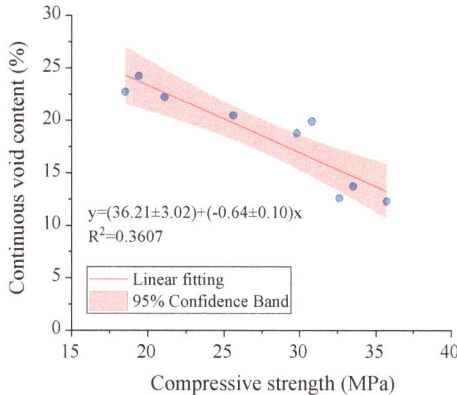

(b) Compressive strength vs. continuous void content

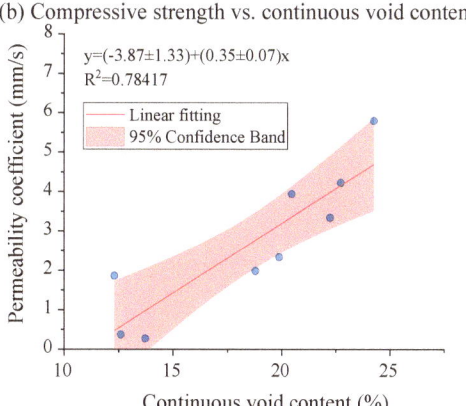

(c) Permeability coefficient *vs.* continuous void content

References

1. Gu, L. L., Feng, J. J., & Xiao, B. F. (2021). Roles of mortar volume in porosity, permeability and strength of pervious concrete. *Journal of Infrastructure Preservation and Resilience, 2*, 19.
2. Zhang, F., Bai, Y., & Chen, B. (2020). Research on the matching relationship between permeability and compressive strength of pervious concrete. *Construction Technology, 49*, 16–21.
3. Xie, X. G., Zhang, T. S., & Wang, C. (2020). Mixture proportion design of pervious concrete based on the relationships between fundamental properties and skeleton structures. *Cement and Concrete Composites, 113*, 103693.
4. Zhang, F., Bai, Y., & Chen, B. (2021). Influence of enhancement technical measures on the performance of pervious concrete. *New Building Materials, 48*, 36–41.
5. Martin, I. I. I., William, D., Kaye, N. B. & Putman, B. (2014). Impact of vertical porosity distribution on the permeability of pervious concrete. *Construction and Building Materials, 59*, 78–84.
6. Alessandra, B., Filippo, G., & Maurizio, C. (2015). Experimental study on the effects of fine sand addition on differentially compacted pervious concrete. *Construction and Building Materials, 91*, 102–110.
7. Costa, F. B. P. D., & Haselbach, L. M. (2021). Pervious concrete for desired porosity: Influence of w/c ratio and a rheology-modifying admixture. *Construction and Building Materials, 268*, 121084.
8. Pradhan, S. K., & Behera, N. (2022). Performance assessment of pervious concrete road on strength and permeability by using silica fume. *Materials Today: Proceedings, 60*, 559–568.
9. Mulu, A., Jacob, P. & Dwarakish, G. S. (2022). Hydraulic performance of pervious concrete based on small size aggregates. *Advances in Materials Science and Engineering, 2022*.
10. Mulyono, T., Anisah. (2019). Properties of pervious concrete with various types and sizes of aggregate. In MATEC Web of Conferences (Vol. 276).

Analysis of Impact Resistance and Optimization of Protective Capacity for Steel Reinforced Concrete Columns

Zongbo Hu

Abstract Through simulated horizontal impact tests on steel-reinforced concrete column components, the dynamic response of steel-reinforced concrete columns under different impact heights, boundary conditions, and impact velocities was obtained. The results indicate that with an increase in impact energy, the peak impact force shows an upward trend. As the impact height increases, the impact force gradually decreases, with higher forces near the end. When the mass of the impacting object is 2580 kg and the minimum velocity is 4.58 m/s, the maximum residual displacement occurs in the impact zone. Based on the simulated test results, the impact failure modes of steel-reinforced concrete columns can be classified as slight damage, moderate damage, severe damage, and critical damage. A model reflecting the changes in mechanical characteristics of steel-reinforced concrete column structures was established based on the damage states and impact resistance mechanisms at each loading stage of the column components. Active and passive protective structures were proposed.

Keywords Steel reinforced concrete columns · Impact resistance · Dynamic response · Working mechanism · Optimization of protective capacity

1 Introduction

Steel-reinforced concrete column structures are extensively used in key areas such as public buildings, transportation facilities, and military engineering. As load-bearing components, steel-reinforced concrete columns that resist lateral forces have garnered significant attention from scholars worldwide regarding their impact resistance and protective performance. However, current research on structural impact

Z. Hu (✉)
Institute of Equipment Management and Support, Engineering University of Chinese People's Armed Police Force, Xi'an, Shanxi 710086, China
e-mail: huzongbo_1985@163.com

Postdoctoral Research Station of Civil Engineering College, Xi'an University of Architecture and Technology, Xi'an, Shanxi 710055, China

D. Li and Y. Zhang (eds.), *Advances in Frontier Research on Engineering Structures II*,
Lecture Notes in Civil Engineering 535, https://doi.org/10.1007/978-981-97-6238-5_14

has primarily focused on applications such as concrete bridge piers for railway bridges, evaluation, design, and strengthening of steel-reinforced concrete columns in airport terminals and stations [1–6]. There is a lack of research on the impact resistance performance and protective structure design of commonly used steel-reinforced concrete columns in a large number of civil and military engineering projects. There are two approaches to enhancing the impact resistance performance of column components: first, addressing the issue of structural rationality; and second, addressing the issue of material compatibility. By combining protective structures and compatible materials based on protective mechanisms, the protective performance of column components can be greatly improved. However, few scholars have studied impact-resistant protective structures. Only researchers such as Remennikov have conducted hammer impact tests on traditional steel tubes filled with rigid polyurethane foam components [7]. Yang et al. [8] simulated the impact resistance performance of steel-reinforced concrete columns reinforced with FRP using numerical simulation. Wang et al. [9] performed lateral impact tests on hollow sandwich steel-reinforced concrete columns. Hou et al. [10] conducted lateral impact tests on steel tube composite columns, examining the effects of component shape, boundary conditions, axial force levels, and impact energy, and obtained information on the failure modes, impact force–time curves, overall deformation, and strain of the components.

Therefore, it is necessary to establish a protective working mechanism for steel-reinforced concrete column structures based on the analysis of impact resistance test results. This involves constructing impact-resistant protective structures and enhancing the structural impact resistance performance.

2 Simulation Tests for Impact on Steel-Reinforced Concrete Columns

2.1 Finite Element Model

The finite element model was implemented using the ANSYS/LS-DYNA software. The comprehensive model system comprises various components, such as an impacting hammer, column base, internal steel, steel reinforcement cage, and other elements. Detailed parameters can be found in Table 1, while the model visualization can be observed in Fig. 1a, b. To ensure accuracy, each part of the model was divided into hexahedral meshes with a mesh size of 10 mm. The contact between components was simulated using a face-to-face approach, with both dynamic and static friction coefficients set to 0.3.

Table 1 Comparison of test results and numerical simulation peak impact force

Specimens	Internal components	Original model (kN)	Optimization model (kN)	Upgrade rate (%)
SRCP-1	PS-I	1174	1490.98	27
SRCP-2	SiO$_2$	1192	1561.52	31
SRCA-3	Damping devices	1170	1696.50	45

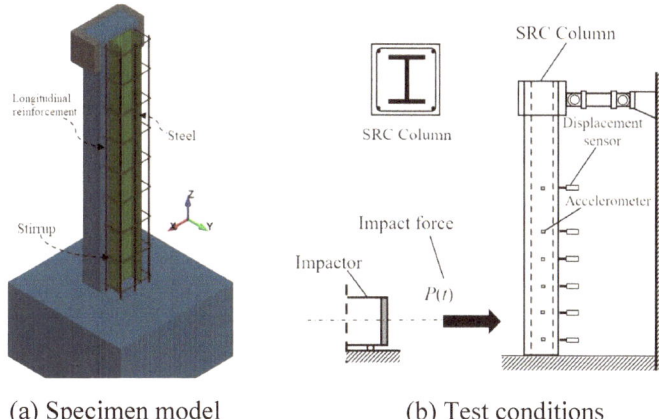

(a) Specimen model   (b) Test conditions

Fig. 1 Experimental mode

2.2 Specimen Parameters

The test objects are steel-reinforced concrete columns, and the variables of the experiment are impact height, impact velocity, and restraint conditions. The main section of the column is rectangular with dimensions of 300 mm × 300 mm. The longitudinal reinforcement of the column consists of HRB400 hot-rolled ribbed steel bars with a diameter of 10 mm, while the transverse reinforcement uses HPB300 light round steel bars with a diameter of 6 mm. The steel configuration is an H-shaped solid-web configuration, using Q235B hot-rolled H-section steel of size 175 mm × 175 mm × 6 mm × 8 mm. Horizontal impact is simulated under three different restraint conditions: fixed-simple, fixed–fixed, and cantilever. There are a total of 14 finite element models, including 12 fixed-simple columns. The mass of the impactor is 1080 kg, 1580 kg, 2080 kg, and 2580 kg, respectively. The impact positions are 400, 600, and 800 mm away from the column heel, and the corresponding impact velocities are 4.3 m/s, 5.1 m/s, and 6.0 m/s, respectively; There is one fixed–fixed constraint column and one cantilever constraint column, with an impact mass of 1080 kg. The impact location is 400 mm away from the column heel, and the corresponding impact velocities are 4.3 m/s, 5.1 m/s, and 6.0 m/s, respectively.

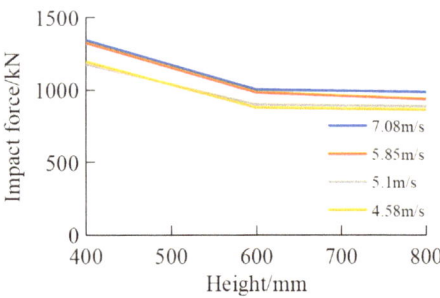

Fig. 2 Maximum impact force

Impact Force

As shown in Fig. 2, with the increase in impact energy, the peak impact force shows an upward trend. As the impact height increases, the impact force gradually decreases. The closer the impact is to the end, the greater the impact force. When the impact position is near the approximate mid-span region, the variation in impact force is not significant. Taking an example of an impact height 400 mm above the bottom of the column, when the velocity of the impacting body increases from 4.58 to 7.08 m/s, the increase in peak impact force is 12.6%. For models with impact heights of 600–800 mm above the bottom of the column, the increase in peak force is 14.7% and 14.6% respectively. The peak value of impact force follows an inverse proportionality with the change in mass.

Displacement

As shown in Fig. 3, with the increase of the initial kinetic energy of the impacting object, the residual displacement in the impacted area of the steel-reinforced concrete column monotonically increases. When the mass of the impacting object is 2580 kg and the minimum velocity is 4.58 m/s, the maximum residual displacement in the impacted area is observed. Therefore, it can be inferred that, with the same impact energy, the deformation of the steel-reinforced concrete column is significantly influenced by the mass of the impacting object.

3   Impact Resistance Working Mechanism

3.1   Failure Mode

As shown in Fig. 4, based on the results of simulation experiments, the failure modes can be classified into four categories: slight damage, moderate damage, severe damage, and critical damage. In the case of slight damage, there are shear inclined cracks, but the concrete remains intact, and there is no significant change in the overall condition of the column. In the case of moderate damage, the protective layer of concrete becomes loose, and wide shear inclined cracks appear in the column. In

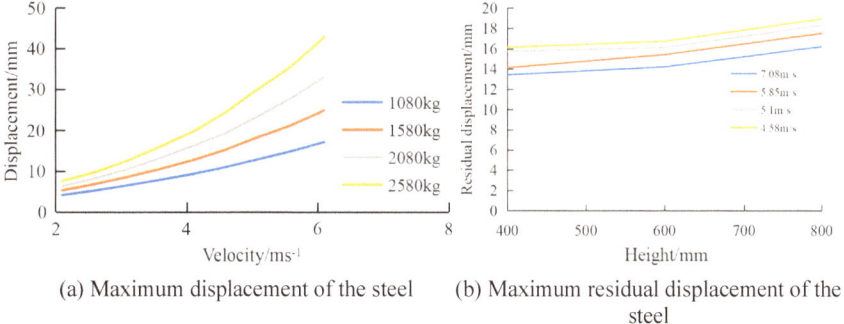

(a) Maximum displacement of the steel   (b) Maximum residual displacement of the steel

Fig. 3 Displacement of column steel

the case of severe damage, concrete spalling occurs, steel reinforcement undergoes non-elastic deformation, and there is noticeable shear misalignment in the impacted area. In the case of critical damage, the concrete becomes detached from the main structure, steel reinforcement fractures, the steel section noticeably yields, and the column collapses due to shear failure.

Fig. 4 Failure mode

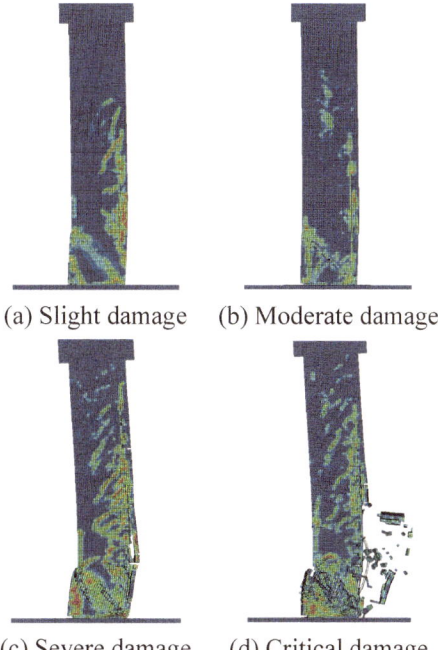

(a) Slight damage   (b) Moderate damage

(c) Severe damage   (d) Critical damage

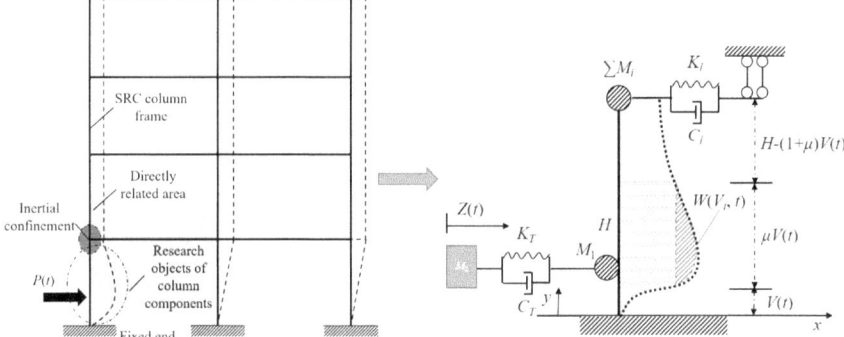

Fig. 5 Working mechanism of steel reinforced concrete column structure under impact

3.2 *Working Mechanism*

According to the simulated test results of the lateral impact on the steel reinforced concrete column frame structure, considering the damage state and mechanical performance indicators of the column components at each loading stage, a "Impact body-structure" coupled model is established to reflect the changes in mechanical characteristics of the steel reinforced concrete column structure under impact. Based on the "impact body-structure" coupling model. As shown in Fig. 5, based on the law of conservation of kinetic energy, the motion equation of the impacted column can be obtained.

$$\sum_{i}^{n} M_i \ddot{W}(V_T, t) + C_l\left[\dot{Z}(t) - \dot{W}(V_T, t)\right] + C_T \dot{Z}(t)$$

$$+ K_l[Z(t) - W(V_T, t)] + K_T Z(t) = -(M_1 + M_2)\ddot{Z}(t) \qquad (1)$$

where, the impact body mass consists of a spring with stiffness K_T and a damper with damping coefficient C_T, supported by the reciprocating motion mass M, and the coupled mass M1 that is coupled with the structure motion. The dynamic deflection of the structure is denoted as $W(V_t, t)$, and the dynamic displacement of mass M is denoted as $Z(t)$, μ is the relative position coefficient of column deflection, which is consistent with the deflection of the structure at its location. These findings aid in understanding the structural response and inform the design of effective protective measures to enhance the impact resistance performance of the columns.

4 Optimization of Impact Protection Performance

The research on protective structures for steel-reinforced concrete columns incorporates the results of simulated lateral impact tests and the mechanism revealed by coupled vibration analysis. There are two methods. Firstly, based on the inertial constraint state of the bottom steel reinforced concrete column, a passive energy dissipation protective structure is adopted (as shown in Fig. 6) to enhance the column ductility. This is achieved by increasing the dynamic displacement of the elastic damping sliding support at the column end, thereby dissipating the instantaneous impact energy. Secondly, based on the conclusions from the coupled vibration analysis of the steel-reinforced concrete column structure, an active energy-absorbing protective structure is employed (as shown in Fig. 7). Materials such as impact-resistant polystyrene (PS-I), filled with reactive SiO_2 particles, or the addition of self-resetting damping devices are used as the contact material of the impact body. This enhances the mass of the coupled motion part of the impact body and reduces the mass of the reciprocating elastic impact part.

Table 1 shows the comparison between the peak impact force of the numerical simulation results of the passive and active impact-resistant column specimens and the original experimental model. From the table, it can be seen that the peak

Fig. 6 Passive impact-resistant column

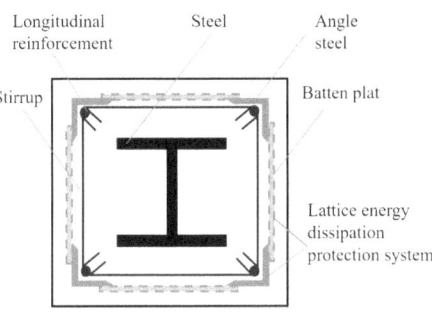

Fig. 7 Active impact-resistant

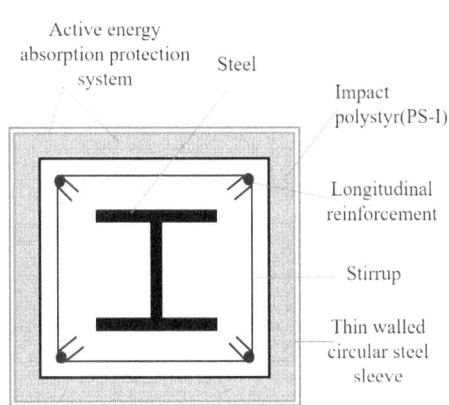

impact force of the column specimens with protective structures has increased by an average of over 35%, which better demonstrates the effectiveness and rationality of the impact-resistant protection optimization model.

5   Conclusion

- The peak impact force exhibits an upward trend as the impact energy increases. Moreover, as the impact height rises, the impact force gradually decreases, with higher forces observed towards the end. Notably, when the impacting object has a mass of 2580 kg and a minimum velocity of 4.58 m/s, the impact zone experiences the maximum residual displacement.
- Based on the simulated test results, the failure modes of steel-reinforced concrete columns under impact can be categorized into four levels: slight damage, moderate damage, severe damage, and critical damage.
- A model reflecting the changes in mechanical characteristics of steel-reinforced concrete column structures was established based on the damage states and impact resistance mechanisms at each loading stage of the column components. Active and passive protective structures were proposed.

Acknowledgements The financial assistance provided by the Shanxi Province Natural Science Basic Research Project No. 2023-JC-YB-419, the China Postdoctoral Science Foundation No. 2020M683432, the Basic Frontier Research Foundation of Engineering University of PAP No. WJY202229. These supports are gratefully acknowledged.

References

1. Zhu, X., Lu, X. Z., & Du, Y. F. (2014). Simulation of running attitude after train derailment. *Vibration and Impact, 33*(23), 145–149.
2. Xu, L. J., Lu, X. Z., & Smith, S. T. (2012). Scaled model to test for collision between an over-height truck and bridge superstructure. *International Journal of Impact Engineering, 49,* 31–42.
3. Lu, Xi. Z., & Lu. X., et al. (2011). Engineering calculation method for impact force on super high vehicles-bridge superstructure. *China Journal of Highway and Transport, 24*(2), 49–55.
4. Remennikov, A., & Kaewunruen, S. (2006). Impact resistance of reinforced concrete columns: experimental studies and design considerations. In: *19th Australasian conference on the mechanics of structures and materials* (pp. 817–824). Taylor & Francis.
5. Wang, H., et al. (2020). Enhancing impact resistance of column structures through SiO2-PS-I composite materials. *Construction and Building Materials, 150,* 453–462.
6. Liu, Q., et al. (2019). Experimental investigation of the seismic performance of SiO2-PS-I composite columns under cyclic loading. *Engineering Structures, 180,* 123–136.
7. Remennikov, A. M., Kong, S. Y., & Uy, B. (2010). Response of foam-and concrete-filled square steel tubes under low-velocity impact loading. *Journal of Performance of Constructed Facilities, 25*(5), 373–381.

8. Yang, Y. F., Zhang, Z. C., & Fu, F. (2015). Experimental and numerical study on square RACFST members under lateral impact loading. *Journal of Constructional Steel Research, 111*, 43–56.
9. Wang, R., Han, L. H., & Zhao, X. L. (2015). Experimental behavior of concrete-filled double steel tubular (CFDST) members under low velocity drop weight impact. *Thin-Walled Structures, 97*, 279–295.
10. Hou, C., Han, L., & Wang, F. (2019). Study on the impact behavior of concrete-encased CFST box members. *Engineering Structures, 198*, 109536.

Research on the Application of Lead Chrome Green in Colored Cement Concrete

Guangrao Fu, Bo Sun, Zhuangzhuang Feng, Hongzhen Li, Quan Yuan, and Xiaohui Cui

Abstract To seek new dyes to improve the performance of colored cement concrete pavement, this paper proposes lead chrome green dyes to replace the traditional iron oxide series dyes, and the color durability test, mechanical test, frost resistance test, compression and flexural test, and matching tests are carried out. The results showed that the dyeing effect of lead chrome green dyes is better than that of traditional iron oxide series dyes; lead chrome green can improve the frost resistance of concrete, and iron oxide green has little effect on the frost resistance of concrete. The research results provide more solutions for improving the preparation process of colored concrete.

Keyword Dye · Colored concrete · Mechanical property · Anti-freezing performance

1 Introduction

With the rapid development of China's transportation, the current colored cement concrete [1–3] pavement has been widely used in specific highway sections such as ETC lanes and expressway service areas. In some developed countries, the research on colored concrete started earlier, and great progress has been made in dyeing agents and construction [4–6]. Positieri [7] studied the effect of iron oxide stains of different colors on concrete properties. The results show that the compressive strength and durability of colored concrete are closely related to the amount of dye, and the performance of colored concrete can be improved through appropriate proportions. Miranda et al. [8] proposed a colored mortar repair technology, which prepared colored mortar by adjusting the coloring agent, to repair concrete defects. Jaroslava and Pavel [9] proposed to use of fine-grained dyes to make colored concrete, researched the amount of dyes, and suggested that the amount of dyes should not exceed 10%. When the amount is below 10%, the color effect of the concrete fades

G. Fu (✉) · B. Sun · Z. Feng · H. Li · Q. Yuan · X. Cui
Shandong Yimeng Transportation Development Group Co., Ltd., Linyi 276002, Shandong, China
e-mail: 569385179@qq.com

© The Author(s) 2025
D. Li and Y. Zhang (eds.), *Advances in Frontier Research on Engineering Structures II*,
Lecture Notes in Civil Engineering 535, https://doi.org/10.1007/978-981-97-6238-5_15

slowly. Compared with traditional colored concrete pavement, the upper colored pavement of the colored composite concrete pavement and the lower layer of ordinary concrete pavement jointly bear the vehicle load, giving full play to the respective advantages of pavements of different strengths, reducing pavement diseases, and improving the strength and service life of the cement concrete pavement.

At present, the preparation methods of colored concrete include the dyeing agent method, surface layer coloring method, and chemical dyeing method. The dyeing agent used for surface coloring has poor durability; the chemical dyeing method has relatively large pollution problems and poor durability. The dyeing agent method was widely used due to its simple preparation process, low price, and good color durability. However, the color of iron oxide series inorganic substances looks dull and faded soon, a much more bright color agent with good durability is needed.

This paper proposes to use lead chrome green dye to replace the traditional iron oxide series dyes, and color durability tests, mechanical properties tests, and antifrost performance tests on them, and the road performance of the cement concrete with lead chrome green are discussed.

2 Selection of Raw Materials and Dyeing Agents

The color fading of cement concrete needs to be considered besides mechanical properties and durability.

2.1 Concrete Raw Materials

Cement. Because ordinary gray cement hurts the color effect of colored concrete, white Portland cement was used in this paper. The St. Durham brand white Portland cement was used (P.W. 42.5), and the basic performance indicators are shown in Table 1.

Coarse aggregate. The gradation composition and physical indicators of the coarse aggregate selected in this test are shown in Tables 2 and 3

Fine aggregate. The fine aggregates used in the test are divided into two types, one is the standard sand used in the mortar test, the other is the ordinary machine-made sand used in concrete, and the detailed parameters are shown in Tables 4 and 5.

2.2 Dyeing Agent

The technical indicators of lead chrome green and iron oxide green are shown in Table 6.

Table 1 Physical properties of cement

Test content	Fineness	Stability	Whiteness/%	Condensation time/min		Compression strength/ MPa		Flexural strength/MPa	
				Initial set	Final set	3 days	28 days	3 days	28 days
Actual measurement	2.6	Qualified	89.67	195	235	29.8	44	5.6	9.3
Standard	≤ 30	Qualified	> 87	≥ 45	≤ 600	≥ 17	≥ 42.5	≥ 3.5	≥ 6.5

Table 2 Gravel grading

Mesh size (mm)	4.75	9.5	16	19	26.5	
Cumulative sieve residue (%)	98.6	80.3	57.2	32.8	0	Qualified
Standard requirement (%)	90–100	70–90	50–70	25–40	0–5	

Table 3 Physical indicators of gravel

Project	Gravel crush indicator (%)	Water absorption (%)	Mud content (% by quality)	Mud content (% by mass)	Apparent density (kg/m³)	Bulk density (kg/m³)
Test results	7.8	0.2	0.4	0	2710	1460
Standard requirement	<15	<2.0	<1.0	<0.2	>2500	>1350

Table 4 Fine aggregate screening results

Mesh size (mm)	4.75	2.36	1.18	0.6	0.3	0.15	Fineness modulus
Cumulative sieve residue (%)	3.8	16.4	38.9	60.1	79.2	98.6	2.85
Standard requirement (%)	0–10	0–25	10–50	41–70	70–92	90–100	

Table 5 Fine aggregate physical properties

Project	Moisture content (%)	Mica content (% by mass)	Mud content (% by quality)	Mud content (% by mass)	Apparent density (kg/m³)	Bulk density (kg/m³)
Test results	0.8	1.3	0.95	0.2	2650	1460
Standard requirement		<2.0	<2.0	<1.0	>2500	>1350

Table 6 Green dye properties

Type of dye	Oil absorption (%)	Relative shading rate (%)	PH value	Light fastness	Alkali resistance	Acid resistant sex	High-temperature resistance
Lead chrome green	25	101	7–8	Pretty good	Good	Good	150°C
Iron oxide green	18.9	98.9	4–7	Good	Pretty good	Good	200°C

2.3 Cement Slurry Color Durability Test

Preparation of cement mortar. The amount of the dyeing agent was set as 0, 2, 4, and 6%. The specific mixing ratio is shown in Table 7.

Evaluation of dyeing effect. The dyeing agent will directly determine the color effect. The color value of the cement mortar samples is shown in Table 8.

- L represents the brightness of the color. The larger the L value is, the brighter the color is;
- a represents the red and green values of the color. A positive number represents red and a negative number represents green;
- b represents the yellow and blue values. Positive numbers represent yellowish and negative numbers represent Bluish.

It can be seen from Table 8 that the color of the samples produced by the same amount of lead chrome green and iron oxide green are almost the same, but in terms of the brightness of the samples, lead chrome green is more colorful.

Color durability test. The durability of lead chrome green in cement concrete is tested by placing cement mortar samples in the air and water respectively.

The maximum value of the color value difference ΔE was taken as the comparison indicator to evaluate the color decay.

As shown in Fig. 1, when the six kinds of samples are placed in the air, the color fading of the test samples using lead chrome green as the dye is proportional to the amount of dye, and with the lead chrome green increased from 2 to 6%, the ΔE of 150 days increased from 6.22 to 9.45; for the test piece using iron oxide green as the dye, with the increase of the dye content, the fading will first increase and then decrease, at 4% The ΔE of 150 days is 9.69, which shows that in the natural environment, the

Table 7 Cement paste mix design

Group	Comment	Water (g)	Dye (g)
1	1500	750	0
2	1500	750	30
3	1500	750	60
4	1500	750	90

Table 8 Color value of cement slurry test piece

Type of dye	L	a	b
2% Iron oxide green	73.89	−13.55	6.23
4% Iron oxide green	62.39	−17.54	6.15
6% Iron oxide green	64.57	−20.36	4.36
2% Lead chrome green	76.44	−12.32	7.15
4% Lead chrome green	73.34	−15.32	8.41
6%Lead chrome green	68.51	−19.52	16.89

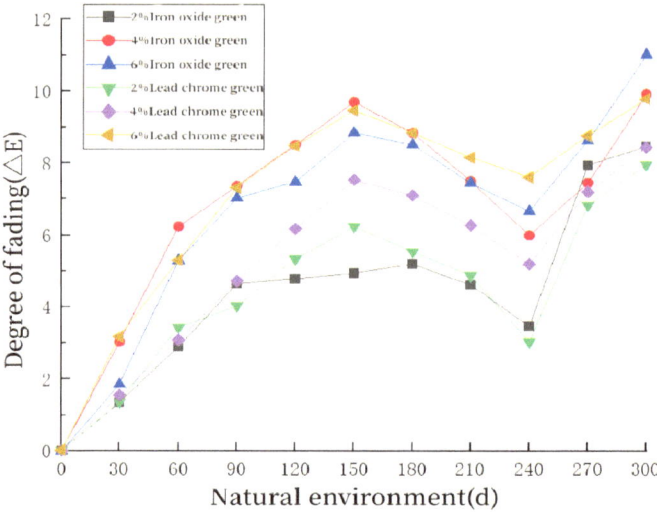

Fig. 1 The color fading of samples in the air

best durability is the lead chromium green test piece with 2% content, and the worst durability is the iron oxide green test piece with 6% content.

As shown in Fig. 2, when the green test piece is in a water bath environment, there is no great difference in the degree of fading of the six test pieces. Among the six proportions, the most serious fading is the 6% iron oxide green test, and the ΔE of 70 days is 3.89; the lightest fading is the 4% lead chrome green test piece, and the ΔE of 70 days is 3.20. It shows that in the water bath environment, the best durability is the 4% lead chromium green test piece, and the worst durability is the 6% iron oxide green test piece.

3   Effect of the Dyeing Agent on the Mechanical Properties of Cement Concrete

By making a green mortar test piece, the optimal dosage of different dyeing agents is determined according to the strength change of the mortar. The optimal dosage of dyeing agent is selected to make concrete samples, and the mechanical properties and frost resistance of concrete are studied.

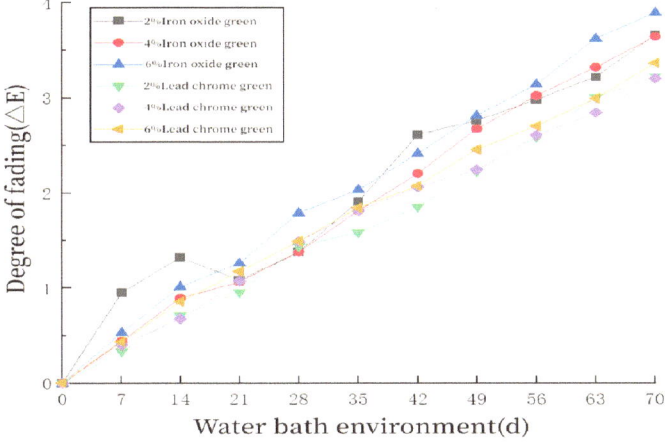

Fig. 2 The color fading of samples in water

3.1 The Effect of Dyeing Agent on the Strength of Cement Sand

Cement mortar test design. The amount of dyeing agent selected is 2%, 4%, and 6%, respectively. The specific mixing ratio is shown in Table 9.

Analysis of test results. Iron oxide green is selected as the iron oxide dye, lead chrome green is selected as the new dye, and the amount of dye is increased from 0 to 6%. The changes in the strength of the samples are shown in Table 10.

It can be seen from Table 10 that the iron oxide green will reduce the strength of the mortar, and with the increase of the dosage, the degree of reduction will gradually increase. At 7 days, the flexural strength dropped from 9.8 to 8.4 MPa, the compressive strength decreased from 34.9 to 33.7 MPa; when the age was 28 days, the flexural strength decreased from 13 to 11.4 MPa, and the compressive strength decreased from 47.5 to 43 MPa. Lead chrome green will cause the strength of the mortar to increase first and then decrease. When the content is 4%, the strength

Table 9 Design of colored cement mortar mixing ratio

Type of dye	Group	Cement (g)	Standard sand (g)	Water (g)	Dye (g)
Not used	1	450	1350	225	0
Iron oxide green	2	450	1350	225	9
	3	450	1350	225	18
	4	450	1350	225	27
New dyes	5	450	1350	225	9
	6	450	1350	225	18
	7	450	1350	225	27

Table 10 Cement sand sample strength

Type of dye	Dosage (%)	Day	Flexural strength (MPa)	Compressive strength (MPa)
Iron oxide green	0	7/28	9.8/13	34.9/47.5
	2		9.2/12.6	34/46.3
	4		8.8/11.6	33.8/45.5
	6		8.4/11.4	33.7/43
Lead chrome green	0		9.8/13	34.9/47.5
	2		10.1/13.3	36.5/47.7
	4		10.9/14	37.7/50.2
	6		10.6/13.8	37/49.9

reaches the maximum value, and the flexural and compressive strengths of 7 days increase by 0.8 MPa and 2.1 MPa, respectively; the folding and compressive strengths increase by 0.8 MPa and 2.4 MPa, respectively.

Selection of dye content. Considering the strength of the concrete mortar and the fading of the color, the optimal content of dyeing agents can be selected. The amount of the iron oxide green is inversely proportional to the strength, but the degree of fading is small, and with the increase of the amount, the color effect is also improved, so the amount is set at 6%; lead chrome green will have a peak strength when the dosage is 4%, and the difference in dyeing effect between the two dosages of 4–6% is not obvious, so the dosage of lead chrome green is set at 4%.

3.2 Mechanical Properties of Colored Concrete

Mix ratio design. Concrete samples with two strengths of C20 and C40 were made according to the amount of dyeing agent, and the ordinary concrete samples were used as the control group to study the effect of dyeing agent on the performance of concrete. We determine the mix ratio of the two kinds of concrete, as shown in Table 11.

Research on the flexural tensile strength of concrete. After the concrete is cured to the specified age in the curing box, the bending tensile test can be performed, and the loading rate is shown in Table 12. Three samples were made for each group, the flexural strength of concrete under 7 days and 28 days curing time was calculated respectively, and the average value was used as the flexural strength of the concrete samples in this group.

It can be seen from Table 13 that 6% iron oxide green will reduce the flexural tensile strength of concrete with a C20 label by 10.4% at 7 days and 11.6% at 28 days; the flexural tensile strength of concrete with a C40 label will be cured for 7 days, and the strength decreased by 10%; the strength decreased by 8.5% when the curing reached 28 days. 4% lead chrome green dye will increase the flexural tensile strength

Table 11 Mix design of colored concrete

Concrete type	Type of dye	Cement (kg/m^3)	Sand (kg/m^3)	Pebbles (kg/m^3)	Water (kg/m^3)	Dye (kg/m^3)
C20	Colorless	380	435	1015	230	0
	Iron oxide green	380	435	1015	230	22.8
	Lead chrome green	380	435	1015	230	15.2
C40	Colorless	510	685	1025	230	0
	Iron oxide green	510	685	1025	230	30.6
	Lead chrome green	510	685	1025	230	20.4

Table 12 Loading rate and size conversion factor of flexural test

The size of the pressure surface of the test piece	Power level	Loading rate (MPa/s)	Size conversion factor
100 mm × 100 mm	f_{cu}<C30	0.02–0.05	0.85
	C30 ≤ f_{cu}<C60	0.05–0.08	
	f_{cu} ≥ C60	0.08–0.10	

of C20 grade concrete by 10.8% and C40 grade concrete by 6.9% after curing for 7 days; after the curing age reaches 28 days, the flexural tensile strength will increase by 8.5% and 7.2% respectively.

Research on the flexural strength of concrete. The concrete can be subjected to compressive tests after curing to the specified age in the curing box, and the loading rate is shown in Table 14.

It can be seen from Table 15 that 6% iron oxide green will cause the compressive strength of C20 grade concrete to decrease by 10.1% and 7.4% after curing for 7 days and 28 days respectively; C40 grade concrete will decrease by 9.9% and 11.2% respectively. 4% lead chrome green will increase the compressive strength of C20 and C40 concrete by 11.2% and 9% when curing for 7 days; the strength will increase by 11.6% and 8.6% when curing for 28 days.

Table 13 Concrete flexural tensile test

Concrete type	Type of dye	Maintenance time (days)	Bending tensile strength (MPa)
C20	Colorless	7	2.3
		28	3.3
	Iron oxide green	7	2
		28	2.9
	Lead chrome green	7	2.5
		28	3.6
C40	Colorless	7	4.3
		28	5.6
	Iron oxide green	7	3.7
		28	5
	Lead chrome green	7	4.6
		28	5.7

Table 14 Compression test loading rate and size conversion factor

The size of the pressure surface of the test piece	Power level	Loading rate (kN/s)	Size conversion factor
100 mm × 100 mm	$f_{cu} < C30$	3.0–5.0	0.9
	$C30 \leq f_{cu} < C60$	5.0–8.0	
	$f_{cu} \geq C60$	8.0–10.0	

Table 15 Concrete compression test

Concrete type	Type of dye	Maintenance time (days)	Compressive strength (MPa)
C20	Colorless	7	17
		28	23
	Iron oxide green	7	15
		28	21
	Lead chrome green	7	18
		28	26
C40	Colorless	7	32
		28	43
	Iron oxide green	7	30
		28	35
	Lead chrome green	7	40
		28	50

3.3 Research on Frost Resistance Performance of Colored Concrete

Low temperature is a common factor that leads to concrete damage, especially in cold regions, most of the concrete damage is caused by freeze–thaw action. At present, the frost resistance of concrete has become an important indicator for evaluating the durability of concrete. This paper adopts the quick freezing method of the concrete frost resistance performance test.

Experimental design. The size of the test mold selected for this test is 100 mm × 100 mm × 400 mm. Each group of tests consists of 3 samples, which can be used continuously. The mass loss rate and dynamic elastic modulus loss rate are measured every 25 times. When the mass loss rate reaches 5% or when the relative dynamic modulus of elasticity drops below 60%, the test can be stopped.

Analysis of test results. The number of freeze–thaw cycles and results of concrete samples of different colors and strengths are shown in Table 16.

It can be seen from Table 16 that when a dyeing agent is added, the dynamic elastic modulus of concrete will increase to a certain extent. When the elastic modulus of concrete shows a downward trend, the rate of decline will gradually increase with the increase of freeze–thaw times; lead chrome green can improve the frost resistance of concrete, and this phenomenon is more obvious in low-grade concrete; iron oxide green has little effect on the frost resistance of concrete.

Table 16 Concrete freeze–thaw cycle test

Concrete type	Type of dye	Cycles	Elastic modulus (MPa)
C20	Colorless	0	26,250
		150	15,500
	Iron oxide green	0	31,000
		150	19,000
	Lead chrome green	0	33,250
		175	18,000
C40	Colorless	0	36,000
		175	21,500
	Iron oxide green	0	34,000
		175	19,600
	Lead chrome green	0	39,000
		175	23,000

3.4 Impact of Colored Concrete on the Environment

The dyes of colored concrete are essentially oxides and have relatively stable chemical properties. When applied to concrete pavements, they will not cause damage to the environment. The only requirement is to avoid ingesting large amounts into the human body during construction.

4 Conclusion

In this paper, different types and dosages of dyes are selected to make colored concrete, and their mechanical properties and durability are studied. The types and optimal dosages of dyes are selected through experiments, and the following conclusions are obtained:

- The dyeing effect of lead chrome green is better than that of traditional iron oxide dyes.
- Iron oxide green dyes generally lead to a decrease in the strength of the mortar.
- The strength increases with the increase of lead chrome green content. Lead chrome green can improve the frost resistance of cement concrete; iron oxide green has little effect on the frost resistance of concrete.
- 4% lead chrome green performed best in durability and mechanical properties tests.

By studying different types and dosages of dyes, it was found that compared with traditional colored concrete pavements, colored composite concrete pavements can give full play to the respective advantages of pavements of different strengths, reduce pavement diseases, and improve the strength and service life of cement concrete pavements. In the future, we can also study the impact of dyes added to cement on the hydration reaction of cement, to conduct microscopic experiments and more in-depth research on various indicators of colored concrete.

References

1. Jingsheng, H., Wenyan, L., & Erdong, F. (2012). Construction of colored concrete pavement in oil and gas field. *Oil and gas fields surface works, 31*(08), 81.
2. Wenyuan, P. (2019). On the construction technology and design application of color concrete pavement. *Smart City, 5*(21), 170–171.
3. Wen, H., & Zheng, W. (2016). Color pavement analysis study. *Tianjin Construction Technology, 26*(05), 61–63.
4. Anonymous. (2018). Lanxess Center takes colored concrete to the next level. *Concrete Products, 121*(11), 32–33.

5. Miranda, J., Valença, J., & Júlio, E. (2019). Colored concrete restoration method: For chromatic design and application of restoration mortars on smooth surfaces of colored concrete. *Structural Concrete, 20*(4), 1391–1401.
6. Craeye, B., Tielemans, T., Lauwereijssens, G., & Stoop, J. (2013). Effect of super absorbing polymers on the freeze-thaw resistance of colored concrete roads. *Road Materials and Pavement Design, 14*(1), 90–106.
7. Physicomechanical properties and durability of structural colored concrete. In *5th ACI/ CANMET/IBRACON International Conference on High-Performance Concrete Structures and Materials 2008* (pp.176–193). American Concrete Institute.
8. Miranda, J., Valenca, J., & Julio, E. (2019). Colored concrete restoration method: For chromatic design and application of restoration mortars on smooth surfaces of colored concrete. *Structural concrete, 20*(4), 1391–1401.
9. Koťátková, J., & Reiterman, P. (2014). Colored concrete with focus on the properties of pigments. *Advanced Materials Research, 36*(03), 1054.

Distribution of Steel Corrosion Products in a Cracked Concrete Beam Exposed to Coastal Atmosphere

Jianfeng Dong, Xuepeng Lin, Xueyuan Liu, Jianzong Guo, and Yuxi Zhao

Abstract This study investigated the distribution of steel corrosion products in a cracked concrete beam, which was in service under coastal atmospheric conditions for more than 40 years. The thickness of corrosion-product-filled paste (CP) and the thickness of corrosion layer (CL) were measured. The migration behaviour of Fe in the concrete cracks and air voids are investigated. The mechanism of CP formation was revealed. The steel corrosion and corrosion-product-filled paste propagates faster in the tension region than in the compressive region of the beam. In the atmospheric environment, the corrosion products neither fill the cracks or air voids nor do they accumulate on the edge of the cracks or inside the air voids.

Keywords Steel corrosion · Corrosion-product-filled paste · Corrosion layer

1 Introduction

One of the main causes of deterioration of reinforced concrete structures is the corrosion of reinforcement [1]. Because the volume of the corrosion products is approximately two to six times the volume of the original steel [1], corrosion products can create expansion pressure at the steel/concrete interface, which eventually causes cracking to the concrete cover. The cracks offer paths for the rapid ingress of aggressive agents to the reinforcement, which in return accelerate the corrosion process of the steel bar [2, 3]. Therefore, the investigation of corrosion-induced cracking is important for predicting the serviceability and durability of reinforced concrete structures.

J. Dong (✉) · X. Lin · X. Liu · J. Guo
Southwest Design and Research Institute Co. Ltd., China Airport Construction Group Corporation, Chengdu 610200, Sichuan, China
e-mail: djf_1991@126.com

J. Dong · Y. Zhao
Institute of Structural Engineering, Zhejiang University, Hangzhou 310058, Zhejiang, China
e-mail: yxzhao@zju.edu.cn

© The Author(s) 2025
D. Li and Y. Zhang (eds.), *Advances in Frontier Research on Engineering Structures II*,
Lecture Notes in Civil Engineering 535, https://doi.org/10.1007/978-981-97-6238-5_16

Numerous studies have dealt with the quantitative prediction of critical steel corrosion when concrete surface cracking occurs. The corrosion products are generally divided into two parts: (1) the corrosion products that fill into the porous zone of concrete around the steel, thus generate the corrosion-product-filled paste (CP) [4–9], and (2) the corrosion layer (CL), the corrosion products that produces expansion pressure on the surrounding concrete [2–13]. Obviously, both CP and CL are important for quantitative prediction of critical steel corrosion.

To quantitatively estimate CP, Michel et al. [4, 5] measured the thickness of CP in cementitious matrix specimens, based on the X-ray attenuation measurement technique and digital image correlation. Subsequently, they proposed a conceptual time-dependent model to describe the propagation of CP from steel corrosion initiation to the cover cracking of the mortar specimen. Based on Scanning Electron Microscopy (SEM) and Energy Dispersive X-Ray Spectroscopy (EDS), previous studies by the authors [6–8] confirmed that the CP and CL generate simultaneously, and the thickness of CP (T_{CP}) increases with the thickness of CL (T_{CL}), and once the average T_{CL} reaches a certain value, T_{CP} no longer shows significant growth.

However, all specimens in previous studies [6–8] were corroded without loading. Obviously, loads could change the pore structure in the concrete [14], which may influence the forming of CP. Moreover, all specimens in previous studies were artificially corroded by the NaCl solution. No study has been carried out on actual concrete components that were in service under natural environment.

Therefore, the objective of this study was to investigate the distribution of steel corrosion products in a concrete beam, which had been in service for more than 40 years, under coastal atmospheric conditions. The steel/concrete interface was observed and the formation of the CP is discussed. The penetration behaviour of Fe in the concrete cracks and air voids are investigated. This study helps to better understand the formation mechanism of CP.

2 Experimental Program

2.1 Concrete Specimens

In this study, a beam, which had been in service in a coastal atmosphere for more than 40 years, was considered. The beam was segmented from the roof of a building, which was located on an island off the east coast of China. The average temperature of this island is 16.2 °C (maximum 38.3 °C; minimum −7.5 °C), the average rainfall is 1624 mm/year, and the rain frequency is 120 days/year. The building was constructed in the 1970s and abandoned in the 1990s.

As shown in Fig. 1a, the East China Sea was south of the beam, while the building and hill were to the north. Thus, the south surface of the beam is the windward surface and is labelled S2, while the north surface is the leeward surface and is labelled S1; the bottom surface was labelled B. The appearance of the beam is shown in Fig. 1b.

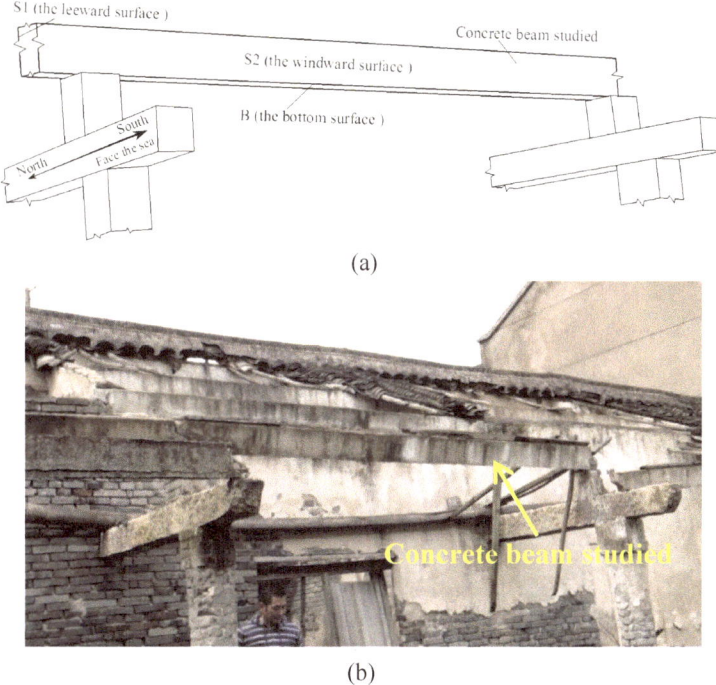

(a)

(b)

Fig. 1 The studied beam; **a** schematic diagram of the beam, **b** the beam

The beam had a size of 60 mm in width and 190 mm in depth; the total length of the beam was 3000 mm, and only the right half part was used in this investigation, since the left half part was severely damaged. The compressive strength (f_c) of the concrete was 22.6 MPa (tested by the rebound method). Two layers of steel bars were used as tension reinforcement, as shown in Fig. 2. Each layer consisted of two bars, and the diameters of the first layer (bottom layer) and the second layer were 5 mm and 4 mm, respectively. Two steel bars, with a diameter of 3 mm, were used as the top-layer steel bars. The six bars were labelled based on their locations, as shown in Fig. 2. The concrete cover thickness of the S1 series bars (i.e. S1-T1, S1-T2, and S1-C) and the S2 series bars (i.e. S2-T1, S2-T2, and S2-C) was 12 mm and 17 mm, respectively, which may have been due to construction error.

2.2 Sample Preparation

The 100-mm-length part in the right end of the beam was abandoned, considering the damage incurred during the segmentation of the beam. In this study, the 200-mm-length part of the beam, shown in Fig. 3, was used, while the other part was stored in order to be studied after several years. The part considered in this study was cast

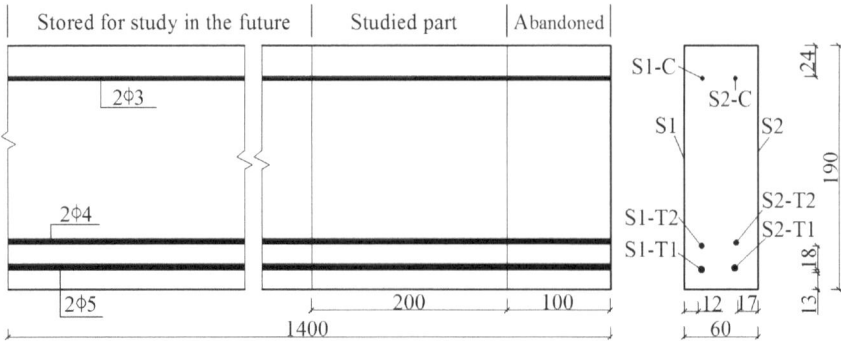

Fig. 2 Reinforcement and concrete cover information of the beam

into a low-viscosity epoxy resin in order to minimize any artificial damage that may have occurred. After the hardening of the epoxy resin, the part was separated from the beam with a concrete cutting machine (Φ335). The part was then poured again into low-viscosity epoxy resin and cut into six blocks with a sectional dimension of 25 mm × 25 mm. Each block consisted of one steel bar, as shown in Fig. 3. Then, the six blocks were further cut into 12-mm-thick slices using an abrasive cutter (SYJ-200) and a diamond blade suitable for hard and brittle materials. The blocks were sliced at low speed, less than 0.05 mm/s, and water was used as coolant in order to minimize damage. 14 slices were obtained for each block; however, some slices were damaged during the cutting process, and approximately 8 slices remained from each block. These slices were labelled as Steel bar label–N, with the third slice of the steel bar S2-T1, for example, being labelled as S2-T1-3. Subsequently, the slices were polished by a UNIPOL-1502 polishing machine for scanning electron microscopy (SEM) observation. After being cleaned by an ultrasonic cleaner, they were placed in vacuum-sealed bags in order to prevent the occurrence of further corrosion.

2.3   Measurement Program

A scanning electron microscope (QUANTA FEG650), operated in backscattered electron (BSE) mode, was applied in order to observe most of the samples; another scanning electron microscope (Hitachi S-3400) was applied to the left samples. Figure 4 shows a BSE image at the steel/concrete interface observed in sample S1-T2-6. The interface was clearly divided into five regions; namely, steel, corrosion layer (CL), millscale (MS), corrosion-product-filled paste (CP), and concrete. The measuring points were set at intervals of 500 μm along the perimeter of the steel bar. A previous study [6] demonstrated that MS forms before the initiation of corrosion, but not during the rust expansion process. Therefore, only CL and CP are discussed in this study. Image-Pro Plus 6.0 software was used to measure T_{CL} (thickness of

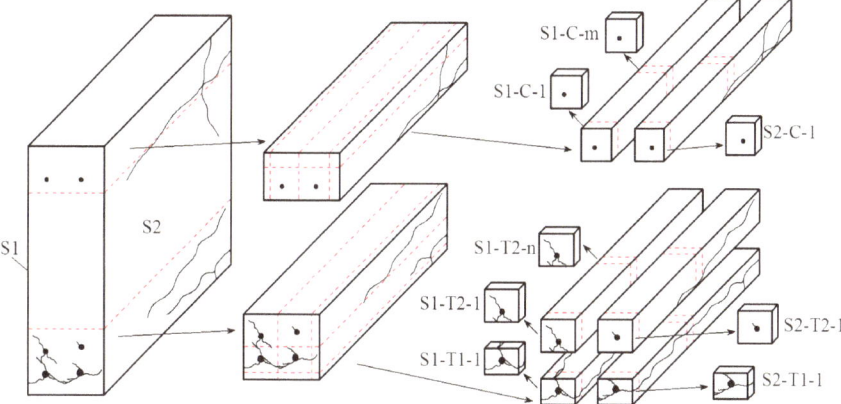

Fig. 3 Schematic diagram of sectioned concrete blocks and slices

CL) and T_{CP} (thickness of CP), as illustrated in Fig. 4. The CL or CP thickness of each sample is the average value of the measurement points, for each sample.

Fig. 4 Rust distributions at steel/concrete interface in sample S1-T2-6

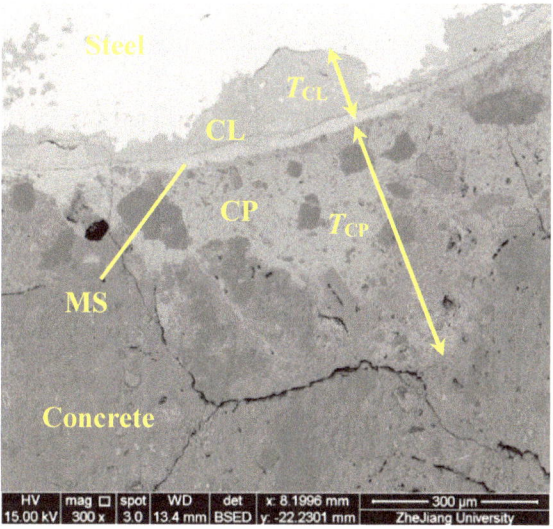

3 Experimental Results

3.1 Steel Corrosion and Concrete Cracking

The crack maps of the beam are shown in Fig. 5. It can be observed that the cracks are mainly distributed near the tension reinforcement, and some of the concrete spalled from the beam. The cracking of S1 was more severe than that of S2 near the tension reinforcement. This is due to the concrete cover thickness, on the side surface of the S1 series bars, being smaller than that of the S2 series bars (possibly due to construction error), as shown in Fig. 2.

The average T_{CL} of each slice are shown in Fig. 6. It needs to be noted that the average T_{CL}, for the S2-T1 point in the red cycle, is much larger than that of the other data, as shown in Fig. 6a. This may be due to the transverse cracks in the beam near these slices, as shown in Fig. 5. This leads to a sudden increase of tested data. Therefore, the point was removed when the average T_{CL} of S2-T1 was calculated. Similarly, the two points in the red cycles of S1-T2, shown in Fig. 6b, were not considered when the average T_{CL} of S1-T2 was calculated.

It can be seen that the average T_{CL} of Series S2 is larger than that of Series S1. This is because the sea wind, which contains a lot of chloride ions and water, acts directly on S2, and results in severe ingression of S2 chloride. Chloride ions reduce the resistance of concrete and directly affect the ionic current flow [15], and it influences galvanic corrosion activity [15–17]; therefore, the corrosion of steel bars in S2 is more severe than that of S1. The average T_{CL} of S2-T1 is slightly larger than that of S1-T1. This may be explained by the fact that both S2-T1 and S1-T1 suffered chloride ingression from the bottom of the beam, in addition to that suffered from the lateral face. As can be seen from Fig. 6, the bottom concrete cover is smaller than that of S2. Additionally, the chloride that penetrated from the bottom surface may play a more important role on the corrosion of Series T1.

It can be observed that, except for the S2C, the average T_{CL} decreased in the order of T1, T2, and C, according to Fig. 6. This is because T1 and T2 are in

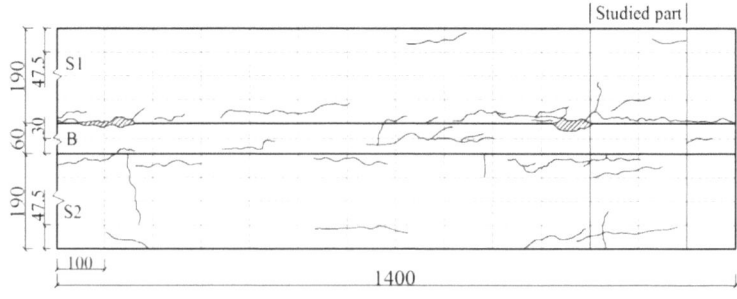

Fig. 5 Crack maps of the beam

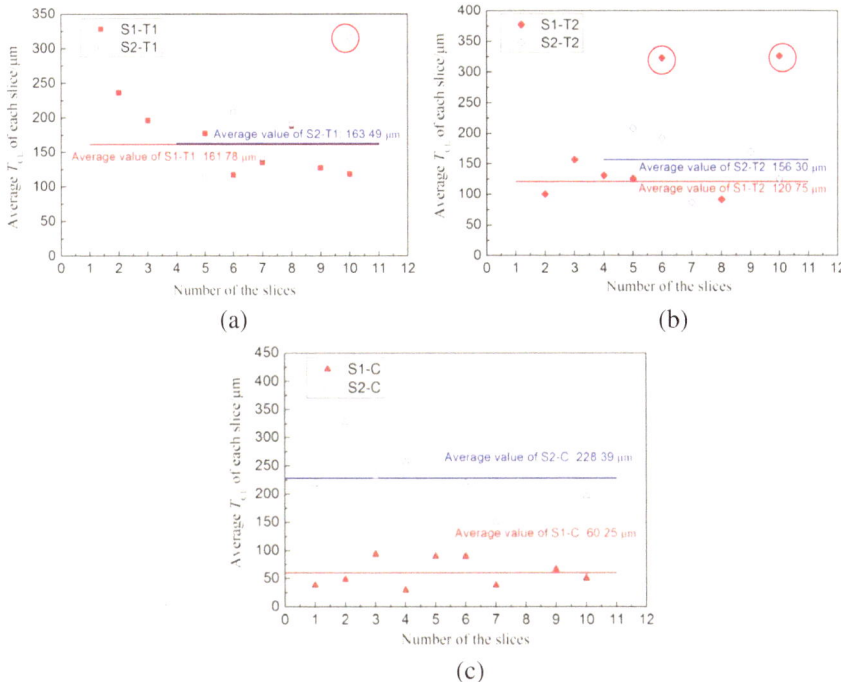

Fig. 6 Average thickness of corrosion layer (T_{CL}) of each slice; **a** Series T1, **b** Series T2 and **c** Series C

the tensile region, while C is in the compressive region. Compressive stress could reduce the number of pores, change the distribution of pore diameters, and reduce the average diameter of micro-pores, which in turn reduced the water transport capacity of concrete [14]. Tensile stress will expand the micro-pores and connect some pores. This accelerates the ingression of chloride ions, water [18], and steel corrosion. With regard to S2-C, some local weaknesses, such as corrosion-induced cracking, were found in the surrounding concrete. This might result in the faster ingression of chloride ions into the concrete, and to the earlier corrosion of the bar. The cracks near S2-C (shown in Fig. 6) accelerated the corrosion of S2-C.

3.2 Migration of Fe Ions in the Concrete Cracks

To investigate the distribution of corrosion products in the concrete, the Fe across the CP region was analysed by EDS along the analytical line shown in Fig. 7a. The results are shown in Fig. 7b, where the horizontal axis represents the distance from the starting point of the analytical line, and the vertical axis is the count of photoelectrons per second (CPS), which reflects the contents of Fe. In the CP region,

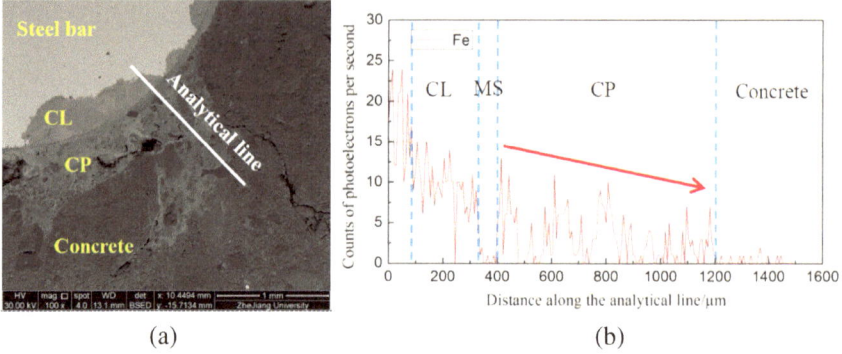

Fig. 7 Distribution of Fe in CP region; **a** BSE image of Sample S2-T2-10 and analytical line, and **b** distribution of Fe in the analytical line analyzed by EDS

it can be observed that the Fe content decreases with the increase of distance from the steel bar. Similar results were also observed in previous studies [5, 6]. After the corrosion of the steel bar, Fe ions accumulate in the steel/concrete interface. Then, the gradient of Fe ions between the steel/concrete interface and the surrounding concrete, induces the migration of Fe ions from the steel/concrete interface to the surrounding concrete, and forms CP. The gradient of Fe ions, between the CP region and the surrounding un-migrated concrete, drives the further migration of Fe ions.

Previous studies [4, 7, 9] have revealed that the corrosion products do not fill the corrosion-induced cracks. Furthermore, more corrosion products can be found on the concrete near the edge of the cracks [7, 9]. Figure 8a illustrates the BSE images of corrosion products in the corrosion-induced cracks obtained from sample S2-T2-10. Again, it can be seen that the corrosion products do not fill the crack. The distribution of Fe across the crack, analysed by EDS along an analytical line, is shown in Fig. 8b. Fe was not found in the crack, and this is consistent with the results of previous studies [4, 7, 9]. However, the corrosion products near the edge of the cracks do not increase according to Fig. 8b. The reason will be discussed in Sect. 3.3.

3.3 Migration of Fe Ions in the Air Voids of the Concrete

In previous studies [8, 11], it was observed that the corrosion products could accumulate in the air voids close to the steel/concrete interface. Figure 9a shows the BSE images of corrosion products in the air voids, which were obtained from sample S2-T2-10. It can be seen that the corrosion products did not fill the air voids, even though the CP was distributed on the surrounding concrete. The distribution of Fe across the air void is shown in Fig. 9b. It can be seen that Fe was distributed in the surrounding concrete; however, little Fe was found in the air void.

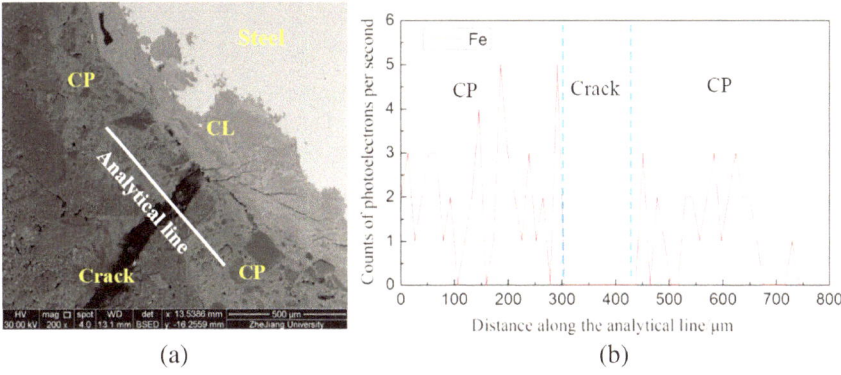

(a)   (b)

Fig. 8 Corrosion product distributions in cracks; **a** BSE image of Sample S2-T2-10 and analytical line, and **b** distribution of Fe across the crack analyzed by EDS

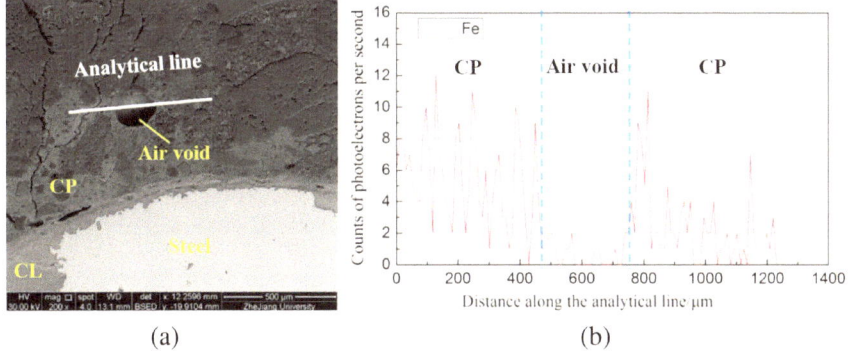

(a)   (b)

Fig. 9 Corrosion product distributions in air voids; **a** BSE image of Sample S2-T2-10 and analytical line, and **b** distribution of Fe across the air voids analyzed by EDS

According to the above discussion, corrosion products did not accumulate on the edges of cracks nor did they fill the air voids. Similar phenomena can be observed in most samples considered in this study. This seems to disagree with previous studies [7–9, 11] and might have been caused by different environments to which the specimens were exposed. In previous studies [7–9, 11], the steel bar in the specimens was corroded in the NaCl solution. Some corrosion products dissolved in the solution and were carried away from the rebar by the solution. Thereby, the corrosion products penetrated the edges of cracks and filled the air voids. In this study, the concrete beams were in service under coastal atmospheric conditions. Thus, they were subjected to chloride ions from the air, but not to the NaCl solution. Thereby, the corrosion products could not be easily carried away. Some rain water may have penetrated into the cracks; however, it did not carry the corrosion products away to penetrate the edges of cracks or fill the air voids, as shown in Figs. 8 and 9.

4 Conclusions

A beam exposed to the coastal atmospheric environment for more than 40 years was used to study corrosion distribution in concrete. The following conclusions can be drawn from this study.

(1) The steel corrosion and corrosion-product-filled paste propagates faster in the tension region than in the compressive region of the beam.
(2) The migration of Fe ions from the steel/concrete interface to the surrounding concrete forms CP.
(3) In the atmospheric environment, the corrosion products neither fill the cracks or air voids nor do they accumulate on the edge of the cracks or inside the air voids.

References

1. Marcotte, T. D., & Hansson, C. M. (2007). Corrosion products that form on steel within cement paste. *Materials and Structures, 40*, 325–340.
2. Asami, K., & Kikuchi, M. (2003). In-depth distribution of rusts on a plain carbon steel and weathering steels exposed to coastal–industrial atmosphere for 17 years. *Corrosion Science, 45*, 2671–2688.
3. Duffo, G, S., Morris, W., Raspini, I., Saragovi, C. (2004). A study of steel rebars embedded in concrete during 65 years. *Corrosion Science, 46*, 2143–2157.
4. Michel, A., Pease, B. J., Geiker, M. R., Stang, H., & Olesen, J. F. (2011). Monitoring reinforcement corrosion and corrosion-induced cracking using nondestructive X-ray attenuation measurements. *Cement and Concrete Research, 41*, 1085–1094.
5. Michel, A., Pease, B. J., Peterová, A., Geiker, M. R., Stang, H., & Thybo, A. E. (2014). Penetration of corrosion products and corrosion-induced cracking in reinforced concrete materials: Experimental investigations and numerical simulations. *Cement and Concrete Composites, 47*, 75–86.
6. Zhao, Y., Wu, Y., & Jin, W. (2013). Distribution of millscale on corroded steel bars and penetration of steel corrosion products in concrete. *Corrosion Science, 66*, 160–168.
7. Zhao, Y., Ding, H., & Jin, W. (2014). Development of the corrosion-filled paste and corrosion layer at the steel/concrete interface. *Corrosion Science, 87*, 199–210.
8. Zhao, Y., Zhang, X., & Jin, W. (2017). Influence of environment on the development of corrosion product-filled paste and a corrosion layer at the steel/concrete interface. *Corrosion Science, 124*, 1–9.
9. Zhao, Y., Yu, J., Wu, Y., & Jin, W. (2012). Critical thickness of rust layer at inner and out surface cracking of concrete cover in reinforced concrete structures. *Corrosion Science, 59*, 316–323.
10. Kim, K. H., Jang, S. Y., Jang, B. S., & Oh, B. H. (2010). Modeling mechanical behavior of reinforced concrete due to corrosion of steel bar. *ACI Structural Journal, 107*, 106–113.
11. Wong, H. S., Zhao, Y. X., Karimi, A. R., Buenfeld, N. R., & Jin, W. L. (2010). On the penetration of corrosion products from reinforcing steel into concrete due to chloride-induced corrosion. *Corrosion Science, 52*, 2469–2480.
12. Care, S., Nguyen, Q. T., L'Hostis, V., & Berthaud, Y. (2008). Mechanical properties of the rust layer induced by impressed current method in reinforced motar. *Cement and Concrete Research, 38*, 1079–1091.

13. Petre-Lazar, I. (2020). *Aging assessment of concrete structures submitted to steel corrosion (PhD Thesis)*, Laval University, Quebec, Canada.
14. Wu, J., Li, H., Wang, Z., & Liu, J. (2016). Transport model of chloride ions in concrete under loads and drying-wetting cycles. *Construction and Building Materials, 112*, 733–738.
15. Song, G., Shayan, A. (1998). *Corrosion of steel in concrete: Causes, detection and prediction: A State-of-the-art review*. ARRB Transport Research, Limited.
16. Gokce, A., Nagataki, S., Saeki, T., & Hisada, M. (2011). Identification of frost-susceptible recycled concrete aggregates for durability of concrete. *Construction and Building Materials, 25*, 2426–2431.
17. Arredondo-Rea, S. P., Corral-Higuera, R., Neri-Flores, M. A., Gómez-Soberón, J. M., Almeraya-Calderón, F., Castorena-González, J. H., & Almaral-Sánchez, J. L. (2011). Electrochemical Corrosion and electrical resistivity of reinforced recycled aggregate concrete. *International Journal of Electrochemical Science, 6*, 475–483.
18. Dong, J., Zhao, Y., Gan, Y., Ding, C., & He, Q. (2017). Comparison of chloride penetration into surface-treated concrete in artificial and natural environments. *Advances in Structural Engineering, 20*, 1315–1324.

Service Life Prediction of Fatigue Damaged Reinforced Concrete Beams Under Seawater Erosion and Freeze–Thaw Cycles

Zijian Liu and Huiyun Cao

Abstract A reasonable and applicable chloride ion diffusion model for fatigue damaged concrete structures exposed to seawater freeze–thaw environments was proposed, And the service life of reinforced concrete beams with fatigue load levels of 0.2 and 0.3, fatigue loading times of 200,000–400,000 under seawater immersion alternating freeze–thaw cycle environment were predicted. Results show that the expected service life of concrete beams without fatigue damage under seawater immersion alternating freeze–thaw cycle was 6.0 years, respectively, which was far less than the requirements of existing specifications for the service life of concrete structures. The service life of concrete structures further decreased as the fatigue load increased. For concrete beams with a fatigue load level of 0.3 loaded 400,000 times, under the alternating freeze–thaw cycles of seawater immersion, the expected service life is only one year.

Keywords Service life prediction · Freeze–thaw · Chloride diffusion · Fatigue damaged

1 Introduction

According to statistical research, steel corrosion was the main cause of failure in reinforced concrete (RC) components [1]. The steel corrosion of RC bridge structures in coastal areas was mainly caused by chloride ion erosion. For the steel bars in concrete, there was a limit concentration of chloride ions required to induce pitting corrosion of steel. When the chloride ion reached the limit concentration on the steel surface, the aggressive chloride ion would activate the passivation film on the steel surface. At the same time, the oxygen-containing environment would promote pitting corrosion of steel with higher redox potential. Therefore, it was of great significance

Z. Liu (✉) · H. Cao
College of Civil and Architectural Engineering, North China University of Science and Technology, Tangshan 063210, Hebei, China
e-mail: sunsword@163.com

© The Author(s) 2025
D. Li and Y. Zhang (eds.), *Advances in Frontier Research on Engineering Structures II*,
Lecture Notes in Civil Engineering 535, https://doi.org/10.1007/978-981-97-6238-5_17

to establish a chloride ion diffusion model in concrete that accurately reflected the effect of external environment for predicting the service life of concrete structures in chloride erosion environment and the durability design of concrete structures.

The study on the diffusion behavior of chloride ions in concrete was first carried out by Collepardi in 1970. He first published the results of chloride ion diffusion based on Fick's second law in 1972 [2]. It was not until the 1990s that researchers gradually realized that the diffusion coefficient D was affected by many factors such as time. The results originally predicted by D as a certain value became too conservative and could not accurately reflect the actual diffusion of chloride ions in concrete [3]. Due to the diverse working environments of bridges in coastal areas, predicting their service life also requires considering specific influencing factors. Many scholars [4, 5] have established a modified model of Fick's second law by considering one or several influencing factors on the diffusion coefficient of chloride ions.

In this paper, the service life of concrete structures will be predicted based on the chloride ion concentration and diffusion coefficient [6, 7] obtained from the seawater erosion and freeze–thaw cycle tests of fatigue damaged reinforced concrete beams completed by the author.

2   Chloride Ion Diffusion Model Based on Fick'S Second Law

Fick's second law assumed that the pores in concrete were uniformly distributed, and the diffusion of chloride ions in concrete was one-dimensional diffusion. The concentration gradient only changed along the direction from the exposed surface to the surface of the steel bar. The surface concentration of the concrete was constant, and the concrete was a semi-infinite medium. The chloride ion concentration formula was shown in Eq. (1).

$$\frac{\partial C}{\partial t} = D\frac{\partial^2 C}{\partial x^2} \tag{1}$$

where C is the concentration of chloride ion which was generally expressed by the mass percentage of chloride ion in cementitious materials, t is exposure time of concrete structures to chloride environment, D is chloride diffusion coefficient, x is depth of chloride ion erosion. Assuming that the chloride ion concentration on the concrete surface reaches a constant saturation after a certain period of time, and the chloride ion diffusion coefficient is a certain value, the boundary conditions and initial conditions of Eq. (1) can be obtained:

Boundary conditions:$C_{(x = 0, t)} = C_s$, initial conditions: $C_{(x, t = 0)} = C_0$.

Thus the analytic solution of (1) can be obtained.

$$C_{(x,t)} = C_0 + (C_s - C_0)[1 - erf(\frac{x}{2\sqrt{D_c t}})] \tag{2}$$

where $C_{(x,t)}$ is the total content of chlorides at a depth x and exposure time t, C_0 is initial chloride ion concentration of concrete, C_s is surface chloride ion concentration, D_c is the diffusion coefficient of concrete in time period t.

It can be obtained from Eq. (2) that:

$$t = \frac{X^2}{4D_c}\left[erf^{-1}\left(\frac{C_S - C_{(X,t)}}{C_S - C_0}\right)\right]^{-2} \tag{3}$$

Substituting $C_{(a_s,t)} = C_c$ into Eq. (3), the time of corrosion can be obtained.

3 Chloride Ion Diffusion Model for Fatigue Damaged Concrete Structures Exposed to Seawater Freeze–thaw Environments

It can be known from many research results [4, 5] that fatigue damage and freeze–thaw cycles will increase the chloride ion diffusion coefficient of concrete, and these two factors have a coupling effect. Therefore, the fatigue damaged reinforced concrete structure under the action of seawater freeze–thaw cycle, the chloride ion diffusion coefficient of concrete:

$$D(t) = F_d \cdot F_t \cdot D_0 \cdot \left(\frac{t_0}{t}\right)^m \tag{4}$$

where F_d is amplification factor of chloride ion diffusion in concrete due to fatigue damage, Ft is amplification coefficient of chloride diffusion in concrete due to freeze–thaw cycles, D_0 is the chloride ion diffusion coefficient of concrete $(m^2 s^{-1})$ when the exposure time of chloride erosion environment is 0, t_0 is the age of concrete at the end of curing (s), m is a time factor, constant, which is related to the mix ratio of concrete and the temperature of the environment. The time factor m determines the change rate of the instantaneous diffusion coefficient $D(t)$ with time t. The value range of m is (0, 1).

The formula for calculating the time factor m specified in the 'Durability Evaluation Standard for Concrete Structures' (CECS.220:2007)is:

$$m = 0.2 + 0.4(\%FA/50 + \%SG/70) \tag{5}$$

where %FA is percentage of fly ash in cementitious materials, %SG is percentage of slag in cementitious materials. This paper was finally determined according to the provisions of 'Concrete Structure Durability Evaluation Standard'.

$$m = 0.30 \tag{6}$$

Defines the diffusion coefficient of chloride ions in concrete Da:

$$D_a = \frac{\int_{t_0}^{t} D_{(t)} dt}{\int_{t_0}^{t} dt} = \frac{T}{t - t_0} \tag{7}$$

where D_a is the average diffusion coefficient of concrete in time period t. It can be known from the author's past work[6, 7] that the average diffusion coefficient D_a of uncracked concrete in fatigue flexural members is approximately linearly related to the fatigue bending moment at the location of concrete and increases with the increase of fatigue loading times. Since the initial value D_0 of the chloride diffusion coefficient of concrete does not change with the time of chloride erosion, it can be obtained by fitting the data of the average diffusion coefficient D_a of concrete with the change of fatigue bending moment.

$$F_d = \begin{cases} 1 + a.b^{\frac{N}{N_0} - 1}.\theta_f & (w < 0.08\,\text{mm}) \\ 1 + a.b^{\frac{N}{N_0} - 1}.\theta_f + c & (w \geq 0.08\,\text{mm}) \end{cases} \tag{8}$$

where N is the number of fatigue loading, N_0 is 200,000 times, θ_f is the ratio of fatigue bending moment to static ultimate bending moment, a, b and c are constants fitted from experimental data. In this paper, $a = 6.73, b = 3, c = 2.69$. w is the fatigue crack width.

Substituting Eq. (4) into Eq. (7):

$$D_a = \frac{F_d \cdot F_t \cdot D_0 \cdot \int_{t_0}^{t} \left(\frac{t_0}{t}\right)^{\alpha} dt}{t - t_0} \tag{9}$$

Substituting the diffusion coefficient of the reference specimen into Formula (9), it can be obtained that $D_0 = 9.787E - 13$ From the test results, it can be seen that in the process of seawater dry–wet cycle, with the hydration of concrete, the fatigue cracks of reinforced concrete beams gradually heal; under the alternating action of seawater immersion and freeze–thaw cycles, the fatigue cracks have a tendency to gradually widen[6]. The beam specimens with the same degree of fatigue damage under different erosion environments were selected for comparison, and the effect of 300 freeze–thaw cycles on the chloride ion diffusion coefficient was obtained:

$$F_t(300) = \begin{cases} 1.83 \ (w < 0.08\,\text{mm}) \\ 3.98 \ (w \geq 0.08\,\text{mm}) \end{cases} \tag{10}$$

Because the damage of freeze–thaw cycle to concrete accumulates with time, for the convenience of calculation, it is assumed that the damage of each freeze– thaw cycle to concrete and the influence on chloride ion diffusion coefficient are the same.

$$F_t(t) = 1 + \frac{F_t(300) - 1}{300}.n.\frac{t}{365 \times 24 \times 3600} \tag{11}$$

Where $F_t(t)$ is the influence of freeze–thaw cycles in time period t in natural environment, n is the number of freeze–thaw cycles in one year in the natural environment.

4 Service Life Prediction of Reinforced Concrete Structures with Fatigue Damage Under Seawater Erosion and Freeze–thaw Action

In the concrete structure under the environment of chlorine erosion, in addition to the chloride ions contained in the concrete itself, other invaded chloride ions need to gather on the surface of the concrete, and then invade the concrete through diffusion, convection and other ways. The change of Cs value of chloride ion concentration on the surface of concrete is mainly related to the age of concrete, the adsorption performance of chloride ion by the material properties of concrete itself, and the environmental conditions (such as the chloride concentration of seawater, the location of the structure, etc.). During a certain period of time at the beginning of chlorine erosion, the chloride ion concentration on the surface of concrete increases gradually with time. However, with the increase of erosion time, the surface concentration value should be basically constant after a certain time.

In the marine environment, the concrete structures in the underwater area, the water level fluctuation area, the splash area and the atmospheric area have different external chloride ion pathways. The chloride ion in the underwater area mainly comes from seawater, and its concentration is relatively constant. The chloride ion sources in the water level fluctuation zone and the splash zone are from waves or sprays, and the maximum chloride ion concentration on the surface of the structure is basically 0.7% -0.9% (mass ratio to concrete). The chloride ion source in the atmospheric area of the seaside is mainly the salt fog of the surrounding marine environment. The maximum chloride ion concentration on the surface of the structure is also relatively stable, basically below 0.4% [8].

At the same time, the composition of the cementitious material of the concrete itself also has an effect on the surface chloride ion concentration. Because the cementitious materials such as slag and fly ash have higher adsorption capacity for chloride ions than cement, the surface chloride ion concentration of concrete mixed with slag and fly ash is generally higher than that of ordinary concrete.

Shi Huisheng [9] et al. sampled concrete structures in the marine environment of the Hangzhou Bay area. The results show that the chloride ion concentration on the surface of concrete in the splash zone and the water level fluctuation zone is about 0.6% ~ 0.7% (mass ratio to concrete), while the chloride ion concentration of concrete in the atmospheric zone is basically below 0.4%. Combined with the test results of this paper, the chloride ion concentration on the surface of concrete is taken as a fixed value of 0.7%.

That is:

$$C_s = 0.7\% \times 2475.1 = 17.3 \, \text{kg/m}^3 \tag{12}$$

Many experimental studies and practical engineering investigations [10, 11] had confirmed the relationship between the risk of steel corrosion caused by Cl-and the concentration of Cl-, and pointed out that the critical chloride ion concentration of steel corrosion (referred to as the critical concentration of chloride ion) is not constant, which is affected by the thickness of concrete cover and environmental conditions. For the service life prediction or durability design of concrete, the critical concentration of chloride ion is generally 0.4% ~ 1.0% of the mass of cementitious material or 0.07% ~ 0.18% of the mass of concrete. For the completely submerged environment and atmospheric environment, a high value can be taken, and the critical concentration of chloride ions in the water level fluctuation zone and the splash zone should be taken as a lower value. In this paper, 0.10% of the mass of concrete is selected as the critical concentration. According to the concrete mix ratio in this paper, the concrete density is 2475.1 kg/m^3. The critical chloride ion content is:

$$C_c = 0.10\% \times 2475.1 = 2.48 \, \text{kg/m}^3 \tag{13}$$

The initial chloride ion concentration on the surface of the specimen at the beginning of seawater erosion is:

$$C_0 = 0.075\% \times 2475.1 = 1.86 \, \text{kg/m}^3 \tag{14}$$

At the beginning of seawater erosion, the initial chloride ion concentration of 0.03 m below the surface of the specimen is:

$$C_{(x,0)} = 0.019\% \times 2475.1 = 0.47 \, \text{kg/m}^3 \tag{15}$$

The time when the steel bar begins to rust is:

$$T = \int_{t_0}^{t} D(x) \cdot dx \tag{16}$$

Replacing $D_c \cdot t$ in Eq. (2) with T in Eq. (16), it can be obtained:

$$C_{(x,t)} = C_{(x,0)} + (C_s - C_0)\left[1 - erf\left(\frac{x}{2\sqrt{T}}\right)\right] \tag{17}$$

When the chloride ion concentration on the surface of the steel bar reaches the critical concentration C_c, Substituting $C_{(a_s,t)} = C_c$ into Eq. (17), the time of corrosion can be obtained. For the concrete structure in the seawater dry–wet cycle environment, the diffusion model $D(t)$ can be regarded as a continuous function, and the time when the corrosion occurs can be directly obtained by Eq. (17) so as to obtain the service life of the reinforced concrete structure.

Table 1 Life prediction results of concrete beam specimens subjected to seawater freeze–thaw cycles

Number	Fatigue bending moment (Mu)	Fatigue times	Fatigue cracks Width (mm)	Expected service life (a)	Life decline (%)
C1	0	0	0	6.0	0
C3	0.20	200,000	0	3.2	46.7
	0.20	200,000	0.11	1.4	76.7
C6	0.30	200,000	0	2.6	56.7
	0.30	200,000	0.11	1.3	78.3
C7	0.28	400,000	0	1.5	75.0
	0.30	400,000	0.10	1.0	83.3

Different from the laboratory simulation of freeze–thaw environment, the concrete structure under the action of freeze–thaw cycle in the actual environment is discontinuous. According to the statistics of meteorological records, between 2011 and 2015, the number of days when the temperature changed around 0 degrees Celsius in Tianjin Port was 269 days, with an annual average of 67.25 days. The freeze–thaw cycle effect in 1 year is equivalent to the whole year, so as to facilitate the calculation. Taking the meteorological conditions of Tianjin Port as an example, the service life of concrete structures under seawater freeze–thaw cycles is predicted in Table 1

It can be seen from Table 1 that under the action of alternating freeze–thaw cycles of seawater immersion, the service life of the uncracked part of the fatigue-damaged concrete member decreases with the increase of fatigue bending moment. The same fatigue bending moment decreases with the increase of fatigue loading times, and the service life of the cracked part is less than that of the uncracked part. The influence of fatigue damage on the service life of the structure under chlorine erosion environment is very obvious.

5 Conclusion

A reasonable chloride ion diffusion model based on the Fick's second law, is proposed for the fatigue damage concrete structure and the influence of seawater freeze–thaw environment. Results show that the expected service life of concrete beams without fatigue damage under seawater immersion alternating freeze–thaw cycle was 6.0 years, respectively, which was far less than the requirements of existing specifications for the service life of concrete structures. The service life of concrete structures further decreased as the fatigue load increased. For concrete beams with a fatigue load level of 0.3 loaded 400,000 times, under the alternating freeze–thaw cycles of seawater immersion, the expected service life is only one year. Therefore, the influence of fatigue damage on the service life of structures under chlorine erosion environment must be taken seriously.

References

1. Ni, J., & He, J. (1990). *Corrosion and protection of concrete and reinforced concrete.* Chemical Industry Press.
2. Collepardi, M., Marcialis, A., & Turriziani, R. (1972). Penetration of chloride ions into cement pastes and concrete. *Journal of the American Ceramic Society, 55*(10), 534–535.
3. Nilsson, L. O., Poulsen, E., Sandberg, P., Srensen, H. E., Klinghoffer, E. (1996). Chloride penetration into concrete, state-of-the-art report, HETEK, Report no.53.
4. Liu, D., & Wang, C. (2023). A review of concrete properties under the combined effect of fatigue and corrosion from a material perspective. *Construction and Building Materials, 369*, 130489.
5. Zhang, T., & Zhang, X. (2023). Experimental research on fatigue performance of reinforced concrete T-shaped beams under corrosion-fatigue coupling action. *Materials, 16*(3), 1257.
6. Liu, z., Diao, B., Zheng, X. (2015). Effects of seawater corrosion and freeze-thaw cycles on mechanical properties of fatigue damaged reinforced concrete beams. *Advances in Materials Science and Engineering.*
7. Wu, J., Diao, B., Zhang, W., Ye, Y., Liu, Z., & Wang, D. (2018). Chloride diffusivity and service life prediction of fatigue damaged RC beams under seawater wet-dry environment. *Construction and Building Materials, 171*, 942–949.
8. Fluge, F. (2001). *Marine chlorides-A probabilistic approach to drive provisions for EN206-1,* 3rd Workshop on Service Life Design of Concrete Sturctures-from Theory to Standardisation, Troms-Φ, Norway.
9. Shi, H., & Wang, Q. (2004). Study on service life prediction of marine concrete. *Journal of Building Materials, 7*(2), 161–167.
10. Maage, M. (1993). *Chloride penetration in high performance concrete exposed to marine environment.* 3rd International Symposium on Utilization of HPC. Li llehammer Norway.
11. Bormforth, P. B. (1999). The derivation of input data for modelling chloride ingress from eight-year UK coastal exposure trials. *Magazine of Concrete Research, 51*(2), 87–96.

Research on Fatigue Rigidity Degradation of Prestressed Concrete Box Girders

Hongyu Zhou, Haoda Wang, Quanzhou Ma, and Shuai Han

Abstract The intercity high-speed railway is developing towards an operating mode of ultra high speed, large capacity, and high frequency, and the frequency of driving loads borne by structures has also significantly increased, which is bound to accelerate the accumulation and development process of fatigue damage on elevated bridge vehicles. Therefore, this article takes the 32 m prestressed concrete box girder of Tianjin High Speed Railway as the research object, and uses similarity theory to design and manufacture three model test beams. To study the fatigue stiffness degradation of prestressed concrete box beams.

Keywords Box girder · Accumulated damage · Stiffness degradation

1 Introduction

In recent years, some large-tonnage fatigue loading equipment has been constructed in China, which has made continuous progress in loading control technology, boundary condition simulation, data collection, and other aspects, directly promoting research work in the field of fatigue damage. Cao Xiang and Zou Xiaobo [1–3] studied the fatigue performance of complex concrete box girder components under vehicular loads, Li Jinzhou and others [4–13] studied the fatigue performance of the lower beam of heavy-duty railways. The study showed that under the vehicle load spectrum, as the axle load of the vehicle increases, the impact force borne by the beam increases. The fatigue life of the bridge depends on the fatigue life of the concrete material. This experiment produced three scaled model beams, numbered XLF1, XLF2, and XLF3. The three model beams are identical, and the three models are loaded with different fatigue upper limits and the same fatigue lower limits.

H. Zhou (✉) · H. Wang · S. Han
Faculty of Architecture, Civil and Transportation Engineering, Beijing University of Technology, Beijing 100124, China
e-mail: ZHYktztgyx@163.com

Q. Ma
Tianjin Intelligent Rail Transit Research Institute, Tianjin 301700, China

© The Author(s) 2025   213
D. Li and Y. Zhang (eds.), *Advances in Frontier Research on Engineering Structures II*,
Lecture Notes in Civil Engineering 535, https://doi.org/10.1007/978-981-97-6238-5_18

2 Calculation of Dynamic and Static Stiffness

2.1 Calculate Stiffness Back from Deflection

The mid span deflection of a simply supported beam is related to bending moment, axial force, and torsion effects, as shown in formula (1), but the contribution of axial force and torsion to the deflection is relatively small and can be ignored. Simplify the stiffness calculation formula for prestressed concrete simply supported beams, as shown in formula (2).

$$f = \sum \int \frac{M_0 M_P}{B} dx + \sum \int \frac{N_0 N_P}{EA} dx + \sum \int \frac{k V_0 V_P}{GA} dx \tag{1}$$

$$B_\delta = \alpha \frac{ML^2}{f} \tag{2}$$

Taking into account the experimental loading plan and the loading form of the box girder structure, α Take as 23/216. The initial stiffness is calculated back from the corresponding deflection value when subjected to fatigue loading of 100,000 times. The remaining stiffness of n fatigue cycles is calculated from the fatigue deflection under the fatigue cycles, as shown in formula (3).

$$B_{\delta n} = \frac{23}{216} \cdot \frac{PL^3}{4 f_n} = 0.524 \frac{P}{f_n} \tag{3}$$

Calculate the fatigue stiffness value, as shown in. Table 1.

Table 1 Calculation results of test static stiffness (unit: $M^2 \cdot Nm$)

Beam number	XLF1					
Number of cycles	0	30	50	100	150	200
Static stiffness	7.61	3.83	3.61	3.04	2.80	2.47
Beam number	XLF2					
Number of cycles	0	30	50	100	150	200
Static stiffness	7.23	5.14	3.61	2.87	2.78	2.03
Beam number	XLF3					
Number of cycles	0	30	50	100	150	200
Static stiffness	7.18	3.89	2.96	2.45	1.63	1.34

2.2 Backward Calculation of Stiffness from Fundamental Frequency

After research, it was found that the natural frequency of a given bridge beam is only related to the stiffness EI of the structure. Therefore, the damage changes in structural stiffness can be determined by measuring the natural frequency of the bridge structure. Select experimental data and calculate the dynamic stiffness values of prestressed box beams under different fatigue cycles based on the fundamental frequency back calculation of stiffness, as shown in Table 2.

The dynamic stiffness of test beams XLF1, XLF2, and XLF3 is shown in Fig. 1

Table 2 Calculation results of test dynamic stiffness (unit: $M^2 \cdot Nm$)

Beam number	XLF1					
Number of cycles	0	30	50	100	150	200
Dynamic stiffness	22.21	21.57	21.6	21.58	21.29	20.99
Beam number	XLF2					
Number of cycles	0	30	50	100	150	200
Dynamic stiffness	20.86	20.4	20.4	20.36	20.3	20.04
Beam number	XLF3					
Number of cycles	0	30	50	100	150	200
Dynamic stiffness	20.31	19.55	19.48	19.59	19.34	19

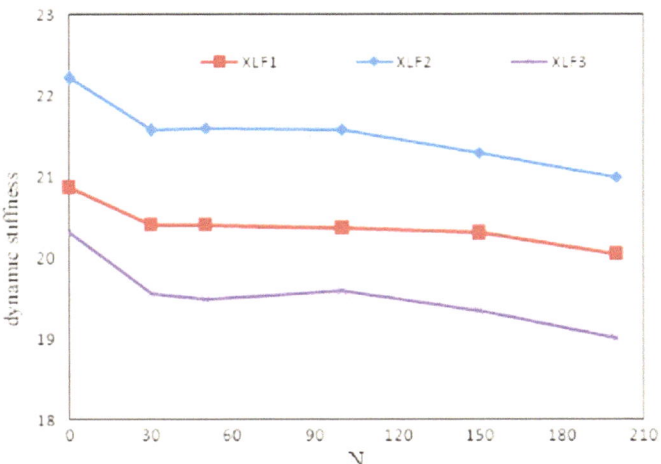

Fig. 1 Comparison of dynamic stiffness damage changes of box girder

Table 3 Comparison of dynamic and static stiffness results

Beam number	XLF1					
Number of cycles	0	30	50	100	150	200
Static stiffness	7.61	3.83	3.61	3.04	2.80	2.47
Dynamic stiffness	22.21	21.57	21.6	21.58	21.29	20.99
Beam number	XLF2					
Number of cycles	0	30	50	100	150	200
Static stiffness	7.23	5.14	3.61	2.87	2.78	2.03
Dynamic stiffness	20.86	20.4	20.4	20.36	20.3	20.04
Beam number	XLF3					
Number of cycles	0	30	50	100	150	200
Static stiffness	7.18	3.89	2.96	2.45	1.63	1.34
Dynamic stiffness	20.31	19.55	19.48	19.59	19.34	19

2.3 Comparative Analysis of the Two

The comparison between the static stiffness value calculated from deflection inversion and the dynamic stiffness value calculated from fundamental frequency inversion is shown in Table 3. It was found that the dynamic stiffness of the prestressed concrete box model beam is greater than the static stiffness, but the degradation law of the two is the same. As the fatigue number increases, both the static and dynamic stiffness gradually decrease.

3 Determination of Fatigue Stiffness Degradation Equation

The stiffness degradation of prestressed concrete beams is an important manifestation of structural damage, which can reflect the damage degree of bridge structures. The degradation law of residual stiffness of beams during fatigue is as follows:

$$D_E = \frac{B_0 - B_n}{B_0 - B_N} \tag{4}$$

Obviously, B_n is related to the cycle ratio n/N, so it can be assumed that DE is a function of the cycle ratio n/N, that is:

$$D_E = \frac{B_0 - B_n}{B_0 - B_N} = g\left(\frac{n}{N}\right) \tag{5}$$

Organized:

$$\frac{B_n}{B_0} = 1 - \left(1 - \frac{B_N}{B_0}\right)g\left(\frac{n}{N}\right) \tag{6}$$

Function $g\left(\frac{n}{N}\right)$ should satisfy that when the cyclic ratio $n/N = 0$, the damage of the bridge component is 0, which can result in $g\left(\frac{n}{N}\right) = 0$.

At the same time, when $n/N = 1$, it is generally considered that the bridge is fatigue damaged, then $g\left(\frac{n}{N}\right) = 1$. When BN data cannot be obtained through experiments, it is recommended to use formula (7) to theoretically calculate the BN value.

$$B_N = \frac{\beta_P \beta_1 \beta_2 M E_c I_0}{\beta_2 M_f + \beta_P (M - M_f)} \tag{7}$$

Use trial calculations to select a function that meets the above requirements, and the function formula is as follows:

$$g\left(\frac{n}{N}\right) = u\left(\frac{n}{N}\right)^v + (1-u)\left[1 - \left(1 - \frac{n}{N}\right)^w\right] \tag{8}$$

Use MATLAB to fit the experimental data in Table 1 and determine each fitting parameter. The fitted curves are shown in Figs. 2, 3 and 4, and the stiffness degradation function parameters obtained from the fitting are shown in Table 4

Take the mean of each parameter in the table and substitute it into formula (8) to obtain the stiffness degradation function as

$$g\left(\frac{n}{N}\right) = 0.8577\left(\frac{n}{N}\right)^{0.1985} + 0.1423\left[1 - \left(1 - \frac{n}{N}\right)^{0.0889}\right] \tag{9}$$

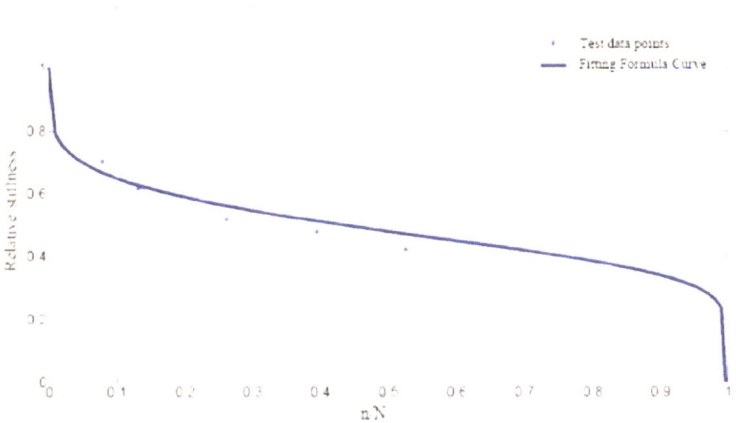

Fig. 2 Fitting curve of model beam xlf1 rigidity

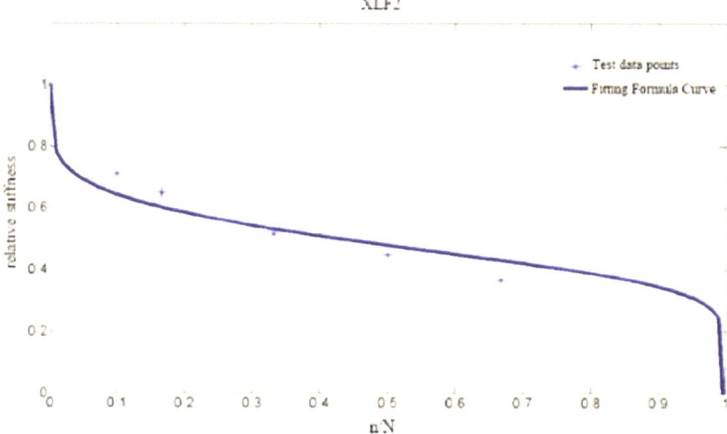

Fig. 3 Fitting curve of model beam xlf2 rigidity

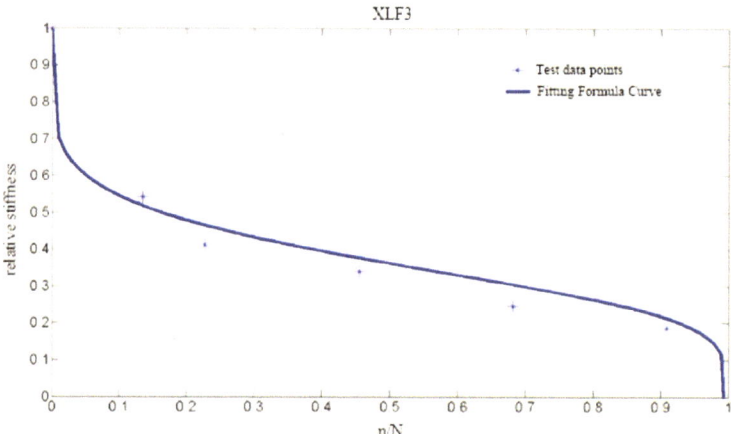

Fig. 4 Fitting curve of model beam xlf3 rigidity

Table 4 Table of fitting parameters of fatigue test beam

XLF1	Fit parameters	u	v	w	R^2
	/	0.8012	0.2119	0.0903	0.9752
XLF2	Fit parameters	u	v	w	R^2
	/	0.7963	0.2011	0.0873	0.9841
XLF3	Fit parameters	u	v	w	R^2
	/	0.9756	0.1824	0.0891	0.9803
Average		0.8577	0.1985	0.0889	/

Substitute the degradation function formula (9) into the stiffness degradation formula (6) to obtain the stiffness degradation formula

$$\frac{B_n}{B_0} = 1 - \left(1 - \frac{B_N}{B_0}\right)\left\{0.8577\left(\frac{n}{N}\right)^{0.1985} + 0.1423\left[1 - \left(1 - \frac{n}{N}\right)^{0.0889}\right]\right\} \quad (10)$$

4 Fatigue Stiffness Degradation Verification

Substitute the data in Table 2 to verify the fitting degree of the stiffness degradation formula (10), as shown in Figs. 5, 6, and 7.

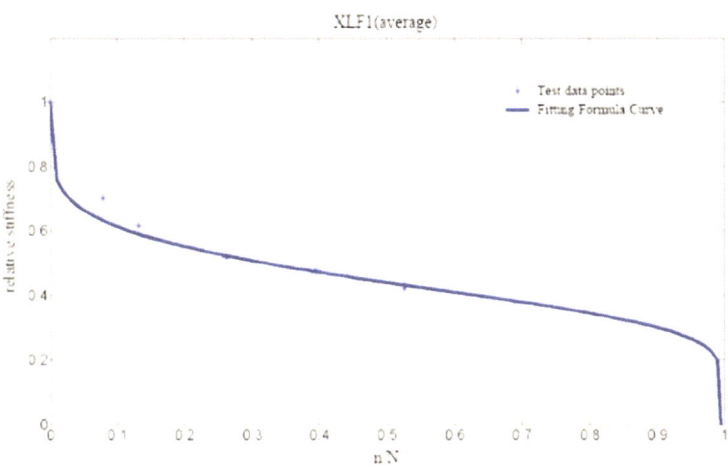

Fig. 5 Theoretical formula fits with xlf1 test data

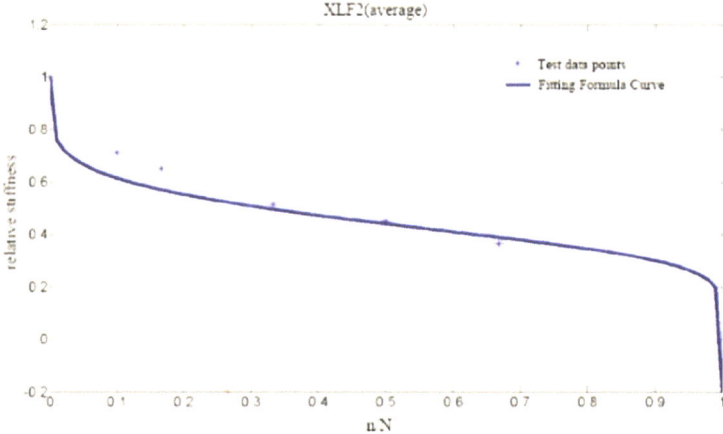

Fig. 6 Theoretical formula fits with xlf2 test data

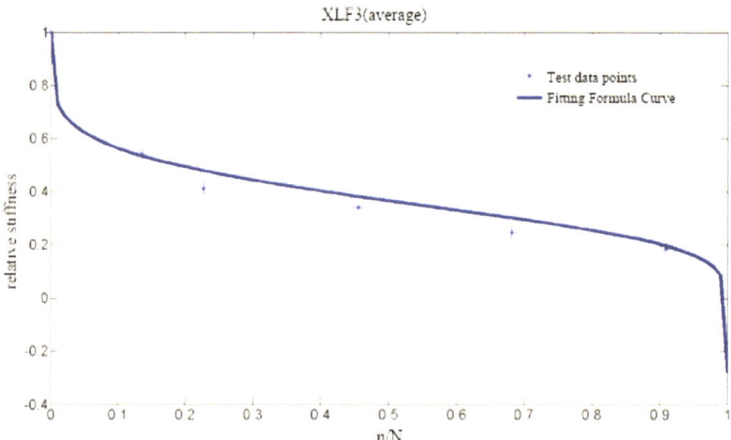

Fig. 7 Theoretical formula fits with XLF3 test data

5   Conclusion

From the above figure, it can be seen that using the mean values of each parameter as the representative values of this parameter is substituted into the stiffness degradation function curve, and it is found that the fitting is good. The stiffness degradation of prestressed concrete box girder components exhibits an S-shape. In the early stage of fatigue, the stiffness degradation is rapid, accounting for 10–20% of the entire cycle. As fatigue loading progresses, stiffness degradation enters a stable stage, accounting for 60–70% of the entire cycle. When approaching fatigue failure, the fatigue stiffness deteriorates rapidly, accounting for about 10% of the entire cycle. In the fatigue test, the test beam did not reach the final fatigue failure stage, but the stiffness degradation law was well matched in the first two stages.

References

1. Cao, X. (2014). Study on the fatigue performance of prestressed concrete box girders under the coupling of dynamic load and environment.
2. Xiaobo, Z. (2010). Experimental study on fatigue performance of highspeed railway prestressed concrete box girders.
3. Hua, J. M., Zou, X. B., & Wang, J. Y. (2011). Experimental study on fatigue performance of prestressed concrete beams with complex box sections on high-speed railways. *Journal Civil Archi Environment Engineering, 33*(03), 8–12.
4. Bai, L. (2014). Research on fatigue life analysis methods for prestressed and reinforced concrete bridges.
5. Li, J. (2013). Experimental research and theoretical analysis on fatigue mechanical behavior of prestressed concrete bridges.
6. Qi, Y. (2015). Experimental study on fatigue performance of prestressed concrete bridges under heavy traffic conditions.
7. Yu, Z. W., Li, J. Z., & Song, L. (2012). Research on fatigue test of heavy duty railway bridges. *China Civil Engineering Journal, 45*(12), 115–126.
8. Li, Y. S., An, L. P., Wei, S. L. Fatigue test research on heavy-duty railway bridges dynamic response analysis and fatigue life evaluation of railway steel truss bridges under the action of heavy-duty trains. *Journal of Shijiazhuang Tiedao University.*
9. Wang, L. (2014). Research on fatigue characteristics of prestressed concrete continuous beam bridges under vehicle load.
10. Liu, Y., Zhang, H. P., Deng, Y. (2014). Research on the influence of vehicle load operation state on fatigue damage of prestressed box beam bridges. *Journal of China and Foreign Highway.*
11. Shua, J. (2010). Research on fatigue reliability design method for highspeed railway prestressed RPC bridges.
12. Wang, B., Shi, Y., Chen, K. (2020). Research on fatigue safety analysis method for highway steel box girders. *Journal Highway Transport Research Development, 16*(10), 226–229.
13. Zhou, L., Zou, l., Zhao, l. (2021). Fatigue test of prestressed steel bars for simply supported box girder bridges on ballastless tracks of high-speed railways. *Journal of Central South University, 52*(02), 509–518

Prediction of the Lateral Ultimate Bearing Capacity of the RC Pipe Pile with Corrosion-Damaged Partial Length

Hongyang Yu and Xiaohui Wang

Abstract For reinforced concrete (RC) pipe pile used in the high-pile wharf in the marine environment, due to the dry–wet cycling in the splash/tidal zone, corrosion of the reinforcing bars in the pipe pile often occurs in this zone. In the present paper, the lateral ultimate bearing capacity of the RC pipe pile with corrosion-damaged partial length is predicted via the finite element method. Corrosion-induced concrete cover cracking, the reduction of effective cross section and strength of steel bars, as well as the bond degradation between steel bars and concrete are considered to predict the lateral ultimate bearing capacity of the corroded RC pipe piles under different axial compression ratios and corrosion levels. The modelling results show that, the ultimate capacity of the corroded RC pipe piles reduces due to the corrosion-damaged partial length. The failure location of the corroded RC pipe piles transfers from the bottom region of the uncorroded pile to the middle-corroded length. The increased axial compression ratio, corrosion level and corrosion-damaged length all affects the ultimate capacity and failure mode of the corroded RC pipe piles.

Keywords Reinforced concrete pipe pile · Finite element analysis · Corrosion-damaged partial length · Lateral ultimate bearing capacity · Failure mode

1 Introduction

In the port engineering, reinforced concrete (RC) pipe piles are widely used in the high-pile wharf. However, due to the frequent dry–wet cycling in the splash/tidal zone, corrosion of the steel reinforcement is usually more severe than the other zones [1]. In the past years, the chloride-induced corrosion problems of the RC pipe piles in the wharves have been extensively studied. Among those research works, the effects of transverse cracks on the corrosion of reinforcement in concrete[2] and

H. Yu · X. Wang (✉)
College of Ocean Science and Engineering, Shanghai Maritime University, Shanghai 201306, China
e-mail: xiaohwang@shmtu.edu.cn

© The Author(s) 2025
D. Li and Y. Zhang (eds.), *Advances in Frontier Research on Engineering Structures II*, Lecture Notes in Civil Engineering 535, https://doi.org/10.1007/978-981-97-6238-5_19

the effects of original incomplete cracks on the service life prediction of RC pipe piles [3, 4] are studied respectively. Effect of the exposure environment [5] and crack depth, crack width [6] on chloride diffusion and permeability into the concrete were evaluated; the time-dependent lateral bearing behaviour of corrosion-damaged RC pipe piles in marine environments is also investigated [7]. For most published papers focusing on the corroded RC pipe piles [3, 4, 7], the full length of the pipe pile is assumed to be uniformly corroded.

According to current literatures, research on the ultimate bearing capacity of the RC pipe pile with partial corrosion-damaged length is very limited. Mechanical properties of full length corroded RC piles are considered in most cases. In the present paper, the lateral ultimate bearing capacity of the RC pipe pile with corrosion-damaged partial length in the dry–wet cycling zone is predicted via the finite element method. Corrosion-induced concrete cracking, reduced cross section and the strength of the corroded steel bars, as well as the bond degradation between the corroded steel bar and concrete are considered. The ultimate lateral bearing capacity of the corroded RC pipe piles under different axial compression ratios and corrosion levels is predicted.

2 Corrosion-Damage Models of the Materials

2.1 Corrosion-Damage Models of the Concrete

In the process of reinforcement corrosion, concrete cover is cracked by the expansion of the corrosion products and concrete compressive strength will be reduced by the corrosion cracking. In the finite element simulation, this process is described by reducing the concrete compressive strength, calculated as follows [8]

$$f_{ck'} = \frac{f_{ck}}{1 + R \cdot \frac{\varepsilon_1}{\varepsilon_{c0}}} \tag{1}$$

where f_{ck} is the peak compressive strength of $150 \times 150 \times 300$ mm sound and uncracked concrete specimens; $f_{ck'}$ is the reduced concrete compressive strength; R is a coefficient related to the roughness and diameter of the reinforcement, the suggested value $R = 0.1$; ε_{c0} is the compressive strain at maximum stress f_{ck}, the suggested $\varepsilon_{c0} = 0.0033$; ε_1 is the average tensile strain of the cracked concrete, given by

$$\varepsilon_1 = \frac{b_f - b_1}{b_1} \tag{2}$$

where b_1 is the cross-sectional width of the uncorroded RC beam, and b_f is the cross-sectional width of the corroded RC beam, which was increased by corrosion cracking. The increment of section width is given by

$$\left(b_f - b_1\right) = n_{bars} \cdot \omega_{cr} \tag{3}$$

where n_{bars} is the number of the bars under compression in the top layer, ω_{cr} is the total corrosion crack width at a certain corrosion depth, given by [9]

$$\omega_{cr} = 2\pi\,(v_{rs} - 1)X \tag{4}$$

where v_{rs} is the volume expansion factor, $v_{rs} = 2$; X is the corrosion depth of the reinforcing bar.

For RC pipe pile, b_1 is taken as the outer diameter of the pipe pile while the number of the compressed bars n_{bars} is approximately taken as half number of the longitudinal rebars of the pipe pile.

2.2 Mechanical Properties of the Corroded Steel

Due to the tension stiffening effects in the cracked concrete, the yield strength of the tensile steel bar embedded in the concrete is calculated by [10]

$$\frac{f_y^*}{f_y} = (0.93 - 2B) \tag{5}$$

where f_y^* is the yield strength of the steel embedded in the cracked concrete, f_y is the steel yield stress, B is the parameter, given by the following equation:

$$B = \frac{\left(\frac{f_t}{f_y}\right)^{\frac{3}{2}}}{\rho_1} \tag{6}$$

where f_t is the tensile strength when concrete cracking strain reaches 0.00008, ρ_1 is the ratio of steel in the concrete.

At a constant flexural load capacity, the yield stress of the corroded rebar is given by [10]

$$f_{sy} = \frac{A_{sn}}{A_s}f_y^* \tag{7}$$

where f_{sy} is yield stress of corroded steel bars; f_y^* is the yield strength of the steel embedded in the cracked concrete; A_s is the cross section of the initial intact rebar; A_{sn}

is the cross section of corroded steel bar. $A_{sn} = (1 - \eta)A_s$, where η is the percentage of the mass loss of the corroded rebar (%).

The yield strain of the corroded rebar is then calculated by $\varepsilon_{sy}{'} = f_{sy}/E_s$, where E_s is the steel elastic modulus. Where ε_{su} is the ultimate strain of uncorroded rebar.

2.3 Bond Strength of the Corroded Steel bar

Degradation of bonding properties is a key factor affecting the performance of RC elements. The nonlinear model of the bond stress-slip developed by CEB-FIP [11] is modified to consider the corrosion effect, see Fig. 1. For the uncorroded bar, the bond stress increases then reaches the maximum bond stress $\tau_{0,max}$. For corroded bar, the reduction in bond strength is defined using the bond strength reduction factor β[8]. In the bond-slip relationship, for both uncorroded and corroded bars, it is assumed that the slip S_1 and the rate of reduction of bond strength in their softening branches arc the same (Fig. 1).

Based on the experimental results, Bhargava [12] proposed a bond strength model to reflect the relationship between the bonding strength and the corrosion level (%)

$$\tau_c(s) = \beta \tau_0(s) \tag{8}$$

$$\beta = \begin{cases} 1 & \eta \leq 1.5\% \\ 1.192e^{-11.7\eta} & \eta > 1.5\% \end{cases} \tag{9}$$

where β is the reduction factor of the bond strength; η is the corrosion level (%); $\tau_0(s)$ is the bond-slip relationship according to CEB-FIP Code 2010[11]; $\tau_c(s)$ is the bond-slip relationship considering the effect of reinforcement corrosion.

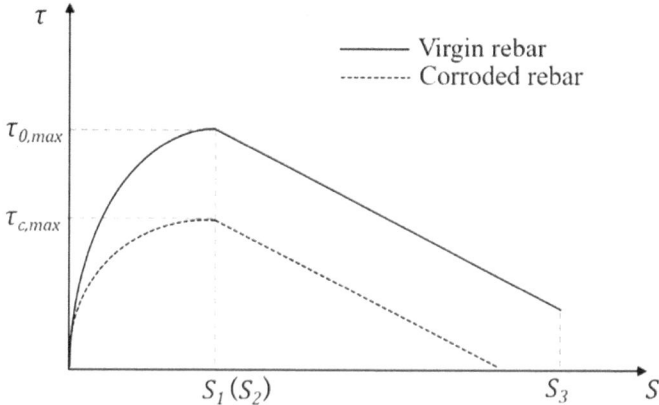

Fig. 1 Bond-slip relationship [8]

In this paper, to consider the degradation of the bond strength of the corroded bar, the methodology proposed by Li [13] is adopted. The flexural stiffness EI of the corroded RC pipe pile is given as [13]

$$EI = \frac{B_s}{\kappa M(\beta)} \tag{10}$$

$$B_s = 0.85 E_c I_0 \tag{11}$$

where B_s is the flexural stiffness of the uncorroded RC pipe pile, E_c is the concrete elastic modulus, I_0 is the equivalent sectional inertia moment the pile. The factor κ and integrated strain factor $M(\beta)$ are given as[13]

$$\kappa = \begin{cases} -1.05\cos(5\pi\beta) + 0.85 & 0 \le \beta < 0.2 \\ 1.0 & 0.2 \le \beta \le 1.0 \end{cases} \tag{12}$$

$$M(\beta) = \begin{cases} 471.016\beta^3 - 135.469\beta^2 - 0.331\beta + 2.777 & 0 \le \beta < 0.2 \\ 1.075 - 0.075\beta & 0.2 \le \beta \le 1.0 \end{cases} \tag{13}$$

3 Finite Element Modeling of the Corroded RC Pipe Piles

3.1 Modelling of the Concrete and Steel bar

Numerical simulation of the RC pipe pile is carried out. The concrete damaged plasticity (CDP) model in Abaqus is used to describe the concrete' compressive and tensile behavior. The stress–strain relation recommended in China concrete structure design code is adopted [14].The constitutive laws for steel reinforcements used in Liu [8] is adopted. For the corroded rebar, assuming the elastic modulus of the steel E_s is the same as the uncorroded one while the yield stress f_{sy} of corroded bars decreases, as given in Eq. (7). In terms of meshing, an 8-node 3D solid element (C3D8R) is adopted for the concrete. The steel reinforcement is modeled by a 2-node linear beam element (B31).

3.2 Equivalent Length of RC Pipe Pile

For an elastic long pile used in the high-pile wharf [15], the under-mud depth of the bending embedded point of the pile is determined firstly by the hypothetical embedded point method as follows

$$t = \upsilon T \tag{14}$$

$$T = \sqrt[5]{\frac{E_p I_p}{m b_0}} \tag{15}$$

When the under-mud depth t of the bending embedment point from the mud surface is determined, the equivalent pile length of the pile is the pile length above the mud surface plus t.

3.3 Setting of Loading and Boundary Conditions

In the finite element modeling, the pipe pile bottom is assumed to be completely fixed in the soil and the top of it is free, see Fig. 2. Set the reference point at the top of the RC pipe pile and the coupling constraint is set between the reference point and the top surface of the pipe pile. A constant axial force is applied to this reference point. Then a horizontal displacement load is applied on the top of the pile until failure.

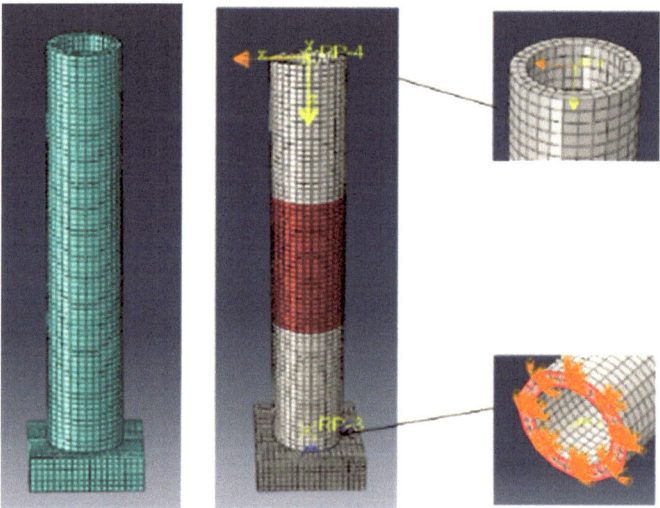

Fig. 2 Boundary modeling and loading setting of the RC pipe pile

3.4 Corrosion Damaged Partial Length in the RC Pipe Pile

Since the lateral ultimate capacity of the corroded RC pipe pile is mainly studied in the present paper and the corrosion-damaged length is mainly located in dry–wet cycling zone (see Fig. 2). The corrosion-damaged length is assumed to be localized in the middle zone of the RC pipe pile. After the equivalent length of pipe pile is determined, the material parameters of the concrete and steel bar in the undamaged and corrosion-damaged lengths are set respectively in the finite element simulation analysis.

3.5 Validation of the Finite Element Model for Uncorroded RC Pipe Pile

A uncorroded RC pipe pile was designed to validate the finite element model. Assume that the top of RC pipe pile is unconstrained and subjected to a horizontal force P_o only. The theoretical horizontal force is predicted by the following formula proposed by Li[13]

$$P^c = \frac{1}{2.429} E_p I_p \lambda^3 u^* \tag{16}$$

$$\lambda = \sqrt[5]{\frac{mb_0}{E_p I_p}} \tag{17}$$

where P^c is the normalized parameter; λ is flexibility factor (m^{-1}), m is the proportional coefficient (kN/m^4); b_0 is the converted width of the pile (m); u^* is the yielding deflection (m), which corresponds to the yielding stress. In this study, u^* is taken as 10 mm [13]. The load–displacement relationship equation is as follows:

$$\begin{cases} \frac{P_o}{P^c} = \frac{u}{u_u^*} & u < u^* \\ \frac{P_o}{P^c} = \frac{u^*}{0.15u} + 0.82 & u > u^* \end{cases} \tag{18}$$

where u is the displacement of pile top (m).

The calculated horizontal force F^*_{max} by Eqs. (16–18) is 89.49 kN while the predicted lateral ultimate capacity F_{max} by finite element model is 76.86 kN. Good agreement is shown in the calculated and predicted results.

Table 1 Material parameters of pipe pile model

Material	Young's modulus (MPa)	Poisson's ratio	Density(kg/m^3)
Concrete (C80)	38,000	0.2	2500
Steel bar (HRB400)	200,000	0.3	7800

Table 2 Concrete damage plasticity parameters

Dilation angle	Flow potential eccentricity	f_{b0}/f_{c0}	K	Viscosity parameter
30°	0.1	1.16	0.6667	0.0008

4 Predicted Results and Discussion

4.1 Information of the RC Pipe Pile Serviced in Suzhou Port

The RC pipe pile is selected from a wharf in Suzhou port, where the pile has a total length of 25.08 m and the length below the mud surface 20.1 m. The outer and inner diameter of the pipe pile 1000 mm and 740 mm, respectively. C80 concrete was selected, where the axial compressive strength of $150 \times 150 \times 300$ mm concrete specimen is 50.2 MPa. The concrete cover of the longitudinal reinforcement is 65 mm. 12 pieces of 16 mm diameter HRB400 steel bars were used as longitudinal bars and the HRB400 steel bars with 8 mm diameter was used for stirrups. The main parameters considered are axial compression ratio and the corrosion level of longitudinal reinforcing bars. The calculated under-mud depth of the bending embedment point by Eqs. (14, 15) is $t = 1.12$ m and the equivalent length of the pipe pile is 6.19 m. Detailed material parameter settings are shown in Tables 1 and 2. Where f_{b0}/f_{c0} is the compressive strength ratio; K is the ratio of the second stress invariant on the tensile meridian to that on the compressive meridian.

4.2 Simulation Results

RC pipe piles with different corrosion levels and axial compression ratios are considered, where the corroded partial length is assumed as 2000 mm.

For uncorroded and corroded RC pipe piles, when the axial compression ratio is 0.6, the plastic strain clouds of the concrete at the ultimate capacity of the piles is shown in Fig. 3. It can be seen from Fig. 3 that, for uncorroded pipe piles, at the ultimate capacity state, the maximum equivalent plastic (MEP) strains of the concrete occur at the bottom region of the pile, see Fig. 3a; while for the corroded pipe pile, the MEP strains occur in the middle part (corroded zone) of the pipe pile, see Fig. 3b. For both pipe piles, the failure is caused by compressive crushing of the concrete. For

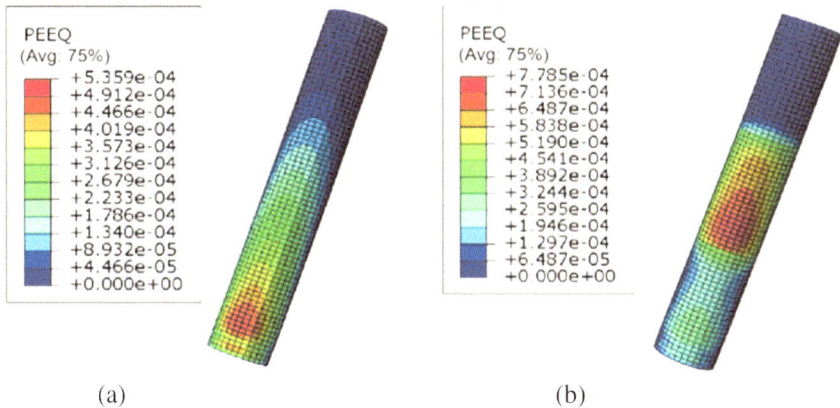

(a)   (b)

Fig. 3 Equivalent plastic concrete strain cloud of RC pipe piles: **a** uncorroded RC pipe pile; **b** RC pipe pile with corroded length 2000 mm

the RC bridge pier specimens with partial corrosion in the middle length [16], the similar failure zone transferring phenomenon was reported. For the pier specimens were subjected to the axial compression loading, the failure zone transferred from the loading end of the uncorroded specimen to the severe corrosion zone. This will result in the middle region of the pile being more susceptible to damage, thus reducing the bearing capacity of the pile.

4.3   Influence of the Axial Compression Ratio on the Lateral Ultimate Bearing Capacity of the Corroded RC Pipe Pile

In the study of this section, the corroded length of the RC piles is assumed 2000 mm while the equivalent length of the pipe pile is 6.19 m, as mentioned above. Figure 4. Shows the lateral load–displacement curves of intact and corroded pipe piles at axial compression ratios of 0.6, 0.65 and 0.7, respectively. When the axial compression ratio exceeds 0.6, the horizontal ultimate capacity of the member gradually decreases with the increase of axial load. When the axial force is the same, the lateral bearing capacity of the pipe piles gradually decreases as the corrosion level of the reinforcement increases. A staged drop in ultimate lateral capacity of 17.7%, 18.7%, 19.2%, is observed for the corroded RC pipe piles subjected to the combination of vertical load, i.e., n = 0.6, 0,65, 0.7, respectively and 4% corrosion level. It shows that the damage to the wave splash zone will affect the overall lateral ultimate capacity of the RC pipe piles.

For corroded RC pipe piles, when the axial compression ratio is 0.6 and the partial corrosion length is 2000 mm, the plastic strain clouds at the ultimate capacity of the piles with different corrosion levels are shown in Fig. 5a, b. It can be observed that,

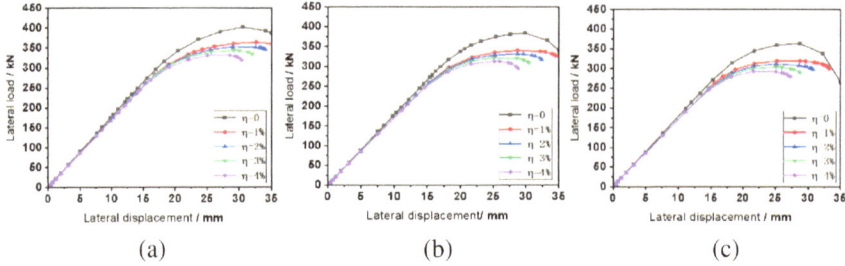

Fig. 4 Lateral load–displacement curves of RC pipe piles at different axial compression ratios: **a** n = 0.6; **b** n = 0.65; **c** n = 0.7

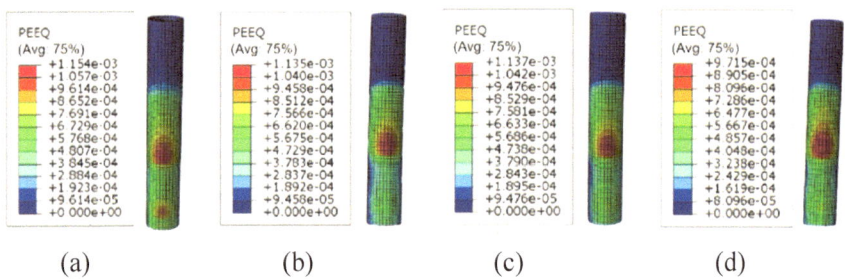

Fig. 5 Equivalent plastic concrete strain cloud of corroded RC piles with different corrosion levels and axial compression ratios: **a** corrosion level = 1%; **b** corrosion level = 3%; **c** axial compression ratio = 0.6; **d** axial compression ratio = 0.7

for corroded RC pipe pile with the same partial corrosion length, when the corrosion level of the bar is small, at the ultimate capacity state, the MEP strains occur at the bottom and middle regions of the pile, see Fig. 5a; the MEP strains of the concrete transfers to the corroded zone the pipe pile with the with the increased corrosion level, see Fig. 5b.

For corroded RC pipe piles, when the partial corrosion length is 2000 mm and the corrosion level is 2%, the equivalent plastic strain clouds at the ultimate capacity of the piles with different axial compression ratios are shown in Fig. 5c, d. It can be observed that, for RC pipe pile with the same corrosion length and corrosion level, the increased axial compression ratio increases the range of the MEP strains of the concrete.

4.4 Influence of the Partial Length Corrosion on the Lateral Ultimate Bearing Capacity of the RC Pipe Pile

In this section, three corrosion-damaged partial lengths, i.e., 2000 mm, 2250 mm and 2500 mm are considered. The axial compression ratio is taken as 0.6 and the corrosion

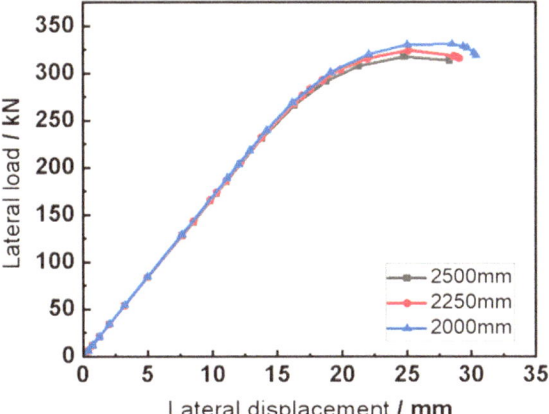

Fig. 6 Load–displacement curves of RC pipe piles with different corrosion-damaged lengths 2250 mm

level of the longitudinal reinforcing bars is 4%. Figure 6 shows the horizontal load–displacement curve of the top of the pile with different corrosion-damaged lengths. For the 2000, 2250, 2500 mm corrosion lengths, the horizontal ultimate capacity of reinforced concrete piles are 329.70 kN, 324.00 kN, 317.29 kN, respectively. Considering the ultimate horizontal bearing capacity 400.52 kN of the sound pipe pile under the same type of the loading, the decreasing trend of the lateral ultimate load capacity is evident with the increased corrosion-damaged length.

For corroded RC pipe piles, when axial compression ratio is 0.6 and the corrosion level is 4%, the plastic strain clouds at the ultimate capacity of the piles with different partial corrosion lengths are shown in Fig. 7. It can be observed from Fig. 7 that, for RC pipe pile with the same axial compression ratio and corrosion level, the increased corrosion length also increases the range of the MEP strains of the concrete. From this it follows that the increased corrosion length will lead to a more significant decrease in the mechanical properties of the pipe pile, making it more susceptible to damage.

5 Conclusions

In the present paper, the lateral ultimate capacity of the RC pipe pile with corrosion-damaged length in the splash/tidal zone is predicted. Corrosion-induced concrete cracking, reduced cross section and strength of the corroded steel bars, as well as the bond degradation are considered in the finite element simulation. The following conclusions can be drawn:

(1) Compared with the ultimate capacity of the uncorroded RC pipe pile, reduced ultimate capacity is shown in the pipe pile with partial corrosion in the middle length. The equivalent plastic concrete strain cloud of uncorroded and corroded

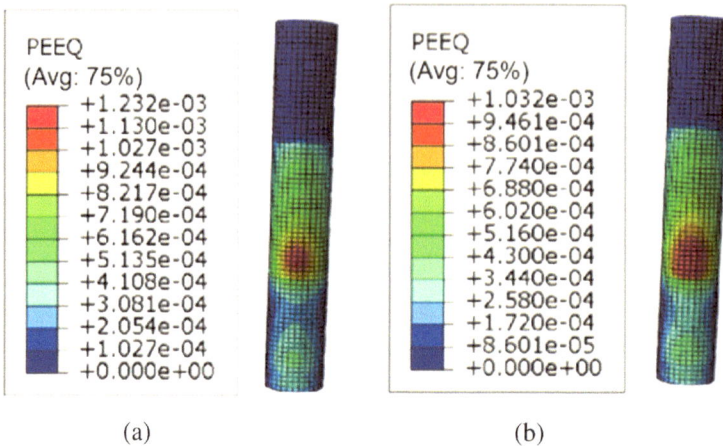

Fig. 7 Equivalent plastic concrete strain cloud of RC piles with different corrosion partial lengths:
a length = ; **b** length = 2500 mm

RC pipe piles indicates a failure mode transfers from the bottom region of the uncorroded pile to the middle-corroded length of the corroded pile.

(2) For corroded RC pipe piles with the same partial length and corrosion level, their ultimate capacity decreases with the increased axial compression ratio or the axial load.

(3) For corroded RC pipe piles with the same corrosion level in the partial length, under the same axial load, the ultimate horizontal bearing capacity decreases with the increased corrosion-damaged length.

(4) For corroded RC pipe piles with partial corrosion in the middle length, the failure mode transfers from the bottom of the uncorroded pile to the corrosion zone with the increased corrosion levels. The range of the MEP concrete strains of the corroded RC pipe piles increases with the increased axial compression ratios, corrosion levels and the corrosion partial lengths.

(5) In this study, the corrosion of the steel reinforcements in RC columns is assumed to be uniform, and simplified the degradation of bonding performance between steel bars and concrete. In practical engineering, the corrosion distribution is more complex, which will influence the bond strength of the corroded steel bar and the mechanical behavior of the member. Future research should consider the effect of uneven corrosion of reinforcement on the mechanical properties of corroded members.

Acknowledgements This research was financially supported by the Shanghai Natural Science Foundation (No. 21ZR1426800).

References

1. Yuan, W., Wu, X., Wang, Y., Liu, Z., & Zhou, P. (2023). Time-dependent seismic reliability of coastal bridge piers subjected to nonuniform corrosion. *Materials, 16*(3), 1029.
2. Jin, X., Chen, J., Li, T., Shen, J., Liu, Q., & Jin, W. (2023). Modelling the corrosion mechanism of steel bars in chloride-contaminated concrete with transverse cracks. *Magazine of Concrete Research, 75*(11), 580–594.
3. Li, L., Yang, C., & Li, J. (2021). Corrosion initiation life of laterally loaded PHC pipe piles served in marine environment: Theoretical prediction and analysis. *Construction and Building Materials, 293*, 123457.
4. Li, L., Li, J., Yang, C. (2019) Theoretical approach for prediction of service life of RC pipe piles with original incomplete cracks in chloride-contaminated soils. *Construction and Building Materials, 228*, 116717.1–116717.14.
5. Bao, J., Wei, J., Zhang, P., Zhuang, Z., & Zhao, T. (2022). Experimental and theoretical investigation of chloride ingress into concrete exposed to real marine environment. *Cement and Concrete Composites, 130*, 104511.
6. Yang, Z., Tang, M., & Zhang, J. (2018). The influences of chloride ion transmission on cracking and chemical corrosion resistance of PHC pipe piles. *Chemical Engineering Transactions, 71*, 1183–1188.
7. Shao, W., Shi, D., Jiang, J., & Chen, Y. (2017). Time-dependent lateral bearing behaviour of corrosion-damaged RC pipe piles in marine environments. *Construction and Building Materials, 157*, 676–684.
8. Liu, X., Zhong, T., Jiang, H., Liu, Y., & Guo, Z. (2023). Deformation limits of corroded reinforced concrete columns. *Structure, 49*, 903–917.
9. Molina, F. J., Alonso, C., & Andrade, C. (1993). Cover cracking as a function of rebar corrosion: Part 2-Numerical model. *Materials and Structure, 26*, 532–548.
10. Baniasad, E., & Dehestani, M. (2019). Incorporation of corrosion and bond-slip effects in properties of reinforcing element embedded in concrete beams. *Structure, 20*, 105–115.
11. CEB-FIP (2010) Fib model code for concrete structures. Ernst & Sohn, Berlin, Germany.
12. Bhargava, K., Ghosh, A. K., & Mori, Y. (2008). Suggested empirical models for corrosion-induced bond degradation in reinforced concrete. *Journal of the Structural Engineering. American Society of Civil Engineers, 134*(2), 221–230.
13. Li, J., Liu, Y., & Zhou, Y. (2013). Service life prediction of horizontal bearing capacity of PHC pipe pile in marine environment. *China Civil Engineering Journals, 46*(12), 109–117.
14. Ministry of Housing and Urban-Rural Development of the People's Republic of China (2010) Code for design of concrete structures (GB50010-2010). China Architecture & Building Press, Beijing.
15. Chinese Standard (2012) Code for pile foundation in port engineering JTS 167-4-2012. China communication press, Beijing.
16. Zhou, H., Xu, Y., Peng, Y., Liang, X., Li, D., & Xing, F. (2020). Partially corroded reinforced concrete piers under axial compression and cyclic loading: An experimental study. *Engineering Structures, 203*, 109880.

Structural Seismic Resistance and Energy Consumption Analysis

Basic Characteristics and Engineering Treatment of Dam Foundation Fault Zone of Gushan Navigation and Power Junction in Hanjiang River

Qiguo Wang and Yi Hu

Abstract Affected by regional structure and multi-stage metamorphism, the dam site area of Gushan Navigation-Power Junction in Hanjiang River has developed faults in dam foundation, with 97 faults in total, among which 9 faults are wider than 1 m, and the maximum width of fracture zone is 10 m, which has great influence on the project. Through research, the dam foundation fault zone mainly affects the dam's anti-sliding stability, settlement deformation, leakage and seepage stability. According to the scale, spatial distribution, material composition and engineering properties of the fault, and considering the layout characteristics of the hub, measures such as trench excavation and backfilling concrete plugs, consolidation grouting and local increase of anchor rods are adopted for engineering treatment. The monitoring data of more than three years show that the deformation and leakage of dam foundation are within the normal range, which shows that the engineering measures of dam foundation for fault zone are effective.

Keywords Avionics engineering fault · Geological features · Deformation of the dam foundation · Leakage of the dam foundation · Processing effects

1 Introduction

The load of gravity dam directly acts on the dam foundation, which mainly depends on the self-weight and sufficient friction between the foundations to ensure stability, which requires higher requirements for the dam foundation. Fault is a kind of geological defect in the rock foundation of gravity dam, which mainly affects the stability of dam foundation against sliding, settlement deformation, seepage and seepage, and the stability of dam abutment slope, etc. If it is not handled well, it will easily lead to

Q. Wang · Y. Hu (✉)
Changjiang Geotechnical Engineering Co., Ltd., Wuhan 430010, Hubei, China
e-mail: huyi408@qq.com

Q. Wang
National Dam Safety Engineering Technology Research Center, Wuhan 430010, Hubei, China

© The Author(s) 2025
D. Li and Y. Zhang (eds.), *Advances in Frontier Research on Engineering Structures II*,
Lecture Notes in Civil Engineering 535, https://doi.org/10.1007/978-981-97-6238-5_20

dam accidents and even dam accidents. Therefore, the fault treatment of gravity dam foundation is a major technical problem in the process of engineering construction, which must be handled in place to eliminate potential geological risks and ensure the safety of dam engineering.

The Gushan Navigation and Power Junction of Hanjiang River is a large-scale (2) project. The construction of this junction is the need to build a high-grade waterway of Hanjiang River and realize the national high-grade waterway planning goal, and also the need to promote the leap-forward strategic development of the water transport industry in the central region of the Yangtze River system. The main tasks of the navigation and power hub are shipping and power generation, with a class VI waterway and a power station with an installed capacity of 180 MW. The dam type is concrete gravity dam with a maximum dam height of 57.2 m. In December 2016, the dam foundation was officially excavated, and the dam foundation pouring was completed at the end of 2022. At present, the project construction has entered the final stage.

A set of shallow metamorphic rock series of the Cambrian Middle Yuejiaping Formation ($\in_{2y}$) is mainly distributed in the dam site area. The main rocks are mica marble and Quartz mica schist, with a small amount of carbonaceous mica schist and quartz schist. However, due to the influence of regional geological structure and multi-stage metamorphism, the dam foundation faults are well developed, and 97 faults are revealed in the construction excavation, including 9 faults with a width greater than 1 m, and the maximum width of the fracture zone is 10 m. The rock mass in the fracture zone, the intersection zone and the fault-affected zone is relatively weak [1], and some fault zones are easy to soften when meeting water, with poor mechanical properties, and the mechanical strength such as deformation modulus, shear strength and compressive strength is significantly different from that of the surrounding intact rock mass. And most of them have weak to moderate water permeability, which affects the stability of dam foundation against sliding, settlement deformation, leakage and seepage, etc. Because of the large number of faults, poor mechanical properties and large width of some broken zones, it brings great technical challenges to the engineering construction [2], which is a major engineering technical problem of this hub, and it is necessary to carry out special reinforcement and anti-seepage engineering treatment for dam foundation fault zones. After research and engineering treatment, the safety monitoring data of the dam during the construction period and the initial operation period show that the engineering treatment effect is good. The research results provide a good reference for similar projects.

Table 1 Formatting sections, subsections and subsubsections

Stage	Stratigraphic code	Thickness (m)	Brief description of lithology
Stage 1	$\in_{2y}^{1}$	175	Gray-black carbonaceous mica schist, gray quartz mica schist
Stage 2	$\in_{2y}^{2}$	15~90	Gray banded mica marble
Stage 3	$\in_{2y}^{3}$	30~70	Gray-white, light gray thick layered mica marble
Stage 4	$\in_{2y}^{4}$	120~210	Light gray and gray quartz mica schist with gray quartz schist
Stage 5	$\in_{2y}^{5}$	360	Gray quartz mica schist and gray banded mica marble

2 Basic Engineering Geological Conditions of Dam Site Area

The dam site area is of medium–low mountain landform, and the Hanjiang River traverses this area from west to east, with a riverbed width of 300 ~ 500 m generally. The valley at the dam site is a U-shaped wide valley with a longitudinal valley, and the left bank is a reverse slope with a bank slope of about 30; The right bank is a straight slope, while the bank slope is relatively steep, and the slope is generally greater than 40. The bedrock in the dam site area is mainly Cambrian metamorphic rocks (see Table 1). The Quaternary is mainly composed of alluvium, alluvium, diluvium, colluvium, landslide accumulation and artificial accumulation.

The cracks in the dam site area are relatively developed. According to the statistical results (see Fig. 1), there are mainly three groups developed, namely: L1-200° ∠55°; L2-305° ∠58°; L3-75° ∠63.

Schmidt net, upper hemisphere projection, number of cracks: 2468.

1: 1%~2%; 2: 2%~3%; 3: 3%~4%; 4: > 4%; 5: Extremely dense point and its number.

3 Development Characteristics of Dam Foundation Faults

According to the preliminary investigation and construction excavation, there are 97 faults in the dam foundation, which are mainly distributed in the dam section between the powerhouse and the installation room, the dam section of the left and right sluice, the dam section of the ecological sluice (longitudinal cofferdam) and the ship lock. The strike of faults is mainly NW ~ NWW, followed by NE ~ NNE, and a few faults are close to EW. The faults are mainly with medium-steep dip angles, among which there are 9 faults with an average width of more than 1 m in the fault fracture zone. See Fig. 2 for the plane distribution of main faults, and Table 2 for the basic characteristics. The remaining faults are small in scale.

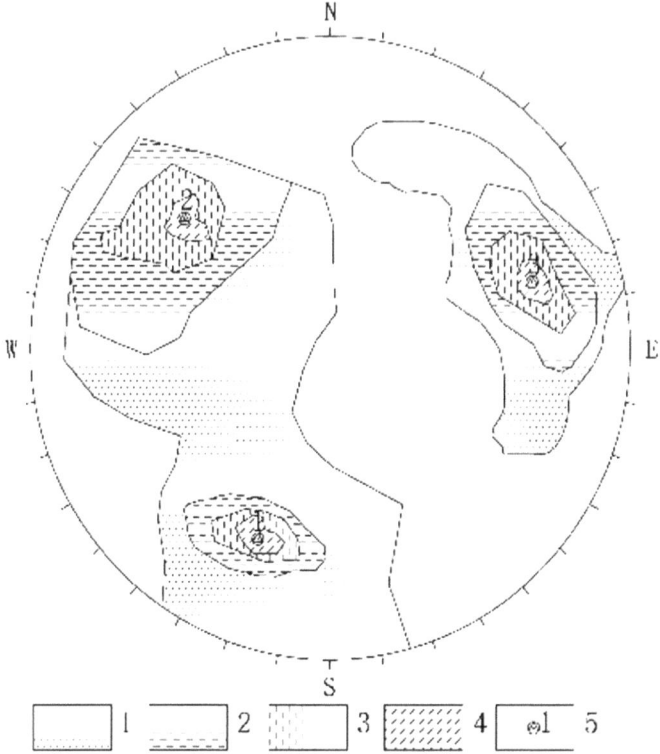

Fig. 1 Crack density map of dam site area

4 Physical and Mechanical Properties and Permeability Characteristics of Dam Foundation Fault Zone

In the early stage of investigation, in-situ test and borehole acoustic wave test were carried out on the fault fracture zone in the dam site. The results of in-situ test are shown in Tables 3 and 4, and the wave velocity of the fault zone is 1820 ~ 2590 m/s, and the results are shown in Table 5. The results show that the deformation modulus, shear strength and wave velocity of the fault zone are small, and the mechanical properties of the fault zone are generally poor. The characteristics, properties and physical and mechanical properties of faults revealed by construction excavation are basically consistent with the previous investigation results. The excavated fault fracture zone has the characteristics of water absorption, softening and disintegration after being soaked in water, which is easy to cause structural damage and become scattered debris particles [3].

In addition, the water permeability tested by the borehole water pressure test in the local fault fracture zone is 3Lu~26 Lu, which is weak-medium. The overall water permeability in the fault zone is quite different, which is mainly related to

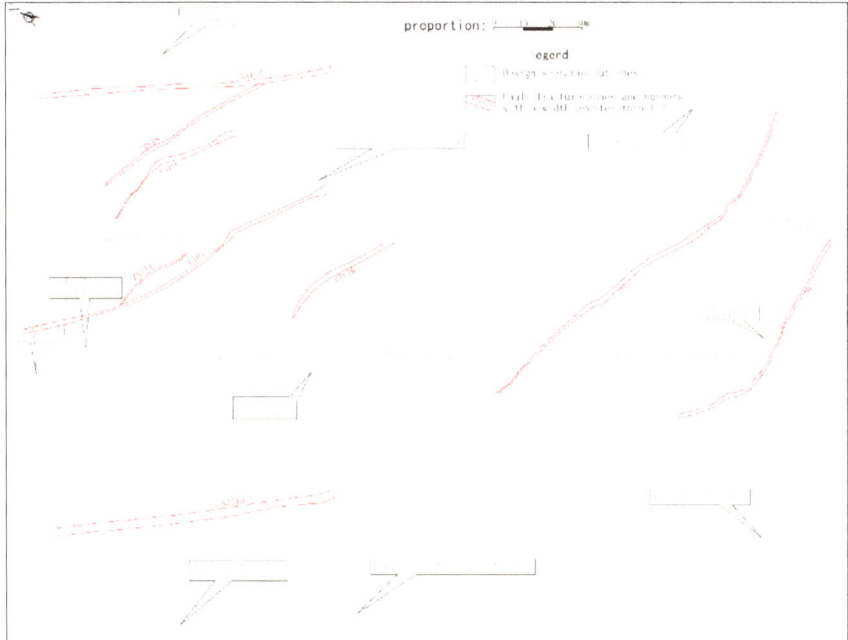

Fig. 2 Schematic diagram of plane distribution of faults with dam foundation width greater than 1 m

the material properties of the fault fracture zone. Generally, the water permeability in the compression-torsion fault zone is relatively small because of the high carbon shale content, while the water permeability in the tensile fault zone is relatively large because of the high broken particles and broken breccia content. Statistics of water pressure test in fault fracture zone are shown in Table 6. The recommended values for the physical and mechanical parameters of the fault fracture zone are shown in Table 7.

5 Engineering Treatment Measures of Dam Foundation Fault Zone

Generally, the treatment methods for the fault fracture zone of concrete gravity dam foundation are different according to the fault dip angle. For steep dip faults, trench excavation and concrete plug reinforcement are generally adopted, and appropriate crack-limiting steel bars can be arranged in the concrete plug according to the actual geological conditions. For the long and slow dip fault fracture zone, measures such as concrete replacement, concrete deep tooth wall and concrete hole plug can be adopted respectively according to the buried depth to increase the anti-sliding stability of the

Table 2 Development characteristics of fault zone with dam foundation width greater than 1 m

No	Attitude of rock		Scale		Fault characteristics	Distribution
	Strata inclination (°)	Dip angle (°)	Length (m)	Width of crushing zone (m)		
SZf10	245	45~58	20	2~4	The reverse fault and fault zone are gray-black, compressive cataclastic rocks, angular-subangular, soft, containing more carbon, and the contact parts of local fault planes are muddy	Upper longitudinal section, upstream approach channel and left non-upstream slope
F_{107}	30~45	70~80	60	4~10	A reverse fault is the contact boundary between the $\in_{2y}^{2}$ and $\in_{2y}^{3}$ strata. The fractured zone consists of compressed sheet-like rocks and fractured rocks	Lower longitudinal section, lock chamber section and left non-downstream slope
XZf12	45	42	19	1.2~2.5	The material in the fault zone is fragmented rock, partially interbedded with quartz fragments. The rock mass undergoes compression deformation in a sheet-like or fragmented manner, with visible scratches on the cross-section	Lower longitudinal section, downstream cutoff wall and downstream approach channel
YZf19	210	80	31.0	0.25~1.8	The material inside the fault zone is grayish black fractured rock, with smooth cross-sections and visible scratches. The fragmented structure is mainly interbedded with a 1 cm thick mud film attached	Right area sluice and lower longitudinal cofferdam

(continued)

Table 2 (continued)

No	Attitude of rock		Scale		Fault characteristics	Distribution
	Strata inclination (°)	Dip angle (°)	Length (m)	Width of crushing zone (m)		
CFF1	215~255	60~75	150.0	0.3~4.5	The material in the fault zone is gray black fragmented rock interbedded with block fractured rock, with smooth cross-sections and visible scratches. The thickness of the mud film is about 0.5-15 cm, and strong wrinkling and compression cause the material in the fault zone to break	Downstream section of sluice gate and dam guide wall in right area, factory building and fishway
F109	210~230	65~80	195	0.5~4.0	The fault surface is smooth with scratches and dark gray mud film, about 3-5 mm thick, and the material in the zone is gray black compressed fractured rock	Workshop, installation site, right non-dam section and right abutment slope
ZXf11	220	65	50	0.5~1.5	Normal fault, mainly blocky rock, partially mixed with cataclastic rock, with fault gouge of 20 ~ 30 cm attached to the fault plane	Lock chamber
F104	40	75	60	0.5~2.0	The fault material is crumpled and compressed fractured rock, with visible compression surfaces and scratches on the cross-section. The rock mass in the affected zone of the fault is relatively fragmented	Lower lock head and drain box culvert of ship lock
ZXf18	40	55	65	0.5~1.5	The fault material consists of crumpled and compressed fractured rocks and block fractured rocks, which are gray black in color, with obvious scratches visible on the cross-section	Waterproof wall at the head and downstream of lower lock of ship lock

Table 3 In-situ deformation test results of fault zone

Content	Serial number	Test site	Loading direction	Modulus of deformation E_0 (Gpa)		Modulus of elasticity E_e (Gpa)		Description of test rock mass
				Experimental value	Average value	Experimental value	Average value	
Fault F_{106}	Ejc-101	The lateral wall of the branch tunnel excavated along the fault at the depth of 13 m in PD6 tunnel on the left bank	Vertical fault plane	0.44	0.30	1.41	1.41	It is mainly composed of extrusion foliation, structural lens and fragments
	Ejc-102			0.17		1.07		
	Ejc-103			0.30		1.76		

Table 4 Field shear test results of faults

content	Rock mass or structural plane	Test site	Number of test points (points)	Shear resistance		Shear resistance (friction)	
				Friction factor f'	Cohesion C'(Mpa)	Friction factor f'	Cohesion C'(Mpa)
Structural plane direct shear	Faulted zone	Fault F_{104} crushed zone	6	0.56	0.05	0.51	0.04
	Fault plane	Fault F_{108} Gently dipping fault plane	3	0.53	0.05	0.50	0.05

Table 5 Statistical table of longitudinal wave velocity results of boreholes in fault parts

Content	Longitudinal wave velocity (m/s)			Number of test segments
	Large average value	Small average value	Average value	
Fault fracture zone	2590	1820	2130	70

Table 6 Statistical table of test results of borehole water pressure in fault fracture zone

Content	Weathering state	Number of test segments	Classification of permeability q(Lu) of rock mass			
			Very slightly		Slightly	
			$q < 0.1$		$0.1 \leq q < 1$	
			Tier number	Percentage (%)	Tier number	Percentage (%)
Fault fracture zone	Slight	11	0	0	5	45.5
			Weakly			
			$1 \leq q < 5$		$5 \leq q < 10$	
			Tier number	Percentage (%)	Tier number	Percentage (%)
			5	45.5	0	0
			moderately		strongly	
			$10 \leq q < 100$		$100 \leq q$	
			Tier number	Percentage (%)	Tier number	Percentage (%)
			1	9	.	0

slow dip fracture zone. The fault fracture zone developed in Gushan Hub is dominated by steep dip, and only a relatively long gentle dip fault is developed in the dam body of local longitudinal cofferdam.

After the excavation of the fault fracture zone is revealed, according to the hydro-geological and engineering geological characteristics of the fault fracture zone in

Table 7 Suggested values of physical and mechanical parameters of fault fracture zone

Type	Content	Natural density (g/cm^3)	Saturated compressive strength of rock (MPa)	Possion ratio	Deformation modulus of rock mass (GPa)	Structure of rock mass			Concrete/rock mass			Permeability coefficient (cm/s)
						Shear strength		shear strength	Shear strength		shear strength	
						f'	C'(MPa)	f	f'	C'(MPa)	f	
V	Fault fracture zone	2.35	0.5~0.8	0.38	0.2~0.3	0.42~0.45	0.05~0.15	0.35~0.40	0.40~0.50	0.05~0.10	0.30~0.40	5×10^{-4}~4.5×10^{-3}

the dam area, in order to ensure the engineering quality and safety, the construction disturbance of the fault fracture zone and the fault intersection zone should be reduced as much as possible before the fault engineering treatment, and the fault zone should not be soaked by engineering water and rain for a long time. According to the occurrence, extension characteristics, width and other properties of the fault fracture zone and the fault intersection zone, drainage, guidance and drainage measures should be taken to guide the engineering water and rain. Then, according to the material composition characteristics of fault fracture zone and fault intersection zone, the scale of fault, the influence degree of fault on buildings and so on, targeted engineering treatment measures are taken, such as trench excavation, backfilling concrete plug, consolidation grouting and local increase of anchor rods. Typical descriptions are given as follows: steep-dip faults which are easy to cause settlement and deformation of buildings at the lock, gentle-dip faults which are easy to cause anti-sliding stability problems at the dam body of longitudinal cofferdam, and steep-dip faults which are easy to cause seepage stability problems at the dam axis:

In view of the large-scale F107 and XZf18 fault parts distributed at the lower lock head of the ship lock, as shown in Fig. 3, because the ship lock part belongs to an important building, the fault part is firstly trenched by 1.5 times of the width until it reaches the undisturbed rock mass, then the concrete plug is backfilled, and the dam structure concrete is poured at the fault and the affected zone with a thickness of not less than 2 m and the design strength reaches 50%, and consolidation grouting is adopted [4]. The consolidation grouting depth is 8 m for the fault part and 8 m for the affected zone part.

As shown in Figs. 4 and 5, the relatively long gently dipping fault in the longitudinal cofferdam dam body is about 10 ~ 50 cm wide, with a visible extension of about 30 m and an overall dip angle of about 15 ~ 20. Firstly, all the gently dipping faults in the water collecting well of the longitudinal cofferdam dam body are excavated, and

Fig. 3 Schematic diagram of F107 and ZXf18 faults of shiplock

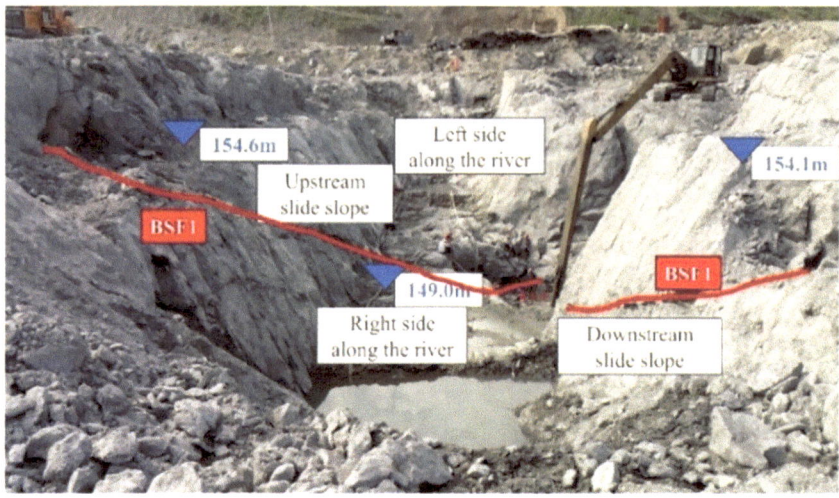

Fig. 4 Schematic diagram of gently inclined BSf1 fault of longitudinal cofferdam dam body

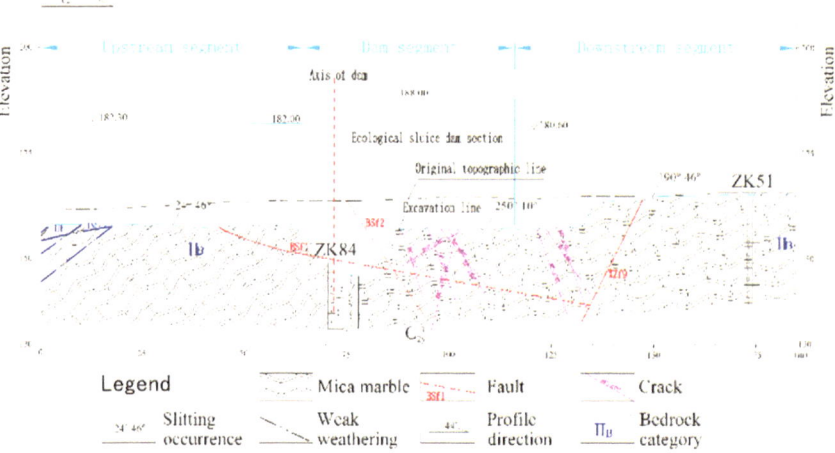

Fig. 5 Schematic diagram of BSf1 fault section with gentle dip angle of longitudinal cofferdam dam body

the gently dipping faults around the water collecting well are treated with consolidation grouting, and the consolidation grouting goes deep into the gently dipping fault for 5 m; Another example is that for the secondary building parts such as the upper and lower longitudinal sections of the longitudinal cofferdam, only 1 ~ 1.5 times of the fault width is adopted for trench excavation, and then the concrete plug is backfilled [5].

Table 8 Treatment measures for faults with a width greater than 1 m

Fault number	Engineering treatment measures
SZf10	Trench digging + cracking limit steel bar + concrete plug backfilling
F_{107}	Trench digging + concrete plug backfilling + consolidation grout
XZf12	Trench digging + concrete plug backfilling
YZf19	Trench digging + concrete plug backfilling
CFF1	Trough digging + concrete plug backfilling + anti-seepage curtain
F109	Trough digging + concrete plug backfilling + bolt + drainage hole + anti-seepage curtain
ZXf11	Trench digging + concrete plug backfilling
F104	Trench digging + concrete plug backfilling + consolidation grout
ZXf18	Trench digging + concrete plug backfilling + consolidation grout

According to the permeability of the fault zone and the seepage control standard of the lower seepage limit of dam foundation, the curtain depth of the fault zone should be appropriately deepened, and it should be 5 ~ F109 meters below the permeability line of 5Lu.

Specific engineering treatment measures for faults with fracture zone width greater than 1 m are shown in Table 8.

6 Engineering Treatment Effect

After the dam foundation fault engineering treatment, the typical monitoring data of bedrock deformation meters at the foundation surface of each dam section are shown in Figs. 6, 7 and 8, among which 4 bedrock deformation meters are buried at the foundation surface of the dam section of the left sluice. By the end of August 2023, the bedrock deformation is between-0.86 mm and 0.83 mm, and the monthly variation is between 0.00 mm and 0.05 mm; Six bedrock deformation meters were buried in the foundation surface of the lock dam section. By the end of August 2023, the bedrock deformation was between-2.40 mm and-0.41 mm, and the monthly variation was between-0.07 mm and 0.04 mm.: Six bedrock deformation meters were buried in the foundation surface of the sluice dam section in the right area. By the end of August 2023, the bedrock deformation was between-0.72 mm and 0.34 mm, and the monthly variation was between-0.02 mm and 0.09 mm; Three bedrock deformation meters were buried in the foundation surface of the dam section of the power plant. By the end of August 2023, the bedrock deformation was between -0.47 and -0.22 mm, and the monthly variation was between -0.09 and 0.04 mm. The longest monitoring data of bedrock deformation meter in each dam section has been 6 years, and the shortest is 3 years. The deformation monitoring data shows that the deformation value changes little with time, and the cumulative displacement is between 0 and 3 mm, all of which fluctuate within the normal range, indicating that the settlement

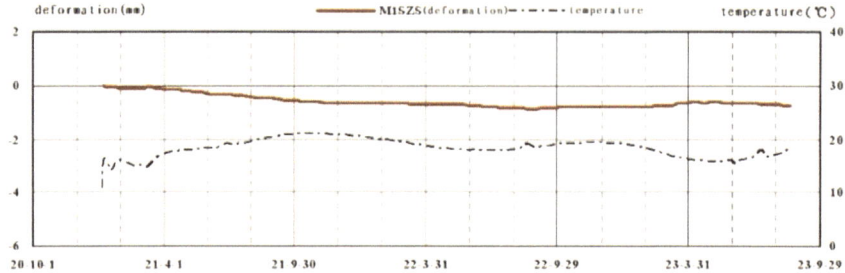

Fig. 6 Typical measurement value change process line of rock deformation meter in ship lock dam section

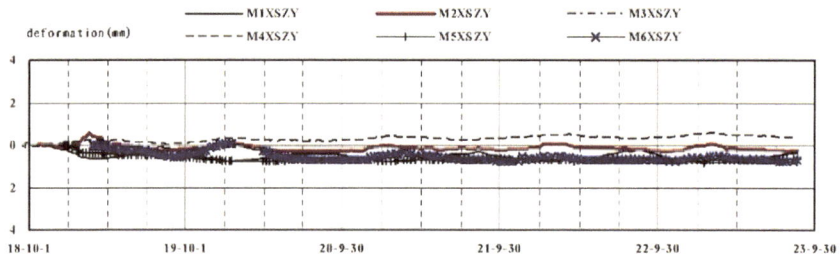

Fig. 7 Typical measurement value change process line of bedrock deformation meter in the right area sluice dam section

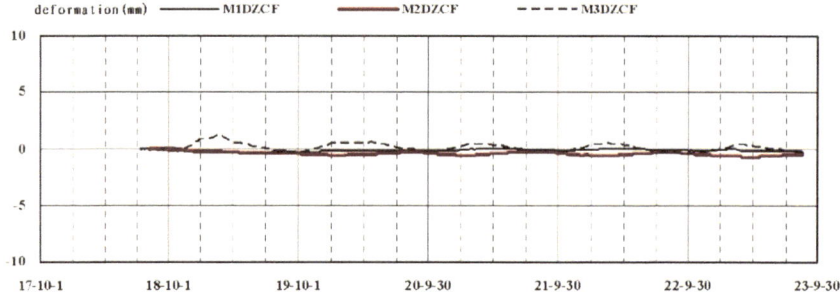

Fig. 8 Typical measurement value change process line of bedrock deformation

and deformation problem of the dam at the fault position has been completely solved, ensuring the safety of the project.

For the CFF1 and F109 faults crossing the seepage control axis of the dam, after the curtain grouting is deepened, the quality of the fault parts is verified by drilling and water pressure test. The test results show that the water permeability of the treated fault zones is less than 5Lu, which shows that the problems of leakage and seepage stability at the fault parts have been completely solved and meet the engineering safety requirements.

7 Conclusion

Faults are well developed in the dam foundation of Gushan Hub, and there are 9 large-scale fault fracture zones. In view of the problems that the fault fracture zone and fault intersection zone are easy to cause dam foundation settlement, deformation, leakage and seepage stability, in the process of construction, firstly, the construction water, rainwater, etc. are guided to the non-engineering parts to avoid deteriorating the mechanical properties of the fault fracture zone, which not only fully ensures the engineering quality, but also minimizes the excavation works of geological defects, engineering measures such as trench excavation and backfilling concrete plug, consolidation grouting and local increase of anchor rods are adopted for treatment [6]. Curtain grouting was adopted to deepen the CFF1 and F109 faults crossing the seepage control axis of the dam. After the engineering treatment, the dam foundation deformation monitoring data and the water permeability of the fault zone are within the scope of safety standards, indicating that the engineering treatment measures for the fault defect parts are appropriate and meet the design requirements. It should be pointed out that although the engineering treatment measures for the dam foundation fault fracture zone are ideal, the project has not passed the test of operation period. The aforementioned bedrock deformation monitoring data and fault water permeability are all detection means during the construction period [7]. In the operation period after the normal water storage of the project, some information-based automatic deformation monitoring and leakage monitoring methods should be carried out for the fault treatment parts, so as to accumulate more detailed basic data for the research on the settlement, deformation and leakage of the fault parts in the subsequent operation period of the project.

References

1. Weiwei, W., Cheng, T., Xijun, L. (2021). Treatment design of wide faults in high seismic intensity gravity dam foundation. *Journal of Engineering Geology, 52*(2), 119–123
2. Gaofeng, L., Shixiang, L., Nengwu, X., et al. (2021). Analysis and treatment of long structural plane of dam foundation rock mass of Baihe Hydropower Station. *People's Yangtze River, 52*(12), 122–125.
3. Qihua, Z., Hongwen, K., & Sheqin, P. (2023). Multi-fault combination model of foundation rock mass of Yebatan Hydropower Station and its engineering influence. *Acta Engineering Geology, 31*(2), 562–573.
4. Shurong, F., Zhongming, J., Huiya, Z., et al. (2015). Experimental study on deformation characteristics of broken rock mass under the dam foundation on the left bank of Xiangjiaba Hydropower Project. *Geotechnical Mechanics, S2*, 539–544.
5. Chelidze, T., Matcharashvili, T., Abashidze, V., et al. (2019). Complex dynamics of fault zone deformation under large dam at various time scales. *Geomechanics and Geophysics for Geo-Energy and Geo-Resources., 5*(4), 437–455.
6. Kim, H.-J., Lyu, Y.-G., Kim, Y.-G., et al. (2012). A study on optimization for location and type of dam considering the characteristic of large fault. *Tunnel and underground space., 22*(4), 227–242.

7. Ariga, Y., Kashiwayanagi, M., & Mizuhashi, Y. (2008). 3-D dynamic analysis method for coupled dam-foundation-fault system by utilizing progressive wave analysis. *Journal of applied mechanics, 11*, 633–640.

Flutter Control Effect and Mechanism of Closed Railing Aerodynamic Measure for Bridge

Xiujuan Jiang, Junyan Wu, Xiaoxia Ning, and Xu Liu

Abstract Through the wind tunnel test, the flutter control effect and mechanism of partly closed railing aerodynamic measure were studied. After taking effective pneumatic measures, the aerodynamic derivative A_2^* was changed from positive to negative forming aerodynamic damping, which increases the total damping of the bridge. The flutter frequency reduced and the critical wind speed of the flutter increased. The flutter type of the bridge is a damp driving in which the reversal of implicated movement is dominant. The performance of the main beam flutter was improved through increasing torsional movement of pneumatic positive damping A and reducing torsional coupling movement implicated generated aerodynamic negative damping term D. The degree of participation of the vertical bending freedom and the bending-torsion coupling was increased when the flutter occurs of the partly closed railing aerodynamic measures beam, and the fluttering shape was transferred from the single degree of freedom torsional vibration to the bending and torsional coupled vibration. Thereby, the flutter frequency was reduced and the critical flutter wind speed was improved.

Keywords Partly closed railing · Aerodynamic measure · Flutter control · Flutter-driving mechanism · Flutter modality

X. Jiang · X. Ning · X. Liu
CCCC Infrastructure Maintenance Group Co., Ltd, Beijing 100011, China
e-mail: 83855480@qq.com

X. Ning
e-mail: 362934301@qq.com

X. Liu
e-mail: yingzi_312@163.com

J. Wu (✉)
Research Institute of Highway Ministry of Transport, Beijing 100088, China
e-mail: wujunyan128@163.com

© The Author(s) 2025
D. Li and Y. Zhang (eds.), *Advances in Frontier Research on Engineering Structures II*,
Lecture Notes in Civil Engineering 535, https://doi.org/10.1007/978-981-97-6238-5_21

255

1   Introduction

The most critical aspect of long-span suspension bridges is the flutter stability of the bridge. Flutter is a self-excited, divergent vibration that, once it occurs, will cause destructive damage to bridges [1]. In order to improve the flutter stability of bridges, it is the most economical and effective safety control method to take corresponding aerodynamic measures, such as adding air nozzles, diversion plates, stabilization plates, and closed railings [2]. The Akashi Strait Suspension Bridge in Japan [3] is equipped with a lower central stabilizing plate, and the Runyang Yangtze River Bridge in China with an upper central stabilizing plate to increase the critical flutter wind speed of the bridge. Sometimes a single aerodynamic measure cannot meet the requirements of the bridge's wind resistance stability, and a comprehensive measure of a combination of multiple aerodynamic measures is required. The Aizhai Bridge [4] adopts the combined measures of a closed central slot, lower central stabilization plate, and 1 m airtight central anti-collision guardrail to increase the critical flutter wind speed of the bridge.

Larsen [5] studied the flutter mechanism of the Tacoma bridge based on the discrete vortex method in CFD. Zhang et al. [6] studied the flutter mechanism of long-span bridges with different main girder section shapes. Rong and Liao [7] studied aerodynamic optimization measures and Yang et al. [8–10] studied the flutter mechanism of several typical bridge sections and some aerodynamic control measures using the two-dimensional three-degree-of-freedom coupled flutter analysis method. Although many scholars have studied the flutter mechanism, there are relatively few studies on the aerodynamic mechanism under the action of aerodynamic control measures.

In this paper, under the background of improving the flutter stability of the bridge, the influence of various aerodynamic measures on its aerodynamic performance is studied, and the combined measures of the lower central stabilization plate, the horizontal deflector, and the partially closed railing are finally determined. The flutter analysis method is the theoretical basis to analyze the flutter control mechanism of the combined aerodynamic measures.

2   Engineering Situations and Pneumatic Measures

It is a suspension bridge, with a main cable span of $150 + 536 + 115$ m. The vertical-span ratio is 1: 11, and the center distance of the main cable is 15.6 m. Steel truss stiffening beams and orthotropic slab decks are used, and the tower adopts a light-tube concrete portal frame structure. Figure 1 shows the standard cross-section of the stiffening beam of the bridge.

The wind speed of the bridge flutter test was 53.1 m/s, and the original design scheme could not meet the flutter stability requirements. For this reason, the wind tunnel laboratory of Chang'an University conducted a wind resistance study of the

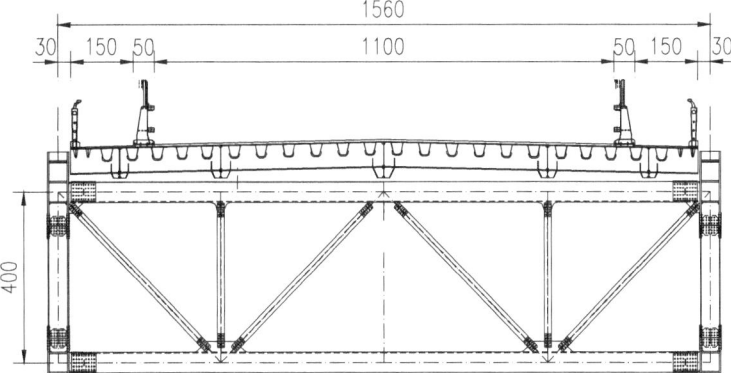

Fig. 1 Standard cross section of stiffening beam of the bridge

bridge and stiffened the steel truss through the segmental model wind tunnel test. The beam carried out aerodynamic performance optimization measures and tested various aerodynamic measures such as the central stabilizer plate. The lower central stabilizer plate and deflector are, the lower center stabilizer plate, the horizontal deflector, and the partially enclosed railing are. Table 1 shows the aerodynamic measures scheme and flutter critical anemometer.

It can be seen from Table 1 that the upper and lower central stabilizing plates of Scheme 3 can effectively improve the critical flutter wind speed. However, the bridge has two lanes and the layout is separated from the upper and lower sides. The upper central stabilizing plate divides the bridge deck into two independent parts that are not connected to each other, which brings inconvenience to the use of the bridge and the handling of possible traffic accidents. After a variety of program tests, the combined aerodynamic measures of the lower central stabilizer plate, horizontal deflector, and partially closed railing were finally determined to improve the flutter stability of the bridge. Figure 2 is a schematic diagram of the eight-part closed railing of the scheme.

3 Influence of Aerodynamic Measures of Central Stabilizer on Aerodynamic Derivatives

Through the wind tunnel test, the aerodynamic derivatives were identified for the flutter data under each working condition. Figure 3 shows the effect of the combined measures of the partially closed balustrade at $0°$ angle of attack on the flutter derivative.

It can be seen from Fig. 3 that at $0°$ angle of attack, the aerodynamic derivative A_2^* of the original section first decreases to a negative value with the increase of wind speed, and then becomes a positive value and gradually increases. The derivatives

Table 1 Aerodynamic measures and flutter critical anemometer

Conditions	Aerodynamic schemes		Flutter critical wind speed (m/s)					Test wind speed (m/s)
	Schematic diagram	Parameters	−5°	−3°	0°	3°	5°	
Original section				48.2	43.8	43.8		53.1
Condition I		$h_1 = 1.12$ m		>70.1	46.0	32.9		
Condition II		$h_1 = 1.12$ m, $h_2 = 0.8$ m	>70.1	>70.1	>70.1	>70.1	48.2	
Condition III		$h_1 = 1.12$ m, $h_2 = 1.28$ m	>70.1	>70.1	>70.1	>70.1	>70.1	
Condition IV		$h_1 = 1.12$ m, $L_1 = 1.28$ m		>70.1	66.0	13.1		
Condition V		$h_1 = 1.12$ m, $L_1 = 1.28$ m, $\alpha = 30°$		>70.1	39.4	30.7		
Condition VI		$h_1 = 1.12$ m, $L_2 = 1.28$ m, $\alpha = -30°$		48.2	39.4	30.7		
Condition VII		$h_1 = 1.12$ m, $L_2 = 1.28$ m, $\alpha = 30°$		39.4	30.7	30.7		
Condition VIII		$h_1 = 1.12$ m, $L_1 = 2.00$ m, (Fig. 2)	65.7	>70.1	65.7	>70.1	61.3	

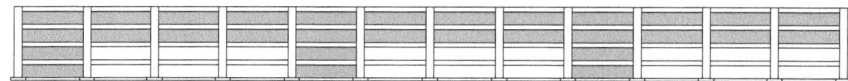

Fig. 2 Diagram of enclosed railing in condition VIII

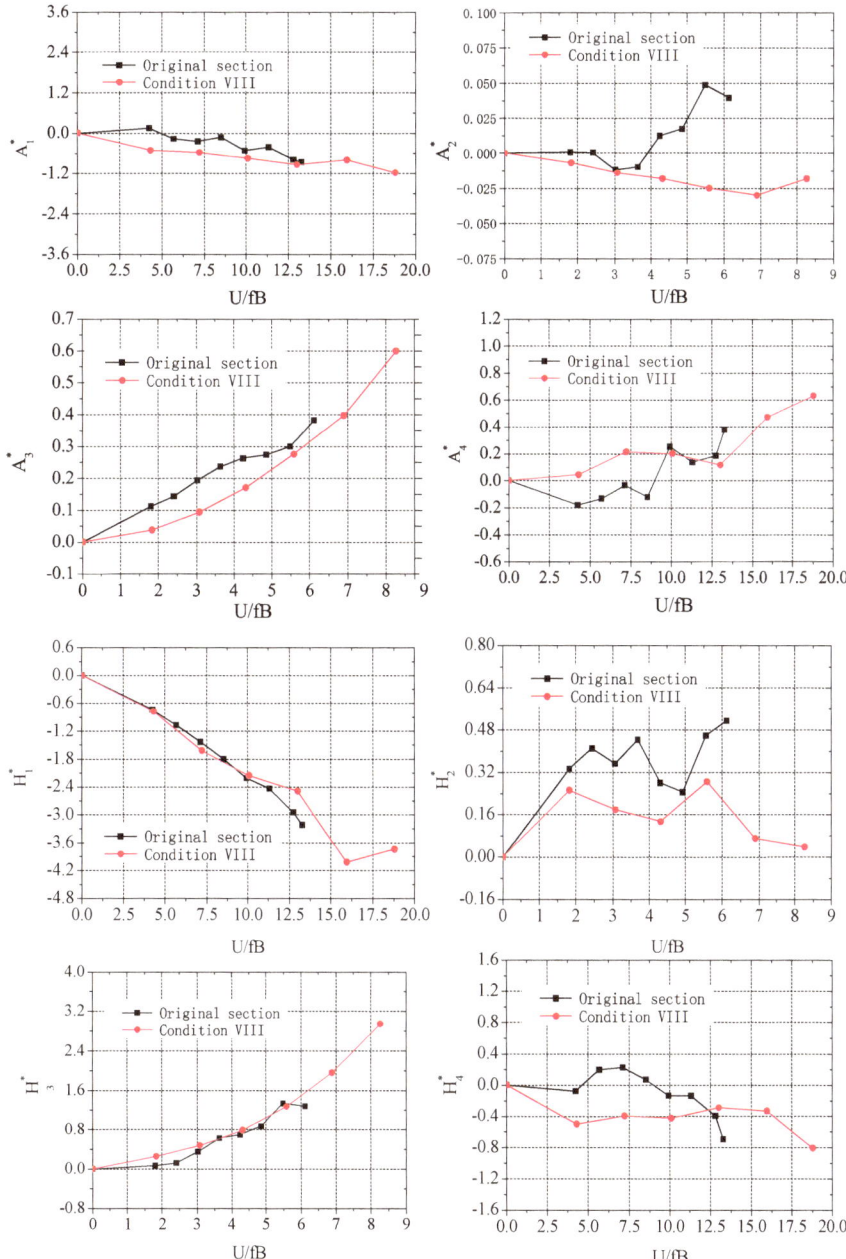

Fig. 3 Effect of combined measures of partially enclosed railings on flutter derivative (0°)

A_2^* are all negative values, and with the increase of wind speed, the absolute value first increases and then decreases. The aerodynamic derivatives A_1^*, A_3^*, H_1^*, H_2^* and H_3^* are basically the same as the development trend of the original section after the combination of partially closed railings, central stabilization plates, and deflectors are adopted, and the values are not much different. The measure section A_4^* is positive, and the original section changes from negative to positive with the increase in wind speed. The measure section H_4^* is negative, and the original section changes from positive to negative with the increase in wind speed.

4 Effect of Combined Pneumatic Measures of Partially Enclosed Railings on Aerodynamic Coefficients

Figure 4 shows the comparison of aerodynamic coefficients between the combined aerodynamic measures section and the original truss section of some closed railings. It can be seen from Fig. 4 that the aerodynamic measures affect the aerodynamic coefficient to varying degrees. In Scheme 8 (adding a lower stabilizer plate, a combined aerodynamic section of a horizontal deflector, and a partially closed railing), the lift moment coefficient C_M is larger than the original design section, and it is more obvious in the negative attack angle area than the positive attack angle area. In addition, compared with the original design section, the slope $dC_M/d\alpha$ of the lift moment coefficient with the angle of attack is always greater than 0, and $dC_M/d\alpha$ gradually increases, indicating that the torsional vibration is more stable, which is beneficial to improving the aerodynamic performance of the truss section.

5 Study on the Mechanism of Flutter Drive

5.1 Frequency and Damping Ratio

Figure 5 shows the variation law of the torsion and vertical bending frequencies of the system with the wind speed, and Fig. 6 shows the variation law of the damping ratio of the system's torsional motion and vertical bending motion with the wind speed. It can be seen from Figs. 5 and 6 that under the attack angles of 0°, 3°, and −3°, the vertical bending frequency and torsional frequency of the main beam do not change much with the wind speed, and they are not coupled to the same frequency. The vibration is damping-driven, and the dominant torsional-bending coupled flutter is the system torsional-involved motion flutter [7]. The torsional damping ratio first increases and then decreases with the increase of wind speed, and reaches the maximum value when the wind speed is 6 m/s, then decreases with the increase of wind speed, and gradually approaches 0, and flutter occurs.

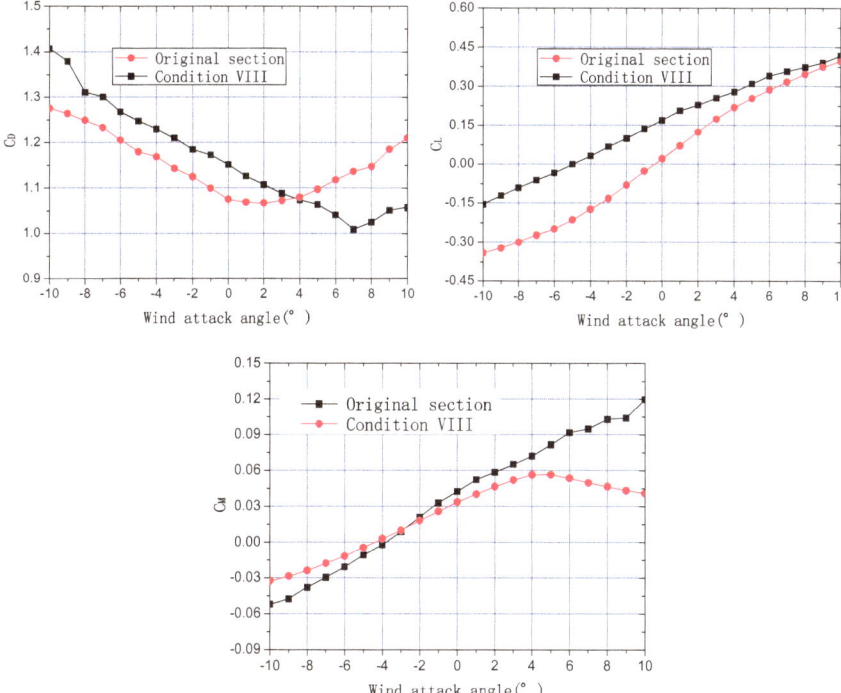

Fig. 4 a Drag coefficient. **b** Lift coefficient. **c** Lift moment coefficient

5.2 Torsional Motion Pneumatic Damping

The bridge truss section is the flutter instability phenomenon caused by the divergence of the torsional motion of the system. Therefore, the in-depth analysis of the development law of the aerodynamic damping of the torsional motion of the system is the key to studying its flutter driving mechanism. The torsional motion of the system includes the main motion of the torsional degrees of freedom and the coupled vertical motion based on the torsional frequency of the system. The coupled vertical motion based on the torsional frequency of the system is excited by the main torsional motion in the vertical degree of freedom, and it contains two components: one is the aerodynamic lift (the H_2^* term aerodynamic lift) generated by the speed of the torsional motion. The second is the coupled vertical motion excited in the vertical degree of freedom by the aerodynamic lift (the H_3^* term aerodynamic lift) generated by the displacement of the torsional motion. These two coupled vertical motions will feedback on the main torsional motion through the coupled aerodynamic lift moment, namely the A_1^* aerodynamic lift moment and the A_4^* aerodynamic lift moment, and together with the torsional main motion, jointly determine the motion law of the system torsional motion.

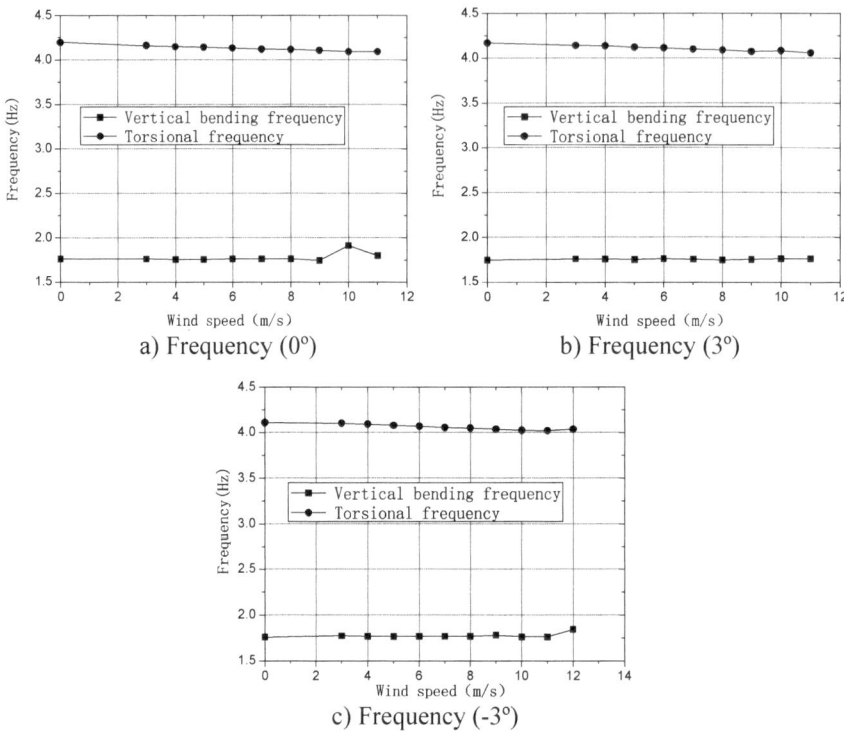

a) Frequency (0°)

b) Frequency (3°)

c) Frequency (-3°)

Fig. 5 The variation law of the torsion and vertical bending frequencies

According to the excitation-feedback principle, there are five ways to generate the aerodynamic damping of the torsional motion of the system [8].

(1) The aerodynamic damping is formed by the aerodynamic lift moment generated by the torsional motion speed and can be simply expressed as $-A_2^*$ (A);

(2) The aerodynamic damping is formed by the coupled aerodynamic lift moment, which is caused by the aerodynamic lift and the vertical velocity generated by torsional velocity, which can be expressed by $-A_1^* H_2^* \cos \theta_1$ (B);

(3) The aerodynamic damping is formed by the coupled aerodynamic lift moment, which is caused by the aerodynamic lift and the vertical displacement generated by torsional velocity, which can be expressed by $A_4^* H_2^* \sin \theta_1$ (C);

(4) The aerodynamic damping is formed by the coupled aerodynamic lift moment, which is caused by the aerodynamic lift and the vertical velocity generated by torsional displacement, which can be expressed by $-A_1^* H_3^* \cos \theta_2$ (D);

(5) The aerodynamic damping is formed by the coupled aerodynamic lift moment, which is caused by the aerodynamic lift, and the vertical displacement generated by torsional displacement, which can be expressed by $A_4^* H_3^* \sin \theta_2$ (E).

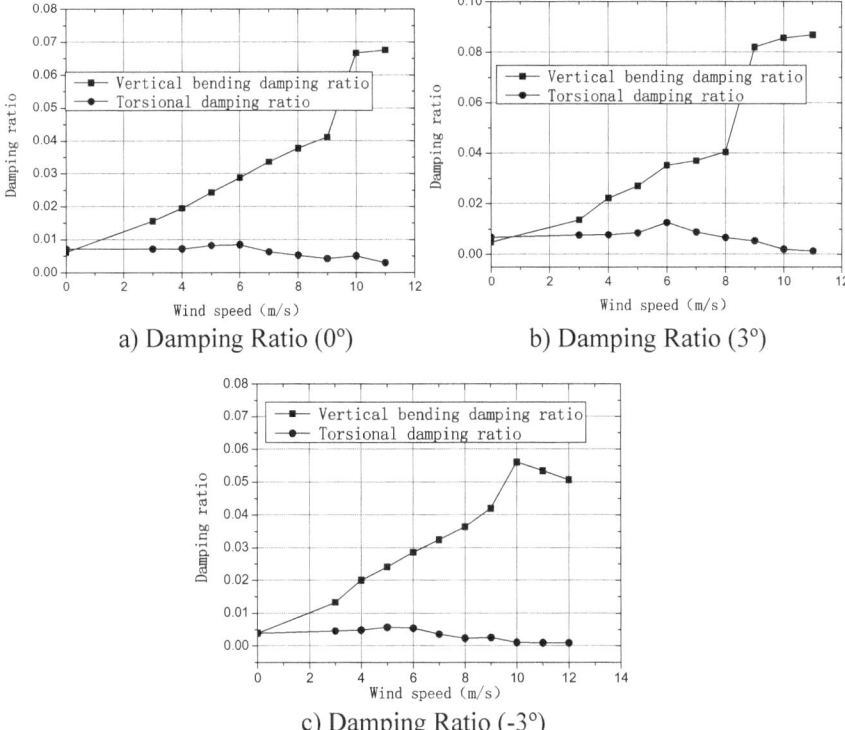

a) Damping Ratio (0°)   b) Damping Ratio (3°)

c) Damping Ratio (-3°)

Fig. 6 The variation law of the damping ratio

$$A \Leftrightarrow -1/2 \cdot \rho B^4/I \cdot A_2^* \quad B \Leftrightarrow -\rho^2 B^6/2m_h I \cdot \Omega_{h\alpha} \cdot A_1^* H_2^* \cos\theta_1$$
$$C \Leftrightarrow \rho^2 B^6/2m_h I \cdot \Omega_{h\alpha} \cdot A_4^* H_2^* \sin\theta_1 \quad D \Leftrightarrow -\rho^2 B^6/2m_h I \cdot \Omega_{h\alpha} \cdot A_1^* H_3^* \cos\theta_2$$
$$E \Leftrightarrow \rho^2 B^6/2m_h I \cdot \Omega_{h\alpha} \cdot A_4^* H_3^* \sin\theta_2$$

Figure 7 is the curve of the aerodynamic damping of the torsional implicated movement in the cross-section of the combined measures of the partially closed railing with the wind speed. After taking the combined measures of partially closed railings, the aerodynamic damping A generated by the torsional motion of the $-3°$, $0°$, and $+3°$ wind angle of attack itself increases with the increase of wind speed and always remains a positive value, which plays a role in stabilizing the system, D-term aerodynamic damping, which always remains negative, is the main reason for the divergence of torsional motion of the drive system.

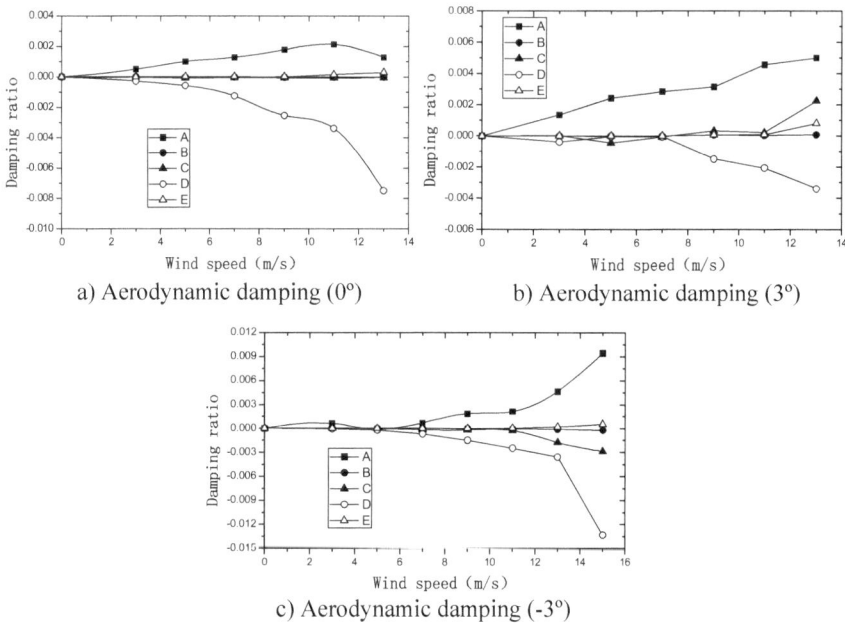

a) Aerodynamic damping (0°)   b) Aerodynamic damping (3°)

c) Aerodynamic damping (-3°)

Fig. 7 The curve of the aerodynamic damping

5.3 Flutter Modality

In addition to the study of the flutter driving mechanism, it is also an important way to explore the flutter mechanism to determine the flutter morphology, that is, to understand the participation of torsion and vertical degrees of freedom in the occurrence of flutter. For two-dimensional bridge segments with two degrees of freedom, the flutter shape vectors of torsional implicated motion and vertical bending implicated motion [7–9] in the flutter critical state can be calculated by Eqs. (1, 2).

$$V_\alpha = \left| \frac{\frac{\rho B^2}{m_h} \Omega_{h,\alpha} \sqrt{H_2^{*2} + H_3^{*2}}}{C_\alpha}, \frac{1}{C_\alpha} \right| \tag{1}$$

$$V_h = \left| \frac{1}{C_h}, \frac{\frac{\rho B^4}{I} \Omega_{\alpha,h} \sqrt{A_1^{*2} + A_4^{*2}}}{C_h} \right| \tag{2}$$

In the formula:

$$\sqrt{\left| \frac{\rho B^2}{m_h} \Omega_{h,\alpha} \sqrt{H_2^{*2} + H_3^{*2}} \right|^2 + 1}, \; C_h = \sqrt{1 + \left| \frac{\rho B^4}{I} \Omega_{\alpha,h} \sqrt{A_1^{*2} + A_4^{*2}} \right|^2}$$

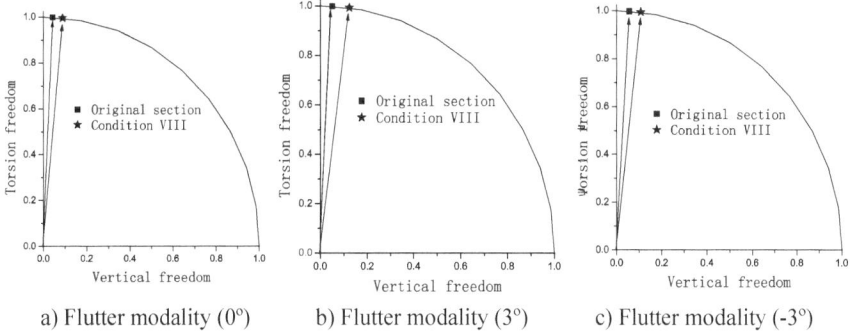

  a) Flutter modality (0°) b) Flutter modality (3°) c) Flutter modality (-3°)

Fig. 8 The flutter shape vector

Figure 8 shows the flutter shape vector of the original section and working condition 8 when the flutter critical state is reached. At 0°, 3°, −3° wind attack angles, the coupling degree of vertical degrees of freedom in case eight reaches the critical state of flutter, and the degree of coupling of vertical degrees of freedom is significantly enhanced compared with the original section, and the critical flutter wind speed of the section has also been significantly improved. The improvement of the degree of freedom participation is beneficial to improve the flutter stability. With the increase in the degree of torsional degree of freedom participating in the flutter, the critical flutter wind speed of the section decreases, and the increase in the degree of participation in the torsional degree of freedom is not conducive to improving the flutter stability.

6 Conclusion

In this paper, the influence of combined aerodynamic measures of partially enclosed railings on aerodynamic performance is studied through wind tunnel tests. The influence of aerodynamic performance is mainly reflected in the change law of aerodynamic derivatives. The direct aerodynamic derivative A_2^* related to torsional motion has the most significant change. After taking effective aerodynamic measures, A_2^* turns from positive to negative, forming aerodynamic damping. The total damping of the bridge section increases, the flutter frequency decreases, and the critical flutter wind speed increases.

The lift moment coefficient of the combined aerodynamic measures of the partially closed railing is always greater than 0, and the slope of the curve $dC_M/d\alpha$ is always greater than 0, and $dC_M/d\alpha$ increases gradually, indicating that the torsional vibration is more stable, which is beneficial to improving the aerodynamic performance of the truss section.

The flutter of the bridge is of the damping-driven type, and the torsional motion of the system is dominant. The flutter control mechanism of the combined aerodynamic measures with partially closed railings was analyzed by using the two-dimensional three-degree-of-freedom coupled flutter analysis method. The motion-generated D-term aerodynamic negative damping can improve the flutter performance of bridge sections in two ways.

The action mechanism of the combined aerodynamic measures of the partially enclosed railing is to increase the degree of participation of the vertical bending degree of freedom and the degree of torsional bending coupling when the flutter occurs, so that the flutter form is transferred from the single-degree-of-freedom torsional vibration to the bending-torsional coupling vibration, thereby reducing the flutter. The vibration frequency is increased to increase the critical flutter wind speed.

References

1. Simiu, F., & Scanlan, R. H. (1996). *Wind effects on structures (3rd Edition) [M]*. John Wiley & Sons.
2. Qinghai, G., Xin, C., Fangliang, W., Jiawu, L., & Jianxin, L. (2014). Wind tunnel test study of vertical vortex-induced vibration of bluff box girder suppressed by aerodynamic measure[J]. *Bridge Construction, 44*(1), 56–62.
3. Ueda, O., Tanaka, H., & Matsushita, Y. (1998). Aerodynamic stabilization for super long-span suspension bridges. In [C] *IABSE. Proc IABSE symposium long-span and high-rise structures*, pp. 721–728.
4. Zhengqing, C., KeJian, O., Huawei, N., & Xugang, H. (2009). Aerodynamic mechanism of improvement of flutter stability of truss-girder suspension bridge using central stabilizer [J]. *China Journal of Highway and Transport, 22*(6), 53–59.
5. Larsen, A. (2000). Aerodynamics of the Tacoma narrows bridge- 60 years later [J]. *Journal of Structural Engineering International, 10*, 243–248.
6. Xinjun, Z., Airong, C., Haifan, X., & Wei, P. (2002) Study on flutter mechanism of long-span bridges with different main girder sections. *China Journal of Highway and Transport[J], 15*(2), 52–56.
7. Rong, X. I. A. N., & Haili, L. I. A. O. (2008). Wind tunnel test study of aerodynamic optimization measures for flutter stability of closed flat steel box girder[J]. *World Bridges, 3*, 44–47.
8. Yongxin, Y., Yaojun, G., & Haifan, X. (2006). Research on the coupled bending-torsion flutter mechanism for thin plate section [J]. *Engineering Mechanics, 23*(12), 1–8.
9. Yongxin, Y. (2002). Two-dimensional flutter mechanism and its application for long-span bridges[D]. Shanghai: Doctoral Dissertation of Tongji University
10. Yongxin, Y., Yaojun, G., & Haifan, X. (2006). Flutter control effect and mechanism of central-slotting for long-span bridges[J]. *Chinese Journal of Civil Engineering, 39*(7), 74–80.

Seismic Performance of a Prefabricated Pre-embedded Steel–Concrete Column Base Connection

Hongyuan Gao, Jie Liu, Changsha Liu, Qun Xie, Jinpeng Gao, Kecheng Liu, and Gang Zhang

Abstract A prefabricated pre-embedded steel–concrete column base was presented which aimed to be applied in a petroleum pipeline gallery and three full-scale specimens were tested under quasi-static cyclic loading to investigate the seismic performance. Failure patterns, hysteretic characteristics, skeleton curves, displacement ductility, and stiffness of columns have been studied. The results showed that concrete spalling failure in the cast-in-place connection part of all specimens occurred and there was no visible crack in the column. The hysteretic curves of specimens indicated relatively good energy dissipation performance. The additional anchor rebar can effectively enhance the connection between the precast column and cast-in-place concrete. The strength along the strong axis and the weak axis was similar based on the comparison of seismic parameters. The specimens still maintained relatively high residual strength even after cast-in-place concrete failure which presented a potential possibility of engineering application in petroleum pipeline gallery.

Keywords Fabricated concrete column · Base connection · Embedded steel · Seismic performance

1 Introduction

Petroleum pipeline gallery is an important infrastructure for petroleum transportation and compared to traditional cast-in-place construction technology, prefabricated petroleum pipe gallery has many advantages, including higher construction quality, lower construction costs, and convenient maintenance [1–3]. Therefore, different types of prefabricated steel–concrete composite structures have been widely studied

H. Gao · J. Liu · C. Liu
China Construction Installation Group Co.Ltd, Ji'nan 250000, China

Q. Xie (✉) · J. Gao · K. Liu · G. Zhang
School of Civil Engineering and Architecture, University of Jinan, Ji'nan 250022, China
e-mail: 2868122374@qq.com

D. Li and Y. Zhang (eds.), *Advances in Frontier Research on Engineering Structures II*,
Lecture Notes in Civil Engineering 535, https://doi.org/10.1007/978-981-97-6238-5_22

in recent years. Zhang et al. [4–8] presented the seismic performance of steel–concrete composite columns and steel-reinforced concrete columns. The embedded circular steel tube generated a positive effect on displacement ductility and bending curvature characteristics. EI-Tawil et al. [9] conducted an experimental study on SRC special-shaped columns considering the inelastic behavior of steel and concrete. Wang et al. [10] conducted quasi-static tests on eight steel–concrete composite columns to study the influence of steel types and loading path on the seismic performance and analyzed the strain development of longitudinal rebar and steel. Gao et al. [11] conducted quasi-static tests on three types of concrete beam-column joints, and the results showed that the prefabricated concrete beam-column joints with sleeve connections presented optimal ultimate bearing capacity and energy dissipation capacity. Cao et al. [12, 13] designed a beam-column joint based on embedded steel into the end of a concrete beam which exhibited good ductility and energy dissipation.

The studies mentioned above mainly focus on steel–concrete composite structures and prefabricated beam-column connection joints. A new type of prefabricated pre-embedded steel–concrete column base has been proposed in this study which is designed for application of petroleum pipeline gallery engineering. Three full-scale specimens were designed and tested under quasi-static cyclic loading to investigate the seismic performance.

2   Experimen Test

2.1   Specimens Design

Two specimens with/without anchor rebar in the pre-embedded steel were designed. The precast concrete column was assembled to the base by bolts, and then concrete was cast in place to cover the steel. The length of the steel was 500 mm including 300 mm embedded part and 200 mm exposed part. The upper part of the steel was welded with the longitudinal rebar in the column, and the shear studs were welded on the steel flange to enhance the robustness. The welded additional rebar extended from the steel flange to ensure the integrity between steel and cast-in-place concrete. The effect of horizon load direction on seismic performance has been considered because of the stiffness difference in the strong axis and weak axis. The detailed information of specimens was shown in Table 1 and Fig. 1.

The compressive strength of precast column concrete is 43.5 MPa and high strength cement-based composite was used for the cast-in cover of the base with a strength of 50.5 MPa. HRB400 E and Q345 steel were adopted and the mechanical properties of steel bars were shown in Table 2.

Table 1 Detailed information on specimens

No.	Column section size/mm	Column height/mm	Steel size/mm	Steel length/mm	Connection mode	Loading direction
PC1	600 × 600	2000	HW400 × 400 × 21 × 24	500	Bolt + cast-in	Parallel flange
PC2	600 × 600	2000	HW400 × 400 × 21 × 24	500	Bolt + cast-in	Vertical flange
PC3	600 × 600	2000	HW400 × 400 × 21 × 24	500	Bolt + cast-in + anchor rebar	Parallel flange

2.2 Test Procedure

The axial load on the specimen was a constant of 1200 kN. The cyclic loading was completed by the MTS actuator, and the loading equipment was shown in Fig. 2. The loading system of combined force–displacement control was conducted. In the force-controlled stage, each load cycle was applied once while the load was repeated three times in the displacement-controlled stage. The loading setup has been shown in Fig. 3.

3 Experiment Results and Discussions

3.1 Failure Modes

For all specimens, the ultimate failure occurred in the zone of cast-in-place concrete as shown in Fig. 4. Compared with PC3, PC1 and PC2 were more seriously damaged. For PC3 the number and width of cracks in the cast-in-place concrete area were less than those of PC1 and PC2 since the restraining of additional anchor rebar. Similar crack development and failure process for all specimens could be seen in Fig. 5. Taking PC1 as an example, the failure process could be briefly described as follow:

(1) At the initial stage of loading, the strains of longitudinal rebar and concrete linearly developed and the specimen was basically in an elastic working state. The first oblique crack with a width of 0.05 mm appeared on the right side of the cast-in-place concrete area under the load of 30 kN as shown in Fig. 5a.

(2) With the load increase, two main diagonal cracks gradually formed which developed along both sides of the cast-in-place concrete, as shown in Fig. 5b. Additionally, cracking occurred at the interface between the cast-in-place concrete bottom and the base top surface. A triangular cone-shaped concrete isolated from the cast-in concrete zone with a maximum crack width of 15 mm.

Fig. 1 Specimens design

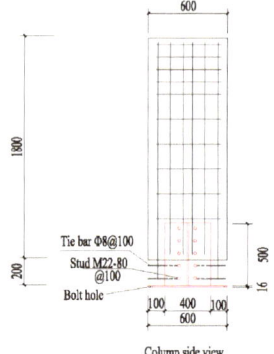

(a) Columns

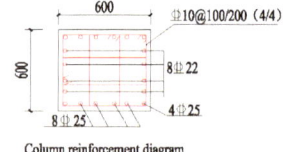

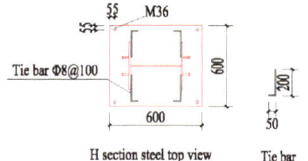

(b) Reinforcement diagram

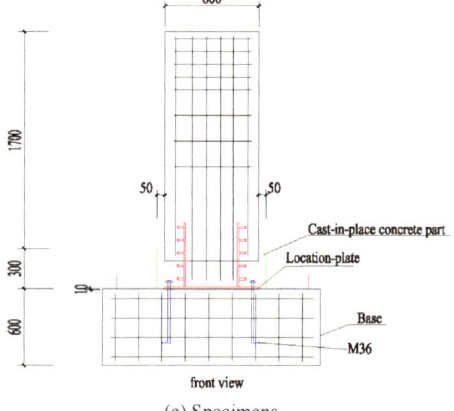

(c) Specimens

Table 2 Mechanical properties of steel bar

Type	Diameter/mm	Yield strength/MPa	Ultimate strength/MPa	Elongation/%	Elastic modulus/GPa
HRB400E	10	437	568	15.6	200
HRB400E	22	460	610	15.5	200
HRB400E	25	435	595	14.5	200

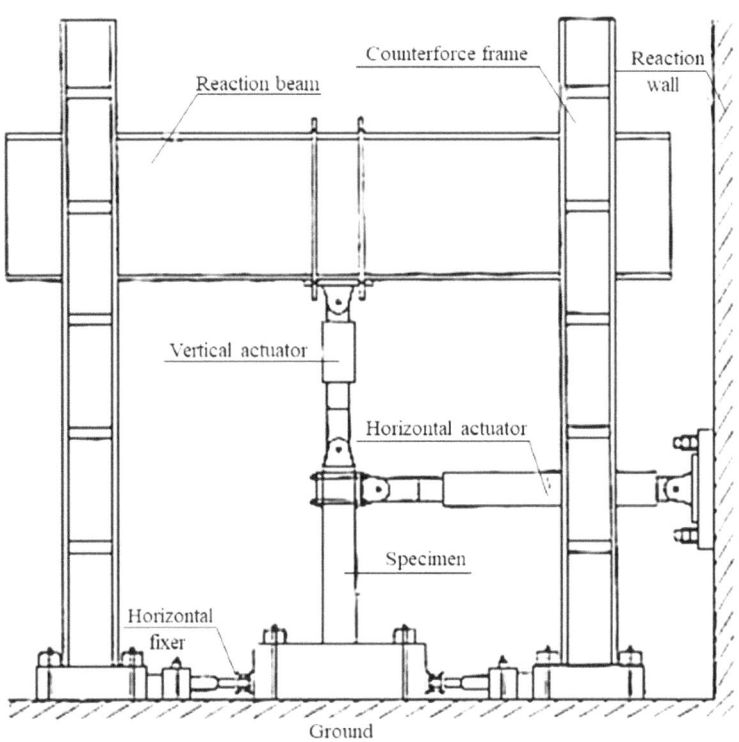

Fig. 2 Loading setup

(3) Near the ultimate state, a vertical crack appeared in the cast-in concrete region as shown in Fig. 5c. Subsequently, the cracks propagated rapidly with a sharp width increase. The specimen exhibited high deformation capacity even after reaching the peak load. In its ultimate state, the cast-in-place concrete region has been completely damaged with exposed steel. However, there was no apparent damage to the precast column surface.

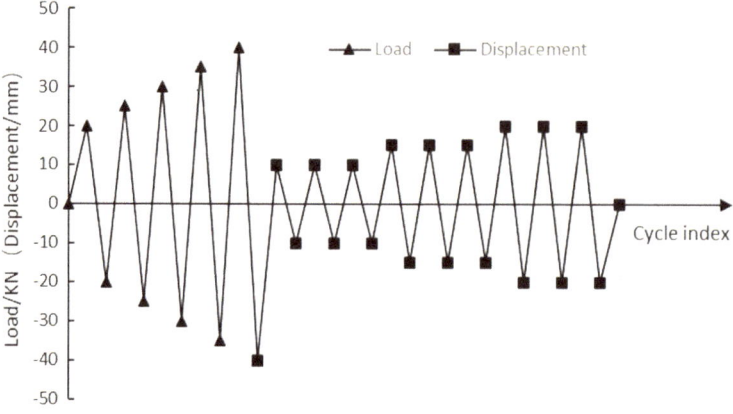

Fig. 3 Loading system

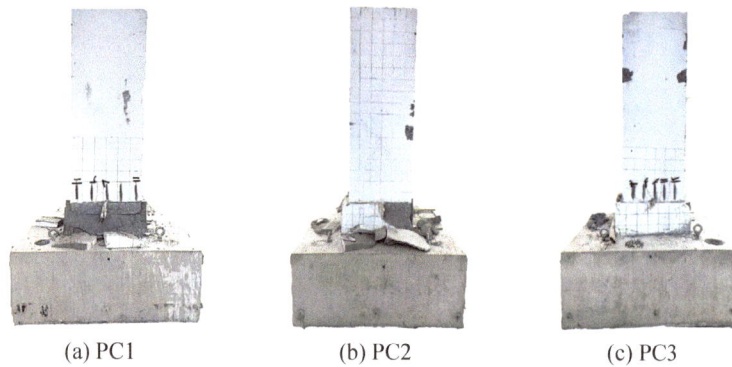

(a) PC1   (b) PC2   (c) PC3

Fig. 4 Failure modes

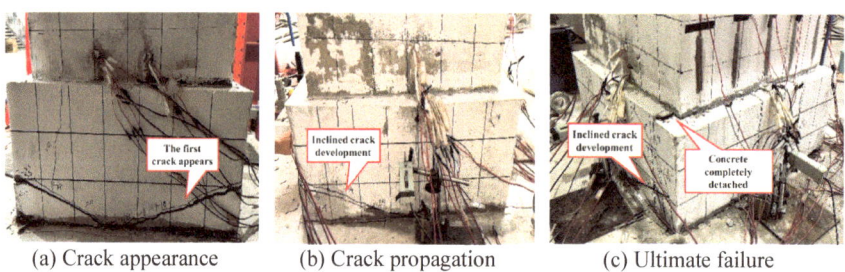

(a) Crack appearance   (b) Crack propagation   (c) Ultimate failure

Fig. 5 Failure process

3.2 Hysteretic Curve

The hysteretic curves were shown in Fig. 6. At the initial stage of loading, the hysteretic curves of specimens were linear, and the residual deformation was small. After the crack appeared in the cast-in concrete region, the hysteretic curves presented nonlinear growth. After peak load, the specimens kept a relatively high residual bearing capacity. At the bottom of PC3, due to the local concrete damage near the anchor bolt of the base, the bearing capacity decreases obviously.

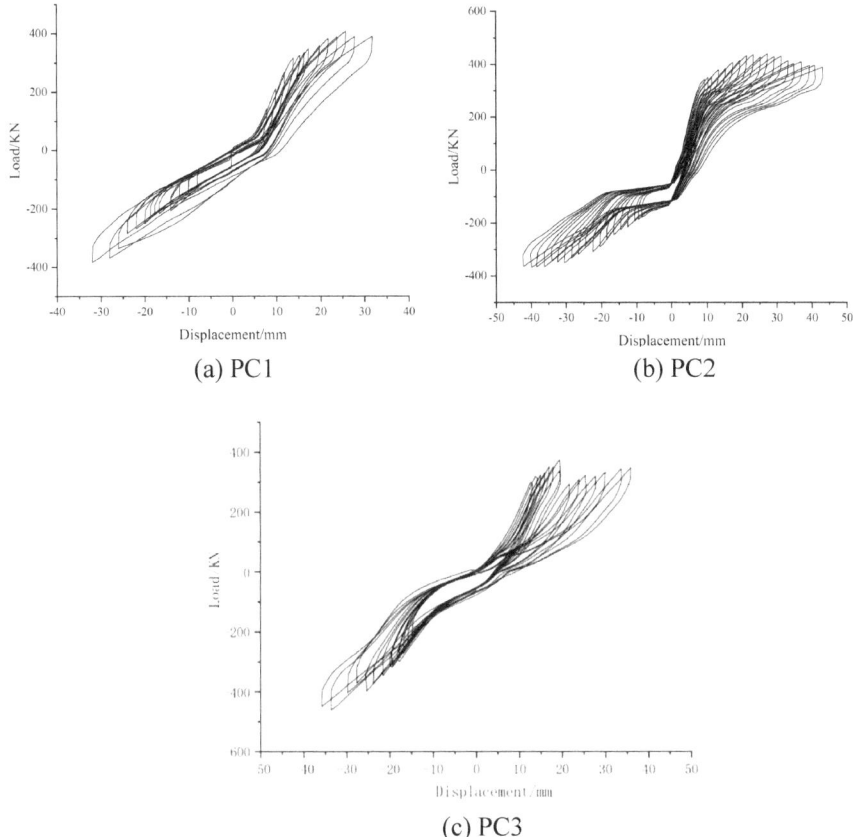

(a) PC1   (b) PC2

(c) PC3

Fig. 6 Hysteretic curves

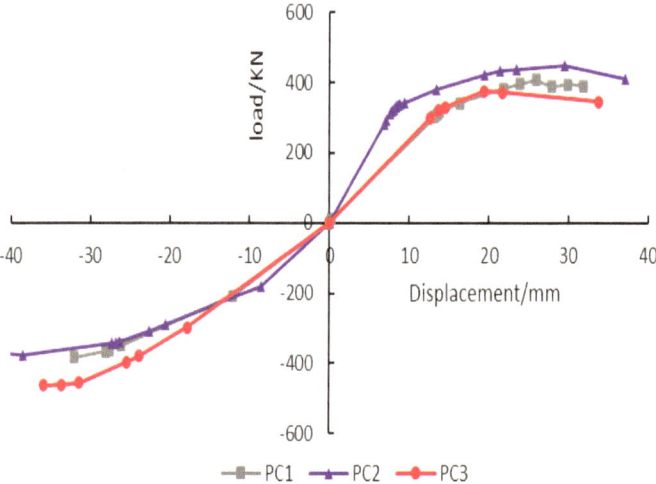

Fig. 7 Skeleton curves

3.3 Skeleton Curves

The skeleton curves of specimens were compared as shown in Fig. 7. PC3's ultimate displacement in both positive and negative directions is slightly higher than that of PC1, which indicated that the existence of additional anchor rebar could improve the strength and deformation capacity of the connection. The comparison showed that the stiffness of PC2 is larger than that of PC1.

3.4 Stiffness Degradation

The stiffness degradation curves were presented in Fig. 8. In the early stage of loading, the stiffness degradation was rapid due to the cracking of concrete and bolt slip. When the cast-in-place concrete completely lost load-bearing capacity, the stiffness degradation of specimens developed in a smooth mode. It could be found that the stiffness degradation rate of PC1 under each load cycle was smaller than that of PC2, meanwhile, the descending section of the stiffness degradation curve was relatively steeper because of the insufficient anchorage length of longitudinal reinforcement in the bottom of the column which resulted in a weaker connection between the precast column and the base.

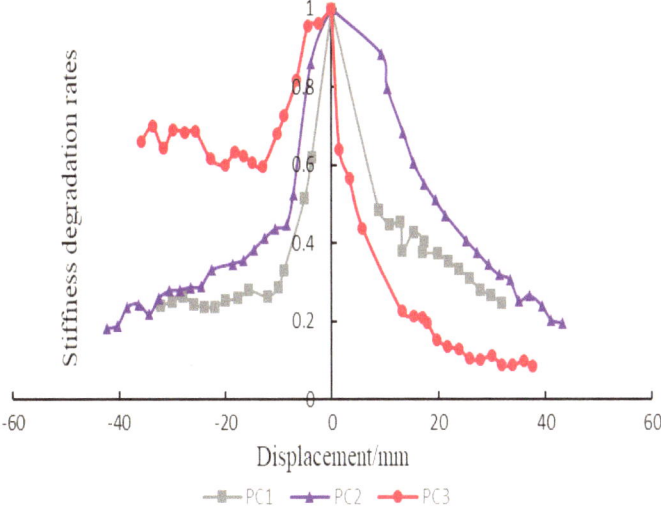

Fig. 8 Stiffness degradation curves

3.5 *Displacement Ductility*

The displacement ductility coefficients of specimens were determined by the energy equivalence method. As shown in Table 3, the displacement ductility coefficient of PC2 is greater than that of the other two specimens. The additional anchor rebar in PC3 generated a positive effect on crack control, but little contribution to the ultimate displacement.

Table 3 Displacement ductility coefficients of specimens

No.	Yield		Peak		Ultimate		Ductility	Average ductility
	P_y/kN	Δ_y/mm	P_m/kN	Δ_m/mm	P_u/kN	Δ_u/mm		
PC1	362.3	19.13	406.5	25.86	389.3	31.87	2.75	2.00
	−345.9	−25.67	−383.3	−32.05	−383.3	−32.05	1.25	
PC2	383.7	14.05	447.0	29.41	394.00	43.14	3.48	3.45
	−328.8	−25.03	−376.8	−38.44	−371.77	−42.25	3.23	
PC3	345.6	16.39	374.1	19.43	318.0	33.71	2.06	1.65
	−430.9	−28.82	−463.0	−35.85	359.71	35.98	1.24	

Note The positive values in the table represent the push loads, and the negative values represent the pull load

4　Conclusion

This study investigated the seismic performance of a prefabricated pre-embedded steel–concrete column base. The main conclusions were presented as follows.

(1) For all specimens the concrete damage occurred in the cast-in-place concrete zone with a similar failure process. The column base could still maintain good ductility and strength even after concrete failure and no visible crack in the column surface appeared.
(2) After the crack appeared, the hysteretic curves presented nonlinear growth. The specimens kept a relatively high residual bearing capacity even after peak load.
(3) The seismic performance along the major axis and the minor axis respectively was similar based on the comparison of PC1 and PC2. The measurement of additional anchor rebar on the steel flange could ensure a reliable connection between the prefabricated column and cast-in-place concrete.

References

1. Gang, W., & Decheng, F. (2018). Research progress on the fundamental performance of precast concrete frame beam-to-column connections[J]. *Journal of Building Structure, 39*(2), 1–16.
2. Deheng, Z., Aiqun, L., & Hong, J. (2014). Progress of investigation on seismic behavior of precast concrete frame structures (I): Study of joint property. *Industrial Construction, 44*(6), 95–100.
3. Congxiao, W., Weishan, L., & Yun, Z. (2015). Experimental study on seismic behaviors of new energy-dissipative prefabricated concrete frame structure joints. *China Civil Engineering Journal, 48*(9), 23–30.
4. Liu, Y., Guo, Z., Ding, J., Wang, X., & Liu, Y. (2020) Experimental study on seismic behaviour of plug-in assembly concrete beam-column connections[J]. *Engineering Structures, 221.*
5. Zhang, Y., Liu, Y., Ruan, J., Liu, Y., Zhong, P., & Yu, T. (2022). Experimental study on seismic performance of prefabricated core steel tube reinforced concrete columns [J]. *Journal of Building Structures, 43*(12), 135–144.
6. Wang, P., Yu, B., & Shi, Q. (2023). Experimental research on seismic performance of new-type steel reinforced concrete composite column [J/OL]. *Industrial Construction* 1–12[2023–03–08].
7. Wang, Q., Shi, Q., Jiang, W., Zhang, X., Hou, W., & Tian, Y. (2013). Experimental study on seismic behavior of steel reinforced concrete columns with new-type cross sections [J]. *Journal of Building Structures, 34*(11), 123–129.
8. Wang, J., & Zhang, H. (2017). Seismic performance assessment of blind bolted steel-concrete composite joints based on pseudo-dynamic testing[J]. *Engineering Structures, 131*(Jan. 15), 192–206.
9. Ei-Tawil, S., & Deierlein, G. G. (1999). Strength and ductility of concrete encased composite columns [J]. *Journal of Structural Engineering, ASCE, 125*(9), 1009–1019.
10. Wang, P., Shi, Q., Wang, F., & Zhang, T. (2018). Experimental study on seismic behavior of steel-concrete composite columns under bidirectional loading paths [J]. *Journal of Building Structures, 39*(01), 52–60.
11. Gao, X., Xu, L., Li, J., & Cao, Y. (2016). Experiment and mechanical behavior analysis on the precast concrete beam-column joint [J]. *Journal of Huazhong University of Science and Technology (Natural Science Edition),44*(10), 47–52.

12. Cao, Y., Sun, Q., Gong, W., Wei, H., Ding, W., & Wang, G. (2016). Experiment on new section steel joints or prefabricated concrete frames [J]. *Journal of Architecture and Civil Engineering, 33*(02), 15–23.
13. Liu, C., Wang, Q., Wang, Y., & Shen, D. (2013). Flexural behavior of precast reinforced concrete beams with steel end connectors [J]. *Journal of Building Structures, 34*(S1), 208–214.

The Influence of Nonlinear Characteristics of Diagonal Rod Insulators on the Seismic Performance of Converter Valve

Lihong Zhang, Jin Xiao, Di Zeng, and Wei Lü

Abstract Seismic safety of the converter valve is crucial for the safe operation of the entire flexible DC transmission system. In order to enhance the seismic resistance of the supported converter valve, composite diagonal rod insulators were designed to be added between the support insulators, which have a nonlinear property that can only be subjected to tension but not compression. Using finite element method, modal analysis of the valve structure was first conducted to study the impact of diagonal rod insulators on the dynamic characteristics of the valve structure. Secondly, artificial seismic waves were input for dynamic time history analysis, and the differences in dynamic response between two valve structure models with nonlinear and linear characteristics of diagonal rod insulators were compared and analyzed. Finally, according to the current specification requirements, three seismic waves were input to conduct seismic safety analysis and evaluation of the valve structure with nonlinear property of diagonal rod insulators. The results indicate that the acceleration and stress response results of the nonlinear valve structure model are greater than those of the linear model, indicating that using a linear model for seismic safety evaluation is more risky. Therefore, it is necessary to consider the nonlinear characteristics of diagonal rod insulators for seismic performance analysis of valve structure.

Keywords Converter valve · Diagonal rod insulator · Nonlinear characteristics · Transient dynamic analysis · Seismic performance

L. Zhang (✉) · J. Xiao · D. Zeng · W. Lü
China Institute of Water Resources and Hydropower Research, Beijing 100048, China
e-mail: zhanglihong08@tsinghua.org.cn

J. Xiao
Xuji Electric Co., Ltd., Xuchang 461000, Henan, China

© The Author(s) 2025
D. Li and Y. Zhang (eds.), *Advances in Frontier Research on Engineering Structures II*,
Lecture Notes in Civil Engineering 535, https://doi.org/10.1007/978-981-97-6238-5_23

1 Introduction

Flexible DC transmission, as a new type of DC transmission system, has become a way for offshore wind farms to connect to onshore power grids. This is because it can independently regulate active and reactive power, provide synchronous AC power for wind farms, and has strong fault crossing ability [1]. The converter valve, as the core component of the flexible DC transmission system, is very expensive, and its safe operation is closely related to the safe and stable operation of the entire flexible DC system. During the long-term operation of the converter valve, extreme situations such as earthquakes may occur, so the seismic safety of the converter valve has attracted high attention from researchers and engineers [2, 3].

According to the different fixing methods of the converter valve in the valve chamber, there are two structural types of valve structures: supported and suspended [4]. The seismic performance of converter valves has been studied by many domestic and foreign researchers. Wu et al. studied the dynamic characteristics and seismic response of suspended valve structures through shaking table test [5–7]. Nakagaki, Nakajima F, and Gao B conducted shaking table tests on the scaled model or partial equivalent component model of the supported converter valve to obtain the natural vibration frequency, displacement, acceleration, and other dynamic responses of the valve [8–10]. In addition to the seismic simulation shaking table test method, finite element numerical simulation technology is also widely used in the seismic research of valve structures. Scholars such as Yu H and Wu X, as well as Huang and Xu respectively, used response spectrum method and time history integration algorithm to analyze the dynamic response of suspended converter valves under seismic loads [11–14]. Zhu used the response spectrum method to study the vibration response law of a supported valve structure [15]. Tu used the time history method to study the dynamic characteristics of the supported valve structure and verified the effectiveness of lead rubber isolation bearings [16]. The support type converter valve is more commonly used in converter stations due to its simple installation and low requirements for the strength of the valve hall roof truss. In order to enhance the seismic resistance of the supported converter valve tower structure, composite diagonal rod insulators, which have properties of light weight, high strength, and pollution flashover resistance, have been designed to be added between the bottom or interlayer support insulators. The end of the diagonal rod insulator is equipped with an adjustment device to solve the on-site safety problem of difficult control of insulator length tolerance [10]. However, the mechanical properties of diagonal insulators are different from those of support insulators, as they can only withstand tension but not pressure, which poses difficulties in numerical simulation due to their nonlinear characteristics. Scholars such as Zhang Q used their own programs to study the dynamic characteristics and seismic resistance of rod type lightning arresters [17], but there are few reports on the seismic performance of rod type converter valves in existing literature.

This article establishes a finite element model based on an actual engineering converter valve. Firstly, modal analysis was conducted on the valve structure to obtain its dynamic characteristics. Then, time history integration algorithm was used to compare and analyze the differences in the dynamic response of the valve tower structure when the diagonal rod insulator only has tensile properties and when it also has tensile and compressive characteristics. Finally, according to the current specifications, three seismic motion time histories were input to conduct seismic safety analysis and evaluation of the valve structure considering the nonlinear characteristics of the diagonal rod insulator.

2 Valve Tower Structure

The research object of this article is the DC energy consumption device valve tower, which consists of five valve layers and is a layered double row support structure. In order to improve seismic resistance, diagonal rod insulators are added between the bottom support insulators, forming a staggered network structure as a whole. The diagram of the valve structure is shown in Fig. 1. The size of the valve structure is 11.5 m long, 4.75 m wide, 12.88 m high, and the total mass is about 75 t. In the finite element model of the valve tower, the vertical direction is defined as the positive direction of the Z-axis, and the plane is horizontally oriented along the long and short sides in the X and Y directions, respectively.

In the finite element model, beam elements are used to simulate components such as supporting insulators, aluminum beam brackets, and insulation crossbeams. The diagonal rod insulator is simulated using nonlinear spring elements, and its stiffness in

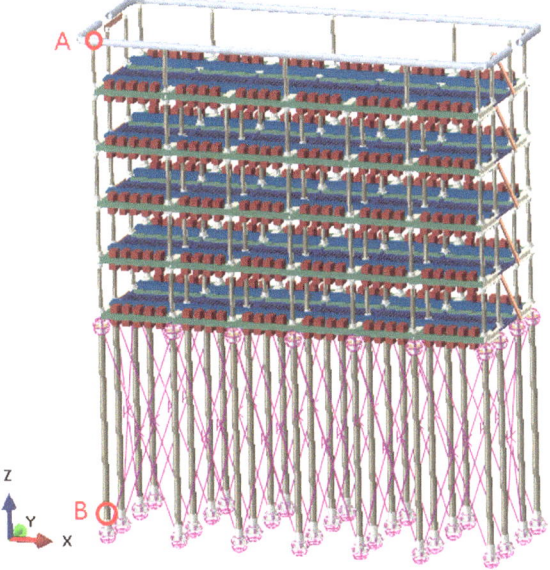

Fig. 1 Valve tower structure diagram

Table 1 Material properties of the valve structure

Material name	Elastic modulus (GPa)	Poisson's ratio	Density (kg/m^3)	Allowable stress (MPa)
Aluminum beam	71	0.34	2800	270
Insulation beam	17.6	0.34	2000	100
Cast steel flange	200	0.3	7800	350
Cast aluminum flange	69	0.34	2700	205
Diagonal insulator	25	0.34	1800	550
Support insulator	40	0.34	1800	155

the tensile stress zone is obtained from material properties and geometric information, while the stiffness in the compressive stress zone is close to 0 [4]. Linear diagonal rod insulators are simulated using linear elastic springs, and the spring stiffness is the same as the tensile stress zone stiffness of nonlinear springs. The material properties in the finite element model of the valve structure are shown in Table 1.

3   Modal Analysis

Modal analysis was conducted on three types of valve structure models, including those without diagonal rod insulators, those with diagonal rod insulators, and those with diagonal rod insulators with stiffness increased by one time. The first mode of vibration of the valve structure is translational motion along the Y-axis, which is the short side direction of the structure. The second mode of vibration is translational motion along the X-axis, which is the long side direction of the structure. And the third mode of vibration is in-plane rotation. The results of the first 10 frequency values of the valve structure are shown in Table 2. It can be seen that compared with the valve structure without diagonal rod insulators, the first order frequency of the model with diagonal rod insulators increases from 0.843 to 1.324 Hz, and the first order frequency of the valve structure increases with the increase of the elastic modulus of the diagonal rod insulators.

Table 2 The first 10 frequency values of the valve structure

Mode number	No diagonal rod insulators (Hz)	With diagonal rod insulators (Hz)	Twice the elastic modulus of diagonal rod insulators (Hz)
1	0.843	1.324	1.400
2	1.235	1.511	1.668
3	1.338	1.709	1.901
4	1.647	1.991	2.113
5	3.589	4.532	5.085
6	3.799	4.715	5.263
7	5.888	5.927	5.965
8	7.298	7.390	7.470
9	7.466	7.921	8.454
10	7.616	8.087	8.635

4 Transient Dynamic Results

4.1 Influence of Nonlinear Diagonal Rod Insulators on the Dynamic Response of Valve Structure

Newmark-β time integration algorithm is used for the transient dynamic analysis. And Rayleigh damping is used to consider the impact of structure damping. The acceleration time history of the artificial seismic waves used in the transient dynamic analysis are shown in Fig. 2. The peak horizontal seismic acceleration is 0.2 g, and the peak vertical acceleration is 0.13 g.

At the bottom of the valve structure, both Y-direction and Z-direction artificial seismic waves are input simultaneously. Two types of valve structure models were analyzed using the transient dynamic method. One model has linear characteristics for diagonal rod insulators, while the other model has nonlinear characteristics that can only withstand tension but not pressure (referred to as linear model and nonlinear

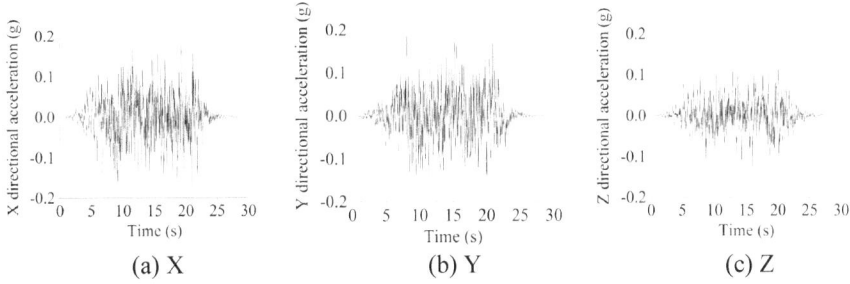

(a) X   (b) Y   (c) Z

Fig. 2 Acceleration curves of artificial seismic wave

model). Figures 3, 4 and 5 show the results of transient dynamics for two models. Figure 3 shows the y-direction acceleration time history curve at the top grading ring of the valve structure (as shown in inspection point A in Fig. 1), Fig. 4 shows the axial stress time history curve of the valve structure support insulator root (as shown in point B in Fig. 1), and Fig. 5 shows the curve of the axial force of the diagonal rod insulator between two rows of towers, which clearly shows that in the nonlinear model, the diagonal rod insulator only has a positive tensile force and no pressure. The results from Figs. 3, 4 and 5 show that the maximum values of the acceleration response at the top of the valve structure, the root stress of the support insulator, and the axial force of the diagonal insulator calculated by the nonlinear model are greater than the corresponding results calculated by the linear model. This indicates that the dynamic response results of nonlinear models are generally greater than the calculation results of linear models. Therefore, it is necessary to consider the tensile and incompressible nonlinear characteristics of diagonal rod insulators when conducting seismic safety analysis of valve structure, as using the calculation results of linear models for seismic safety evaluation is biased towards danger.

Fig. 3 Time history of y-direction acceleration of the top inspection point

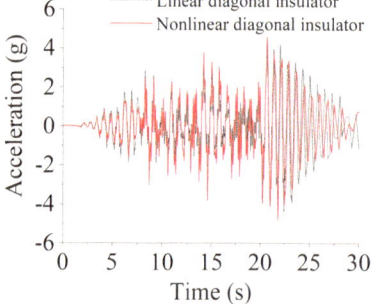

Fig. 4 Axial stress time history of the root of support insulator

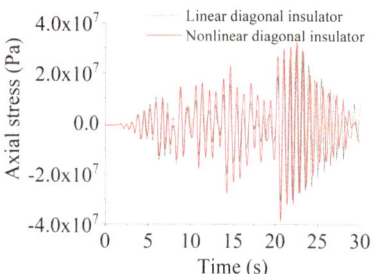

Fig. 5 Time history of axial force of diagonal rod insulators

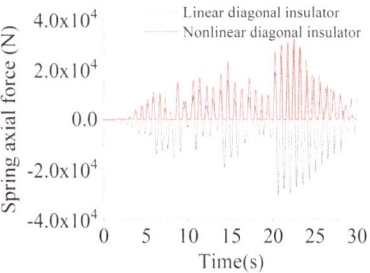

4.2 Seismic Performance Analysis of Valve Structure Based on Nonlinear Models

According to the "Code for Seismic Design of Electric Power Facilities" (GB 50,260–2013), when using the dynamic time history method for seismic analysis, the input seismic motion time history should not be less than three, and at least one artificially synthesized seismic motion time history should be included. Therefore, for the nonlinear model with only tension characteristics of diagonal rod insulators, in addition to the artificial seismic waves described in Sect. 4.1, two natural seismic waves were input, namely Elcentro wave and Taft wave. Elcentro wave belongs to near field seismic wave, with wide frequency band and relatively uniform distribution, suitable for Class II sites, while Taft wave belongs to medium and far field seismic wave, suitable for Class III sites. All seismic wave amplitudes were adjusted to the peak horizontal acceleration of 0.2 g, and the peak vertical acceleration was taken as 65% of the horizontal acceleration. When conducting transient dynamic analysis of each seismic wave, the two horizontal seismic actions in the X and Y directions are combined with vertical seismic actions, respectively. Referred to as "X + Z" working condition and "Y + Z" working condition respectively, in addition, the self weight effect is also considered in both working conditions. The displacement peak values of the inspection point at the top of the valve structure (as shown in point A in Fig. 1) under different working conditions are listed in Table 3. It can be seen that the displacement value at the top of the valve structure along the horizontal direction is relatively large, with a maximum of about 0.086 m, and the vertical displacement value is relatively small. The peak stress results of the main components of the valve structure are shown in Table 4. From the results in the table, it can be seen that the stress results of the grading ring, aluminum beam, insulation beam, support insulators and other components of the valve structure do not exceed the allowable stress of the material, meeting the strength requirements of GB50260-2013 "Code for Seismic Design of Electric Power Facilities" [18].

Table 3 Peak displacements of the valve (unit: m)

	Artificial seismic wave		Elcentro wave		Taft wave	
	X + Z	Y + Z	X + Z	Y + Z	X + Z	Y + Z
X direction	0.075	0.004	0.086	0.004	0.065	0.005
Y direction	0.001	0.061	0.001	0.062	0.001	0.086
Z direction	0.001	0.001	0.001	0.001	0.001	0.001

Table 4 Peak stress results of the main components of valve (unit: MPa)

	Artificial wave		Elcentro wave		Taft wave		Allowable stress	Whether the requirements are met or not
	X + Z	Y + Z	X + Z	Y + Z	X + Z	Y + Z		
Grading ring	1.30	1.84	1.55	2.08	1.35	2.61	260	Yes
Aluminum beam	8.21	55.83	9.41	65. 23	7.69	89.28	260	Yes
Insulation beam	34.64	32.70	38.14	27.74	30.00	43.27	100	Yes
Cast steel flange	12.71	99.92	13.60	102.73	11.12	144.85	350	Yes
Aluminum flange	4.26	17.64	4.93	18.81	3.72	26.35	205	Yes
Support insulator	2.41	31.85	4.15	40.43	4.03	56.74	150	Yes
Diagonal insulator	67.42	83.24	74.53	89.34	56.55	124.55	550	Yes

5 Conclusions

Through modal analysis and transient dynamic analysis of the valve structure, the following conclusions are drawn in this article:

(1) The presence of web arranged diagonal rod insulators will increase the natural vibration frequency of the valve structure, and the natural vibration frequency of the valve structure will increase with the increase of the stiffness of the diagonal rod insulators.

(2) The dynamic response results of the valve structure model when the diagonal rod insulator attribute is nonlinear are generally greater than the results of the valve structure model when the diagonal insulator attribute is linear. This means that using the calculation results of the linear model for seismic safety evaluation is biased towards danger. Therefore, it is necessary to consider the tensile and incompressible nonlinear characteristics of diagonal rod insulators.

(3) By inputting three seismic waves, the peak calculated stress of the key components of the energy dissipation valve structure studied in this article is less than the allowable stress of the material, and its seismic performance meets the strength requirements in the GB50260 specification.

Due to limited time and space, some of the research work in this article needs to be further deepened and improved. In the future, a model considering the nonlinear characteristics of diagonal rod insulators can be used to further analyze the impact of parameters such as stiffness of diagonal rod insulators on the seismic performance of valve structures.

References

1. Xu, Z. (2017). *Flexible DC transmission system (second edintion)*. China Machine Press, Beijing. *ISBN978711401858.*
2. Larder, R. A., Gallagher, R. P., & Nilsson, B. (1989). Innovative seismic design aspects of the intermountain power project converter stations. *IEEE Transactions on Power Delivery, 4*(3), 1708–1714. https://ieeexplore.ieee.org/
3. Yang, Z., Xie, Q., Zhou, Y., & Mosalam, K. M. (2018). Seismic performance and restraint system of suspended 800 kV thyristor valve. *Engineering Structures, 169,* 179–187. https://doi.org/10.1016/j.engstruct.2018.05.022
4. Zhang, L., Xiao, J., Lv, W., & Zeng, D. (2021). Research on seismic analysis of converter valve of offshore flexible high voltage direct current considering the effect of initial displacement. E3S Web of Conferences. *EDP Sciences, 01037,* 1–5. https://doi.org/10.1051/e3sconf/202125 201037
5. Wu, J. X. (2012). Study on the seismic collapse capacity of large converter station valve hall structure with electric equipments. Guangzhou.Guangzhou University. https://kns.cnki.net
6. Wei, W. H., Zhang, D., Yu, M., & Zhung, D. (2014). Seismic response of structure with suspended mass. *Journal of Harbin Institute of Technology, 46,* 98–104. https://kns.cnki.net
7. Zhang, W. W., Yu, H. B., & Gao, B. (2022). Experimental research on shaking table for seismic simulation of UHV DC suspended converter valves. *High Voltage Apparatus, 58*(08), 142–149. https://kns.cnki.net
8. Nakagaki, S., Uchida, T., Koyama, M., Ikegame, H., Kobayashi, A., Takeda, H., & Ohkubo, H. (1991). Studies on aseismic measure for multistage thyristor valves. *IEEE Transactions on Power Delivery, 6*(4), 1819–1824.
9. Nakajima, F., Yamazaki, T., Itoh, K., Matsumoto, T., & Sakai, T. (1993). Design and type test of a light-triggered thyristor valve for back-to-back systems. *IEEE Transactions on Power Delivery, 8*(1), 31–37. https://doi.org/10.1109/61.180316
10. Gao, B., Zhang, W., Yu, H., & Liu, G. (2018). Anti seismic design of a supported converter valve structure. *Machine Design and Manufacturing Engineering, 47*(11), 35–39.
11. Yu, H. B., & Liu, B. (2016). Application of FEA to seismic design of uhv converter valve tower structure. *High Voltage Apparatus, 52*(8), 184–188. https://doi.org/10.13296/j.1001-1609.hva.2016.08.031
12. Wu, X. F., & Sun, Q. G. (2011). Seismic analysis of an UHV suspended converter valve structure. *World Earthquake Engineering, 27*(2), 207–210. https://doi.org/10.1080/17415993.2010.547197
13. Huang, M. Z. (2021). Method of damping ratio calculation and seismic analysis of suspended converter valve. Beijing: North China University of Technology. https://kns.cnki.net
14. Xu, J., Xie, Q., Yang, Z., Zhuo, R., Hu, R., & Sun, B. (2020). Seismic performance analysis of±800 kv uhvdc converter valve hall System [J]. *High Voltage Apparatus, 56*(1), 96–103.

15. Zhu, X. H., Zhang, C., & Xie, Y. L. (2014). Research on anti-seismic performance of IGBT valves of Flexible HVDC. *China Electric Power (technology), 08*, 19–22. https://kns.cnki.net

16. Tu, M. H. (2022). *Study on parameter identification and isolation performance of supported flexible DC converter valve structure.* Guangzhou: South China University of Technology. https://kns.cnki.net

17. Zhang, Q. H., Yang, Y. D., & Su, W. Z. (1989) Seismic analysis of insulator rod electrical equipment. *World Earthquake Engineering, 02*, 30–38. https://kns.cnki.net

18. China Ministry of Construction. (2013). *GB50260--2013 code for seismic design of electrical installation.* Beijing: Chinese Plan Stress. http://www.nssi.org.cn/

A Project-Level Pavement Performance Prediction Framework Based on Machine Learning and Time Series Data

Xinyu Wei and Hui Wang

Abstract The numerical deviations of the existing pavement performance prediction models for project-level forecasting make it not suitable for maintenance and repair (M&R) funding planning. Project-level forecasting lacks multi-feature parameters, and high-frequency maintenance makes it lack sufficient time series data. We have constructed a prediction framework to overcome the problem of insufficient data and large forecast bias. The performance of the time series status data-based prediction model is stable. Time series status data play important roles in improving the prediction accuracy, and the effect is greater than that of the dataset size. The difference in prediction accuracy between the machine learning algorithms is not significant. The variable selection conclusions obtained by the mean decrease impurity (MDI) sorting can effectively support random forest (RF), gradient boosting decrease tree (GBDT), and extreme gradient boosting (XGboost) prediction models. The dimensionality of feature data is greatly reduced. The 1–5 year prediction deviations of ride quality index (RQI) and rutting depth index (RDI) are basically within ± 2, which demonstrates the framework is an effective project-level forecasting method for pavement performance.

Keywords Project-level management · High-frequency maintenance · Feature parameter selection · Pavement performance prediction · Time series status data · Machine learning

X. Wei · H. Wang (✉)
Key Laboratory of New Technology for Construction of Cities in Mountain Area, School of Civil Engineering, Chongqing University, Chongqing 400045, China
e-mail: mickysophy@cqu.edu.cn

School of Civil Engineering, Chongqing University, Chongqing 400045, China

© The Author(s) 2025
D. Li and Y. Zhang (eds.), *Advances in Frontier Research on Engineering Structures II*,
Lecture Notes in Civil Engineering 535, https://doi.org/10.1007/978-981-97-6238-5_24

1 Introduction

The choice of benefit indicators and the forecasting accuracy significantly impact the results of project maintenance and repair (M&R) planning [1–3]. The prediction accuracy of the established rutting depth index (RDI) is usually lower compared to the international roughness index (IRI) [4, 5], and pavement condition index (PCI) [5]. Studies [15] based on a large amount of road network data (e.g., the LTPP dataset) show that PCI and IRI are highly correlated and even substitutable for each other. So we can select metrics with higher predictive accuracy for budget planning in the project. The prediction accuracy (R^2) can be substantially improved when the prediction bias is large, such as 10 points on a percentage scale [6]. This accuracy is enough for network management but cannot support the forecast of expressway performance and M&R planning. Markov Mediterranean prediction model [7] and deep learning models [2, 4] were constructed to solve this problem. However, the models are usually better suited to estimate the number of road sections in a certain state at a given moment [8] than to predict the performance of a particular road section. The biased results for specific road sections are still significant, and deep learning algorithms perform even worse than random forest (RF), gradient boosting decrease (GBDT), and other machine learning models. Elsayed [9] explored the superiority of GBDT algorithm in solving time series prediction problems by comparing various deep learning algorithms with GBDT.

Prediction accuracy is affected by the relevant feature data, and the significant effects of climate and heavy load change on pavement performance decay models have been proven many times [8, 10]. So, even with project-level management, we need to pay attention to the impact of these characteristics on decay trends. Detail distresses incorporated into feature parameters have been found to greatly improve the short-term prediction accuracy of pavement performance [11, 12]. Machine learning algorithms can be used to develop accurate time series forecasts, and forecasting models based on five years of data were found to perform better than models based on ten years of data [13]. The accuracy of machine learning in time series prediction has been constantly revised by the number of features simplified to improve short-term forecasting [14, 15]. Rutting depth and IRI predictions are greatly influenced by recent values, age, and maintenance information, while traffic load has little effect on short-term prediction [13, 16]. We also cannot ignore the huge impact of the level of maintenance [7] or construction [17]. However, research on the accuracy of project-specific time-series forecasting is somewhat lacking. Project-level predictions often rely on supervised learning or deterministic physical models [18] and the anisotropy of road sections and lanes is not considered. A genetic algorithm (GA) enhanced hybrid neural network (HNN) model reached an R^2 of 0.74 in single-project performance prediction, but the deviation of the prediction results was more than 5 [19] and there was a lack of time dimensional prediction accuracy exploration.

The high-frequency maintenance of daily distress also challenges the representativeness of performance data. Moreover, the M&R budget cycle of 3–5 years has a low tolerance for forecast deviation. Li [20] realized a precise prediction of asphalt

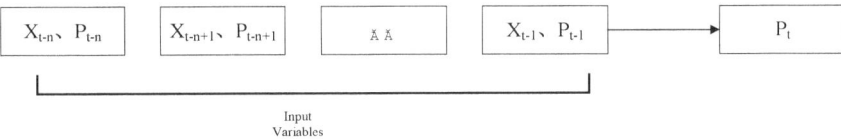

Fig. 1 Prediction framework

pavement performance one year ahead, which cannot support 3–5 years of management. Project-level management is much less rich in data than net-level management, so it is more important to obtain a framework for selecting effective feature parameters and accurately forecasting in the event of insufficient data. This paper will build and verify a time series-based machine learning model for project-level asphalt pavement performance prediction.

2 Data and Methods

2.1 Multivariate Time Series-Based Machine Learning Framework

Framework. It is multi-step forecasting with a direct strategy, using all variables up to the forecast year (year t) as input variables and building a separate time series model for each forecast period, as shown in Fig. 1.

Since longer historical datasets tend to have fewer complete observations, we selected various past observations, and several datasets have been created to explore the feasibility of the framework.

2.2 Prediction Algorithm Selection and Evaluation

Machine learning algorithms. Generally speaking, there are three main characteristics of project-level pavement performance prediction: (1) many input variables and high data dimensionality, (2) data types that include both numerical and categorical data, and (3) less data volume than net management. The neural networks are difficult to be met with insufficient sample size. The SVM algorithm is less effective with larger data dimensions, and prediction accuracy is much worse than RF, extreme gradient boosting (XGBoost) [3]. Therefore, we choose three modeling approaches: extreme gradient boosting (XGBoost), gradient-boosted decision trees (GBDT), and random forests (RF) [21] for the following study. The optimization objectives of the three algorithms are all average deviation-like metrics, and the machine learning algorithms comparison is used to determine the impact of the algorithm. The mean square error (MSE) was used as the training objective of the regression models.

Evaluation Methods. The coefficient of determination (R^2), root mean square error (RMSE), and mean absolute error (MAE) were used to evaluate the model performance and were calculated as shown in Eqs. (1–3).

$$R^2 = 1 - \frac{\sum_{i=1}^{n}(Y_i - \hat{Y}_i)^2}{\sum_{i=1}^{n}(Y_i - \overline{Y}_i)^2} \tag{1}$$

$$RMSE = \sqrt{\frac{1}{n}\sum_{i=1}^{n}(Y_i - \hat{Y}_i)^2} \tag{2}$$

$$MAE = \frac{1}{n}\sum_{i=1}^{n}|Y_i - \hat{Y}_i| \tag{3}$$

where Y_i is the actual value, n is the sample size of Y_i, $\hat{Y}_i$ is the prediction value, $\overline{Y}_i$ is the mean of the measured values.

2.3 Data and Methods

Analysis of Influencing Factors. Before predicting pavement performance for a specific project, it is necessary to determine valid variables based on influence factor analysis. Table 1 summarizes the essential factors affecting the performance of different pavement performance metrics in previous studies.

Given that the project data collections were based on JTG 5210–2018, we introduce three metrics: the pavement condition index (PCI), the ride quality index (RQI), and the rut depth index (RDI). PCI is a 100-system value based on the distresses' calculation areas weighted, RQI is a 100-system value conversion based on the average IRI, and RDI is a hundred-system value conversion based on the maximum depth of the maximum rut depths of all profiles. Table 1 shows that the previous variables significantly affect the predicted pavement performance decay, regardless of the performance associated with PCI, RQI, or RDI. Furthermore, different pavement performance metrics can affect each other [11, 21]. However, only historical data for a specific time is utilized. The correlation of different indicators and the non-apparent information given by the historical state are not valued. Since rutting is the accumulation of permanent pavement deformation, the road age factor has significant effects on the prediction of both RDI and RQI-related metrics. However, road age has a low impact on predicting distress-related metrics, and the degree of effect varies for different projects. It should also be noted that road type (soil base or bridge) was found to be the most sensitive factor in the PDCI model [16]. Additionally, structural

Table 1 Summary of research on the importance of variables selected for asphalt pavement performance prediction

No.	Methods	Metrics	Variable importance ranking
1 [16]	Average impact value of neurons of input variables (MIV)	PDCI (Comprehensive indicators of pavement distress)	Pavement or bridge, previous PDCI, the material of subbase, base material, distress treatment type, road age, the material of asphalt layer, maintenance type
2 [22]	BPN network performs sensitivity analysis	PCI	Flexible arterial: previous PCI, road age; Flexible collector: previous PCI, AADT, ESALs
3 [16]	Average impact value of neurons of input variables (MIV)	IRI	Distress treatment type, previous IRI, maintenance type, maintenance material, road age, the material of asphalt layer
6 [11]	Feature contribution degree in RF	IRI	Age, crack, AADTT, RD
4 [3]	Feature contribution degree in XGBoost and RF	IRI	XGBoost based: material property (NO_200_Passing), hydraulic conductivity, a variable of traffic (KESAL_YEAR), modified thickness; RF-based: modified thickness, material property (NO_200_Passing), a variable of traffic (KESAL_YEAR), age
5 [13]	Sensitivity analysis	IRI	previous IRI
6 [16]	Average impact value of neurons of input variables (MIV)	RD	Previous RD, road age, treatment age, maintenance type, the material of subbase, distress treatment type

factors (surface, base and, subbase, etc.), distress treatment & maintenance types, environmental & climatic factors, and traffic factors cannot be ignored in predicting pavement performance. Besides, environmental & climatic factors are not considered as they are non-variables for 3–5 years of project-level management. For the method of input variables selection, the feature contribution selection model is superior to correlation analysis, according to the previous study [11]. The mean decrease impurity (MDI) [7] as a metric implying the contribution of the RF algorithm was chosen for this study. The MDIs are calculated by Eq. (4).

$$\text{MDI} = \frac{N^t}{N} * \left(IMP - \frac{N_R^t}{N^t} * IMP_R - \frac{N_L^t}{N^t} * IMP_L \right) \tag{4}$$

where IMP, IMP_R, and IMP_L are the impurity value, the impurity value in the right child, and the impurity value in the left child, respectively. N^t is the number of

Fig. 2 Project locations. Left: Qingyin expressway (G20, G2001); Right: Chongqing bypass expressway (G5001)

samples at the current node, N_R^t is the number of samples in the right child, N_L^t is the number of samples in the left child, and N is the sample size.

Project and data description. This study collected data from two expressway projects, the Qingyin Expressway (G20, G2001) and the Chongqing Bypass Expressway (G5001). The locations are shown in Fig. 2. Both projects are expressways with similar engineering designs and pavement structures.

We summarize five main categories of input variables: roadway information, maintenance and repair (M&R), distress data, performance data, and road age. The dataset was processed and performed a 7:3 cut, with 70% as the training data and 30% as the test data.

3   Results

3.1   Selection of Variables Based on MDI

We explore machine learning prediction models built on three performance metrics: PCI, RQI, and RDI. The input variables are shown in Table 2 and ranked by MDI to select candidate variables to obtain a better set of data features. The ranking results are shown in Figs. 3, 4 and 5.

Except for the large fluctuations and the low overall MDIs for PCI, there is a certain similarity in the influence trend of parameters in different projects. Figure 3 shows that T, previous pavement condition metrics, and previous distress status (patching, longitudinal crack, and transverse crack) all had some effect on PCI prediction but were not highly significant. Figure 4 shows that PrevRQI and PrevIRI significantly affect the prediction of RQI, followed by T, T0, PrevRD, and PrevRDI. Lane, patching, PrevPCI, and PrevDR also have some influence. Figure 5 shows

Table 2 Description of input variables

Type	Name of the input variable	Description
Basic information	Pavement or tunnel (section type, ST)	The road section, including tunnel or not
	Lane	1st, 2nd, 3rd, 4th
Previous Distress condition	Alligator crack (ALC, m^2)	The corresponding distress area at age T_0
	Transverse crack (TC, m^2)	
	Longitudinal crack (LC, m^2)	
	Pothole (m^2)	
	Block patching (BP, m^2)	
	Patching (P, m^2)	
Maintenance information	Maintenance type (MT)	None, 4 cm in-situ thermal regeneration, MS-3 micro-surfacing, low-noise anti-slip super-surfacing (Type II), high-permeability fog seal, milling top layer re-paving, milling top and middle surface layer re-paving, milling top, middle and bottom layer re-paving
Pavement condition	Previous DR (PrevDR, %)	The corresponding pavement condition index at age T_0
	Previous IRI (PrevIRI, m/km)	
	Previous RD (PrevRD, mm)	
Pavement performance	Previous PCI (PrevPCI)	The corresponding pavement performance index at age T_0
	Previous RQI (PrevRQI)	
	Previous RDI (PrevRDI)	
Age	T_0 (years)	The interval between road construction and input variable detection
	T (years)	The interval between road construction and predicted performance detection

that T significantly affects RDI prediction, PrevRD, PrevRDI, and lane also have considerable effects.

Lane has no particular influence on PCI prediction, some effect on RQI, and a significant impact on RDI, indicating that traffic structure has a more considerable influence on RDI. The section type affected little on the pavement performance predictions, suggesting that whether or not it is a tunnel does not affect the decay

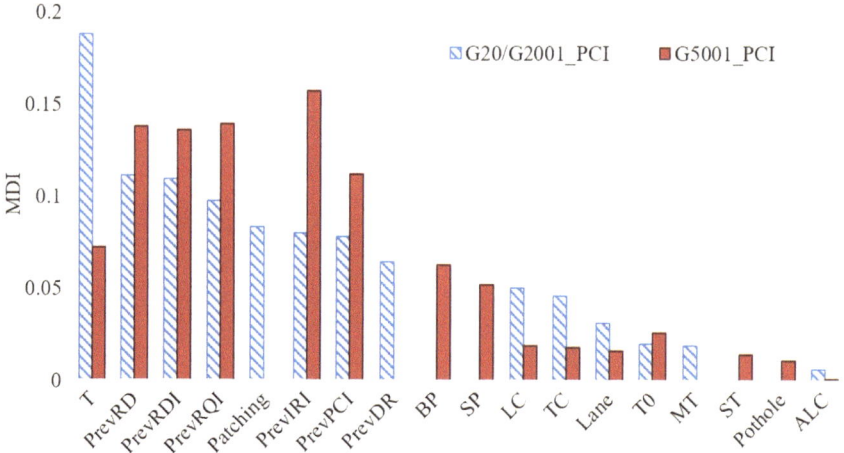

Fig. 3 Sorting of MDI results for PCI

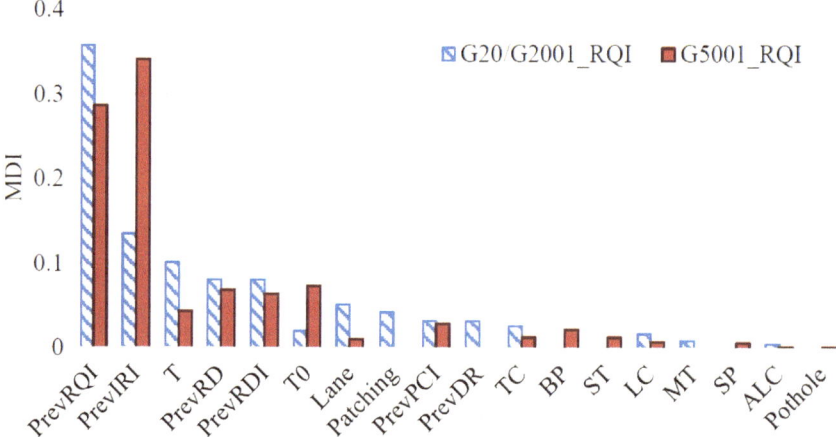

Fig. 4 Sorting of MDI results for RQI

degree of the three performance metrics. It is somewhat surprising that the maintenance type has no significant effect on all pavement performance predictions. This
may be because the maintenance conditions of a project are largely similar, and the
differences in their selection of maintenance measures are often taken for localized
idiosyncratic sections. Based on the MDI ranking, we selected T, T0, and other variables with MDI greater than 0.03 to obtain the variable feature sets to build PCI,
RQI, and RDI prediction models, as shown in Table 3.

The independent variables constructed by the decay model of different benefit
indicators contribute differently to the same project. The main contributing variables
of both RQI and RDI are their specific monitoring data. RDI and RQI-based metrics

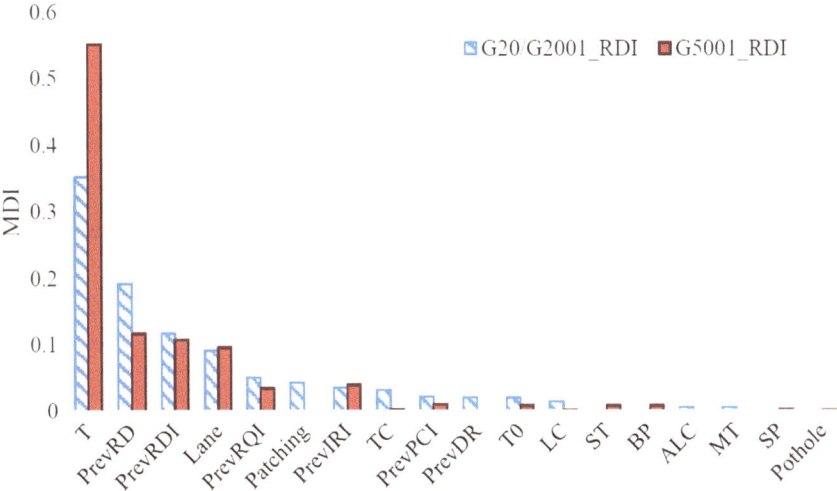

Fig. 5 Sorting of MDI results for RDI

Table 3 Data set information

Dataset ID	Predicted performance index	Input variables	Number of input variables
G20/G2001_PCI	PCI	PrevRD, prevRDI, prevRQI, patching, prevIRI, prevPCI, prevDR, LC, TC, T, T_0	11
G20/G2001_RQI	RQI	PrevRQI, prevIRI, prevRD, prevRDI, Lane, Patching, prevPCI, prevDR, T, T_0	10
G20/G2001_RDI	RDI	PrevRD, prevRDI, lane, prevRQI, patching, prevIRI, TC, T, T_0	9
G5001_PCI	PCI	PrevIRI, prevRQI, prevRD, prevRDI, prevPCI, BP, SP, T, T_0	9
G5001_RQI	RQI	PrevIRI, prevRQI, prevRD, prevRDI, T, T_0	6
G5001_RDI	RDI	PrevRD, prevRDI, lane, prevIRI, prevRQI, T, T_0	7

influence PCI, as PCI is a comprehensive condition indicator based on multi-type distress statistics. The established PCI indicators are not suitable for the M&R plan management of pavements with high-frequency maintenance. The influence variables of the RQI prediction model presented some differences for different projects (different data sets), while the influence variables of the RDI prediction model are more consistent. This may indicate that RQI is more influenced by daily maintenance than RDI, and the significant effect of absolute road age on rutting decay also indirectly proves this inference.

3.2 Selection of Performance Indicators Based on Machine Learning Forecast Results

Based on the feature parameters selected in Table 3, three machine learning models, GBDT, XGBoost, and RF, were used to evaluate the prediction models for different pavement performance metrics. The evaluation results are shown in Fig. 6.

As seen in Fig. 6, the prediction results of the three machine algorithms are similar for the same pavement performance metrics. The RDI model has the best overall prediction accuracy (R^2), the RQI model followed, and the PCI model has achieved unsatisfactory prediction results. The previous selection of variables also found that the MDI values of the variables that influenced the PCI prediction were generally low. Improving the prediction accuracy of PCI may need to incorporate other variables, such as routine maintenance and repair records, and PCI may be better suited to a net-level prediction model. It can be seen the improvement of XGBoost for GBDT is not obvious in this type of project (with fewer features). The two performance metrics (RQI and RDI) and the GBDT algorithm were chosen for further study on time series. Predictions based on the other two algorithms (XGBoost and RF) were also calculated, and the results are reflected in the prediction scatter plots figures.

The three different types of machine learning models do not show significant differences in prediction bias in terms of accuracy. This is in contrast to established studies [23] that concluded that the choice of prediction models is critical. All three models are suitable for the prediction framework proposed in this paper, and there is no significant difference (Fig. 6).

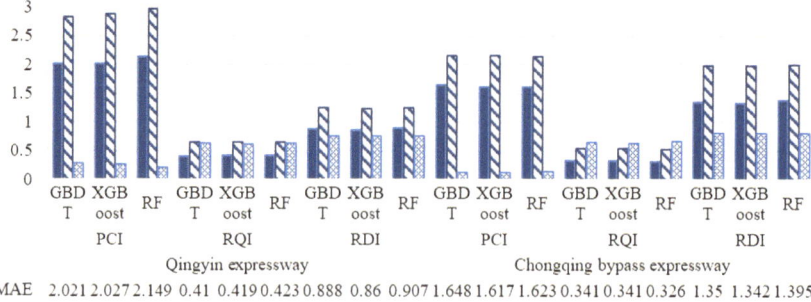

Fig. 6 Evaluations of machine learning prediction models with different indicators

3.3 Prediction Evaluation Results

Since the M&R cycles are usually five years, we only explore the effect of 1–5 years of performance prediction. The results of the RQI and RDI prediction models for the Qingyin and Chongqing bypass highways are shown in Figs. 7 and 8. Figure 7 shows that for the RQI forecast using time series data, MAE and RMSE are insignificant. The larger the historical period, the higher the prediction accuracy. In addition, a relatively high R^2 can be achieved when forecasting the performance metrics 1–2 years ahead. As seen in Fig. 8, the pattern of the prediction model evaluation results for RDI is generally consistent with that of RQI. Still, the RDI prediction errors are typically higher than those of RQI. However, the R^2 results for RDI are much better than those of RQI. To further observe the precision of the prediction, we selected the optimal combination of time series prediction models to draw the scatter plots in Figs. 9, 10, 11 and 12.

Figures 9, 10, 11 and 12 show that RQI and RDI forecast deviations increase with the forecast year. Within the error line of ± 2, there is a small amount of over-error range data for RDI predictions. However, the trend is not clear for the Chongqing bypass. Supported by sufficiently long time-span data, the framework (Figs. 9, 10, 11 and 12) performs well with few prediction results outside the error line. The prediction results of the framework (Figs. 7 and 8) reveal there is a degree of correction of the historical data on the prediction models, and its effect is greater than the effect of the dataset size. When the period of historical data is longer than the time of the forecast, the deviation of the forecast is acceptable. Furthermore, it should be noted that RDI prediction bias decreases over time, which may reflect its suitability for long-term pavement performance prediction.

Fig. 7 Evaluations of RQI prediction

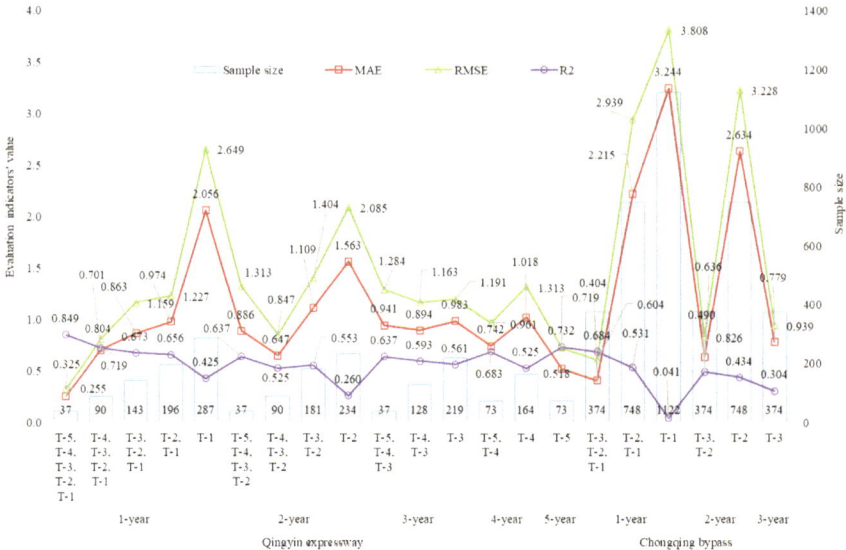

Fig. 8 Evaluations of RDI prediction

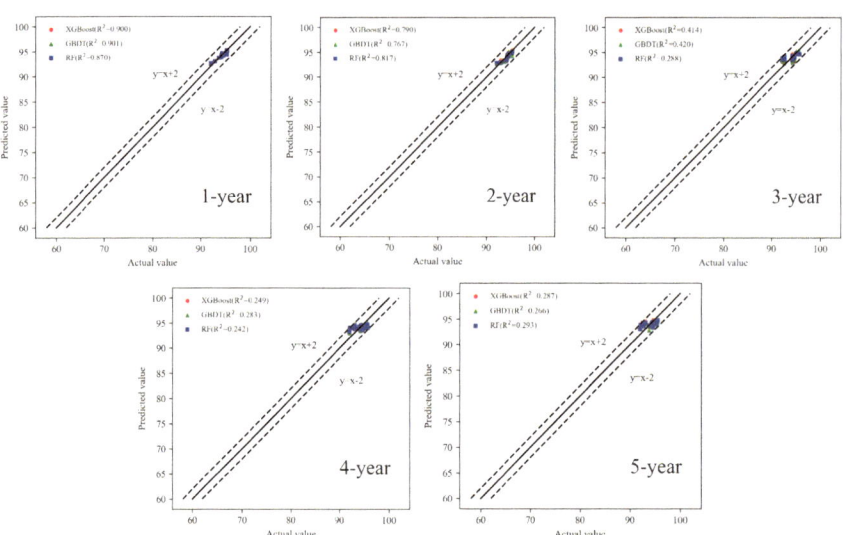

Fig. 9 RQI prediction scatter plots of Qingyin expressway

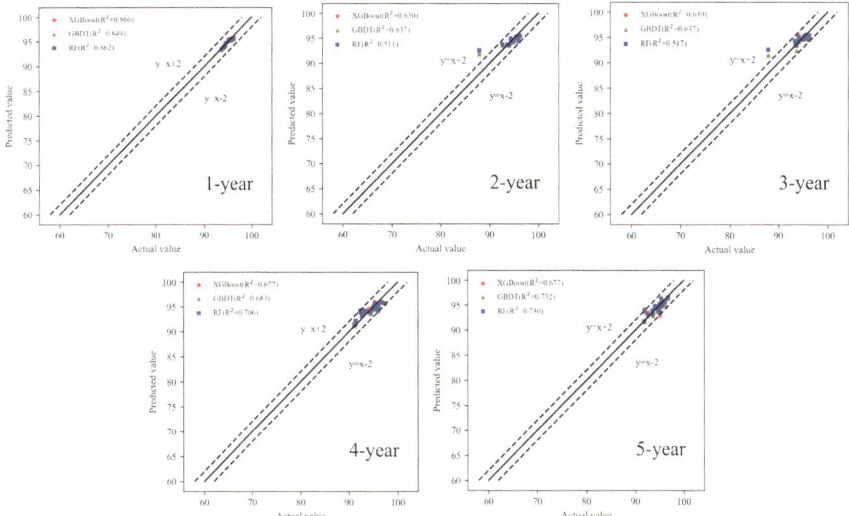

Fig. 10 RDI prediction scatter of Qingyin expressway

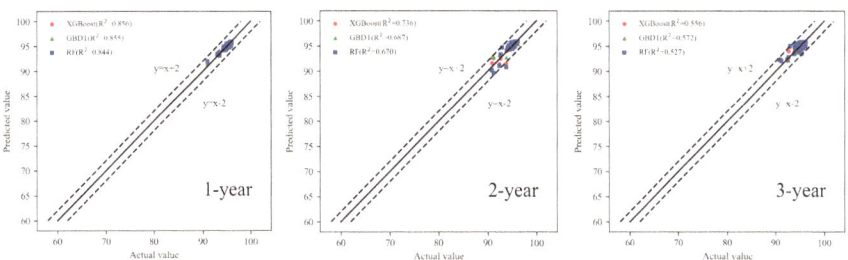

Fig. 11 RQI prediction scatter plots of Chongqing bypass

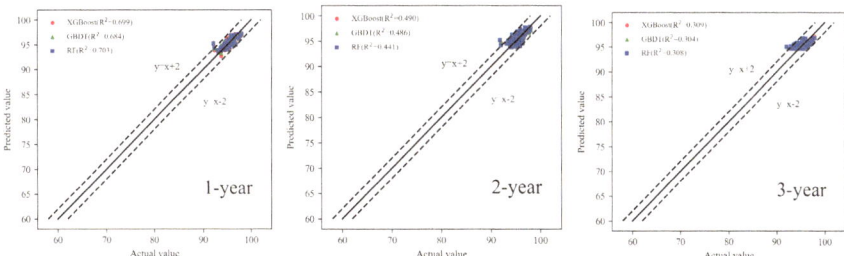

Fig. 12 RDI prediction scatter plots of Chongqing bypass

4 Conclusions

This paper conducts a comparative study based on a machine learning prediction framework. The data sets of two expressways in different areas were used to train and test the methods. The main contributions of the paper are as follows.

- A framework is proposed to effectively predict short-term pavement performance and control deviations, which can be used for the M&R planning of high-grade roads.
- With the selection of MDI-based input variables, machine learning models such as GBDT, RF, and XGBoost have no difference in prediction accuracy after simple tuning.
- PCI has the worst predictability and is not suitable for short-term M&R planning of expressways. The 1–5 year prediction deviations of RQI and RDI are within $\pm$ 2, and the mean errors of RQI are usually lower than those of RDI prediction.
- The time span-based prediction model based on the framework performs stable.

The representative statistical index of rutting and its prediction need to be further studied. Besides, more projects can be further validated.

References

1. Chen, W., & Zheng, M. L. (2021). Multi-objective optimization for pavement maintenance and rehabilitation decision-making: A critical review and future directions. *Automation in Construction, 130*, 1–23. https://doi.org/10.1016/j.autcon.2021.103840
2. Choi, S., & Do, M. (2020). Development of the road pavement deterioration model based on the deep learning method. *Electronics, 9*(1), 1–15. https://doi.org/10.3390/electronics9010003
3. Damirchilo, F., Hosseini, A., Mellat Parast, M., & Fini, E. H. (2021). Machine learning approach to predict international roughness index using long-term pavement performance data. *Journal of Transportation Engineering, Part B: Pavements, 147*(4), 1–14. https://doi.org/10.1061/JPEODX.0000312
4. Gong, H., Sun, Y., Mei, Z., & Huang, B. (2018). Improving accuracy of rutting prediction for mechanistic-empirical pavement design guide with deep neural networks. *Construction and Building Materials, 190*, 710–718. https://doi.org/10.1016/j.conbuildmat.2018.09.087
5. Sidess, A., Ravina, A., & Oged, E. (2021). A model for predicting the deterioration of the pavement condition index. *International Journal of Pavement Engineering, 22*(13), 1625–1636. https://doi.org/10.1080/10298436.2020.1714044
6. Barzegaran, J., Dezfoulian, R. S., & Fakhri, M. (2021) Estimation of IRI from PASER using ANN based on k-means and fuzzy c-means clustering techniques: A case study. *International Journal of Pavement Engineering*, 1–15. https://doi.org/10.1080/10298436.2021.2000988
7. Osorio-Lird, A., Chamorro, A., Videla, C., Tighe, S., & Torres-Machi, C. (2018). Application of Markov chains and Monte Carlo simulations for developing pavement performance models for urban network management. *Structure and Infrastructure Engineering, 14*(9), 1169–1181. https://doi.org/10.1080/15732479.2017.1402064
8. Moreira, A. V., Tinoco, J., Oliveira, J. R., & Santos, A. (2018). An application of Markov chains to predict the evolution of performance indicators based on pavement historical data. *International Journal of Pavement Engineering, 19*(10), 937–948. https://doi.org/10.1080/10298436.2016.1224412

9. Elsayed, S., Thyssens, D., Rashed, A., Jomaa, H. S., & Schmidt-Thieme, L. (2021). *Do we really need deep learning models for time series forecasting*? Available online at http://arxiv.org/pdf/2101.02118v2.

10. Gudipudi, P. P., Underwood, B. S., & Zalghout, A. (2017). Impact of climate change on pavement structural performance in the United States. *Transportation Research Part D: Transport and Environment, 57*, 172–184. https://doi.org/10.1016/j.trd.2017.09.022

11. Luo, Z. Y., Wang, H., & Li, S. L. (2022). Prediction of international roughness index based on stacking fusion model. *Sustainability, 14*(12), 1–13. https://doi.org/10.3390/su14126949

12. Wang, C. B., Xu, S. Z., & Yang, J. X. (2021). Adaboost algorithm in artificial intelligence for optimizing the IRI prediction accuracy of asphalt concrete pavement. *Sensors (Basel, Switzerland), 21*(17), 1–16. https://doi.org/10.3390/s21175682

13. Marcelino, P., de Lurdes Antunes, M., Fortunato, E., & Gomes, M. C. (2021). Machine learning approach for pavement performance prediction. *International Journal of Pavement Engineering, 22*(3), 341–354. https://doi.org/10.1080/10298436.2019.1609673

14. Dong, Y., Shao, Y., Li, X., Li, S., Quan, L., Zhang, W., & Du, J. (2019). Forecasting pavement performance with a feature fusion LSTM-BPNN model. In Proceedings of the 28th ACM international conference on information and knowledge management (pp. 1953–1962).

15. Gharieb, M., Nishikawa, T., Nakamura, S., & Thepvongsa, K. (2022). Modeling of pavement roughness utilizing artificial neural network approach for laos national road network. *Journal of Civil Engineering and Management, 28*(4), 261–277. https://doi.org/10.3846/jcem.2022.15851

16. Yao, L., Dong, Q., Jiang, J., & Ni, F. (2019). Establishment of prediction models of asphalt pavement performance based on a novel data calibration method and neural network. *Transportation Research Record, 2673*(1), 66–82. https://doi.org/10.1177/0361198118822501

17. Onayev, A., & Swei, O. (2021). IRI deterioration model for asphalt concrete pavements: Capturing performance improvements over time. *Construction and Building Materials, 271*, 1–11. https://doi.org/10.1016/j.conbuildmat.2020.121768

18. Justo-Silva, R., Ferreira, A., & Flintsch, G. (2021). Review on machine learning techniques for developing pavement performance prediction models. *In Sustainability, 13*(9), 5248. https://doi.org/10.3390/su13095248

19. Li, Z., Zhang, J., Liu, T., Wang, Y., Pei, J., & Wang, P. (2021). Using PSO-SVR algorithm to predict asphalt pavement performance. *Journal of Performance of Constructed Facilities, 35*(6), 04021094. https://doi.org/10.1061/(ASCE)CF.1943-5509.0001666

20. Li, J., Yin, G., Wang, X., & Yan, W. (2022). Automated decision making in highway pavement preventive maintenance based on deep learning. *Automation in Construction, 135*, 104111. https://doi.org/10.1016/j.autcon.2021.104111

21. Nadar, A., Abd El-Hakim, R. T., El-Badawy, S. M., & Afify, H. A. (2020). International roughness index prediction model for flexible pavements. *International Journal of Pavement Engineering, 21*(1), 88–99. https://doi.org/10.1080/10298436.2018.1441414

22. Amin, S. R., & Amador-Jiménez, L. E. (2017). Backpropagation neural network to estimate pavement performance: Dealing with measurement errors. *Road Materials and Pavement Design, 18*(5), 1218–1238. https://doi.org/10.1080/14680629.2016.1202129

23. Shtayat, A., Moridpour, S., Best, B., & Rumi, S. (2022). An overview of pavement degradation prediction models. *In Journal of Advanced Transportation, 2022*, 1–15. https://doi.org/10.1155/2022/7783588

Research on Non-destructive Identification Technology of Hidden Pavement Damage Based on Radar

Chunliang Li, Heguang Fang, Xian Li, and Gongning Zhai

Abstract Using 3D ground-penetrating radar to detect the structural layer is a non-destructive and efficient detection method, but analyzing the collected radar data is highly complex work, and there is a lack of adequate data processing methods for analyzing the structural damage of the pavement layer using radar data. To solve the above problems, this study starts from the propagation characteristics of the radar in the pavement layer, analyses the influence of structural damage of the pavement layer on the radar signal, then proposes the corresponding eigenvalues, and develops an algorithm that can effectively extract the eigenvalues, and then carr.

Keywords Radar · Non-destructive identification · Hidden disease · Wavelet transforms

1 Introduction

Geological radar has become a powerful tool in physical exploration with its high resolution and working efficiency and has been widely used in many engineering fields [1, 2]. In recent years, it has only begun to be applied to a new technology for highway detection in China. At present, China's road engineering detection in the application of short-pulse radar, there are three main problems: (1) a single emission frequency; (2) detection area is small, each time only a cross-section can be measured; (3) only the thickness of the structural layer of the pavement can be measured. Therefore, the current short-pulse radar can not be realized to reflect the detection of the pavement structure's internal condition, and the thickness detection is only one section, which can not obtain the thickness distribution of the entire

C. Li · H. Fang · G. Zhai
Qingdao Highway Development Center, Qingdao 266061, Shandong Province, China

X. Li (✉)
Shandong Transportation Institute, Jinan City 250031, Shandong Province, China
e-mail: xianli0124@163.com

© The Author(s) 2025

D. Li and Y. Zhang (eds.), *Advances in Frontier Research on Engineering Structures II*,
Lecture Notes in Civil Engineering 535, https://doi.org/10.1007/978-981-97-6238-5_25

internal pavement structure. The single transmitting frequency prevents the ability to obtain the internal conditions at different depths [3, 4].

To solve the above problems, this study starts from the propagation characteristics of radar in the pavement layer, analyses the influence of structural damage of the pavement layer on the radar signal, then proposes the corresponding eigenvalue, and develops an algorithm that can effectively extract the eigenvalue to achieve the full-coverage, high-precision and rapid detection of hidden pavement diseases.

2 Characterization of Pavement Structural Damage and Ground-Penetrating Radar Signals

Ground Penetrating Radar (GPR) works by transmitting electromagnetic waves to the pavement, receiving the reflected waves and analyzing them to obtain information about the pavement [5]. The interaction of electromagnetic waves with an object: reflection, incidence and scattering [6]. The propagation of ground-penetrating radar signals through the pavement structure is relatively simple in undamaged pavement. A road pavement can be simplified as a layered structure consisting of a homogeneous, isotropic medium. The propagation of the electromagnetic field of ground-penetrating radar in a road pavement consists of two main processes: propagation within the exact medium and partial reflection and partial incidence at interfaces formed by different media. Thus, the received signal generally contains several distinct pulse waveforms. Between the pulse waveforms, there are flat, straight lines. In the case of structurally damaged paved structures, the layers are no longer a homogeneous medium due to cracks within the layers, separation of the layers, water accumulation due to trials, etc. Scattering is formed in the middle of the original pulse waveforms. The distribution of electromagnetic waves is also haphazard due to the irregularity of the cracks, voids, and water patterns [7]. Propagation of ground-penetrating radar signals in intact paving layers is shown in Fig. 1.

Damage to the interior of the structural paving layer can be determined based on whether or not scattering of radar signals occurs. The degree of damage to the paving layer can also be calibrated based on the scattering intensity of the radar signal. When the paving structure is intact, the scattering intensity between the radar signal pulses is almost zero; when the paving layer is cracked, dehollowed or waterlogged, the scattering between the radar signal pulses occurs, and the scattering intensity is positively correlated with the degree of damage. Therefore, the scattering power can be extracted as an eigenvalue to identify the damage inside the pavement. Of course, according to the relationship between the wavelength and the relative size of the scattering body, the above conclusion is only valid within a specific range of wavelength and frequency. Therefore, when extracting this eigenvalue, care must be taken to distinguish between electromagnetic waves of different frequencies.

After determining the use of scattering intensity as a criterion for judging the degree of disease, how to effectively extract the scattering intensity from the complex

Fig. 1 Propagation of ground-penetrating radar signals in intact paving layers

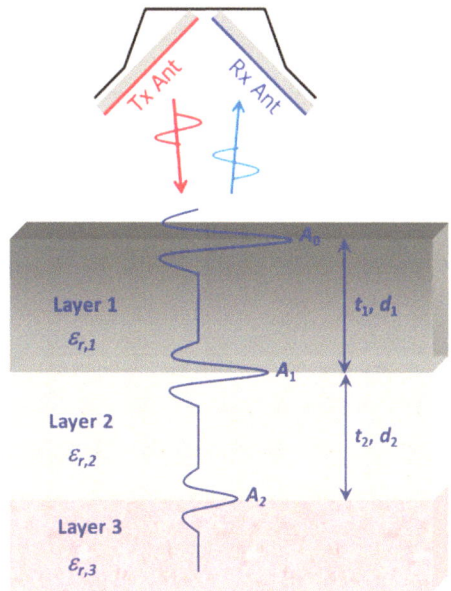

radar signals becomes the difficulty and focus of the research. This project uses the wavelet transform, an advanced digital signal processing method, for feature extraction.

3 Structural Damage Determination of Pavement Based on Wavelet Transforms

3.1 Principle of Wavelet Transform

In digital signal processing methods, a signal in the time domain can be transformed into a sign in the frequency domain using the Fourier transform. The movement in the frequency domain can be converted into a signal in the time domain by the inverse Fourier transform. However, the disadvantage is that viewing the information in the time and frequency domains is impossible [8]. As an advanced method in signal processing, the wavelet transform has become an essential and indispensable tool for analyzing signals and images. The advantage of wavelet transform over Fourier transform is that the analyzed signal can be arbitrarily zoomed and shifted, and its features can be extracted. The formula for continuous wavelet transform is shown in Eq. (1).

$$\psi_{\alpha,\tau}(t) = \frac{1}{\sqrt{\alpha}} \psi\left(\frac{t_1 \tau}{\alpha}\right), \alpha_1 R^1, \tau R \tag{1}$$

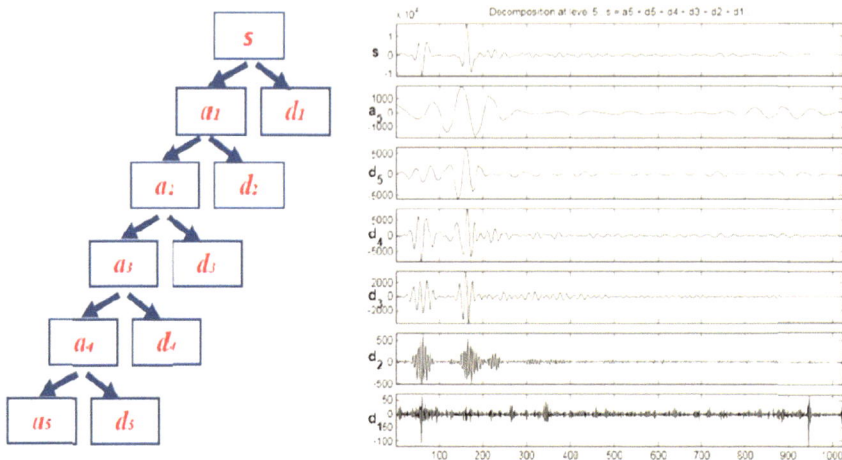

Fig. 2 Decomposition of the signal using wavelet transform

where, $\psi\left(\frac{t-\tau}{\alpha}\right)$ is the transform of scaling and shifting of the mother wave $\psi(t)$, a is the scaling coefficient, τ is the shifting coefficient, $Cf(a,\tau)$ is the continuous wavelet transform of the signal f(t), and the superscript "*" indicates the complex conjugate.

As shown in Fig. 2, using wavelet transform for signal decomposition, the original signal is decomposed into approximation coefficient a1 and detail coefficient d1. A wavelet decomposition is performed again for approximation coefficient a1, and the approximation coefficient a1 is decomposed into approximation coefficient a_2 and detail coefficient d_2, and so on with five layers of wavelet decomposition, and the original signal s is decomposed into one approximation coefficient a_5 and five detail coefficients d_1, d_2, d_3, d_4, d_5. d_5. Summing the approximation coefficients a_5 and the five detail coefficients yields the original signal s. An example of wavelet decomposition is shown on the right side of the figure. Each detail coefficient has a Pseudo Frequency, which decreases from d1 to d_5.

3.2 Application of Wavelet Analysis for Situational Discrimination of Structural Damage on Pavements

Selection of Mother Wave in Wavelet Analysis

In wavelet analysis, the selection of the mother wave is essential. It has been found that for sound research, the shape of the mother wave should be as similar as possible to the form of the original wave emitted. Figure 3 shows the shape of the original waveform of the ground-penetrating radar with different wavelet mother waves.

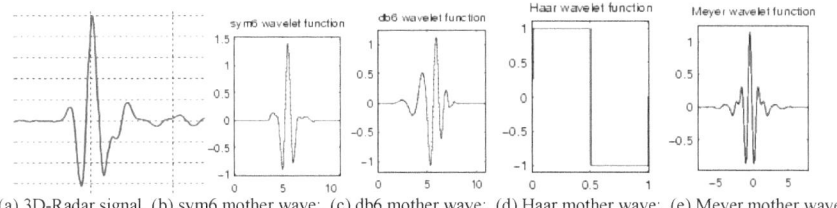

(a) 3D-Radar signal (b) sym6 mother wave; (c) db6 mother wave; (d) Haar mother wave; (e) Meyer mother wave

Fig. 3 Comparison of the shapes of the original radar wave and the wavelet mother wave

Wavelet Decomposition

The original signal is decomposed by wavelet transform into three wavelet detail coefficients d_1, d_2, d_3 and one wavelet approximation coefficient a_3. In this segment, the sampling length is 25 ns, and the number of sampling points is 256, so the sampling frequency is 10.24 GHz. Using the sym6 mother wave, the pseudo-frequencies of different detail coefficients are: d_1: 3.724 GHz; d_2: 1.862 GHz; d_3: 0.931 GHz. Based on the different pseudo-frequencies, the wavelength of the electromagnetic field in the medium can be estimated, and feature extraction in other frequency bands can be performed.

Eigenvalue Extraction

After wavelet decomposition of the signal, to get the index of the scattered intensity, the eigenvalues containing the scattered intensity information in the signal need to be extracted and analyzed. The extraction of eigenvalues is divided into two main steps: selecting appropriate wavelet detail coefficients and selecting right eigenvalues to remove the scattering information contained in the wavelet detail coefficients.

The purpose of selecting appropriate wavelet detail coefficients is to retain only certain specific pseudo-frequencies so that the scattering of electromagnetic fields can be controlled in the frequency range sensitive to road damage. For example, for detail coefficients at too low a frequency, their wavelengths are too long, and they become insensitive to cracks, dehiscence, etc. Therefore, detail coefficients at too low a frequency do not contain the required scattering information and must be rejected.

In the radar data collected in this study, the sampling frequency is 10.24 GHz. The pseudo-frequencies of the different detail coefficients for the 5-layer wavelet decomposition using the sym6 mother wavelet are d_1: 3.724 GHz, d_2: 1.862 GHz, d_3: 0.931 GHz, d_4: 0.466 GHz, and d_5: 0.233 GHz, respectively. The corresponding wavelengths are: d_1: 0.036 m, d_2: 0.072 m, d_3: 0.144 m, d_4: 0.288 m and d_5: 0.577 m. Based on the structural dimensions of the road, the electromagnetic waves with wavelengths in the range of d_4: 0.288 m and d_5: 0.577 m and above are insensitive to structural damages on the road, and their scattering intensities are also small. Therefore, the detail coefficients d1, d_2 and d_3 are selected as the basis for eigenvalue extraction in this study.

After selecting the wavelet detail coefficients, the eigenvalue extraction mainly considers the effect of scattering intensity. In two-dimensional signals, the strength of scattering is primarily manifested as the difference in the roughness of the image texture. In one-dimensional signals, the power of distribution is mainly manifested by the difference in the degree of fluctuation. The more significant the change of the one-dimensional motion, the rougher the texture of the superimposed two-dimensional image, the greater the scattering and the greater the degree of road damage. The variance, standard deviation and signal energy can express the degree of fluctuation of a 1D signal. In this study, the standard deviation and signal energy are chosen as the feature values to be compared to select the best feature extraction method.

The greater the difference between the dielectric constant of the substance inside the disease and that of the surrounding substance, the more obvious it is in the radar profile. The greater the conductivity of the medium inside the damage, the greater the energy absorbed, and the greater the attenuation of the amplitude of the electromagnetic wave. When the pavement structure inside the filling material is disordered, in the radar radar image corresponds to show the region of the waveform is more disordered, can see the obvious interference phenomenon.

4   Pavement Structural Layer Disease Classification and Grading

Where conditions permit, some of the diseased locations are selected for verification using the coring method. The internal diseases of the paving layer structure are mainly classified into loose (loose, segregated, broken), poorly bonded between layers, water-rich, and de-voided types [9]. The image description of disease characteristics and grading basis are shown in Tables 1 and 2.

5   Conclusion

(1) Based on a series of steps in wavelet analysis, such as the selection of parent wave, wavelet decomposition and eigenvalue extraction, a non-destructive recognition method of pavement structural layer disease radar based on the principle of wavelet transform is proposed.

(2) The disease characteristics of the radar-based pavement layer structure, such as internal looseness (loose, segregation, fragmentation), interlayer bonding, water rich, hollow, etc., were clarified, and the typical disease classification was realized.

(3) The technical results of this research provide support for rapid, non-destructive and accurate identification of internal damage in the pavement structural layer, realise detection from point to point, improve the representativeness of the

Table 1 Radar image description of disease characteristics

Typical radar image	Radar image anomaly characterization	Properties of the medium inside the structural layer
	The waveforms are subhorizontal to the same phase axis, have continuity and similarity, and are stable along the amplitude segments of the same phase axis	Structural layer interface or layered construction layer interface
	Both sides of the waveform have the same phase axis nearly horizontal distribution, but the same phase axis stagger or interruption; the stagger interruption distance is not more than 10 cm	Internal vertical cracks in the basement
	Localized robust reflected waveforms or reflected waveforms with the same phase axis. In upward convexity, the waveform is long or the sensation of multiple reflected waves	Localized hollowing or loosening within the structural layer
	Reflected waves appear locally, with long wavelengths and attenuated amplitudes, in clouds	Topical water-rich
	Localized subhorizontal reflective bands, or strong reflective bands located at the interface dreamy waves	Localized loosening of structural layers, segregation or loose interlayers along interfaces
	Along the line profile. Reflected waves with high tilt angles are distributed in the same phase	Poor joint bonding of structural layers laid in phases

detection data, realise efficient identification of internal disease in the pavement structural layer, and provide data support for scientific maintenance of pavements.

Table 2 Classification and grading of pavement structural layer diseases

Type of disease	Grade	Disease characteristics image description and grading basis
Loose (loose, segregated, broken)	Light	The amplitude of the radar-reflected wave in the diseased area is more significant than that of the surrounding medium, and the waveform is scattered. There is a mild loss of the homogeneous axis at the interface of the road structure layer below the disease
	Medium	The amplitude of radar-reflected waves in the diseased area is much larger than that reflected from the surrounding medium, the waveform is scattered, and the homogeneous axis of the interface of the road structural layer below the disease is partially missing
	Heavy	The amplitude of the radar-reflected wave in the diseased area is much larger than that reflected from the surrounding medium, the waveform is completely scattered and disordered, and the same phase axis of the interface of the structural layer of the road underneath the disease completely disappears
Poor bonding between layers	–	The radar-reflected wave homoclinic axis is interrupted, producing two parallel axes to the unconfined area with a large reflected wave amplitude
Affluent water	Light	Radar reflected wave amplitude and initial radar emission electromagnetic wave inversion, amplitude than the surrounding medium reflected wave amplitude is small, part of the spectrum of high-frequency signals are absorbed and truncated, mainly for the low and medium frequency spectrum
	Heavy	Radar reflected wave amplitude and the initial radar emission electromagnetic wave inversion, amplitude than the surrounding medium reflected wave amplitude is much smaller, most of the high-frequency signals on the spectrum is absorbed and truncated, mainly for the low-frequency range
Fall through	–	Strong reflections are formed on the data profile with high amplitudes of reflected waves, and due to multiple reflections and bypasses at the edges of the de-embedded region, hyperbolic reflection arcs and bypassed wave tails are formed

References

1. Cui, F., Ning, M., Shen, J., & Shu, X. (2022). Automatic recognition and tracking of highway layer-interface using faster R-CNN. *Journal of Applied Geophysics, 196*, 104477.
2. Wang, Y. (2022). A brief analysis of advance geological prediction technology and application of highway tunnel. *Insight-Information, 4*(2), 49–52.
3. Ali, J., Abdullah, N., Yahya, R., Ullah, I., Das, B., & Jusoh, M. (2020). Ultra-wideband antenna development to enhance gain for surface penetrating radar. *Wireless Personal Communications, 115*, 1821–1838.
4. Cohen, E. B., Horton, K. G., Marra, P. P., Clipp, H. L., Farnsworth, A., Smolinsky, J. A., & Buler, J. J. (2021). A place to land: spatiotemporal drivers of stopover habitat use by migrating birds. *Ecology Letters, 24*(1), 38–49.
5. Li, Y., Liu, C., Yue, G., Gao, Q., & Du, Y. (2022). Deep learning-based pavement subsurface distress detection via ground penetrating radar data. *Automation in Construction, 142*, 104516.
6. Fan, J., Ma, T., Zhu, Y., & Zhang, Y. (2023). Ground penetrating radar detection of buried depth of pavement internal crack in asphalt surface: A study based on multiphase heterogeneous model. *Measurement, 221*, 113531.
7. Wei, S., Li, C., Mao, X., & Ai, D. (2023). Investigation on the multimodal failure characteristics of cement mortar under uniaxial compression loading. *Construction and Building Materials, 392*, 131900.
8. Yu, H., Li, H., & Li, Y. (2020). Vibration signal fusion using improved empirical wavelet transform and variance contribution rate for weak fault detection of hydraulic pumps. *ISA transactions, 107*, 385–401.
9. Hanandeh, S. (2022). Introducing mathematical modeling to estimate pavement quality index of flexible pavements based on genetic algorithm and artificial neural networks. *Case Studies in Construction Materials, 16*, e00991.

Mean-Squared Response Sensitivity Method for Bridge Damage Detection

Ning An, Rui Ma, and Peng Dong

Abstract A new sensitivity approach for detecting bridge deterioration using the mean-squared response is proposed, in which the dynamic responses of the bridge are estimated using a vehicle-bridge dynamic interaction model. The sensitivity of the mean-squared response to bridge damage is investigated and matrices are generated by defining the stiffness variation of the bridge element as an index of damage factor. The mean-squared response sensitivity method detects local bridge damage with significantly fewer iteration times, and the identification accuracy is proportional to the number of measuring points. The mean-squared response sensitivity approach may correctly identify bridge local damage, including single-element and two-element damage.

Keywords Damage detection · Vehicle-bridge system · Mean-squared response · Sensitivity method

1 Introduction

Damage detection for engineering structures has attracted great attention from researchers in recent decades, and numerous detection methods have been developed. As a finite element (FE) update approach, the sensitivity-based model update method is an effective damage detection method that is based on the solution of a first-order Taylor series that minimizes an error function of matrix perturbations. The finite element model is continuously updated in this method until the estimated output of the structure (such as the response at measuring locations, the structure's natural frequencies, or any other value) is close enough to the true one. Natural frequencies [1, 2], mode shapes [3, 4], curvature mode shapes [5], modal strain energies [6, 7], frequency response functions [8], and stiffness or flexibility sensitivities [9] are typical frequency domain parameters used in various sensitivity-based methods. However, for online damage detection of structures in service, these

N. An (✉) · R. Ma · P. Dong
Research Institute of Highway Ministry of Transport, Beijing 100029, China
e-mail: n.an@rioh.com

© The Author(s) 2025
D. Li and Y. Zhang (eds.), *Advances in Frontier Research on Engineering Structures II*,
Lecture Notes in Civil Engineering 535, https://doi.org/10.1007/978-981-97-6238-5_26

frequency-domain metrics frequently necessitate a large number of sensors and human processing. In comparison, time domain methods can make better use of measured information than frequency domain methods and require fewer measuring points [10, 11]. Lu and Law [12] present a time-domain sensitivity-based update approach. The sensitivity matrix of dynamic response of a structure with regard to a perturbation of a system parameter is addressed in their work, and numerical simulation is used to demonstrate the usefulness and accuracy of the proposed method.

In general, the response sensitivity strategies produced good results. However, these methods only use the answers of one or a few measurement points and do not fully exploit the relationships between the measuring points. The mean square of the bridge's responses caused by moving trains is used in this paper to detect bridge degradation. The numerical simulation results demonstrate the procedure and impact of the mean-squared response sensitivity method for bridge damage identification.

2 Bridge Response Caused by Vehicle

Each vehicle in the dynamic model of the bridge-vehicle interaction system is made up of a car-body, two bogies, four wheel-sets, and spring-dashpot suspensions between the three components. The automobile body, bogies, and wheel sets are viewed as rigid components, with elastic deformation during vibration being ignored. A 10-DOF dynamic system can mimic the idealized model for each vehicle.

In the vehicle-bridge interaction dynamics, it is commonly assumed that the wheel-sets maintain contact with the bridge track, so that when track deformation is not taken into account, the movement of any wheel-set can be determined by the movement of the bridge deck and track irregularity at the position of the wheel-set. The coupled motion equations for the bridge-train system can be expressed as follows in this case:

$$\begin{bmatrix} \mathbf{M}_{vv} & \mathbf{0} \\ \mathbf{0} & \mathbf{M}_{bb} \end{bmatrix} \begin{Bmatrix} \ddot{\mathbf{X}}_v \\ \ddot{\mathbf{X}}_b \end{Bmatrix} + \begin{bmatrix} \mathbf{C}_{vv} & \mathbf{C}_{vb} \\ \mathbf{C}_{bv} & \mathbf{C}_{bb} \end{bmatrix} \begin{Bmatrix} \dot{\mathbf{X}}_v \\ \dot{\mathbf{X}}_b \end{Bmatrix} + \begin{bmatrix} \mathbf{K}_{vv} & \mathbf{K}_{vb} \\ \mathbf{K}_{bv} & \mathbf{K}_{bb} \end{bmatrix} \begin{Bmatrix} \mathbf{X}_v \\ \mathbf{X}_b \end{Bmatrix} = \begin{Bmatrix} \mathbf{F}_v \\ \mathbf{F}_b \end{Bmatrix} \quad (1)$$

where $\mathbf{M}$, $\mathbf{C}$, and $\mathbf{K}$ are the mass, damping, and stiffness matrices of the vehicle and the bridge, respectively, and vb and bv represent the train-bridge interaction.

When both sides of Eq. (1) are differentiated with regard to the damage factor of each element, e.g. the damage factor of the kth element α_k in the damage factor vector, the following results are obtained:

$$\begin{bmatrix} \frac{\partial \mathbf{M}_{vv}}{\partial \alpha_k} & \mathbf{0} \\ \mathbf{0} & \frac{\partial \mathbf{M}_{bb}}{\partial \alpha_k} \end{bmatrix} \begin{Bmatrix} \ddot{\mathbf{X}}_v \\ \ddot{\mathbf{X}}_b \end{Bmatrix} + \begin{bmatrix} \frac{\partial \mathbf{C}_{vv}}{\partial \alpha_k} & \frac{\partial \mathbf{C}_{vb}}{\partial \alpha_k} \\ \frac{\partial \mathbf{C}_{bv}}{\partial \alpha_k} & \frac{\partial \mathbf{C}_{bb}}{\partial \alpha_k} \end{bmatrix} \begin{Bmatrix} \ddot{\mathbf{X}}_v \\ \ddot{\mathbf{X}}_b \end{Bmatrix} + \begin{bmatrix} \frac{\partial \mathbf{K}_{vv}}{\partial \alpha_k} & \frac{\partial \mathbf{K}_{vb}}{\partial \alpha_k} \\ \frac{\partial \mathbf{K}_{bv}}{\partial \alpha_k} & \frac{\partial \mathbf{K}_{bb}}{\partial \alpha_k} \end{bmatrix} \begin{Bmatrix} \mathbf{X}_v \\ \mathbf{X}_b \end{Bmatrix}$$

$$+ \begin{bmatrix} \mathbf{M}_{vv} & 0 \\ 0 & \mathbf{M}_{bb} \end{bmatrix} \begin{Bmatrix} \frac{\partial \ddot{\mathbf{X}}_v}{\partial \alpha_k} \\ \frac{\partial \ddot{\mathbf{X}}_b}{\partial \alpha_k} \end{Bmatrix} + \begin{bmatrix} \mathbf{C}_{vv} & \mathbf{C}_{vb} \\ \mathbf{C}_{bv} & \mathbf{C}_{bb} \end{bmatrix} \begin{Bmatrix} \frac{\partial \dot{\mathbf{X}}_v}{\partial \alpha_k} \\ \frac{\partial \dot{\mathbf{X}}_b}{\partial \alpha_k} \end{Bmatrix} + \begin{bmatrix} \mathbf{K}_{vv} & \mathbf{K}_{vb} \\ \mathbf{K}_{bv} & \mathbf{K}_{bb} \end{bmatrix} \begin{Bmatrix} \frac{\partial \mathbf{X}_v}{\partial \alpha_k} \\ \frac{\partial \mathbf{X}_b}{\partial \alpha_k} \end{Bmatrix} = \begin{Bmatrix} \frac{\partial \mathbf{F}_v}{\partial \alpha_k} \\ \frac{\partial \mathbf{F}_b}{\partial \alpha_k} \end{Bmatrix} \tag{2}$$

In the equation, the differentials of $\mathbf{M}_{bb}$, $\mathbf{M}_{vv}$, $\mathbf{C}_{vv}$, $\mathbf{C}_{vb}$, $\mathbf{C}_{bv}$, $\mathbf{K}_{vv}$, $\mathbf{K}_{vb}$, $\mathbf{K}_{bv}$, $\mathbf{F}_v$ and $\mathbf{F}_b$ are all equal to 0, for they are independent of α_k.

Equation (2) can be simplified as

$$\begin{bmatrix} \mathbf{M}_{vv} & 0 \\ 0 & \mathbf{M}_{bb} \end{bmatrix} \begin{Bmatrix} \frac{\partial \ddot{\mathbf{X}}_v}{\partial \alpha_k} \\ \frac{\partial \ddot{\mathbf{X}}_b}{\partial \alpha_k} \end{Bmatrix} + \begin{bmatrix} \mathbf{C}_{vv} & \mathbf{C}_{vb} \\ \mathbf{C}_{bv} & \mathbf{C}_{bb} \end{bmatrix} \begin{Bmatrix} \frac{\partial \dot{\mathbf{X}}_v}{\partial \alpha_k} \\ \frac{\partial \dot{\mathbf{X}}_b}{\partial \alpha_k} \end{Bmatrix}$$

$$+ \begin{bmatrix} \mathbf{K}_{vv} & \mathbf{K}_{vb} \\ \mathbf{K}_{bv} & \mathbf{K}_{bb} \end{bmatrix} \begin{Bmatrix} \frac{\partial \mathbf{X}_v}{\partial \alpha_k} \\ \frac{\partial \mathbf{X}_b}{\partial \alpha_k} \end{Bmatrix} = \begin{Bmatrix} 0 \\ -\frac{\partial \mathbf{K}_b}{\partial \alpha_k} \mathbf{X}_b - \beta \frac{\partial \mathbf{K}_b}{\partial \alpha_k} \dot{\mathbf{X}}_b \end{Bmatrix} \tag{3}$$

By solving Eq. (3), the velocity response vector $\mathbf{u}_i = [u_{i,1}, u_{i,2}, \ldots \ldots u_{i,m}]$ and their response sensitivity with respect to the damage factors can be obtained via $\frac{\partial \dot{u}(x,t)}{\partial \alpha_k}$.

3 Mean-Squared Response Detection Method

When there are p measuring points on the structure, for each measuring point i, assume velocity time-history $u_i(t)$ is discretized at m sampling points, they can be written as vectors $\mathbf{u}_i = [u_{i,1}, u_{i,2}, \ldots \ldots u_{i,m}]$, thus the jth sampling point of the mean square of the response can be expressed as:

$$s_j = \frac{1}{p} \sum_{i=1}^{p} u_{i,j}^2 \tag{4}$$

The damage of an element of the bridge can be represented through its flexural stiffness. In this case, the damage factor α_k for the kth element is defined as:

$$\alpha_k = \frac{(EI)_k - (EI)_{k0}}{(EI)_{k0}} \times 100\% \tag{5}$$

where $(EI)_{k0}$ is the original flexural stiffness of the kth element, and it becomes $(EI)_k$ upon the occurrence of damage $(EI)_k$.

For n elements of the bridge, the damage factor vector can be defined as $\alpha = \{\alpha_1 \ \alpha_2 \ \ldots \ \alpha_n\}^T$. When any element of the bridge is damaged, the response and its mean square are all changed, so they are the functions of the vector α. According to this principle, a mean-squared response sensitivity-based update bridge damage detection method is proposed, which is based on the modification of structural stiffness matrix to reproduce as closely as possible the measured dynamic response from the data. The updated matrices are used to compare with the original correlated matrix and then an indication of damage can be got to quantify the location and extent of the damage.

4 Numerical Illustration of Mean-Squared Response Sensitivity Method

Figure 1 depicts a bridge-vehicle system that is used to demonstrate the application of the mean-squared response sensitivity approach. The bridge features a 32-m-long beam with a uniform cross-section, a Young's modulus of 3.55104 MPa, a sectional area of 3.955 m^2, and an area moment of inertia of 4.341 m^4. The beam is discretized into 16 equal-length segments. The train is composed of 4 vehicles, the parameters of which are stated in Table 1.

Vertical track irregularity is regarded as self-excitation to train-bridge system vibration. The Grade-4 track irregularity spectrum is used in the study based on the track irregularity spectra.

The train runs through the bridge at a speed of 25 m/s. The bridge's dynamic responses are computed. To imitate the polluted measurement, the computed responses are polluted with white noise in the following analysis, expressed as:

$$u_p = u_m + e_p N_0 \sigma(u_m) \tag{6}$$

where u_p and u_m are, respectively, the polluted and the original measured velocity of the bridge; e_p is the noise level, defined as the ratio of noise amplitude to the response amplitude (between 0 and 1); N_0 is a standard normal distribution vector with zero mean and unit standard deviation; $\sigma(u_m)$ is the standard deviation of u_m.

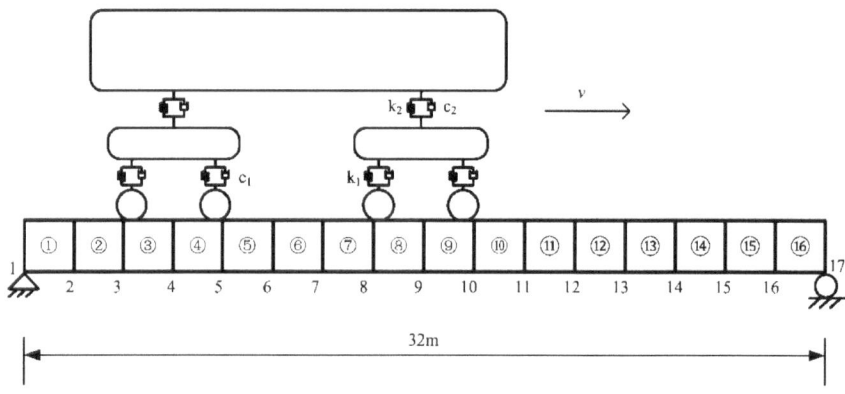

Fig. 1 Analysis model of train-bridge system

Table 1 Parameters of vehicle

Parameter	Unit	Value
Distance between two wheel-sets $2d_1$	m	2.4
Distance between two bogie centers $2d_2$	m	18.5
Full length of the car	m	26.57
Mass of wheel-set	kg	1675
Mass of bogie	kg	3086
Mass of car-body	kg	48,200
Mass moment of inertia of bogie	kg·m^2	4730
Mass moment of inertia of car-body	kg·m^2	2,999,000
Damping of the primary suspension system c_1	kN·s/m	30
Damping of the secondary suspension system c_2	kN·s/m	140
Stiffness of the primary suspension system k_1	kN/m	2538
Stiffness of the secondary suspension system k_2	kN/m	960

Figure 2 shows the the determined damage result only after 8 iterations utilizing the mean-squared velocity of 5 nodes (Node 5, 7, 9, 11, and 13). It demonstrates that the mean-squared velocity may be used to reliably identify damage.

The update bridge damage detection approach based on mean-squared velocity sensitivity is likewise relevant in the case of 2 damages. For example, when 2 damages are placed into the beam's Elements 5 and 12, with damage factors of 12 and 8%, respectively, and noise levels of 3%, the damages are well recognized using the mean-squared response of five nodes (Nodes 5, 7, 9, 11, and 13), as shown in Fig. 3.

Fig. 2 Identified damage of Element 6

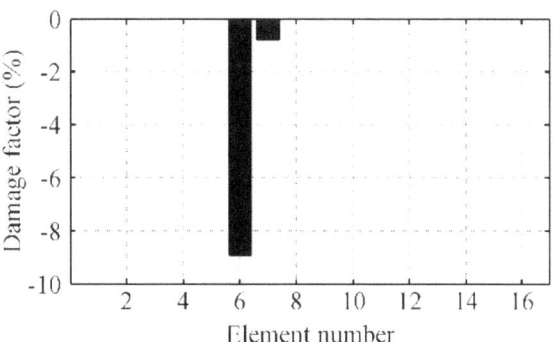

Fig. 3 Identified damages of Elements 5 and 12

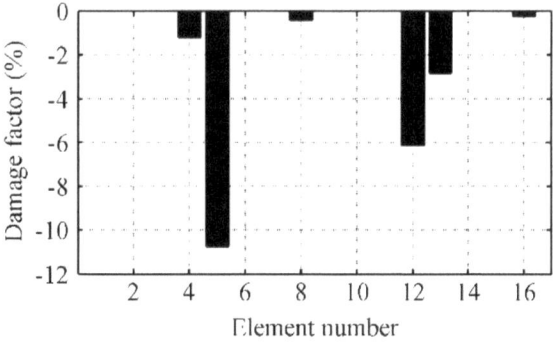

5   Conclusion

The mean-squared response sensitivity method is proposed to detect the damage of bridge using the mean-squared response induced by running trains. The numerical results show that the mean-squared response sensitivity method behaves very well. The following conclusions can be drawn from the results:

(1) Compared with the response sensitivity method that does not fully use the connections among different measuring points and has large amount of calculation, the mean-squared response sensitivity method is effective to detect local damage of bridge using much fewer iteration times.

(2) The mean-squared response sensitivity method has advantages especially in case of multiple measuring points. The accuracy of identified result is related to the number of measuring points: the more the measuring points, the better the results.

(3) The mean-squared response sensitivity method can identify accurately the local damage of bridge, including single-element damage and two-element damage, and is insensitive to measurement noise.

Acknowledgements This work was financially supported by Central Public-Interest Scientific Institution Basic Research Fund (2021-9053a and 2021-9053b) and Open Project of National Engineering Laboratory of Bridge Safety and Technology (Beijing).

References

1. Jian-min, S., Dan, L., & Wang-ji, Y. (2022). Regularization methods for solving modal sensitivity-based damage equations: A comparative study[J]. *Chinesel Journal of Computational Mechanics, 39*(1), 70–79.

2. Yang, Q. W., & Sun, B. X. (2011). Structural damage identification based on best achievable flexibility change[J]. *Applied Mathematical Modelling, 35*(10), 5217–5224.

3. Hou, J., Li, Z., Zhang, Q., Jankowski, Ł., & Zhang, H. (2020). Local mass addition and data fusion for structural damage identification using approximate models. *International Journal of Structural Stability and Dynamics, 20*(11), 2050124.
4. Lee, E. T., & Eun, H. C. (2015). Damage identification of a frame structure model based on the response variation depending on additional mass[J]. *Engineering with Computers, 31*(4), 737–747.
5. Xiang, C., Li, L., Zhou, Y., & Yuan, Z. (2020). Damage identification of beam structures based on modal curvature utility information entropy. *Journal ofl Vibration and Shock, 39*(17), 234–244.
6. Tong, W., Liang, T., & Zhi-xiang, Z. (2021). Deflection curvature area difference method for damage location of bridge structures[J]. *Advanced Engineering Sciences, 53*(6), 165–174.
7. Entezami, A., & Shariatmadar, H. (2014). Damage detection in structural systems by improved sensitivity of modal strain energy and Tikhonov regularization method[J]. *International Journal of Dynamics and Control, 2*(4), 509–520.
8. Majumdar, A., Maiti, D. K., & Maity, D. (2012). Damage assessment of truss structures from changes in natural frequencies using ant colony optimization. *Applied Mathematics and Computation, 218*(19), 9759–9772.
9. Tomaszewska, A. (2010). Influence of statistical errors on damage detection based on structural flexibility and mode shape curvature. *Computers & Structures, 88*(3–4), 154–164.
10. Choi, S., & Stubbs, N. (2004). Damage identification in structures using the time-domain response. *Journal of Sound and Vibration, 275*(3–5), 577–590.
11. Link, M., & Weiland, M. (2009). Damage identification by multi-model updating in the modal and in the time domain. *Mechanical Systems and Signal Processing, 23*(6), 1734–1746.
12. Lu, Z. R., & Law, S. S. (2007). Features of dynamic response sensitivity and its application in damage detection. *Journal of Sound and Vibration, 303*(1–2), 305–329.

Structural Dynamics and Building Bearing Capacity Analysis

Research on Bearing Capacity Size Effect of RC Frame Considering Material Size Effect

Hongyu Zhou, Shuai Han, Yun Zhou, and Xiaohua Zhao

Abstract The size effect correction formulas of peak strength and peak strain of concrete materials are summarized by using regression analysis theory and Bažant size effect law. The finite element software Seismostruct is used to simulate the unidirectional nappe of three groups of plane frame structures with different sizes of modified materials, and the simulation results are compared with the experimental results. The results show that the bearing capacity of the structure decreases with the increase of size, and the simulation results considering the size effect correction are closer to the test results, so it is suggested that the material size effect correction should be taken into account in the structural design analysis.

Keywords Component · Regression analysis · Reinforced concrete frame · Size effect · Numerical simulation

1 Introduction

As one of the most common architectural forms, the mechanical performance of reinforced concrete frame structures has always been the research focus of scholars. Su [1] carried out the quasi-static loading test on a three-story and three-span RC frame. The test results show that reducing the vertical load and beam-column linear stiffness ratio of the structure is beneficial to realize the ideal "beam hinge" failure mechanism. Tang [2] carried out a Pushover analysis of six frame structures. The simulation results show that the bottom shear force is the largest under uniformly

H. Zhou (✉) · S. Han
Faculty of Architecture, Civil and Transportation Engineering, Beijing University of Technology, Beijing, China
e-mail: ZHYktztgyx@163.com

Y. Zhou
Beijing Glory PKPM Technology Co., Ltd, Shanghai Branch, Shanghai, China

X. Zhao
China Academy of Building Research, Beijing, China
e-mail: 459569497@qq.com

© The Author(s) 2025
D. Li and Y. Zhang (eds.), *Advances in Frontier Research on Engineering Structures II*,
Lecture Notes in Civil Engineering 535, https://doi.org/10.1007/978-981-97-6238-5_27

distributed loading, and the first-order vibration mode of the structure is closer to the actual deformation of the structure. According to the research results, it is confirmed that there is a size effect in the mechanical properties such as peak strength and peak strain of concrete material [3] and members [4].

In the above research results, the size effect of the actual frame structure is not considered, the numerical simulation results are not compared with the real test data, and the size effect correction of concrete materials is not taken into account. There is still a gap in the research on the size effect of frame structure. Therefore, in this paper, the frame structure model based on fiber beam-column element is established by using Seismostruct finite element software, and the size effect of concrete material is taken into account to compare the simulation results with the real experimental data to study the bearing mechanical properties of the frame structure.

2 Structural Model and Finite Element Model

2.1 Original Test Frame Structure

In this paper, the frame in the pseudo-static collapse test of the reinforced concrete frame structure of Tsinghua University [5] is selected as the simulation prototype model. The original structure is a three-story and three-span RC frame with a plane size of 4.95×9 m, a story height of 3.3 m, a span of 6 m, a column section size of 200×200 mm, a beam section size of 250×500 mm, concrete grade C30, longitudinal reinforcement HRB335, stirrups HPB235, and the designed axial compression ratio of the first story and middle column is 0.80. The model is enlarged according to the similarity coefficients 1.5, 2.0, and three RC frames with different scales are obtained, numbered KJ1, KJ2, and KJ3. The reinforcement of the frame beam, slab, and column is shown in Table 1.

2.2 Finite Element Model

In this paper, the Seismostruct software is adopted, and the nonlinear concentrated plastic hinge fiber element based on force is adopted. The constitutive model of steel bar is 80% discount line steel, and the constitutive model of concrete in the core area is Mander nonlinear concrete constitutive model. The finite element model of the frame structure is shown in Fig. 1.

Table 1 Reinforcement of frame beam, slab, and column

		KJ1	KJ2	KJ3
Column	Longitudinal bar	8B8	8B12	8B16
	Stirrups	A4@50	A6@75	A8@100
		A4@100	A6@150	A8@200
	Longitudinal bar	4B10 + 4B8	4B16 + 4B12	4B20 + 4B16
	Stirrups	A6@70	A10@105	A12@140
		A6@140	A10@210	A12@280
Beam	Longitudinal bar (top)	4B10	4B16	4B20
	Longitudinal bar (bottom)	4B8	4B12	4B16
	Stirrups	A4@50	A6@75	A8@100
		A4@100	A6@150	A8@200
Slab	Top	6 A4	6 A6	6 A8
	Bottom	6 A4	6 A6	A8

Unit: mm

Fig. 1 Finite element model of RC plane frame

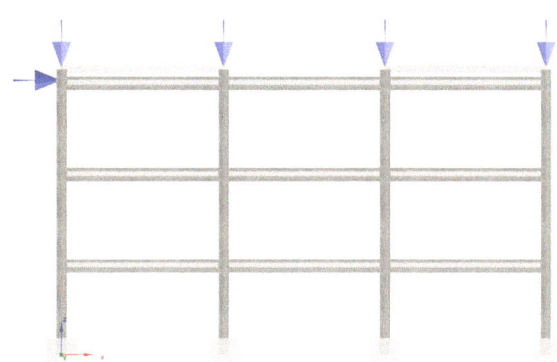

3 Correction of Size Effect of Concrete Material

The research results of many scholars show that the steel bar has no obvious size effect, and this study only modifies the size effect on the peak strength and peak strain of concrete. In this paper, the experimental data are combined with the Bažant size effect law for regression analysis. The theoretical expression of the Bažant size effect is shown in Formula (1).

$$\sigma_N = \frac{Bf_c}{\sqrt{1 + D/D_0}} \tag{1}$$

In the formula, f_c is the compressive strength of the material; B is the dimensionless constant; D is the cross-sectional size of the specimen; D_0 is the geometric correlation constant of the structure.

The mean value in Fig. 2 is the mean value of the data points of this topic and other scholars' experimental data [6–10]. The regression analysis is carried out with reference to the theoretical formula of the size effect of Bažant. The axial compressive strength of the 150 mm cube is taken as the reference strength, and the size effect parameters are obtained by regression analysis of the data points of the mean function, that is, $B = 1.1675$, $D_0 = 467.9715$. The fitting formula of the peak strength size effect is obtained as follows:

$$\sigma = \frac{1.1675 f_c}{\sqrt{1 + D/467.9715}} \quad 70.7 \le D \le 467.97 \tag{2}$$

The data of the peak strain size effect of concrete confined by relevant stirrups are selected as the correction basis. The average value in Fig. 3 is the arithmetic average value of each scholar's experimental data point [11–13], based on the size and side length of the original column 200 mm. The correction value of the axial compressive strength of concrete in each layer of the original test material is shown in Table 2.

The structure is regarded as a cube of concrete confined by 200 mm stirrups, and the modified formula of the peak strain size effect of concrete confined by stirrups is obtained by fitting the mean data points, where D is the side length of the concrete. By substituting different cube sizes D, the calculated results are shown in Table 3.

$$\frac{\varepsilon}{\varepsilon_{200}} = -0.117 \ln(D) + 1.6184 \quad 100 \le D \le 400 \tag{3}$$

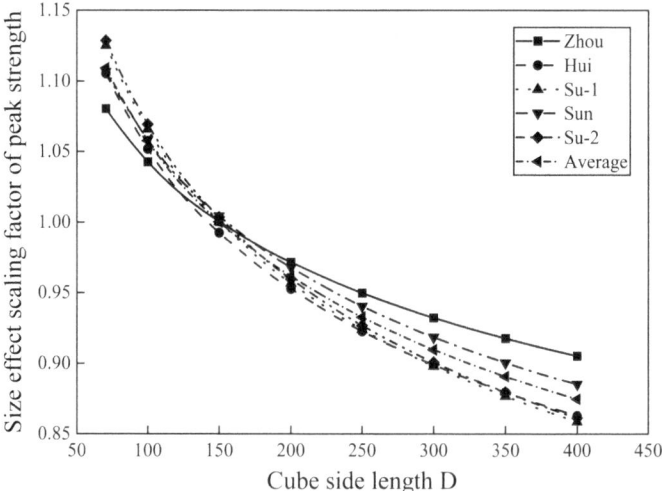

Fig. 2 Size effect coefficient of peak strength

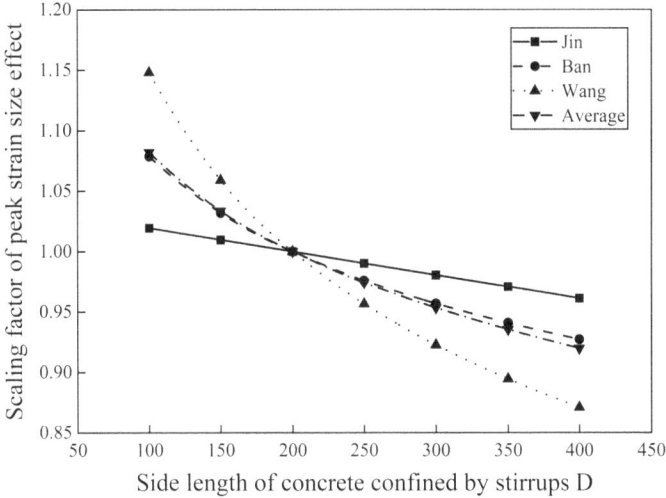

Fig. 3 Peak strain size effect coefficient

Table 2 Correction value of size effect on compressive strength of concrete

Cube side length D/ mm	The first layer correction value/Mpa	Second layer correction value/Mpa	Third layer correction value/Mpa
150	36.20	34.70	33.60
200	34.82	33.38	32.32
300	32.47	31.13	30.14
400	30.54	29.28	28.35

Table 3 Modified value of peak strain size effect of concrete confined by stirrups

Cube side length D/mm	200	300	400
Peak strain correction	0.00400	0.00381	0.00351

4 Numerical Simulation Results and Analysis of RC Plane Frame

4.1 Overview of Basic Operating Conditions

In this paper, the test condition is divided into the following two parts: (1) unidirectional nappe of KJ1, KJ2, and KJ3 under static constitutive model; (2) only considering the size effect, the unidirectional nappe of KJ1, KJ2, and KJ3 is modified. A summary of the analysis conditions is shown in Table 4.

Table 4 Analysis of working conditions

Loading condition	Frame model	Constitutive model	Loading mode
Condition 1	KJ1	Static constitutive model	Unidirectional loading-1
Condition 2	KJ2	Static constitutive model	Unidirectional loading-2
Condition 3	KJ3	Static constitutive model	Unidirectional loading-3
Condition 4	KJ1	Size effect modified constitutive model	Unidirectional loading-1
Condition 5	KJ2	Size effect modified constitutive model	Unidirectional loading-2
Condition 6	KJ3	Size effect modified constitutive model	Unidirectional loading-3

4.2 Loading Scheme Design

While keeping the axial compression ratio constant, the axial pressure ratio applied to the center column and side column is maintained at 1: 2. The equivalent gravity loads on the upper part of the frame structure with different sizes are shown in Table 5.

The displacement loading mode is adopted, the horizontal displacement at the top of the structure is controlled, the left side node of the top floor of the frame is the loading point, 1% of the yield displacement extracted from the hysteretic curve of the frame is loaded as the reference displacement, and 80% of the peak bearing capacity of the frame is the failure load. Among them, the hysteretic loading system of KJ2 and KJ3 is 1.5 times and 2 times the displacement of KJ1 horizontal reciprocating loading system respectively, and the three horizontal loading points are loaded at the loading ratio of 3: 2: 1. The hysteresis curve obtained by loading is shown in Fig. 4. After that, the positive and negative skeleton curve of the hysteresis curve is extracted, and the equivalent efficiency method is used to solve the yield displacement. the calculation results of the characteristic points of the skeleton curve of each frame structure are shown in Table 6. According to the calculation results in the table, KJ1, KJ2, and KJ3 respectively take 0.4319 mm, 0.6457 mm, and 0.8099 mm as the displacement reference values of unidirectional nappe loading, and take the top loading point of the structure as the control point to output displacement and loading force. The reduction of the bearing capacity to 80% of the peak bearing capacity is regarded as frame failure.

Table 5 Equivalent gravity load of superstructure

	KJ1	KJ2	KJ3
Side column	163	366.75	652
Central column	326	733.5	1304

Unit: kN

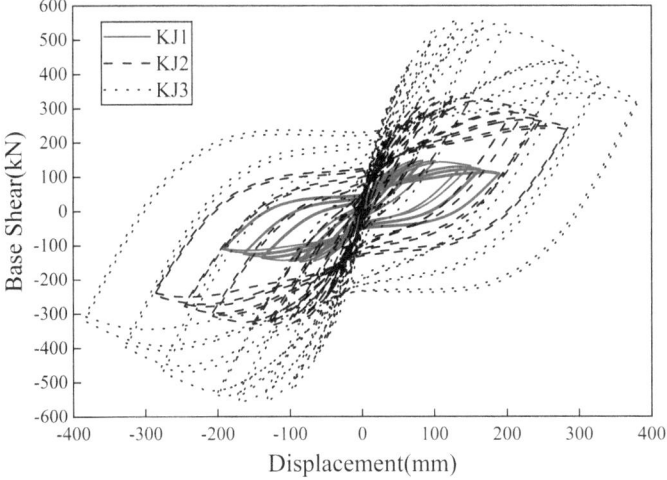

Fig. 4 Loading hysteretic curves of plane frame structures with different scales

Table 6 Feature points of skeleton curve of frame structure

Frame model	Equivalent yield point of the forward skeleton		Equivalent yield point of the negative skeleton		Average equivalent yield point	
	$\triangle$y/mm	Py/kN	$\triangle$y/mm	Py/kN	$\triangle$y/mm	Py/kN
KJ1	43.27	126.55	43.10	125.80	43.19	126.18
KJ2	64.23	285.79	64.91	286.76	64.57	286.27
KJ3	79.86	473.00	82.12	476.45	80.99	474.73

4.3 Simulated Loading Results

The nonlinear static analysis module is used to simulate the loading of each working condition in the Seismostruct software. The bearing capacity and displacement of the frame control point under each working condition are shown in Table 7, and the load–displacement curve at the frame control point is shown in Fig. 5.

4.4 Size Effect Analysis

As can be seen from Fig. 5 and Table 7, after considering the size effect of concrete, the peak bearing capacity and ultimate bearing capacity of KJ1, KJ2, and KJ3 decreased by 2.07%, 3.72% and 5.88%, respectively. After considering the size effect of the constitutive peak strength and peak strain of concrete, the bearing capacity decreases

Table 7 Loading results of frame structure under various working conditions

Loading condition	Peak bearing capacity P_m/kN	Peak displacement Δ_m/mm	Ultimate bearing capacity P_u/kN	Ultimate displacement Δ_u/mm
Condition 1	158.59	116.78	126.87	93.42
Condition 2	366.25	170.07	292.99	136.06
Condition 3	627.49	214.73	501.99	171.78
Condition 4	155.31	117.25	124.24	93.80
Condition 5	352.61	174.20	282.08	139.36
Condition 6	590.56	216.24	472.45	172.99

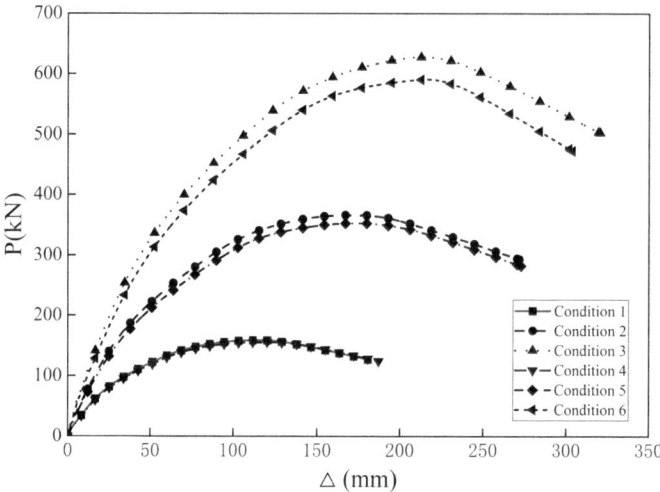

Fig. 5 The load–displacement curve of frame structure under various working conditions

in varying degrees compared with that without modification, and the larger the size is, the greater the size effect is.

As can be seen from Fig. 6, the unmodified KJ1 model, that is, the original test model, the single point simulation loading results are in good agreement with the original structural test data, and its change law is consistent with the test curve, which verifies the correctness of the finite element model established in this paper.

Combined with the analysis of Table 8, the peak bearing capacity of modified KJ2 and KJ3 is 2.27 and 3.80 times higher than that of KJ1, respectively, and the peak bearing capacity of KJ2 is 127% higher than that of KJ1. Compared with KJ2, the peak bearing capacity of KJ3 is 67% higher.

The results show that the increase of peak bearing capacity is affected by the size effect caused by the increase of structural size, and the larger the structural size is, the slower the increase of peak bearing capacity will be.

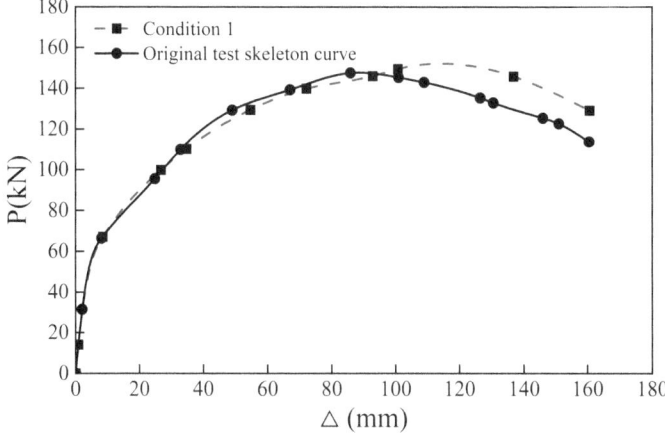

Fig. 6 Comparison of load–displacement curves of frame structures

Table 8 Changes of peak bearing capacity of each frame after correction

Frame model	Improvement of peak bearing capacity	Percentage of adjacent size increase (%)
KJ1	1.00	0
KJ2	2.27	127%
KJ3	3.80	67%

5 Conclusion

In this paper, the frame model based on fiber beam-column element is established by using Seismostruct finite element software, and the monotone push-over simulation is carried out considering the size effect correction of concrete materials. The simulation results are compared with the experimental data. The main results are as follows:

(1) When considering the correction of the material size effect, the numerical simulation results of reinforced concrete frame structures are closer to the actual test results under the action of unidirectional thrust. Therefore, for reinforced concrete frame structures, it is suggested that the size effect of materials should be taken into account in the actual structural design and scale test analysis.

(2) With the increase of the size of the structure, the bearing capacity of the structure increases gradually, but the increasing range of the bearing capacity of adjacent structures of different sizes will gradually decrease with the increase of size, and there is a certain degree of size effect.

References

1. Liu, B., Su, J., & Ma, Y. (2019). Pseudo-static collapse experiment of multi-story multi-span reinforced concrete plane frames. *Journal of China Civil Engineering Journal, 52*(08), 24–39.
2. Tang, J. (2013). *Pushover analysis of reinforced concrete frame structure.* D. Qingdao University of Technology.
3. Du, X., Jin, L., & Li, D. (2017). A state-of-the-art review on the size effect of concretes and concretestructures (I): Concrete materials. *Journal of China Civil Engineering Journal, 50*(09), 28–45.
4. Du, X., Jin, L., & Li, D. (2017). A state-of-the-art review on the size effect of concretes and concretestructures (II): RC members. *Journal of China Civil Engineering Journal, 50*(11), 24–44.
5. Lu, X., Ye, L., Pan, P., Zhao, Z., Ji, X., & Qian, J. (2012). Pseudo-static collapse experiments and numerical prediction competition of RC frame structure I: RC frame experiment. *Building Structure, 42*(11), 19–22.
6. Zhou, H., Liu, H., & Zhao, X. (2019). Research of fatigue compression damage mechanism and size effect of concrete materials. *Journal of Concrete, 12*, 1–5.
7. Hui, H., Li, L., & Yang, H. (2015). Experimental study on impact of strength grade on size effect of concrete strength. *Journal of Concrete, 07*, 31–34.
8. Su, J., Fang, Z., & Yang, Z. (2014). Influence of aggregate mixture and strength grade on dimensional effect of concrete uniaxial compressive behavior. *Journal of Building Structures, 35*(05), 120–127.
9. Sun, Q. (2017). *Experimental Study on size effect of C30 steel fiber reinforced.* D. North China University of Water Resources and Electric Power.
10. Su, J., Fang, Z., & Yang, Z. (2014). Experimental study on impact of aggregate mixture on dimensional effect of concrete cubic compressive strength. *Journal of Building Structures, 35*(02), 152–157.
11. Jin, L., Li, P., & Du, X. (2021). Size effect law of nominal compressive strength for squared concrete columns confined by stirrups. *Journal of Basic Science and Engineering, 29*(04), 986–998.
12. Wang, T. (2013). *Size effect and performance of concrete confined by stirrups under axial compression.* D. Dalian University of Technology.
13. Ban, S. (2013). *Experimental study on size effect on behavior of concrete confined by stirrups.* D. Dalian University of Technology.

Study on the Bond-Slip Mechanism and Mechanical Analysis of Steel Reinforced Recycled Concrete

Guoliang Bai and Tao Ni

Abstract Based on the results of a large number of experimental studies on the bond-slip launch of steel recycled concrete, the group tries to clarify the transformation law of chemical adhesive force, mechanical bite force and friction force in the process of applied load; analyzes the mechanism of bond evolution through fine view theory and peeling wear theory, and explains the principle that the bond force of steel recycled concrete is weaker than that of ordinary steel concrete; takes micro-segment theory for steel recycled concrete and gives the basic Equation of bond-slip from the perspective of elasticity. The basic Equation of bond-slip is given, and the formula of bond-slip is derived from the perspective of elastic mechanics. The results of the study can provide some reference for the design and calculation theory of steel recycled concrete structures.

Keywords Profile recycled concrete · Bond-slip mechanism · Mechanical analysis · Elastodynamics

1 Introduction

Currently, the annual emission of construction waste in China regarding waste concrete type can reach billion tons [1]. The waste concrete is processed and added to concrete in the form of coarse aggregate to prepare recycled concrete, which can realize the reuse of construction waste and achieve the goal of carbon peaking and carbon neutrality [2], which has significant environmental and economic benefits and is an inevitable choice for the future construction industry. However, the residual old mortar at the interface of recycled coarse aggregate and internal microcracks leads to lower mechanical properties of recycled concrete compared to normal concrete [3]. To change the mechanical properties of recycled concrete and increase the range of applications of recycled concrete, steel profiles are embedded in them to form a combined steel-recycled concrete structure [4].

G. Bai · T. Ni (✉)
Xian University of Architecture and Technology, Beilin District, Xi'an, China
e-mail: 1559625894@qq.com

© The Author(s) 2025
D. Li and Y. Zhang (eds.), *Advances in Frontier Research on Engineering Structures II*,
Lecture Notes in Civil Engineering 535, https://doi.org/10.1007/978-981-97-6238-5_28

The profile steel recycled concrete structure is a new type of structure that combines the profile steel and recycled concrete to share the load, and the prerequisite for working together is a good bonding performance between the two [5]. The concept of bond strength is an intuitive indicator of the superiority of bond performance, which is essentially a matter of shear transfer, and bond strength is also an important parameter in determining the principal structure relationship and the code of breaking strength [6]. A large number of studies have been done at home and abroad on section recycled concrete, but most of them are limited to the study of bond strength and seismic performance. Bai and Ma [7] studied the bond performance of the interface of section recycled concrete columns by launching tests. The group of Xue Jianyang of Xi'an University of Architecture and Technology University studied the seismic performance of section steel recycled concrete columns, section steel recycled concrete frame structures, and section steel recycled concrete frame-infill wall structures from various aspects [8], and the shear strength of section steel recycled concrete beam-column nodes were also studied and analyzed [9]. Jing et al. [10] designed eight-section steel recycled concrete column-steel beam nodes, and verified the seismic performance of the combined frame nodes by repeated tests at low perimeters and numerical simulations, which showed that the strength of the recycled concrete and the thickness of the steel web significantly affected the horizontal bearing capacity, but had no significant effect on the ductility and deformation capacity of the combined nodes. Ma et al. [11] studied the bonding performance of the beam structure of section steel recycled concrete and established a formula for calculating the bond strength considering the replacement rate of recycled aggregates.

At present, the establishment and analysis of the bond-slip mechanism and bond-slip mechanics model of section steel recycled concrete at home and abroad are less studied, and there are still some differences and problems. The study of the bond-slip mechanism problem is characterized by complexity, which is due to the special features of the geometry and force properties of the steel sections in the profile recycled concrete, as well as the limitations of the observation means. Most of the previous theoretical studies on the bonding mechanism are analyzed from a macroscopic point of view, i.e., the interpretation of the three influencing factors and change laws of chemical bonding force, mechanical bite force, and sassafras force.

In this paper, we will analyze from macroscopic and fine-scale perspectives, through the mechanism of regulating the interfacial movement of steel and recycled concrete, the theory of peeling layer and the fine-scale weak layer, and establish the mechanical model of steel recycled concrete, and establish the bond-slip calculation Equation from the perspectives of elastic mechanics and energy method, respectively. It provides a reference for the experimental research and application calculation design of section recycled concrete and provides a basis for the finite element simulation of bond-slip of section recycled concrete.

2 Research on the Mechanism of Bond-Slip

The chemical bonding force exists mainly before the relative slip of the two materials, the profile and the recycled concrete, and relies on the bond stress between the cement mortar and the profile to resist the longitudinal shear stress. Therefore, the influence of the chemical bonding force is mainly the shear capacity between the cement crystals and the compressive strength of the cement paste. A large number of studies have shown that for recycled concrete, the proportion of chemical adhesion to the total bond is much higher than the proportion in the light round steel, which is the main component of the bond between the steel and recycled concrete interface.

Physical friction occurs under the following conditions: (1) the objects are in contact with each other and squeezed; (2) the two contact surfaces are rough; (3) there is a relative movement between the objects or the tendency of relative movement. Steel and recycled concrete in contact with each other, in the process of loading, steel or concrete will produce deformation and the formation of mutual extrusion, and the tendency to slip or slide, the interface between the two materials is relatively rough, so the force of sassafras is present, and is always changing. The size of the friction force depends on the positive stress and friction coefficient at the interface, some experts and scholars use the method of increasing the roughness of the interface to improve the bond stress at the interface, such as abrasive treatment of the steel, the use of striped steel, rebar, etc. The same method is also used to increase the grip force between the steel and concrete to improve the bond stress at the interface, such as increasing the lateral reinforcement rate and protective layer thickness.

The formation of mechanical bite force depends on the phenomenon of mutual staggering of rough peaks between the steel and recycled concrete, i.e., the plowing effect [12]. The mechanical bite force is generally considered to occur after the interface slip, together with the friction force to provide longitudinal bond stress, and mechanical bite force in the specimen slip process is changing, the main factors affecting the material properties, interface roughness, etc.

The changes of chemical bonding force, mechanical bite force, and friction force in the loading process are as follows: when the member is in the early stage of loading, the slip between the interface of steel and recycled concrete does not occur, and the chemical bonding force takes the main load, as shown in Fig. 1, with the increase of load, the initial relative slip occurs at the interface of the loading end, indicating that the chemical bonding force is gradually lost near the loading end, and the friction force and mechanical bite force start to participate in the Work, Zheng et al. pointed [13] that the chemical bonding force from generation to loss, the specimen loading end there is a relatively fixed bond diffusion length. For the part where the slip has occurred, the development of bond stress is limited, while the corresponding non-slip part of the bond stress is increased, so the local bond stress increases accordingly, while the interface closer to the loading end will be the earlier the complete loss of chemical bonding force. At this point, the load continues to increase and the slip gradually penetrates along the direction of the anchorage length. The main reason for the loss of chemical bonding force is the formation and development of cracks within the interface between the steel and recycled concrete, which affect the transfer of bond stress and break the relative stability of the steel and recycled concrete. In the rising section of the load, there are only two cases, one is the chemical bonding force, mechanical bite force, and friction force bearing the load at the same time, and the other is only the mechanical bite force and friction force bear the load, this situation generally occurs with the loading end near. When the ultimate load of the specimen is reached, the chemical bonding force, mechanical bite force, and friction force coordinate to reach the maximum load-bearing capacity, and the ultimate load is in equilibrium, but this equilibrium relationship will be broken instantly, and the member enters the load drop section, at this stage, the chemical bonding force bearing the main load is almost completely lost, and the slip also develops from the loading end to the free end. As the cement crystals in the interface between the steel section and the recycled concrete are sheared or squeezed, and then narrow abrasive bodies are produced [14]. According to the mechanism of motion regulation of the contact surface between the steel section and the recycled concrete, these abrasive bodies rough the surface of the smooth steel section to a certain extent and increase the friction between the interfaces, so that the load decreases rapidly after the complete loss of chemical cementation, but the abrasion between the interfaces makes the load. When the abrasive body after a period of wear, the profile steel and recycled concrete interface morphology gradually improved, the surface pressure and friction coefficient decreases, so as to reach a stable wear stage, and finally, the load slip curve becomes an approximately flat straight line, at which time the mechanical bite force is almost lost, leaving only a relatively weak friction force, bonding is flow plastic nature.

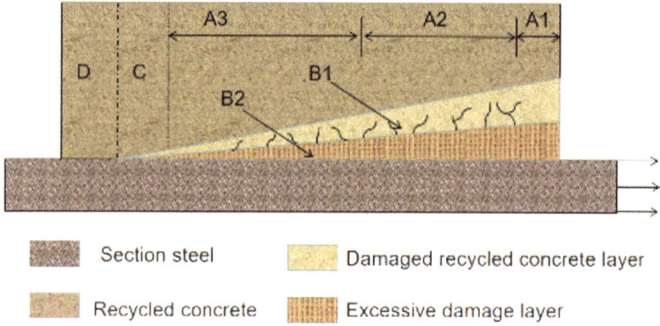

Fig. 1 Diagram of bond stress distribution

3 Theoretical Analysis of Peeling Wear

Steel recycled concrete structure and steel concrete structure has great similarity [15], so the bond-slip mechanism of the two also has similar characteristics. Steel recycled concrete bond-slip launch test, the slip of steel and recycled concrete is precisely caused by the shear deformation of the two, the value of the size of the difference between the deformation of steel and recycled concrete. The peel-and-wear theory is often used for intermetallic surface slip, but due to the high surface stiffness of recycled concrete, which can be approximated as a soft metal compared to steel, the evolution of the interfacial mechanical bite force can be explained based on dislocation theory, peel-and-wear theory, and fracture and plastic deformation of the substrate near the bond surface.

As shown in Fig. 2, the surface of recycled concrete can be regarded as a soft surface compared to the surface of the steel section, and the contact state before the slip is rough peak to rough peak, which is the formation factor of the initial mechanical bite force. When the section steel slips with the recycled concrete, the rough peak of the soft surface is easily deformed, while the soft rough peak is first fractured after the interface under external load occurs a large slip, thus forming a smoother surface. In this way, the contact state is not rough peak to rough peak, but the rough peak of the hard surface to the relatively smooth soft surface, as the rough peak of the hard surface continues to slide on the soft surface, triggering the shear plastic deformation of the soft surface layer, resulting in dislocations within the concrete surface, the phenomenon of plowing, the resistance of the plowing effect is both an important component of friction, but also the main cause of the later mechanical bite force. Under the continuous action of the external load, the shear deformation accumulates during the sliding process of the steel sections and recycled concrete, causing dislocation accumulation on the contact surface of the recycled concrete, which in turn leads to the generation of cracks or pores. This also proves that during the loading process, the mechanical bite force is not constant in one layer due to the relative slip of the section steel and recycled concrete, but wears out continuously and finally tends to stabilize.

Recycled concrete layer

Soft surface roughness peak

adhesive layer

Hard surface roughness peak

Section steel layer

Fig. 2 Schematic diagram of bonding and sliding stripping

4   Linear Elasticity Solution of Bond-Slip

Domestic and foreign experimental studies on the bond-slip performance of recycled concrete with steel sections show that the stress–strain along the anchorage length direction on the steel sections is distributed in an exponential form under the launch load, and the steel sections generally do not reach the yield strength and are still in the elastic phase, and the bond stress is the differential of the stress on the steel sections. Therefore, according to the stress state and equilibrium conditions of the profile regenerated concrete, the interrelationship between bond stress and relative slip can be analyzed from the perspective of elastic mechanics.

(1) $p \leq p_0$

It is assumed that when $P \leq P_0$, the bond stress is exponentially distributed along the anchorage length, and the bond stress is: $\tau_e = \tau_0 e^{kx}$, where k is Changshu, which can be determined through experiments; Recycled concrete is in a linear elastic state.
   Assuming stress function:

$$\phi(x, y) = e^{kx} f(y) \tag{1}$$

Stress function compatibility Equation:

$$\left(\frac{\partial^2}{\partial x^2} + \frac{\partial^2}{\partial y^2}\right)\left(\frac{\partial^2 \phi}{\partial x^2} + \frac{\partial^2 \phi}{\partial y^2}\right) = 0 \tag{2}$$

Equation (1) is brought into Eq. (2) yields

$$f^{(4)}(y) + 2k^2 f''(y) + k^4 f(y) = 0 \tag{3}$$

By solving the complex double root form of the constant coefficient linear differential Equation, we can get:

$$\phi = e^{kx} f(y) = e^{kx}[(c_1 + c_2 y) \cos ky + (c_3 + c_4 y) \sin ky]. \tag{4}$$

The stress component is as follows.

$$\sigma_x = \frac{\partial^2 \phi}{\partial y^2} = e^{kx} f''(y) = e^{kx}[(2c_4 - c_1 k^2$$
$$- c_2 k^2 y) \cos ky - (2c_2 k + c_3 k^2 + c_4 k^2 y) \sin ky] \tag{5}$$

$$\sigma_y = \frac{\partial^2 \phi}{\partial x^2} = k^2 e^{kx} f(y) = k^2 e^{kx}[(c_1 + c_2 y) \cos ky$$
$$+ (c_3 + c_4 y) \sin ky] \tag{6}$$

$$\tau_{yx} = \frac{\partial^2 \phi}{\partial x \partial y} = ke^{kx}f'(y) = ke^{kx}[(-c_1k - c_2ky$$

$$+ c_4)\sin ky + (c_2 + c_3k + c_4ky)\cos ky] \tag{7}$$

According to the physical Equation, it can be concluded that:

$$\left. \begin{array}{l} E\varepsilon_x = E\dfrac{\partial u}{\partial x} = \sigma_x - \mu\sigma_y \\[2mm] E\varepsilon_y = E\dfrac{\partial v}{\partial y} = \sigma_y - \mu\sigma_x \end{array} \right\}. \tag{8}$$

According to boundary conditions, when $y = a$, $\sigma_y = 0$, $\tau_{yx} = \tau_0 e^{kx}$, it can be concluded that:

$$(c_1 + c_2a)\cos ka + (c_3 + c_4a)\sin ka = 0 \tag{9}$$

$$k[(c_4 - c_1k - c_2ka)\sin ka$$
$$+ (c_2 + c_3k + c_4ka)\cos ka] = \tau_0. \tag{10}$$

When $x = l$, $y = 0$, $u = 0$, $v = 0$, it can be concluded that:

$$2c_4 - (k + \mu)c_1 = 0 \tag{11}$$

$$c_2(1 - \mu) - (1 + \mu)kc_3 = 0. \tag{12}$$

The simultaneous Equations of Eqs. (9), (10), (11), and (12) are solved:

$$\left. \begin{array}{l} c_1 = -\dfrac{\tau_0}{k} \bigg/ \{[\dfrac{k+\mu}{2} - k - k^2a\dfrac{1+\mu}{1-\mu}(-\dfrac{\cos ka + \frac{k+\mu}{2}a\sin ka}{\sin ka + \frac{1+\mu}{1-\mu}k\cos ka})]\sin ka+ \\[4mm] \quad [\dfrac{2}{1-\mu}k(-\dfrac{\cos ka + \frac{k+\mu}{2}a\sin ka}{\sin ka + \frac{1+\mu}{1-\mu}k\cos ka}) + ka\dfrac{k+\mu}{2}]\cos ka\} \\[4mm] c_2 = \dfrac{1+\mu}{1-\mu}kc_3 \\[3mm] c_3 = -\dfrac{\cos ka + \frac{k+\mu}{2}a\sin ka}{\sin ka + \frac{1+\mu}{1-\mu}k\cos ka}c_1 \\[4mm] c_4 = \dfrac{k+\mu}{2}c_1 \end{array} \right\} \tag{13}$$

(2) $p > p_0$

Fig. 3 Calculation diagram
of elasticity method

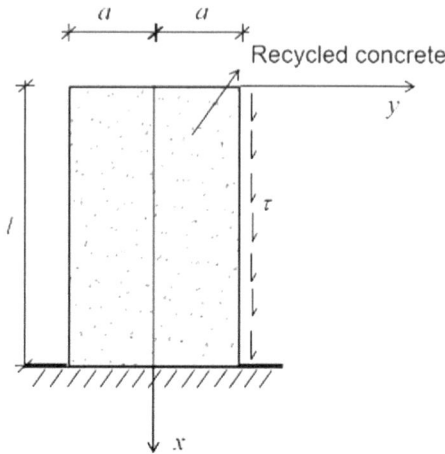

From the bond-slip tests of recycled concrete with steel sections, it can be seen from Fig. 3 that when the load reaches a certain level of P0, the bond stress tends to a constant along the anchoring length direction. Therefore, it is assumed that $\tau_e = \tau_b$, the stress function is:

$$\phi = (x + A)f(y). \tag{14}$$

The stress function is substituted into Eq. (13) yields:

$$(x + A)f^{(4)}(y) = 0. \tag{15}$$

f (y) is solved and substituted into the stress function:

$$\phi = (x + A)(By^3 + Cy^2 + Dy + E). \tag{16}$$

The corresponding stress components are:

$$\sigma_x = \frac{\partial^2 \phi}{\partial y^2} = (x + A)(6By + 2C) \tag{17}$$

$$\sigma_y = \frac{\partial^2 \phi}{\partial x^2} = 0 \tag{18}$$

$$\tau_{yx} = \frac{\partial^2 \phi}{\partial x \partial y} = 3By^2 + 2Cy + D. \tag{19}$$

According to the boundary conditions: $y = a$, $\sigma_y = 0$, $\tau_{yx} = \tau_b$ at that time; When $x = l$, $y = 0$, $u = 0$, $v = 0$, it can be solved as: $A = 0$; $B = \frac{-\tau_b}{4a^2}$; $C = \frac{-\tau_b}{4a}$; $D = 0$.

By substituting it into the above formula, the corresponding stress and displacement components can be obtained.

5 Conclusions

(1) The bonding stress is explained by chemical adhesion, mechanical bite force, and sassafras force, and the factors affecting the bonding stress are introduced. And from the macroscopic bond-slip mechanism analysis, the evolution of the chemical adhesive force, mechanical bite force, and sassafras force in each stage of the three under the launch test are clarified.
(2) The formation and variation of mechanical bite force are explained by theoretical analysis of peel wear.
(3) Using the knowledge of elastodynamics, the relationship between the relative slip of the interface and the bond stress is analyzed to give the solution of bond-slip elastodynamics, and finally the rationality of the elastodynamic solution is analyzed.

References

1. Jianzhuang, X. (2022). Promising recycled concrete. *Science and Technology Vision, 09*, 1–4.
2. Guo, C. X. (2022). The construction industry's carbon peak and carbon neutralization time are tight, and the task is heavy. *Energy Conservation and Environmental Protection, 02*, 10–12.
3. Saravanakumar, P., Abhiram, K., & Manoj, B. (2016). Properties of treated recycled aggregates and its influence on concrete strength characteristics. *Construction and Building Materials, 111*, 611–617.
4. Tie, Z. H., Yong, Y., Yang, X. J., Hong, W. Y., & Yong, L. Y. (2003). Study on mechanical properties of bond and slip of steel reinforced concrete and its basic problems. *Journal of Mechanical Progress, 01*, 74–86.
5. Jianyang, X., Gang, W., & Hui, L. et al. (2014). Experimental research on seismic behaviors of steel reinforced recycled concrete frame structure. *Xi'an University of Architecture and Technology: Natural Science Edition, 05*, 629–634. https://doi.org/10.15986/j.1006-7930.2004.05.003
6. Yang, X. J., Yong, Y., Tie, Z. H., Suo, Z. S., & Yong, L. Y. (2002). State of the art on the bond-slip behaviors of SRC structures. *Structural Engineers, 04*, 52–58+51.
7. Guoliang, B., Jinfeng, M., Biao, L., et al. (2020). Study on the interfacial bond-slip constitutive relation of I-section steel and fully recycled aggregate concrete. *Construction and Building Materials, 238*, 117688.
8. Jianyang, X., Liangjie, Q., & Zheng, L. et al. (2015). Study on seismic performance levels and allowable deformation values of steel reinforced recycled concrete frame structure with infilled wall. *The Journal of Xi'an University of Architecture and Technology: Natural Science Edition, 01*, 15–20. https://doi.org/10.15986/j.1006-7930.2015.01.004
9. Xue, J., Zhai, L., Bao, Y., Ren, R., & Zhang, X. (2018). Cyclic loading tests and shear strength of steel reinforced recycled concrete beam–column inner joints. In The structural design of tall and special buildings, (Vol. 27(6)).

10. Jing, D., Changling, S., Hui, M., Yunhe, L., & Xiaoran, C. (2022). Nonlinear numerical analysis and restoring force model of composite joints with steel reinforced recycled concrete columns and steel beams. *Coatings, 12*(11).

11. Jinfeng, M., Guoliang, B., Haozhao, M., Xueyi, B., & Tao, N. (2022). Beam-type experimental study on interfacial bond-slip behavior of steel reinforcement recycled concrete. *Construction and Building Materials, 351.*

12. Guang, Y. Y. (2017). Experimental study on bond-slip behavior between steel shape and concrete in steel reinforced recycled. Xi'an University of Architecture and Technology Concrete Structure.

13. Suo, Z. S., Yong, Y., Yang, X. J., Hong, Y. M., & Tie, Z. H. (2002). Study on bond-slip behavior of steel reinforced concrete. *Journal of Civil Engineering, 04*, 47–51.

14. Suo, Z. S., Zhuan, D. G., Wei, T., Zhe, W. M., & Lei, Z. (2007). Theoretical study on bond strength between shaped steel and concrete in SRC composite structures. *01*, 96–100+105.

15. Yong, Y. (2003). Study on the basic theory and its application of bond-slip between steel shape and concrete in SRC structures. Xi'an University of Architecture and Technology Concrete Structure.

Mechanical Analysis of Cement Concrete Pavement Paving Layer Based on Abaqus

Suchang Jiao, Hongjuan Wu, Zhengrong Xu, and Dafeng Wang

Abstract Based on the theory of elastic layered system, this paper establishes a three-dimensional finite element model of cement concrete overlay structure on old cement concrete pavement with the Abaqus program, and analyzes the influence of modulus change of old cement concrete pavement, modulus change of composite foundation and various overlay thickness on the load stress at the most unfavorable position of different overlay bottom. The analysis results show that before microcracking the old cement concrete pavement, it is necessary to treat the diseases such as voids at the bottom of the old cement concrete slab to ensure the stability of the strength of the old cement concrete slab after microcracking. The strength of the old cement concrete slab after microcracking is not as high as possible, and the strength of the old cement concrete slab is the best at 4000 Mpa. The process of road reconstruction should be timely, the foundation should be reinforced, and its strength should be kept at 180 MPa; the recommended thickness of cement concrete overlay is 18 cm.

Keywords Road engineering · Stress under load · Cement concrete pavement paving · Finite element analysis

1 Introduction

The renovation of old cement concrete pavements on rural roads has become a hot topic in recent years. The technique of homogenizing and treating micro-cracks in old cement concrete pavements is a new method for the renovation of rural roads. This technique involves using specialized equipment to create micro-cracks uniformly distributed in all directions within the old cement concrete pavement. The blocks interlock and fit together, maximizing the load-bearing capacity of the existing pavement while releasing the internal shrinkage stress of the concrete, thus reducing the overall stiffness [1].

S. Jiao · H. Wu (✉) · Z. Xu · D. Wang
School of Civil Engineering, Northwest Minzu University, Lanzhou 730124, China
e-mail: whj0116@163.com

© The Author(s) 2025
D. Li and Y. Zhang (eds.), *Advances in Frontier Research on Engineering Structures II*,
Lecture Notes in Civil Engineering 535, https://doi.org/10.1007/978-981-97-6238-5_29

This paper is based on the application research project of the homogenization and regeneration technology for old cement concrete pavements on rural roads in Gansu Province. After the micro-crack homogenization treatment of the old cement concrete pavement, a plan is proposed to overlay a new layer of cement concrete considering the specific conditions of the road. The specific plan is as follows: "micro-crack + white", consisting of two discontinuous sections, 1.8 km and 0.6 km in length, respectively [2]. The road deterioration is more severe in these two sections, so the treatment involves micro-crack homogenization and overlaying an 18 cm thick cement concrete layer (white + white). Considering that the load-bearing capacity requirements for the post-micro-crack panels in the "white + white" process are not high, the principle of crack permeation is followed to control crack width and prevent lateral displacement [3].

Based on the theory of elastic layered systems, this study employs the Abaqus software to establish a three-dimensional finite element model for mechanical analysis involving different variables [4].

2 Analysis of the Mechanical Properties of the Overlay Pavement Structure

2.1 Finite Element Analysis Setup

The main steps for performing a computational analysis using the Abaqus software include: creating the part, assigning material properties, assembling the components, defining analysis steps, defining interactions, applying loads, meshing the model, setting output variables, submitting the job, and visualizing the results. To conduct a mechanical analysis of the overlay structure using the current recommended structure for cement concrete pavement issued in 2021, Abaqus finite element software can be utilized. Since the tire's interaction with the pavement is extremely short-lived, resulting in minimal visco-plastic deformation of the pavement structure, and considering that the stress calculation formulas used in China's current "Design Code for Highway Cement Concrete Pavement" (JTG D40-2021) are derived from regression analysis based on finite element solutions of finite-sized thin plates on an elastic foundation, the three-dimensional spatial model of the pavement structure composed of the cement concrete overlay, existing concrete pavement, and subgrade can be treated as an elastic layered system [5, 6]. The finite element numerical method is employed to analyze the stress conditions of the overlay structure on the old cement concrete pavement [7]. The computational model will utilize C3D8R elements (eight-node quadratic hexahedral elements with three-dimensional stress-reduced integration) for stress and displacement analysis [8]. The following specifications are defined for each structural layer [9–12].

1. Each layer of the pavement is assumed to be isotropic, continuous, and uniformly elastic;

Table 1 Main model parameters for each structural layer

	Subgrade dimensions	Old cement concrete pavement and overlay width	Old cement concrete pavement thickness	Overlay thickness
Layer	Length: 8.01 m, Width: 5 m, Depth: 9 m	3.5 m	18 cm	10–20 cm of cement concrete

2. Vertical and horizontal displacements between each layer are continuous;
3. There is no horizontal displacement at the side of the model, and the vertical and horizontal displacements at the bottom are zero;
4. The pavement joints have a width of 1 cm and extend across the entire pavement width, but they do not carry any load;
5. The self-weight of the pavement structure is neglected.

2.2 Model Dimensions

The dimensions of the model should be determined based on the actual conditions of the cement concrete pavement and the proposed overlay structure. Convergence analysis should be conducted considering different subgrade depths. When the subgrade depth is not less than 9 m, the stress tends to reach equilibrium. Therefore, the dimensions of the subgrade can be set as $8.01 \times 5 \times 9$ m. The model parameters for each structural layer are shown in Table 1.

2.3 Meshing

To enhance the accuracy of the computational results, the meshing of the subgrade, old cement concrete pavement, and overlay layer was refined. The mesh size used for the meshing was set at 0.25×0.25 m. This finer meshing helps capture the detailed behavior of the structure and ensures more accurate analysis results [13].

2.4 Model Parameters

The main computational parameters for each structural layer are shown in Table 2.

Table 2 Computational parameters

Structural layer	Thickness (cm)	Young's modulus (MPa)	Poisson's ratio
Composite subgrade	900	60–300	0.35
Old cement concrete pavement	18	800–12,000	0.2
Cement concrete overlay	10–20	31,000	0.2

2.5   Load Scheme

The surface load on the pavement is based on the standard axle load BZZ-100, with a tire pressure of 0.7 MPa. The distance between dual wheels is 32 cm, and the distance between the axles of the two wheels on each side is 182 cm. According to relevant literature, for the purpose of meshing, the wheel load is converted to a rectangular area of 18.9 cm × 18.9 cm using the principle of equal effective area [14]. This model is primarily used to analyze the maximum tensile stress, maximum shear stress, equivalent stress, and maximum deformation of the road overlay layer. The equivalent stress is derived from the fourth strength theory, which states that when the stress state at the critical point of the material reaches simple tensile yield, the material undergoes yield failure.

Research indicates that when the wheel load acts in the middle of the cement concrete overlay layer, it has the greatest impact on the stress distribution, leading to stress concentration and putting the overlay layer in the most unfavorable working state [15]. Therefore, this paper uses finite element analysis to study the effect of the wheel load on the cement concrete overlay layer, specifically when the load is applied to the middle of the cement concrete panel, as shown in Fig. 1.

Fig. 1 Load overlay position diagram

3 Analysis of Load Distribution in the Overlay Layer

3.1 Finite Element Analysis of Elastic Modulus Variation in the Existing Cement Concrete Pavement

Table 3 presents the equivalent stress, maximum tensile stress, maximum shear stress, and deformation experienced by the overlay layer with a thickness of 18 cm for different elastic moduli of the existing cement concrete pavement, including 800 1000, 1500, 2000, 4000, 8000 and 12,000 MPa.

From Table 3 and Fig. 2, it can be observed that at elastic moduli of 800 MPa and 12,000 MPa, the equivalent stress values are 0.5234 and 0.3138, respectively, representing a reduction of 40.0%. The maximum tensile stress values are 0.6051 and 0.4607, respectively, indicating a reduction of 23.9%. The maximum shear stress values are 0.5714 and 0.4532, respectively, showing a decrease of 20.7%. Therefore, it can be concluded that with an increase in the elastic modulus of the existing cement concrete pavement, the equivalent stress, maximum tensile stress, and maximum shear stress of the overlay layer decrease.

These analysis results indicate that under the same vehicle load, the tensile stress at the bottom of the overlay layer decreases with an increase in the elastic modulus of the existing cement concrete pavement. The stress at the bottom of the layer is less than 0.7 MPa, and the reduction is more than 20%. Therefore, the overlay layer, after the microcracks are introduced in the existing cement concrete pavement, meets the requirements of the traffic load.

From Table 3 and Fig. 3, it can be observed that as the elastic modulus of the existing cement concrete pavement increases from 800 MPa to 12,000 MPa, the deformation values are 0.07786 mm and 0.05826 mm, respectively, representing a reduction of 25.2%. Therefore, it can be concluded that with an increase in the elastic modulus of the existing cement concrete pavement, the deformation of the overlay layer decreases, indicating an overall increase in the strength of the road structure. This improvement enhances the road's service life and reduces maintenance requirements.

Table 3 Summary table of finite element analysis results for elastic modulus variation in the existing cement concrete pavement

Elastic modulus (MPa)	Equivalent stress (MPa)	Max. tensile stress (MPa)	Max. shear stress (MPa)	Deformation (mm)
800	0.5234	0.6051	0.5714	0.07786
1000	0.4838	0.5707	0.5532	0.07065
1500	0.4448	0.5516	0.5289	0.06534
2000	0.4112	0.5124	0.5025	0.06275
4000	0.3634	0.5051	0.4914	0.06063
12,000	0.3138	0.4607	0.4532	0.05826

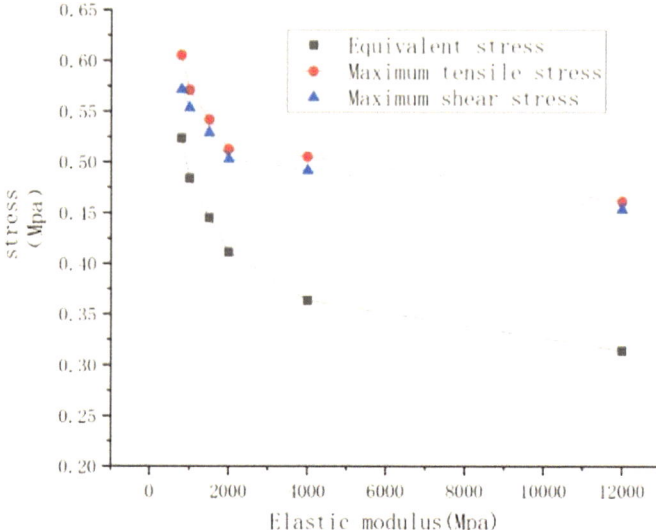

Fig. 2 Variation of equivalent stress, max. tensile stress, and Max. shear stress with different elastic moduli of the existing cement concrete pavement

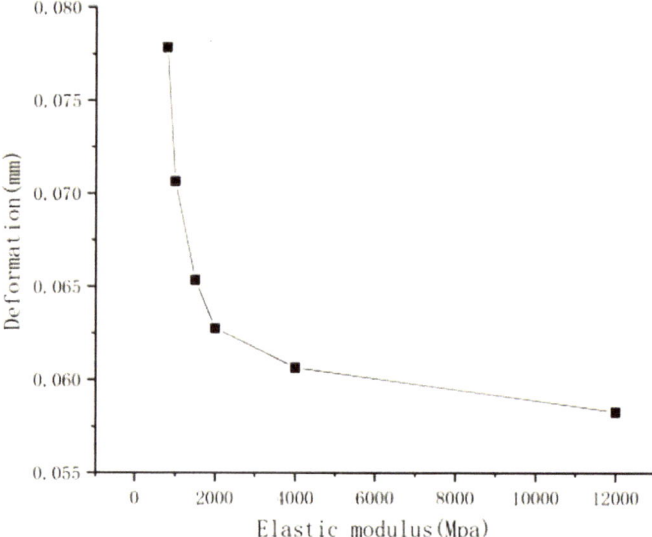

Fig. 3 Variation of deformation with different elastic moduli of the existing cement concrete pavement

However, after reaching an elastic modulus of 4000 MPa in the existing cement concrete pavement, the reduction in equivalent stress, maximum tensile stress, maximum shear stress, and deformation of the cement concrete overlay layer becomes less significant. Therefore, it is recommended to use an elastic modulus of 4000 MPa for the existing cement concrete pavement after microcracks occur.

3.2 Finite Element Analysis of Elastic Modulus Variation in the Composite Subgrade

Based on the above analysis results, considering the elastic modulus of the existing cement concrete pavement as 4000 MPa, the influence of elastic modulus variation in the composite subgrade on the equivalent stress, maximum tensile stress, maximum shear stress, and deformation of the overlay layer is analyzed.

Table 4 presents the results of the analysis, showing the equivalent stress, maximum tensile stress, maximum shear stress, and deformation experienced by the overlay layer under different elastic moduli (60, 120, 180, 240 and 300 MPa) of the composite subgrade when the elastic modulus of the existing cement concrete pavement is 4000 MPa.

From Table 4 and Fig. 4, it can be observed that at different elastic moduli of the composite subgrade, specifically 60 MPa and 300 MPa, the equivalent stress of the overlay layer is reduced by 13.8%, with values of 0.5234 and 0.4513, respectively. Similarly, the maximum tensile stress is reduced by 7.5%, with values of 0.6543 and 0.6051, and the maximum shear stress is reduced by 8.8%, with values of 0.5714 and 0.5208. These results indicate that with an increase in the elastic modulus of the composite subgrade, the equivalent stress, maximum tensile stress, and maximum shear stress of the overlay layer decrease. Therefore, under the same vehicle load, a smaller elastic modulus of the composite subgrade leads to higher stress levels in the overlay layer.

From Table 4 and Fig. 5, it can be observed that as the elastic modulus of the composite subgrade increases from 60 to 300 MPa, the deformation of the overlay

Table 4 Overall analysis data of finite element analysis for variation of elastic modulus in the composite subgrade

Elastic modulus of subgrade (MPa)	Equivalent stress (MPa)	Max. tensile stress (MPa)	Max. shear stress (MPa)	Deformation (mm)
60	0.5234	0.6543	0.5714	0.08363
120	0.4863	0.6358	0.5494	0.08022
180	0.4549	0.6156	0.5244	0.07481
240	0.4526	0.6098	0.5221	0.07347
300	0.4513	0.6051	0.5208	0.07168

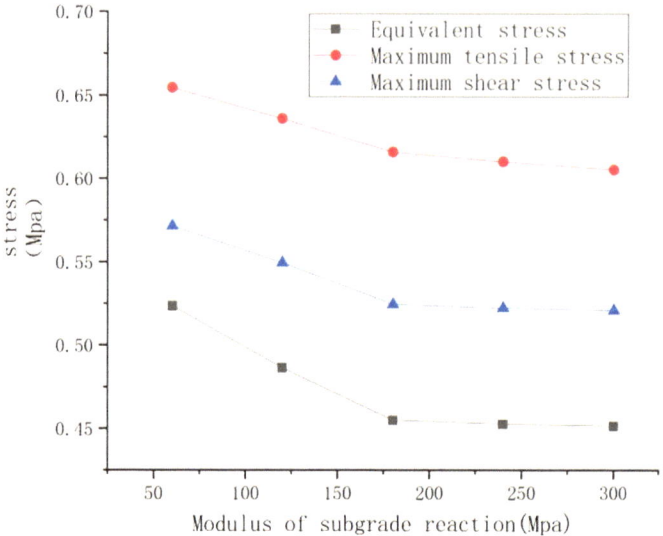

Fig. 4 Variation of equivalent stress, maximum tensile stress, and maximum shear stress with different elastic moduli of the composite subgrade

layer decreases by 14.3%, with values of 0.08363 mm and 0.07168 mm, respectively. This indicates that with an increase in the elastic modulus of the composite subgrade, the deformation of the overlay layer decreases. Furthermore, there is a significant reduction in deformation when the elastic modulus of the composite subgrade increases from 60 to 180 MPa. This suggests that the strength of the composite subgrade has a significant influence on the overall structure of the pavement. Therefore, before performing crack treatment on the cement concrete pavement, it is necessary to address issues such as voids at the bottom of the slabs to maintain uniformity and integrity after cracking.

From Figs. 4 and 5, it can be observed that the reduction in stress and deformation decreases after the elastic modulus of the composite subgrade reaches 180 MPa. Taking into account the economic feasibility and construction difficulties of reinforcing the subgrade, it is recommended to use an elastic modulus of 180 MPa for the composite subgrade.

3.3 Finite Element Analysis of the Effects of Cement Concrete Overlay Thickness Variation

Based on the recommended analysis above, the effects of varying the thickness of the cement concrete overlay on stress and deformation will be analyzed, considering a

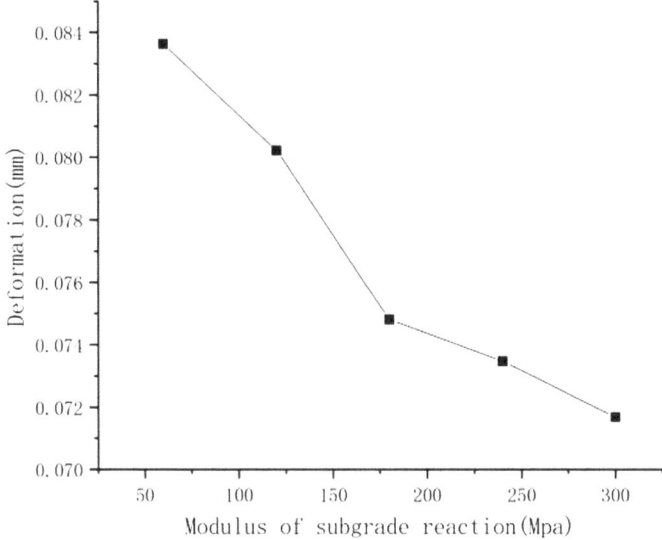

Fig. 5 Variation of deformation with different elastic moduli of the composite subgrade

concrete pavement modulus of 4000 MPa and a composite subgrade elastic modulus of 180 MPa.

According to Table 5 and Fig. 6, it can be observed that as the thickness of the cement concrete overlay increases from 10 to 20 cm, the equivalent stress in the overlay decreases by 14.9% (from 0.3937 to 0.3349), the maximum tensile stress decreases by 46.4% (from 0.2345 to 0.1257), and the maximum shear stress decreases by 25.2% (from 0.225 to 0.1684). This indicates that with an increase in the thickness of the cement concrete overlay, the equivalent stress, maximum tensile stress, and maximum shear stress in the overlay decrease. The analysis results demonstrate that under the same vehicle load, a smaller thickness of the cement concrete overlay results in higher values for the three stress indicators in the overlay.

Table 5 The results of the analysis are presented and the accompanying figure

Overlay thickness (mm)	Equivalent stress (MPa)	Max. tensile stress (MPa)	Max. shear stress (MPa)	Deformation (mm)
10	0.3937	0.2345	0.225	0.08843
12	0.3825	0.1961	0.2113	0.08563
14	0.3711	0.167	0.1975	0.0849
16	0.3599	0.1442	0.1838	0.08304
18	0.349	0.1259	0.1702	0.08148
20	0.3349	0.1257	0.1684	0.08099

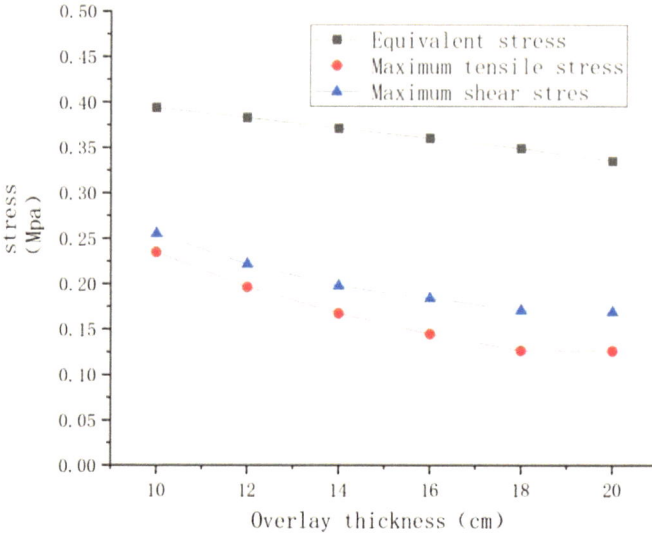

Fig. 6 Variation of equivalent stress, maximum tensile stress, and maximum shear stress with different thicknesses of cement concrete overlay

According to Table 5 and Fig. 7, it can be observed that as the thickness of the cement concrete overlay increases from 10 to 20 cm, the deformation decreases from 0.08843 mm to 0.08099 mm. From the graph, it can be seen that with an increase in the thickness of the cement concrete overlay, the deformation of the overlay decreases. Considering the changes in the three stress indicators shown in Fig. 6, there is a significant reduction in deformation when the thickness of the overlay increases from 10 to 18 cm. However, the decrease in deformation becomes less significant when the thickness increases from 18 to 20 cm. Taking into account cost considerations, it is recommended to have an overlay thickness of 18 cm.

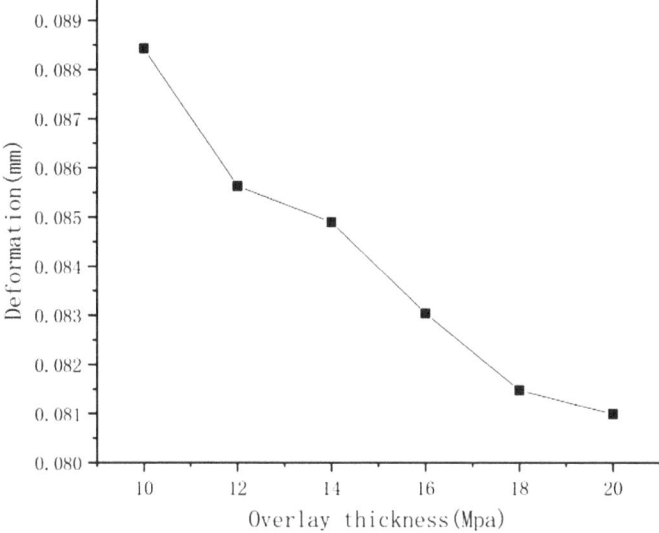

Fig. 7 Deformation variation with different thicknesses of cement concrete overlay

4 Conclusion

Cement concrete pavement is a commonly used road structure in China, and it is of great significance for the development of the transportation industry and environmental improvement to recycle and reuse damaged cement concrete pavement. Based on the results obtained using the Abaqus finite element analysis software, the following recommendations are proposed for overlaying cement concrete pavement on the micro-cracked old pavement:

(1) Before performing microcrack treatment on old cement concrete pavement, it is necessary to address underlying diseases such as voids at the bottom of the old cement concrete panels. This ensures the stability of the strength of the old cement concrete panels after microcrack treatment. The strength of the treated old cement concrete panels should not be excessively high but rather optimized around 4000 MPa or higher, with a focus on eliminating concentrated stress within the panels.

(2) A smaller modulus of the subgrade results in higher internal stresses and deformations within the overlay under the same load. This increases the likelihood of reflective cracking. Therefore, during road reconstruction, it is essential to strengthen the composite subgrade in a timely manner, with an optimal strength of around 180 MPa.

(3) With an increase in the thickness of the cement concrete overlay, the three stress indicators and deformation within the overlay decrease. However, the reduction in stress and deformation becomes less significant when the overlay thickness

increases from 18 to 20 cm. Taking cost into consideration, it is recommended to have an overlay thickness of 18 cm.

Acknowledgements Funding Projects: Fundamental Research Funds for the Central Universities: Grant No. 31920200033; Research Project of Gansu Provincial Department of Transportation: Project No. 2020–03; Special Fund for Fundamental Research of Central Universities: Grant No. 31920230043.

References

1. Duxing, W. (2019). Homogenization treatment and overlay technology for microcracked old cement concrete pavement. *Chang'an University*. https://doi.org/10.26976/d.cnki.gchau.2019. 000129
2. Meiqin, L., Shang, W., & Fei, P. (2023). Simulation and analysis of three-point bending experiment with hollow beam based on Abaqus. *Journal of Physics: Conference Series, 2566*(1).
3. Pavement Management Guide-Executive Summary Report; Bemanian, S., & Sebaaly, P. (1999). Cost-effective rehabilitation of Portland cement concrete Pavement tin Nevada transportation research record, *1684,* 156–164.
4. Zhesheng, G., & Weijun, Z. (2006). Mechanical analysis of asphalt overlay on old cement concrete pavement. *Journal of Chang'an University (Natural Science Edition), 06,* 23–26. https://doi.org/10.19721/j.cnki.1671-8879.2006.06.006
5. Tongbo, S. (2010). Techniques for the breaking and remodeling of old cement concrete pavement and optimization of overlay structure (Master's thesis, Shandong University). https://kns. cnki.net/KCMS/detail/detail.aspx?dbname=CMFD2012&filename=1011059565.nh
6. Junwei, S. (2016). Research on micro-cracking regeneration technology and engineering application of old cement concrete pavement. *Highway and Transportation Science and Technology (Applied Technology Edition),* 73–74. https://kns.cnki.net/kcms/detail/detail.aspx?FileName= GLJJ201604029&DbName=CJFQ2016
7. Xueliang, S. (2011). Study on construction control parameters and overlay structure of old cement concrete pavement breaking (Master's thesis, Chang'an University). https://kns.cnki. net/KCMS/detail/detail.aspx?dbname=CMFD2012&filename=1011186642.nh
8. Peifeng, C., & Hong, L. (2017). Mechanical study on asphalt overlay structure of old cement concrete pavement based on Abaqus. *Highway Engineering, 42*(01), 9–12+30.
9. Hong, L. (2016). Mechanical study on asphalt overlay structure of old cement concrete pavement (Master's thesis, Northeast Forestry University). Available at: https://kns.cnki.net/KCMS/ detail/detail.aspx?dbname=CMFD201701&filename=1016309404.nh
10. Houlan, H., Mingsen, K., & Shijie, M. (2009). Finite element mechanical analysis of crack relieving layer in asphalt overlay of old cement concrete pavement. *Highway, 08,* 15–19. Available at: https://kns.cnki.net/kcms/detail/detail.aspx?FileName=GLGL200908005&DbName= CJFQ2009
11. Danfeng, L. (2013). Study on mechanical behavior of dual-layer cement concrete pavement slabs (Master's thesis, Fuzhou University). Available at: https://kns.cnki.net/KCMS/detail/det ail.aspx?dbname=CMFD201602&filename=1015346705.nh
12. Baosheng, Y., Yuhua, P., & Mingming, G. (2013). Load stress analysis of asphalt concrete overlay on old cement concrete pavement based on finite element method. *Highway, 04,* 85–89. Available at: https://kns.cnki.net/kcms/detail/detail.aspx?FileName=GLGL201304023&DbN ame=CJFQ2013

13. Yanqing, D., Jinhua, S., Tao, W., & Liang, W. (2011). Coupled stress analysis of asphalt overlay structure on old cement concrete pavement based on stress absorbing layer. *Journal of Ludong University (Natural Science Edition), 01*, 86–91. Available at: https://kns.cnki.net/kcms/detail/detail.aspx?FileName=WOOD201101019&DbName=CJFQ2011

14. Hao, Y. (2012). Study on reconstruction technology and overlay structure of old cement concrete pavement (Master's thesis, Chang'an University). Available at: https://kns.cnki.net/KCMS/detail/detail.aspx?dbname=CMFD201402&filename=1013018998.nh

15. Sheng, Z., & Yuhang, H. (2008). Mechanical analysis of asphalt overlay structure on old cement concrete pavement. *Journal of Chang'an University (Natural Science Edition), 02*, 18–21+25. https://doi.org/10.19721/j.cnki.1671-8879.2008.02.005

Effect of Pit System Stiffness on Force and Deformation of Adjacent Monopile

Minmin Jiang, Di Wang, Yu Li, and Zhaoran Xiao

Abstract The soil displacement caused by the foundation excavation will impose additional stresses on the pile foundations within its influence range, resulting in lateral displacement or insufficient bearing capacity of the pile foundations, which will adversely affect the pile foundations and cause uneven settlement of the structures in the long run, triggering local damage to the soil below the pile foundations and even distorting the deformation of the upper structures, affecting their safety in use. In this paper, the two-stage method is used to analyze the lateral displacement of pile foundations. The free displacement field of the soil during the excavation of the foundation pit is used as the input condition, and the lateral displacement and internal force of the monopile are solved by introducing the improved foundation reaction modulus into the Winkler foundation model. Combined with the numerical model, the effects of pit excavation on the surrounding soil, supporting structure, and adjacent pile foundations are investigated. The effects of different factors on the force and deformation pattern of pile foundations are analyzed, the degree of influence of different comprehensive stiffness of pit support on pile foundations is explored, and the comprehensive stiffness is related to the maximum horizontal displacement of pile foundations to propose a method for predicting the maximum horizontal displacement of pile foundations, which can be a reasonable reference for similar pile foundation projects.

Keywords Pit · Adjacent monopile · Combined stiffness · Numerical analysis · Displacement · Internal forces

1 Introduction

In fact, the excavation process is the process of redistribution of soil stress. Due to the self-organization of the soil, the process will drive the soil to deformation, specifically in the form of horizontal displacement of the surrounding soil, surface settlement,

M. Jiang · D. Wang (✉) · Y. Li · Z. Xiao
School of Civil Engineering, Henan University of Technology, Zhengzhou, Henan, China
e-mail: 1348685280@qq.com

© The Author(s) 2025
D. Li and Y. Zhang (eds.), *Advances in Frontier Research on Engineering Structures II*,
Lecture Notes in Civil Engineering 535, https://doi.org/10.1007/978-981-97-6238-5_30

substrate uplift, and other phenomena, and the pit support system, adjacent to the existing monopiles at different depths will be subject to different degrees of force, so that the retaining wall and pile lateral shift to the pit [1–3]. The lateral shift of the soil will exert additional stress on the monopile within its influence, resulting in the monopile lateral shift or insufficient bearing capacity, thus adversely affecting the monopile.

To address this issue, scholars both at home and abroad have conducted relevant research. For theoretical analysis, Stewart [4] and Heyman [5] have proposed empirical formulas for soil pressure around the pile, lateral displacement at the top of the pile, and the bending moment of the pile. Huang et al. [6] used the finite difference method to analyze the mechanical properties of group piles in non-homogeneous foundations and to investigate the shading effect of group piles on the surrounding soil. Zhang et al. [7] used the two-stage analysis method, based on the Winkler foundation model, considered the pile-soil coordinated deformation condition, established the horizontal displacement control differential Equation for the neighboring monopiles under the foundation excavation, and derived the soil transfer coefficient under horizontal motion by the plane Mindin solution, analyzed the shading effect between the group piles, and calculated the response degree of each monopile in the group pile.

Numerical simulations have also been extensively studied by many related scholars. Yao et al. [8] used FLAC 3D to numerically simulate axially loaded piles under soil lateral displacement and analyzed the effects of soil strength, pile stiffness, and different pile top restraint conditions on the force characteristics of monopiles. Liang et al. [9] used LPILE Plus 5.0 to fully discuss the effects of pile top restraint conditions, lateral soil resistance of the monopile, and different soil displacement states on the monopile displacement and internal forces during the excavation of the foundation pit, based on the p-y curve for non-linear load transfer. Liang et al. [10] used ABAQUS to simulate the interaction between different soil lateral displacement modes and the adjacent monopiles and discussed the degree of influence of the maximum soil lateral displacement value and distribution range on the monopiles.

Research on monopile force deformation is progressing rapidly, but there are many ways of foundation support, and it is difficult to quantitatively analyze the effect of different support structure system parameter changes on monopile force deformation during foundation excavation. Therefore, research on this aspect is needed. This paper introduces the comprehensive stiffness of foundation support to systematically analyze the effect law of different support schemes on monopile and investigates the effect of foundation excavation on monopile horizontal displacement and bending moment under the same foundation support stiffness from the perspective of monopile and ground connection wall spacing, monopile stiffness, monopile length, and pile top vertical load.

2 Theoretical Analysis

2.1 Integrated Stiffness Theory

The influence factors of the support structure are systematized to study the effect of integrated stiffness on the force deformation of monopiles. The most widely used comprehensive stiffness calculation method at home and abroad is the dimensionless comprehensive stiffness formula for the pit support system proposed by Clough [11], as shown in Eq. (1). The ratio between the displacement of the support structure and the excavation depth of the pit may appear more discrete and cannot reflect the function relationship between the two more accurately.

$$\eta = \frac{EI}{\gamma_w h^4} \tag{1}$$

In Eq. (1), EI denotes the flexural stiffness of the enclosure wall (pile); E denotes the modulus of elasticity of the wall material; I denotes the moment of inertia of the section per unit width; h denotes the average vertical spacing of the supports; γ_w denotes the water weight.

Zhang et al. [12] proposed a comprehensive stiffness for MVSS based on Clough's consideration of temporal effects, foundation reinforcement, and horizontal spacing of internal supports, as shown in Eq. (2).

$$\eta_{mvss} = k_t k_j \frac{mk_1 + nk_2}{(m+n)k_1} EI / (r_w h^2 Hs) \tag{2}$$

In Eq. (2), E denotes the modulus of elasticity of the material, i denotes the section moment of inertia per unit width, k_t denotes the combined adjustment factors for spatiotemporal effects, insertion ratios, and other influencing factors, k_j denotes the foundation strengthening impact factor, s denotes the support average horizontal spacing, $\frac{mk_1 + nk_2}{(m+n)k_1}$ denotes the influence factor for support stiffness, m denotes the number of steel support courses, n denotes the number of concrete support courses, k_1 denotes the support stiffness, and k_2 denotes the Concrete support stiffness.

Since $h = H/N$ (N is the number of internal support channels), the above Equation is rewritten as Eq. (3).

$$\eta_{mvss} = k_t k_j \frac{mk_1 + nk_2}{(m+n)k_1} EIN / (\gamma_w H^3 s) \tag{3}$$

2.2 Theory of the Effect of Foundation Excavation on Monopile Displacement

Yu et al. [13] derived the foundation model by converting continuous soil displacements into non-linear unit width displacements and deriving the elastic theory method with the foundation beam displacement solution of the Winkler foundation model under the action of unit displacements, comparing the connection between the two solutions so as to find the foundation model, which is an improvement of the Vesic formula, considering the situation where the foundation beam is at a certain depth below the ground surface, which is introduced in this paper to solve for the foundation reaction modulus, The modulus of foundation reaction Equations are shown in Figs. 4 and5.

$$k_y = \frac{3.08}{\eta} \left(\frac{E_s B^4}{E_p I_p} \right)^{0.125} \frac{E_S}{(1 - V_s)^2} \tag{4}$$

$$\eta = \begin{cases} 2.18 & h/B \leq 0.5 \\ \left(1 + \frac{1}{1.7h/B}\right) & h/B > 0.5 \end{cases} \tag{5}$$

In Eq. (4), h denotes the depth of the pile into the soil, the other symbols are the same as in the previous formula. We make $\lambda = \left(\frac{k_y}{4E_p I_p} \right)^{0.25}$ substitute the deflection differential equation for the monopile case, as shown in Eq. (6).

$$E_p I_p \frac{d^4 y_p}{dz^4} k_y \left[y_p - s_y(z) \right] = 0 \tag{6}$$

In Eq. (6), E_p and I_p denote the modulus of elasticity and section inertia distance of the single pile respectively. k_y denotes the modulus of reaction of the foundation on the side of the pile. $S_y(z)$ denotes the free horizontal displacement of the surrounding soil caused by the excavation of the foundation pit.

Then Eq. (6) can be rewritten as Eq. (7):

$$\frac{d^4 y_p(z)}{dz^4} + 4\lambda^4 \left[y_p(z) - s_y(z) \right] = 0 \tag{7}$$

By equating the pile length n along the depth direction, with two virtual equipartition points at the top and bottom of the pile respectively, the monopile calculation model is shown in Fig. 1, then the standard deviation Equation of Eq. (7) is shown in Eq. (8).

$$y_{p,i-2} - 4y_{p,i-1} + (6 + 4\lambda_i^4 h^4)y_{p,i} - 4y_{p,i+1} + \\ y_{p,i+2} = 4\lambda_i^4 h^4 s_{y,i} \tag{8}$$

Fig. 1 Monopile calculation model

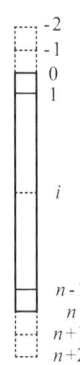

Combined with the monopile boundary condition equation, the effect of horizontal displacement of the surrounding soil on the monopile can be obtained, and the expression is given in Eq. (9).

$$[K_y]\{y_p\} = [K_{sy}]\{s_y\} = \{F_y\} \tag{9}$$

In Eq. (9), $[K_y]$ denotes the monopile horizontal stiffness matrix, $\{y_p\}$ denotes the monopile horizontal displacement vector, $[K_{sy}]$ denotes the horizontal stiffness matrix of the soil around the foundation pit, $\{s_y\}$ denotes the soil displacement field around the foundation pit located at the monopile horizontal soil displacement, and $\{F_y\}$ denotes the monopile horizontal load column vector.

The expression for the horizontal displacement of a monopile is shown in Eq. (10).

$$\{y_p\} = ([K_{sy}]\{s_y\} + \{F_y\})[K_y]^{-1} \tag{10}$$

The load is proportional to the horizontal resistance of the soil on the pile side. When the soil on the pile side of node i increases to the ultimate resistance P_{fi}, Peripheral free soil displacement limit values are $S_{yc,i}$. The critical horizontal displacement value of a single pile is $y_{pf,i}$. The expression of the differential equation is shown in Eq. (11).

$$E_p I_p \frac{d^4 y_p(z)}{dz^4} + k_y [y_{pf}(z) - s_{yc}(z)] = 0 \tag{11}$$

3 3D Model Parameters

In this paper, ABAQUS is used for simulation and the specific parameters of the model are as follows.

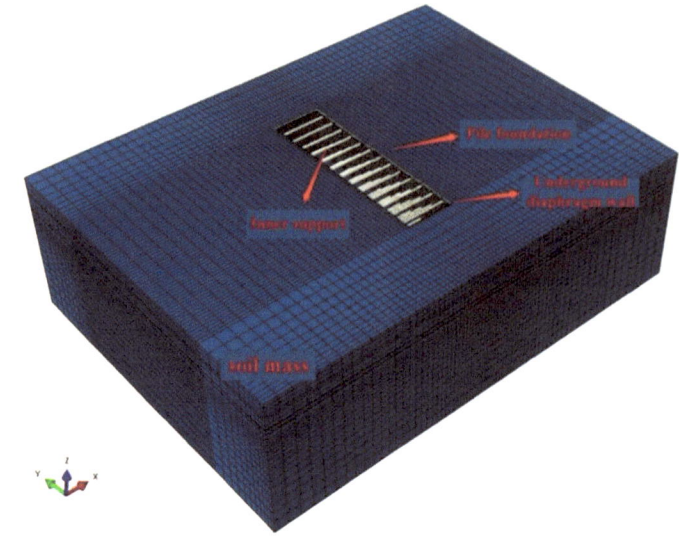

Fig. 2 The overall meshing of the model

3.1 Pit Parameters

The three-dimensional foundation pit model built in this paper is 80 m long and 20 m wide, with an excavation depth of 16 m, using an underground diaphragm wall enclosure structure, with a wall width of 0.6 m and a burial depth of 32 m. The single pile diameter is 1 m, with a diameter length of 46 m, located 3 m behind the diaphragm wall. Three vertical supports are arranged along the depth direction of the foundation pit, with a horizontal spacing of 6 m and a vertical spacing of 5 m. The supports are all 609 mm in diameter and 16 mm in wall thickness, wall thickness of 16 mm steel pipe support. The overall meshing of the model is shown in Fig. 2. The specific excavation processes were: (1) excavation to −1.5 m; (2) application of the first horizontal support at the −1.0 m elevation and excavation to −3.5 m; (3) excavation to −6.5 m; (4) application of the second horizontal support at the −6.0 m elevation and excavation to −9.5 m; (5) excavation to −11.5 m; (6) application of the third horizontal support at the −11.0 m elevation and excavation to −13.5 m; (7) excavation to −16.0 m.

3.2 Material Parameters

The geological conditions of the model soil are soft soil geology and the specific soil physical and mechanical indicators are shown in Table 1.

Table 1 Soil layer parameters

Number	Name of soil layer	Severe /(kN/m³)	Cohesive force /kPa	The angle of internal friction /(°)	Modulus of elasticity / MPa	Poisson's ratio
1	Clay chalk	18.3	5	31.6	4.3	0.34
2	Silty chalky clay	17.5	11	14.5	2.5	0.36
3	Powdery clay	19.4	48	16.7	8.1	0.29
4	Sandy clay	18.8	8	30.3	8.4	0.34
5	Powdery clay with sandy clay	18.2	17	16.8	4.7	0.35
6	Powdery clay with sandy chalk	18.1	19	19.9	6.7	0.35

The ground link wall and monopile models are considered in accordance with C30 reinforced concrete. The specific physical property parameters of the reinforcement, internal support, and monopile are shown in Table 2.

As shown in Fig. 3, when the pit is excavated to − 7.5 m and − 16 m, the maximum horizontal displacement of the monopile measured by Goh et al. is 15 mm and 28 mm, and the maximum displacement of the pit occurs near the excavation surface. The numerical simulation results in this paper are consistent with the results measured by Goh et al. and the theoretical calculation results in this paper, but the specific values differ from them due to different soil conditions and support conditions. The lack of ability to limit the deformation of the monopile below the excavation surface of the foundation pit caused the displacement and bending moment below the excavation surface to be larger.

Table 2 Steel support and concrete physical and mechanical parameters

Structural parameters	Density/(kg/m³)	Modulus of elasticity/(GPa)	Poisson's ratio
Reinforcing steel, steel supports	7800	200	0.17
C30 Reinforced concrete	2550	30	0.2

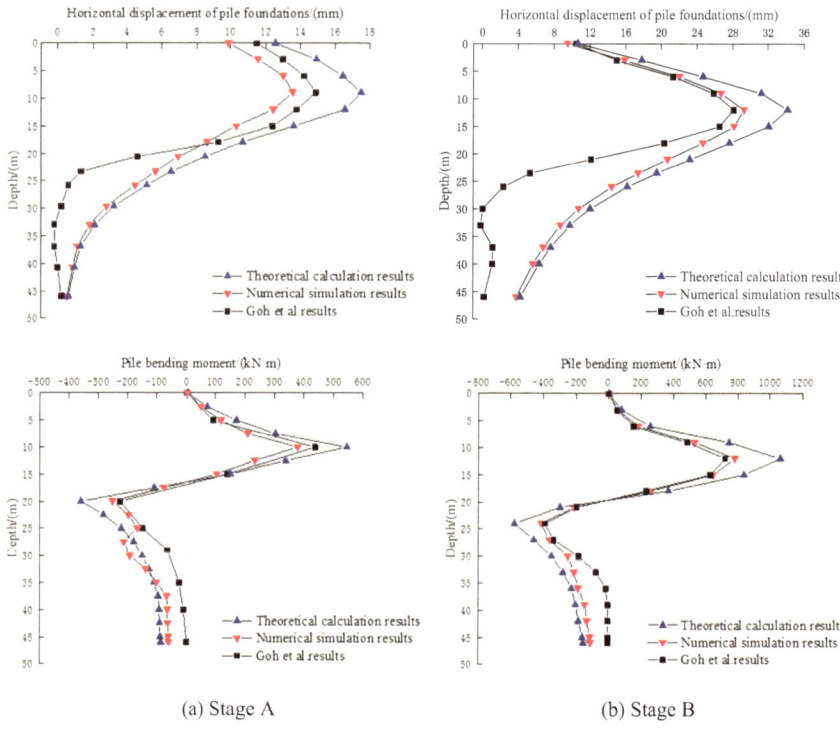

(a) Stage A   (b) Stage B

Fig. 3 Comparison of numerical simulation results with the measured results of Goh et al.

Fig. 4 Diagram of
maximum horizontal
displacement of a single pile
for varying combined
stiffness due to different
ground link wall thicknesses

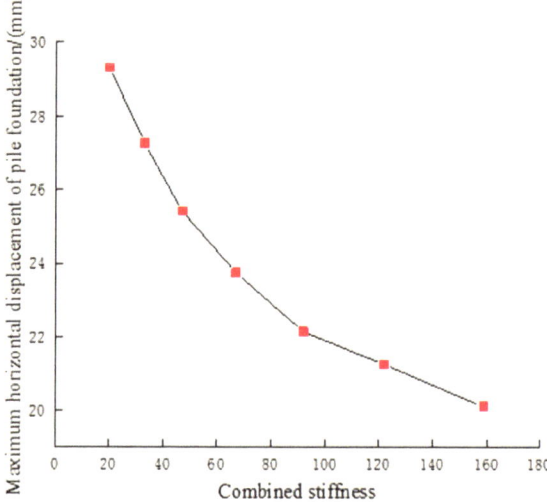

Fig. 5 Variation of single pile displacements for different numbers of vertical supports causing changes in combined stiffness

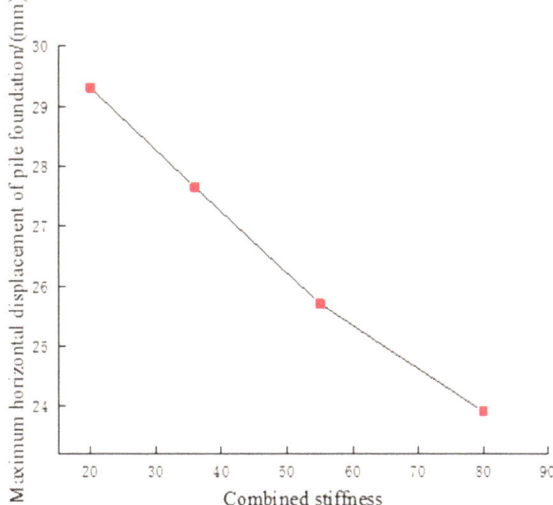

4 Analysis of Monopile Deformation Under Different Combined Stiffness of Support

4.1 Integrated Stiffness Arrangement Scheme for Supports

Using the numerical results of the model as the basis for analysis, the control variable method is used to change the thickness of the ground link wall, the number of vertical bracing courses, and the horizontal bracing spacing one by one so as to change the comprehensive stiffness of the foundation support system and to study the response law of the foundation excavation to the displacement of the monopile, while other parameters, such as the material parameters of the monopile, the excavation depth of the foundation, and the burial depth of the ground link wall, remain unchanged.

As shown in Table 3, the thickness of the ground connection wall was set at 0.6, 0.7, 0.8, 0.9, 1.0, 1.1, and 1.2 m, and the number of vertical internal supports was 3, 4, 5, and 6 (where the spacing between two adjacent internal supports was 5, 4, 3 and 2 m). The horizontal spacing of internal supports was 2, 4, 6 and 8 m respectively, for a total of 112 groups of foundation support solutions with comprehensive stiffness calculated.

Table 3 Parameters for calculating the combined stiffness of the support structure

Depth of foundation/m	Floor-to-wall thickness/m	Number of internal support courses	Horizontal spacing of internal supports/m
16	0.6	3	2
	0.7	4	4
	0.8	5	6
	0.9	6	8
	1		
	1.1		
	1.2		

4.2   Analysis of Monopile Deformation Under the Change in Combined Stiffness Due to the Thickness of the Same Ground Link Wall

The increase in the thickness of the ground link wall will increase the comprehensive stiffness of the pit support system, as shown in Table 4, and the flexural stiffness of the ground link wall itself will also be significantly increased, and its ability to inhibit the displacement of the soil into the pit will gradually increase, which will lead to a certain contraction of the soil in the direction away from the pit, and the maximum settlement value of the ground surface and the horizontal displacement of the soil behind the wall will be gradually reduced.

As shown in Fig. 4, when the thickness of the local connecting wall increases, i.e., the comprehensive stiffness increases, the displacement of the monopile decreases, and the trend of decreasing monopile displacement shows a step weakening, which is due to the process of increasing the flexural stiffness of the wall itself from 540 to 4320 MN·m². The difference between the deformation of the top of the wall and the maximum deformation of the wall gradually decreases, resulting in the weakening

Table 4 Value of change in combined stiffness due to change in thickness of ground link wall

Depth of foundation/m	Floor-to-wall thickness/m	Number of internal support courses	Horizontal spacing of internal supports/ m	Combined stiffness
16	0.6	3	6	20
	0.7			33
	0.8			47
	0.9			67
	1			92
	1.1			122
	1.2			159

Table 5 Value of change in combined stiffness due to change in the number of vertical support courses

Depth of foundation/m	Floor-to-wall thickness/m	Number of internal support courses	Horizontal spacing of internal supports/ m	Combined stiffness
16	0.6	3	6	20
		4		36
		5		55
		6		80

of the displacement field of the soil behind the wall, and the maximum horizontal displacement of the monopile also decreases, and the location where the maximum displacement appears does not change, while the displacement of the top of the pile end shows a trend of gradually increasing.

4.3 Analysis of Single Pile Deformation with Different Numbers of Vertical Bracing Courses Causing the Change in Combined Stiffness

As shown in Table 5, the addition of vertical bracing increases the support density and reduces the average vertical spacing of the pit support, increasing the combined stiffness.

As can be seen from Fig. 5, the effect on the maximum horizontal displacement of the single pile is significant with the increase in the number of vertical supports, with each additional internal support. The maximum horizontal displacement of the single pile is reduced by 1.8 mm on average, and the reduction rate is relatively stable.

4.4 Analysis of Single Pile Deformation with Varying Horizontal Support Spacing Causing Combined Stiffness

As shown in Table 6, similarly, reducing the horizontal support spacing increases the support density and reduces the average spacing of the horizontal supports for the pit support, increasing the combined stiffness.

Figure 6 shows the comparison of the maximum horizontal displacement of a single pile when the horizontal support spacing is increased sequentially to cause comprehensive stiffness changes. It can be seen from the figure that the displacement of a single pile decreases by 4.08 m on average when the horizontal support spacing is increased by one time.

Table 6 Value of change in combined stiffness due to change in horizontal support spacing

Depth of foundation/m	Floor-to-wall thickness/m	Number of internal support courses	Horizontal spacing of internal supports/ m	Combined stiffness
16	0.6	3	2	60
			4	30
			6	20
			8	15

Fig. 6 Variation of single pile displacements for different horizontal support spacing caused by the change in combined stiffness

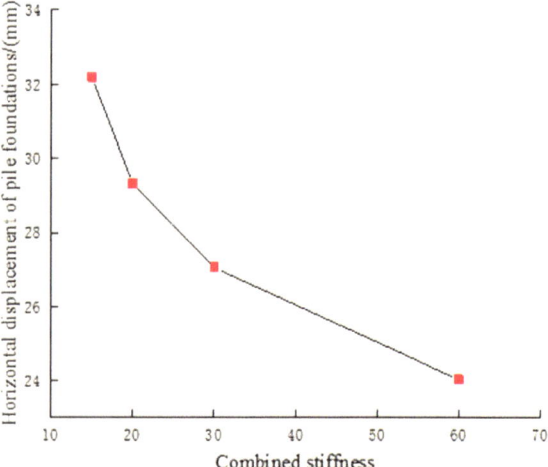

4.5 Combined Stiffness of Different Combination Schemes

By combining the thickness of the ground link wall, the number of vertical bracing courses, and the horizontal spacing of the bracing, the combined stiffness of the 112 different support schemes in Table 3 was calculated, and in order to better evaluate the relationship between the lateral movement of the monopile and the foundation support structure, each group of support schemes kept the monopile at 3 m behind the ground link wall, as shown in Fig. 7. The fitting formula is shown in Eq (12).

$$w_{\max}/H = a\eta_{mvss}^{b} \tag{12}$$

In Eq (12), $w_{\max}/H$ denotes the ratio of the maximum horizontal displacement of a single pile to the depth of excavation, a,b denotes the fitting parameter, and η_{mvss} notes the expression for the MVSS combined stiffness function.

We fit the following Eq (13) as a function of the integrated stiffness of the support structure and the lateral displacement of the monopile.

Fig. 7 Normalized monopile displacement versus integrated stiffness fit diagram

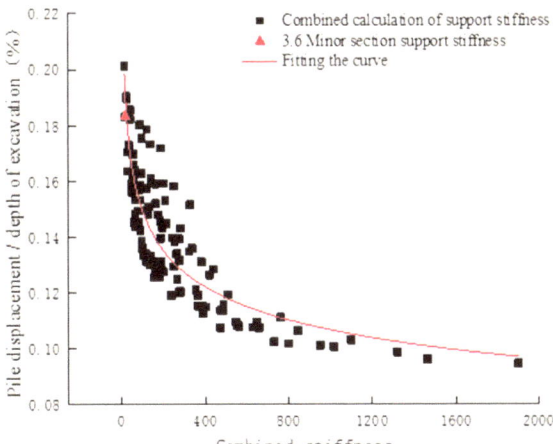

$$w_{max}/H = 0.296\, \eta_{mvss}^{-0.148} \tag{13}$$

The expression for the maximum horizontal displacement of a single pile can be rewritten as Eq. (14)

$$w_{max} = 0.296\, \eta_{mvss}^{-0.148} \tag{14}$$

When the integrated stiffness is less than 446, increasing the support stiffness can significantly reduce the influence of the ground-linked wall on the surrounding soil, and the rate of reduction of the maximum horizontal displacement of the monopile also increases. When the integrated stiffness is greater than 446, we continue to increase the integrated stiffness of the foundation support structure and the deformation trend of the ground-linked wall. At the same time, due to the increase in the stiffness of the ground link wall, the maximum horizontal displacement of the single pile further tends to level off. When designing the pit support structure system, the depth of the pit, the thickness of the ground link wall, and the surrounding structures should be fully considered, so that the comprehensive stiffness of the pit support can be controlled in an economical, reasonable, and safe range and the most suitable pit support scheme can be selected.

5 Deformation Analysis of Monopile Forces at the Same Combined Stiffness

5.1 Effect of Horizontal Spacing of Monopiles and Ground Link Walls

To compare the extent to which the monopiles were affected by distance during the excavation of the pit, the distances between the monopiles and the edge of the ground link wall were taken to be 3, 5, 7, 9, 11, 13, 15, 17, 19, 33 and 48 m respectively, as shown in Fig. 8, with the model parameters of pit excavation size, depth and support conditions remaining unchanged. From Fig. 9, it can be seen that the excavation of the foundation pit causes the displacement of the monopiles towards the direction of the foundation pit, and the lateral displacement shows a gradual weakening phenomenon as the spacing between the ground link wall and the monopiles increases, which is in line with the law of the change of the displacement field of the surrounding soil.

When the distance between the monopile and the ground connection wall is 3 m, the pile body produces the largest horizontal lateral displacement, which is located about 4 m above the excavation surface of the foundation pit, and the lateral displacement shape is overall fish-belly shaped, and the lateral displacement at the top of the monopile is 1/3 of the maximum lateral displacement, so the distance between the monopile and the foundation pit should be avoided too close in real life projects.

When the distance between the two is within the excavation depth of the pit, the horizontal displacement of the monopile decays sharply with increasing distance in a linear relationship until the distance between the two is 15 m, and the maximum lateral displacement of the pile is shifted to the top of the pile end from about 12 m depth. At this time, the influence of the horizontal displacement field of the soil

Fig. 8 Diagram of horizontal displacement of monopiles at the varying spacing between monopiles and diaphragm walls

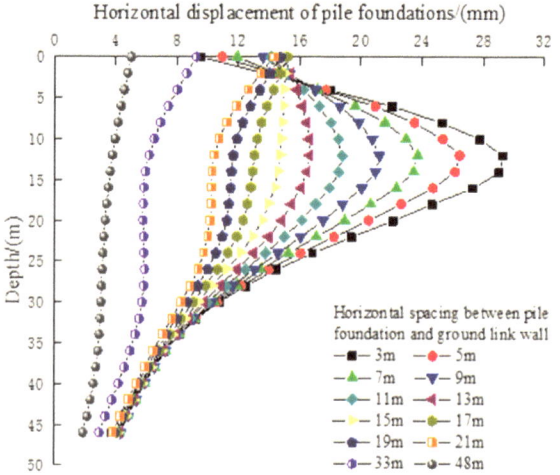

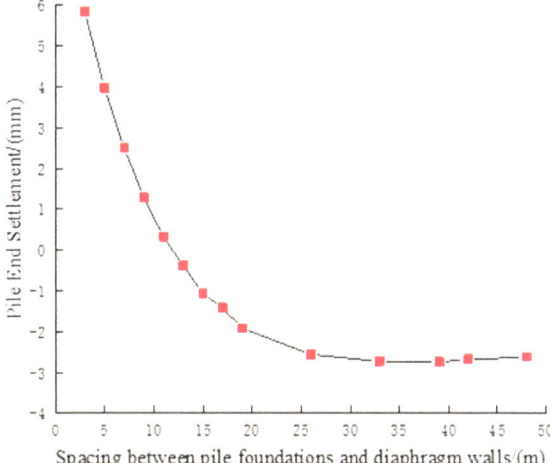

Fig. 9 Pile end settlement diagram for varying spacing between monopile and diaphragm wall

around the pit on the monopile is equal to the influence of the surface settlement, the deformation form of the monopile above the excavation surface of the pit is "1"-shaped, the horizontal displacement at the top of the pile is 15.08 mm, and the displacement value near the excavation surface is 14.38 mm, both of which are basically the same.

When the distance between the monopile and the ground link wall is greater than the excavation depth, the influence of the horizontal displacement around the foundation pit is gradually greater than the influence of the surface settlement monopile, and the pile starts to show a trend of displacement away from the foundation pit direction and the maximum horizontal displacement is always located at the top of the pile end.

When the distance between the monopile and the ground connection wall is twice the excavation depth, the maximum lateral displacement of the monopile is only 1/3 of the distance between the monopile and the ground connection wall of 3 m. When the distance between the monopile and the ground connection wall is greater than 3 times the excavation depth, i.e., 48 m, the additional lateral displacement of the pile is already very small and the displacement values of the top and bottom of the pile end are the same.

The monopile is affected by soil settlement during the excavation of the pit, and various parts of the pile will produce additional settlements of varying sizes. The further away it is from the pit, the weaker it is affected by surface settlement. In this paper, the settlement value at the top of the pile end is taken for analysis.

When the distance between the monopile and the ground connection wall is less than about 0.7 H, the monopile as a whole tends to move upwards. With the distance increasing, the settlement value at the top of the pile end decreases in a "cliff-like" manner, when the distance between the monopile and the ground connection wall is about 0.7 H, the maximum settlement value of the monopile is located at about 30 m below the surface, the settlement value at the top of the pile end is close to 0, and the monopile moves from upwards gradually turns to a downward trend.

When the spacing between the two is greater than twice the excavation depth, the settlement of the monopile tends to level off and the displacement value at the bottom of the pile end is approximately − 2.73 mm ("−" represents the downward movement of the monopile), at which point it can be seen that surface settlement has a weak effect on the monopile.

5.2 Effect of Monopile Stiffness

The elastic model of the monopile and the diameter of the monopile both have a direct effect on the monopile stiffness. This paper analyses the effect of foundation excavation on the force–deformation characteristics of the monopile by varying the value of the elastic modulus to achieve a change in monopile stiffness. As shown in Fig. 10, the elastic modulus of a single pile is 0.3 GPa (cement-soil mixing monopile), 3 GPa (low-strength plain concrete monopile), 30 GPa (reinforced concrete monopile), and 300 GPa (steel monopile) for discussion and analysis, and the model parameters such as excavation size, depth and supporting conditions of the foundation pit remain unchanged.

The combination of the monopile horizontal displacement diagram and bending moment diagram shows that when the monopile modulus of elasticity is less than 30 GPa, the monopile displacement increases with the decrease of modulus of elasticity, the maximum displacement value appears without change in position, while the additional bending moment decreases abruptly, the maximum additional bending moment position also changes with the change of modulus of elasticity, the maximum bending moment of reinforced concrete monopile is 3.56 times of that of plain concrete monopile, due to the cement mix pile Due to the small modulus of elasticity of the pile itself, the difference between the maximum and minimum additional bending moment values of the pile is only 59 kN·m. Both the cement mix pile and the plain concrete pile exhibit a flexible pile trait. When the monopile trait is shown as a rigid pile, the additional bending moment of the monopile increases while the horizontal displacement decreases. From the above analysis, it can be concluded that the change in the modulus of elasticity of the monopile during the excavation of the foundation pit does not greatly affect the lateral displacement of the pile, but greatly affects the monopile bending moment.

5.3 Effect of Monopile Length

The length of the monopile itself during the excavation of the foundation pit also has a certain influence on its displacement and additional bending moment. 8, 16, 24, 32, 46, and 56 m monopile lengths are taken for comparative analysis, and

Fig. 10 Monopile deformation force diagram for changing monopile stiffness

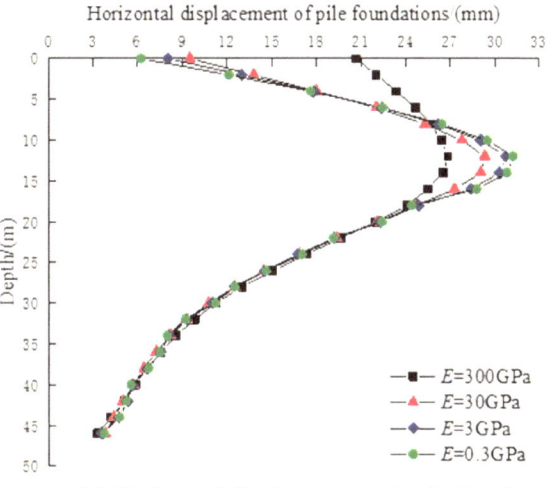

(a) Horizontal displacement of a single pile

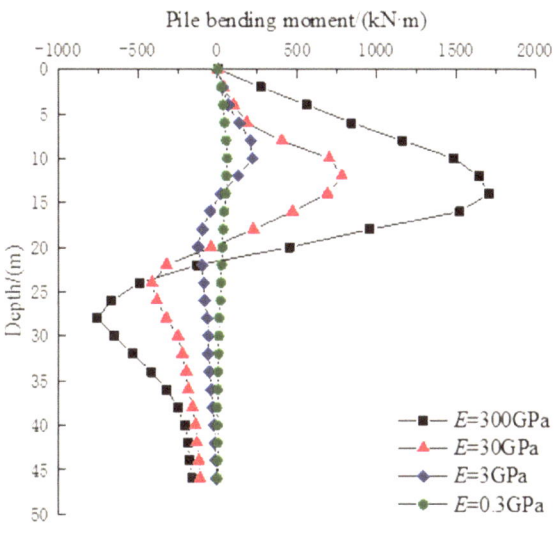

(b) Pile bending moment

the model parameters such as excavation size, depth, and support conditions of the foundation pit remain unchanged, as shown in Fig. 11 for the horizontal displacement and bending moment diagrams when the length of the monopile changes.

When the length of the monopile is smaller than the excavation range of the foundation pit, the monopile produces inclined rotation towards the direction of the foundation pit, and the overall deformation form is diagonal. The overall deformation value of the monopile is proportional to the length of the monopile, but the displacement of the top of the pile end is always smaller than the displacement of the bottom

Fig. 11 Single pile
deformation force diagram
for changing single pile
length

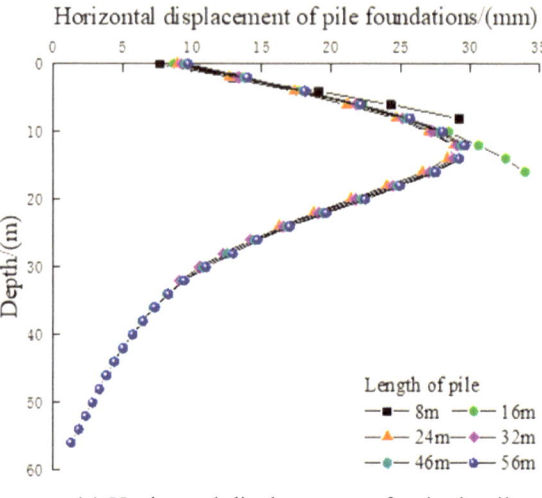

(a) Horizontal displacement of a single pile

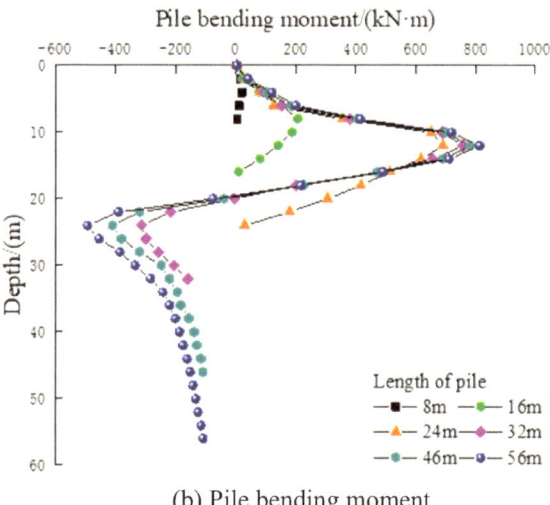

(b) Pile bending moment

of the pile when the length of the monopile is 16 m, and the horizontal displacement of the bottom of the pile is 33.94 mm. At the same time, the additional bending moment of the pile body changes from "1" to fish-belly type, and the maximum bending moment value does not change, i.e., it is located at about 1/2 of the single pile when the length of the single pile is 8 m. The maximum bending moment value of the single pile is only 20.60 kN·m, and at this time the single pile belongs to the short pile.

When the monopile length is 1–1.5 times H (H is the excavation depth of the foundation pit), the deformation form of the monopile is presented as fish-belly type, the maximum horizontal displacement is located at the maximum excavation depth of the foundation pit, the displacement value is 28.81 mm, compared with the monopile length less than H, and 1–1.5 times H length monopile bending moment value suddenly increases. At this time the monopile belongs to a medium-length pile. When the length of the monopile is 3.5 times H, i.e., 56 m, the maximum horizontal displacement of the monopile is 29.61 mm and the maximum additional bending moment value is 813.94 kN·m. It can be seen that when the length of the monopile is greater than 1.5 times H, the maximum horizontal displacement and an additional bending moment of the monopile do not increase significantly with the increase of the length of the monopile, and the monopile belongs to the long pile at this time.

5.4   Effect of Vertical Loads on Top of Piles

In the actual project construction, the top of the pile needs to bear the vertical load transmitted by the superstructure, so it is necessary to consider the response degree of the pile body under different vertical loads and set the vertical load on the top of the pile to 0, 625, 1250 and 2500 kN in turn, with the model parameters such as pit excavation size, depth and support conditions remaining unchanged, and the horizontal displacement and additional bending moment of the single pile as shown in Fig. 12.

The combination of the two figures shows that when the top of the pile is subjected to vertical load, the soil around the pit moves into the pit thus causing the monopile to have a tendency to move sideways. The vertical load on top of the pile evolves step by step from the axial load before the excavation of the pit to an eccentric load and the eccentric distance also starts to change, so the monopile will gradually be affected by the eccentric load and its horizontal displacement and additional bending moment will increase as the vertical load increases.

When the pile top load is 2500 kN, the maximum displacement and additional bending moment of the single pile are 31.50 mm and 958.84 kN·m respectively, and the increase is less obvious than when there is no load at the pile end.

6   Concluding Remarks

(1) Variations in the thickness of the ground link wall, the number of vertical bracing courses, and the horizontal spacing have different degrees of influence on the comprehensive stiffness of the enclosure, and under the same conditions, increasing the horizontal spacing has a greater influence on the control of single pile deformation than increasing the number of internal bracing courses.

(2) The rate of change of the maximum horizontal displacement of the single pile and the comprehensive stiffness of the pit is in a power function relationship. When the comprehensive stiffness is less than 446, the change in stiffness has a

Fig. 12 Single pile
deformation force diagram
for change in vertical load on
top of a pile

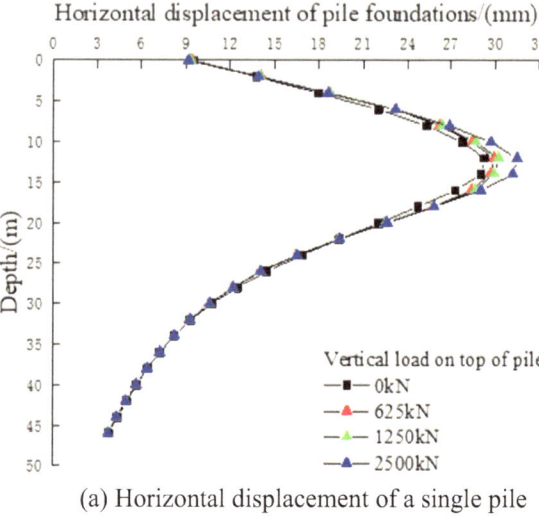

(a) Horizontal displacement of a single pile

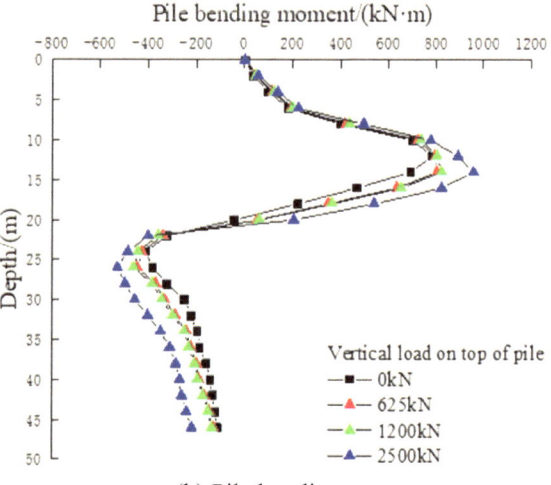

(b) Pile bending moment

greater impact on the maximum horizontal displacement of the pile. When the
comprehensive stiffness is greater than 446, the change of stiffness has a smaller
impact on the maximum horizontal displacement of the pile, the functional
relationship between the comprehensive stiffness of the pit and the maximum
horizontal displacement of the single pile is established, which provides a new
method for predicting the maximum horizontal displacement of the single pile.

(3) When the distance between the pile foundation and the ground connection wall
increases from 3 to 48 m, the maximum horizontal displacement of the pile
foundation tends to decrease. When the distance between the two is about 1
times the excavation depth, the pile deformation above the excavation surface

is in the form of "1". When the distance between the two is greater than 26 m, the change in distance has almost no effect on the settlement value of the pile top.

(4) The increase in pile foundation stiffness has a small effect on the horizontal displacement of the pile foundation, while it has a significant effect on the additional bending moment of the pile foundation. The maximum bending moment value of reinforced concrete pile foundations is 3.56 times that of plain concrete pile foundations. When the length of the pile foundation is less than or equal to the excavation depth of the pit, the horizontal displacement of the pile foundation shows a linear change.

(5) When the vertical load at the top of the pile is continuously increased, the pile foundation is subjected to eccentric loading and the horizontal displacement and bending moment are increased.

References

1. Shi, J. W., Liu, G. B., Huang, P., et al. (2015). Interaction between a large-scale triangular excavation and adjacent structures in Shanghai soft clay. *Tunnelling and Underground Space Technology, 50*, 282–295.
2. Liu, G. B., Huang, P., Shi, J. W., et al. (2016). Performance of a deep excavation and its effect on adjacent tunnels in Shanghai soft clay. *Journal of Performance of Constructed Facilities, 30*(6), 04016041.
3. Xiao, X., Zhang, Y. Q., Li, M. G., et al. (2017). Responses of the strata and supporting system to dewatering in deep excavations. *Journal of Shanghai Jiao Tong University (Science), 22*(6), 705–711.
4. Stewart, D. P., Jewell, R. J., & Randolph, M. F. (1994). Design of piled bridge abutments on soft clay for loading from lateral soil movements. *Geotechnique, 44*(2), 277–296.
5. Heyman, L., & Boerema, F. (1961). Bending moments in piles due to lateral earth pressure. In Proceedings of 5th ICSMFE, Paris (Vol. 2, pp. 425–429).
6. Mao-song, H., Chen-rong, Z., & Zao, L. (2008). Lateral response of passive pile groups due to excavation-induced soil movement in stratified soils. *Chinese Journal of Geotechnical Engineering, 07*, 1017–1023.
7. Chen-rong, Z., & Mao-song, H. (2012). Behavior of pile foundation due to excavation-induced lateral soil movement. *Chinese Journal of Geotechnical Engineering, 34*(S1), 565–570.
8. Guo-sheng, Y., Fa-yun, L., Jing-pei, L., & Feng, Y. (2011). 3D numerical analysis for behavior of axiallyloaded pile subjected to lateral soil movement. *Journal of Tonggi University (Natural Science), 39*(01), 1–6.
9. Fa-yun, L., Jie, H, & Jing-pei, L. (2008). Behavior of single pile subjected to lateral soil displacement induced by excavation. *Chinese Journal of Geotechnical Engineering, 30*(S1), 260–265.
10. Fa-yun, L., & Yan-chu, L. (2011). Numerical analysis for effects of lateral soil movement on adjacent piles. *Chinese Journal of Geotechnical Engineering, 33*(S2), 399–403.
11. Clough, G. W., Smith, E. M., & Sweeney, B. P. (1989). Movement control of excavation support systems by iterative design. In *Proceedings of Foundation Engineering: Current Principals and Practices, ASCE.* (Vol. 2, pp. 869–884). Geotechnical Special Publication.
12. Ge, Z., & Hai-he, M. (2016). A new system stiffness of retaining structure of deep foundation pit in soft soil area. *Rock and Soil Mechanics, 37*(05), 1467–1474.

13. Jian, Y., Chen-rong, Z., & Mao-song, H. (2012). Subgrade modulus of underground pipelines subjected to soil movements. *Chinese Journal of Rock Mechanics and Engineering, 31*(01), 123–132.

Research on the Influence Mechanism of Side Pile Foundation of Shield Tunnel Based on Transparent Soil Test Technology

Hongxiang Gong, Qian Yin, Wanchun Chen, Jiangwei Wang, Yukun Liu, and Yunliang Cui

Abstract Combined with the practical geotechnical problem of shield tunnel side piercing pile foundation, this paper develops a set of test device that can realistically simulate formation loss and post-wall grouting during shield tunneling by introducing transparent soil test technology and digital image processing technology, carries out indoor model tests on the existing pile foundation of double-line tunnel side penetration, obtains the deformation gauge and dynamic development process of soil and pile foundation caused by double-line tunnel, and analyzes the influence of tunnel burial depth and relative distance between pile and tunnel. The results show that the increase of tunnel burial depth will cause the expansion of soil settlement trough range and peak value. The deformation caused by the pile foundation located on both sides of the double-line tunnel accounts for about 90% of the total deformation compared to the tunnel, and the farther the pile foundation on both sides of the double-line tunnel, the less affected it is. This article can provide guidance and suggestions for the construction of similar projects.

H. Gong · Q. Yin
Nanjing Metro Construction Co., Ltd, Nanjing 210018, China
e-mail: 18851190018@163.com

Q. Yin
e-mail: 18851190168@163.com

W. Chen
China Construction Eighth Bureau Rail Transit Construction Co., Ltd, Nanjing 210023, China
e-mail: 2041566154@qq.com

J. Wang (✉) · Y. Liu
Hohai University, 210029, Nanjing 213200, China
e-mail: 1972935166@qq.com

Y. Liu
e-mail: 3306810054@qq.com

Y. Cui
Hangzhou City University, Hangzhou 310015, Zhejiang, China
e-mail: 15990160799@163.com

© The Author(s) 2025
D. Li and Y. Zhang (eds.), *Advances in Frontier Research on Engineering Structures II*,
Lecture Notes in Civil Engineering 535, https://doi.org/10.1007/978-981-97-6238-5_31

Keywords Shield tunnel · Side piercing pile foundation · Transparent soil model test · Deformation of the pile foundation

1   Introduction

The shield machine method can effectively reduce the disturbance of the surroundings during construction [1–3]. However, the disturbance of the surrounding soil deformation during shield tunnel excavation is an extremely complex dynamic three-dimensional change process, so it is urgent to analyze the influence mechanism of the internal force and deformation of adjacent pile foundation during excavation, and summarize the relevant measures, which has great research value and social significance.

The problem of tunnel crossing existing pile foundation has received a lot of attention from many researchers. Morton [4] carried out model experiments on tunnel boring under weak ground, and obtained the relationship between the formation loss rate and the surrounding soil mass and adjacent pile foundation failure areas. Sun [5] based on transparent soil test technology, designed the systematic scheme of shield tunnel boring model test, and analyzed and summarized the mesoscopic mechanism of soil deformation caused by shield construction. Zhao [6] realized the use of physical model test to observe the three-dimensional spatial displacement field with the help of transparent soil model test technology and three-dimensional reconstruction technology.

This paper focuses on the practical and complex geotechnical engineering problem of the side piercing pile foundation of shield tunnel, combines transparent soil test technology, comprehensively analyzes the deformation and force characteristics of the soil mass around the tunnel and the adjacent pile foundation, summarizes the influence law of pile foundation and soil mass under different burial depths and pile-tunnel distances, and obtains the influence law and deformation mechanism of pile foundation and soil under double-line tunnel excavation.

2   Visualized Two-Line Shield Tunnel Construction Test

2.1   Design of the Test Device

The shield tunnel model box is $500 \times 400 \times 300$ mm transparent plexiglass cuboid, the model box is divided into two equal compartments 1 and 2, zone 1 is the tunnel and pile foundation model and transparent soil sample, zone 2 is injected with mixed oil. The grouting system is composed of a small air compressor, a controller, a grouting tank, an air pressure pipe and a grouting pipe. The controller ensures that the grouting is carried out continuously and smoothly; The pneumatic pipe connects the silent

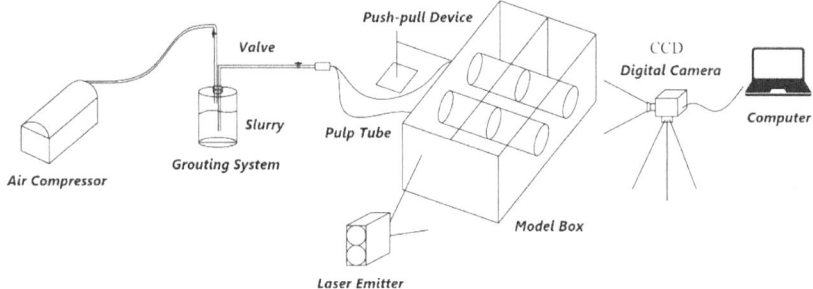

Fig. 1 Model diagram of the test system

air compressor, the control box and the slurry tank through the three-way connector; The grouting pipe adopts a silicone capillary hose with an outer diameter of 3 mm, which connects the grouting tank and the tail of the outer tube of the model box, and obtains the grouting volume by weighing the mass reduction before and after the grouting tank. The digital image measurement system includes laser emitters, CCD cameras and particle image speed processing software PIVlab, which can visualize the whole process of tunnel side pile foundation disturbance. Figure 1 shows the specific model diagram.

2.2 Preparation of Transparent Soil Samples

A transparent soil sample prepared by fused silica sand and mixed oil was used [7, 8]. The transparent soil solid particles are made of fused silica sand with a particle size range of 1-3 mm, and the pore liquid is mineral oil mixed with n-dodecane and 90# white oil.

2.3 Trial Protocol

In this test, a variety of spatial conditions were set by changing the buried depth of different tunnels and the position of the pile foundation, and 16 groups of tests were set, and the specific test group parameters are shown in Table 1.

In this paper, three kinds of tunnel burial depths of 30 mm, 60 mm, and 90 mm correspond to three burial depth ratios of H/D of 0.5, 1.0, and 1.5 respectively. Two pile foundation position layout forms are designed in this experiment, of which the horizontal distance between pile foundation and tunnel boundary line is 30, 45, 60 mm, and the detailed pile foundation layout form is shown in Fig. 2.

Table 1 Test design scheme of side pile foundation of shield tunnel

Number	Burial depth (mm)	Pile foundation layout form	Vertical distance between the pile end and the tunnel axis (mm)
A1	90	Type I	50
A2	60	Type I	50
A3	30	Type I	50
B1	90	Type I	0
B2	60	Type I	0
B3	30	Type I	0
C1	90	Type I	−50
C2	60	Type I	−50
D1	90	Type II	50
D2	60	Type II	50
D3	30	Type II	50
E1	90	Type II	0
E2	60	Type II	0
E3	30	Type II	0
F1	90	Type II	−50
F2	60	Type II	−50

Fig. 2 Pile foundation arrangement

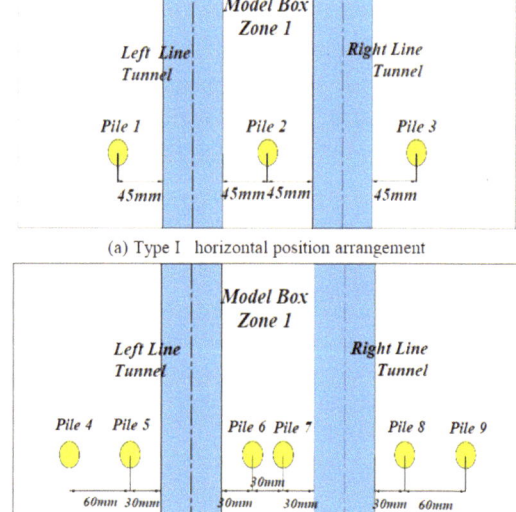

(a) Type I   horizontal position arrangement

(b) Type II   horizontal position arrangement

3 Analysis of Results

3.1 Influence of Tunnel Burial Depth on Adjacent Pile Foundations

Tunnel burial depth is the main factor of shield tunnel excavation on the surrounding soil and adjacent pile foundation. The different burial depth ratios of all test groups are mapped together with the settlement extremes caused by tunnel excavation, as shown in Fig. 3a. When the tunnel diameter, formation loss and soil parameters were consistent, the maximum surface settlement value was negatively correlated with the tunnel burial depth [9, 10]. The corresponding curves of soil depth and settlement extremes under the three burial conditions are plotted in Fig. 3b. It can be seen from the figure that when the buried depth of the tunnel increases, the settlement extreme value area also increases, that is, the increase in the buried depth will lead to a decrease in the surface settlement value and the settlement value at the tunnel vault position, and the deeper the depth of the ground layer, the greater the settlement extreme value of the soil layer, and the larger the settlement extreme value range of different soil layers when the buried depth is deeper.

Fig. 3 Correspondence between tunnel burial depth and settlement extreme

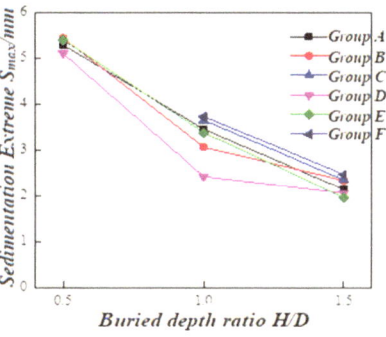

(a) Diagram of the burial depth ratio and the settlement extremum

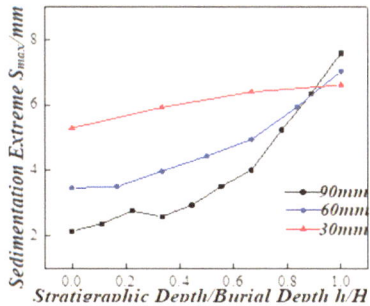

(b) Soil depth corresponds to settlement extreme

3.2 Influence of Pile Foundation Position on the Adjacent Pile Foundations

Shield tunnel excavation has different effects on different areas, so the relative position of the pile foundation and the tunnel is also an important factor affecting the deformation of the pile foundation. The lateral displacement extremes of pile foundations in different positions under different test conditions are extracted in Tables 2 and 3, Table 2 is the lateral displacement extremes of three pile foundations in type I under various test conditions, Table 3 is the lateral displacement extremes of six pile foundations in type II, it can be seen from the table that the farther the relative horizontal distance of the pile tunnel, the smaller the side displacement value of the pile foundation, from the data in the table, it can be seen that the farther the relative horizontal distance of the pile and tunnel, the smaller the lateral displacement value of the pile foundation, and the horizontal distance of the pile tunnel shows an obvious positive correlation.

Table 2 Side shift extremes of Type I pile foundations

Number	Pile 1 extremes (mm)	Pile 2 extremes (mm)	Pile 3 extremes (mm)
A1	0.946	0.016	−0.937
A2	1.108	0.003	−1.105
A3	1.058	−0.012	−1.069
B1	0.974	−0.008	−0.979
B2	0.902	−0.152	−0.822
B3	1.447	−0.021	−1.432
C1	0.912	−0.016	−0.909
C2	1.342	−0.012	−1.333

Table 3 Side shift extremes of Type II pile foundations

Number	Pile 4 extremes (mm)	Pile 5 extremes (mm)	Pile 6 extremes (mm)	Pile 7 extremes (mm)	Pile 8 extremes (mm)	Pile 9 extremes (mm)
D1	0.476	1.131	−0.196	0.198	−1.133	−0.477
D2	0.404	1.103	−0.273	0.258	−1.115	−0.411
D3	0.427	1.397	−0.463	0.459	−1.414	−0.431
E1	0.462	0.909	−0.284	0.288	−0.911	−0.466
E2	0.622	1.509	−0.268	0.264	−1.507	−0.621
E3	0.594	1.923	−0.579	0.583	−1.933	−0.601
F1	0.565	0.961	−0.326	0.341	−0.961	−0.562
F2	0.658	1.565	−0.264	0.228	−1.561	−0.654

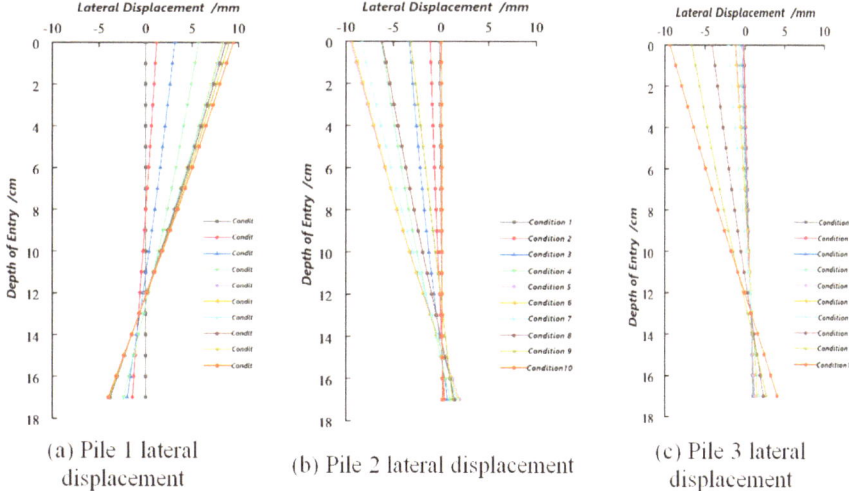

(a) Pile 1 lateral displacement

(b) Pile 2 lateral displacement

(c) Pile 3 lateral displacement

Fig. 4 Lateral displacement of each pile

In order to further study the deformation of pile foundations at different positions with the progress of construction under the same test conditions, the lateral movement data of piles 1, 2 and 3 of group A1 were extracted and plotted in Fig. 4. It can be seen from the figure that the deformation of the pile foundation caused by the closer tunnel accounts for about 90% of the total deformation.

As the tunnel excavation progressed, the tunnel continued to induce the deformation of the pile foundation. The deformation in the middle of the double-line tunnel will first bias towards the excavated tunnel under construction, and the pile foundation will gradually reset when the other side of the tunnel is constructed. The double-line tunnel has a superposition effect on the influence of pile foundation on both sides, and has a countervailing effect on the foundation. The pile foundation rotates at the buried depth of the tunnel axis, the soil body in the upper part of the tunnel will continue to move towards the gap generated by the tunnel construction, and the soil below the tunnel buried depth is basically not deformed, and the moving soil drives the pile foundation to deform together, so that the rotation point of the pile foundation is consistent with the buried depth of the tunnel axis.

The position of the pile foundation not only affects the size of its own deformation, but also the different position of the pile foundation will have a reaction to the deformation of the soil layer, and the single cause control method is used to analyze the reaction of the pile foundation position on the soil deformation, as shown in Fig. 5, the surface settlement of the experimental group with different pile foundation positions and other conditions are drawn in a figure for comparative analysis. The ideal Peck curve is a fairly smooth curve, and under the same buried depth conditions, the surface settlement curve changes in multiple positions, and these inflection points basically correspond to the position of the pile foundation. The analysis reason is that the excavation of shield tunnel causes soil deformation, soil disturbance causes

Fig. 5 Comparative Curves of ground surface settlement at different pile foundation positions

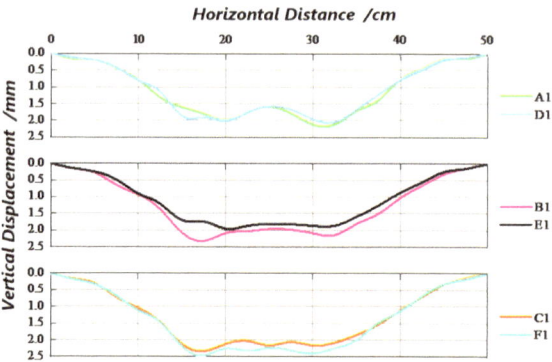

additional stress of pile foundation near pile body, and under the dual influence of additional stress of pile body and stress generated by soil layer disturbance, the vertical displacement of soil around the pile is limited to a certain extent, so the settlement curve has obvious concave and convex changes, and this suppression effect is also weakened with the increase of the relative distance between the tunnel and the pile foundation.

4 Conclusions

This chapter studies the cross-existing pile foundation of double-line shield tunnel through transparent soil model test, simulates the progressive excavation process of shield tunnel, analyzes the influence mechanism of tunnel construction disturbance adjacent pile foundation by using transparent soil test technology of non-destructive intervention and transparent visualization, and conducts influencing research on tunnel burial depth, pile-tunnel spacing and other factors, the main research contents are as follows:

(1) The change of tunnel burial depth will not affect the soil settlement mode, the tunnel burial depth is inversely proportional to the extreme value of ground settlement, and the increase in buried depth will cause the expansion of the soil settlement trough range and peak increase, and the increase in buried depth will lead to a decrease in the surface settlement value and an increase in the settlement value of the tunnel vault position.

(2) During the construction of the double-line shield tunnel, the vertical settlement and lateral displacement of the adjacent pile foundation also continued to increase. The deformation caused by the near-tunnel of the pile foundation on both sides of the double-line tunnel accounts for about 90% of the total deformation, and the influence of the farther tunnel is small, and the deformation of the double-line tunnel is greater than that caused by the single-line tunnel. The total deformation of the pile foundation located in the middle of the double-line

tunnel is small, and the backward tunnel will reset the deformation caused by the pilot tunnel, and it is necessary to avoid fatigue and shear failure of the pile foundation. The relative lateral distance of pile tunnel is an important influencing factor affecting the extreme value of the lateral displacement of the pile body, and the relative vertical distance of the pile tunnel is the main reason for determining the form of lateral displacement of the pile body.

References

1. Hisatake, M. (1994). Ground surface settlements due to shield tunnels. In J. W. Bull (Ed.), Soil-structure interaction: Numerical analysis and modelling. E and FN, (pp. 647–672).
2. Yun, Z., Zongze, Y., & Yongfu, X. (2002). Analysis of surface deformation caused by shield tunneling. *Journal of Rock Mechanics and Engineering, 03*, 388–392.
3. Xinjiang, W., Mobang, Z., Zhi, D., & Xiao, Z. (2020). Research status and prospect of the influence of shield tunneling on existing metro tunnels. *Rock and Soil Mechanics, S2*, 1–20. http://kns.cnki.net/kcms/detail/42.1199.O3.20201006.1253.003.html
4. Morton, J.-D., & King, K.-H. (1979). Effects of tunnelling on the bearing capacity and settlement of piled foundations. 57–68.
5. Jizhu, S., & Wenhui, X. (2011). Research on experimental design of shield tunnel model based on transparent soil. *Journal of Wuhan University of Technology, 33*(05), 108–112.
6. Honghua, Z., Cong, L., Xiaowei, T., et al. (2020). Research on spatial deformation visual measurement system based on transparent soil and 3D reconstruction technology. *Rock and Soil Mechanics, 41*(09), 3170–3179.
7. Sadek, S., Iskander, M. G., & Liu, J. Y. (2002). Geotechnical properties of transparent silica. *Canadian Geotechnical Journal, 39*(1), 111–124.
8. Wanghua, S., Yue, G. (2011) Current situation and prospect of transparent soil experimental technology. *Journal of China Coal Society, 36*(04), 577–582.
9. Peck, R.-B. (1969). Deep excavations and tunnelling in soft ground. proc.int.conf.on smfe.
10. Gang, W., Xinhai, Z., & Xinxin, H. (2017). Study on surface subsidence caused by double-line horizontal parallel shield construction considering multiple factors. *Journal of Disaster Prevention and Mitigation Engineering, 37*(06), 923–930.

Study of Difference in Dynamic Response of Underground Structure at Different Blast Angles

Qindong Lin, Chun Feng, Yundan Gan, Jianfei Yuan, Wenjun Jiao, and Ying Yang

Abstract The dynamic response of underground structures under blast loading is of great military significance to strategically valuable targets, and the difference in the dynamic response of underground structures at different blast angles is investigated based on a FEM-DEM coupled method. First, a full-time numerical simulation under blast loading is carried out. Then, the displacement and fracture characteristics of the underground structure are analyzed quantitatively. The results indicate that the spatial distribution of displacement and initial fracture type is closely related to the position of the explosive. The concrete at the top boundary and right boundary reaches the opposite boundary earliest at 90° and 0° respectively. The time-history curves of crack ratio have obvious stage characteristics, the maximum tensile crack ratio is obtained at 90°, and the maximum shear crack ratio is obtained at 0°.

Keywords Dynamic response · Underground structure · CDEM · Blast loading

1 Introduction

Many strategically valuable targets are moved underground, relying on high-resistance underground protective structures to resist weapon destruction. To improve the ability to destroy underground targets, the application of earth penetrators develops rapidly. Therefore, it is of great military significance to study the destructive effect of blast shock waves on the underground structure [1, 2].

Currently, scholars mainly study the dynamic response of underground structures based on experimental study, theoretical analysis, and numerical simulation. Liu et al. [3] found that the damaging effect is related to overpressure, duration of explosion, surrounding media, and structures in the media through 11 batches of

Q. Lin · Y. Gan (✉) · J. Yuan · W. Jiao · Y. Yang
Xi'an Modern Chemistry Research Institute, Xi'an 710065, Shaanxi, China
e-mail: ganyundan@163.com

C. Feng
Institute of Mechanics, Chinese Academy of Sciences, Beijing 100190, China

© The Author(s) 2025
D. Li and Y. Zhang (eds.), *Advances in Frontier Research on Engineering Structures II*,
Lecture Notes in Civil Engineering 535, https://doi.org/10.1007/978-981-97-6238-5_32

blast experiments. Zheng et al. [4] proposed that cracks, deformation, and damage to structures are the most serious when the cylindrical charge explodes beside the straight wall. Liu et al. [5] proposed that the failure mode develops from the concrete cracks at the back surface of the blast to the concrete spalling and deformation of steel bars. Based on the similarity theory and the dimensional analysis method, Liu et al. [6] determined the key parameters affecting penetrator depth and established the function relation.

Compared with experimental study and theoretical analysis, numerical simulation has the advantages of high efficiency, accuracy, and low cost. Cao et al. [7] analyzed the dynamic response and cumulative damage of underground caverns under cyclic explosion at low levels and a single explosion at high levels and found that the cumulative damage of surrounding rock presents a dramatic nonlinear relationship with explosion times. Huo et al. [8] explored the damaging effect of the underground arched structure under the condition of the 45° side top blast. The results showed that the failure mode of the structure gradually transitions from spalling failure and bending failure to shear failure and overall failure with the increase in the charge mass. Keskin et al. [9] examined the behavior of an underground circle-shaped tunnel under an impact load generated by ground-surface explosions, and the result indicated that the concrete lining and embedded steel rebar cage of the tunnel attain almost the same pressure profiles during the explosion period. Sun et al. [10] simulated the underground arch structures with different spans subjected to the shock wave and found that the structure undergoes partial failure on the vault and the whole concrete structure forms overall cracks.

Many scholars have conducted numerical simulations on the damage evolution process of underground structures under explosion shock waves, and most of the numerical methods belong to the continuum mechanics method. Owing to the disadvantages of mechanical theory and numerical algorithms, these methods cannot accurately simulate the initiation and expansion process of crack rather than deleting elements. Based on a FEM-DEM coupled method, this paper investigates the difference in the dynamic response of underground structures under different blast angles, which benefits the further analysis of destruction evaluation of earth penetrators on underground structures.

2   Numerical Simulation

2.1   Numerical Model

The continuum-discontinuum element method (CDEM) is adopted in the study to conduct the numerical simulation [11–13]. The numerical model of the explosive, underground structure and rock mass is plotted in Fig. 1. The vertical height of the rock mass is 80 m, and the horizontal length is 120 m. The thickness of the concrete wall is 1.8 m, the vertical height of the underground structure is 12 m, and the

Fig. 1 Numerical model of explosive, rock mass and underground structure

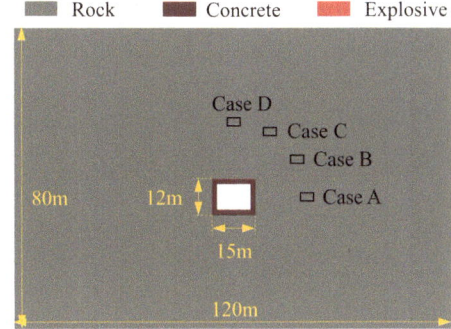

Table 1 Mechanical parameters of concrete and rock

Material	Density (kg/m³)	Elastic modulus (GPa)	Tensile strength (MPa)	Cohesive strength (MPa)
Concrete	2500	35	9	18
Rock	2300	10	3	7

Table 2 Mechanical parameters of explosive

Material	Charge density (kg/m³)	Internal energy (J/m³)	CJ pressure (Pa)	Detonation velocity (m/s)
TNT	1630	7e9	20e9	6930

horizontal length is 15 m. The diameter of the explosive is 0.2 m, and the distance from the center point of the underground structure is 15 m. In four cases, the azimuth angle of the explosive increases from 0° to 90°. The model is meshed by triangular elements, the element size of the explosive is 0.05 m, the element size of the concrete is 0.2 m, and the element size of the rock is 1 m. The mechanical parameters of rock and concrete are listed in Table 1. The JWL equation of state is adopted to simulate the explosion process, and the mechanical parameters are listed in Table 2.

2.2 Numerical Results

To investigate the effect of explosive azimuth angle on the dynamic response of underground structure, the azimuth angle θ_e is set to four values, $\theta_e = 0°$ in case A, $\theta_e = 30°$ in case B, $\theta_e = 60°$ in case C and $\theta_e = 90°$ in case D. The full-time dynamic response of underground structure under blast loading is simulated, and the difference in displacement and fracture characteristics are analyzed.

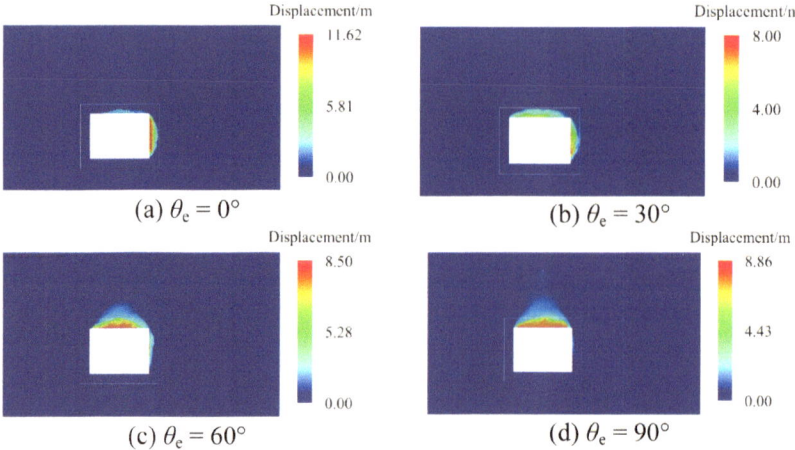

Fig. 2 Displacement nephograms corresponding to different azimuth angles

Displacement characteristic

The displacement nephograms corresponding to different explosive azimuth angles at $t = 0.5$ s are plotted in Fig. 2, and the solid white line represents the interface between the underground structure and the rock mass. It is observed that the value and spatial distribution of displacement at the top boundary and right boundary change with the increase of azimuth angle, while the spatial distribution of displacement at the bottom boundary and left boundary remains unchanged.

In the case of $\theta_e = 0°$, the maximum displacement is located at the right boundary of the underground structure, and the displacement value at the middle part is similar. As the distance from the air surface increases, the displacement gradually decreases. When the distance reaches the thickness of the concrete wall, the displacement of the outer rock mass is much smaller than that of the concrete wall. The concrete at the top boundary also cracks and slips, while the displacement is small.

In the case of $\theta_e = 30°$, the concrete at the right boundary and top boundary undergoes large displacement, and the maximum displacement is located at the right boundary, which is due to the concrete at the right boundary being closer to the explosive. For the right boundary and top boundary, the maximum displacement is located at the midpoint, and the displacement gradually decreases with the increase of distance from the air surface. The spatial distribution characteristic of displacement indicates that the region where large displacement occurs is mainly located at the concrete wall, and the rock around the explosive also undergoes slippage.

In the case of $\theta_e = 60°$, the concrete at the right boundary and top boundary undergoes large displacement. Since the concrete at the top boundary is closer to the explosive, its displacement value is larger than that of the concrete at the right boundary. For the top boundary, the maximum displacement is located at the midpoint, and the displacement gradually decreases as the distance from the air surface increases.

When the distance reaches the thickness of the concrete wall, the displacement of the outer rock mass also undergoes slippage and movement.

In the case of $\theta_e = 90°$, the concrete at the top boundary undergoes large displacement, while the displacement at the right boundary, left boundary, and bottom boundary is small. The displacement at the top boundary has the obvious discontinuous characteristic, which indicates that the concrete undergoes fracture and flyaway, and the displacement value at the middle part is similar. In addition, the rock above the top concrete wall also moves, and the region is larger than that in the case of $\theta_e = 60°$.

To accurately investigate the displacement evolution characteristic of concrete walls, the displacement time-history curves at midpoint A of the top boundary and midpoint B of the right boundary are plotted in Fig. 3. For midpoint A, when the azimuth angle is $0°$ and $30°$, the displacement increases gradually with the growth of time. When the azimuth angle is $60°$ and $90°$, the displacement increases gradually with the growth of time and remains unchanged subsequently, which indicates that the concrete at the midpoint of the top boundary reaches the bottom boundary, and the concrete in the case of $\theta_e = 90°$ reaches the bottom boundary earlier. For midpoint B, when the azimuth angle is $0°$, the displacement increases gradually with the growth of time and remains unchanged subsequently, which indicates that the concrete at the midpoint of the right boundary reaches the left boundary. When the azimuth angle is $30°$, $60°$, and $90°$, the displacement gradually increases with the growth of time, and the growth rate of displacement in the case of $\theta_e = 30°$ and $60°$ is similar.

Crack characteristic

The initial fracture nephograms of the concrete interface at $t = 0.5$ s are plotted in Fig. 4. It is observed that the initial fracture type of concrete interface is composed of tensile fracture and shear fracture, and the spatial distribution of fracture type is closely related to the position of the explosive. The rock near the explosive mainly undergoes shear fracture, and the proportion of rock with tensile fracture gradually increases as the distance from the explosive increases. For the underground structure,

Fig. 3 Time-history curves of displacement

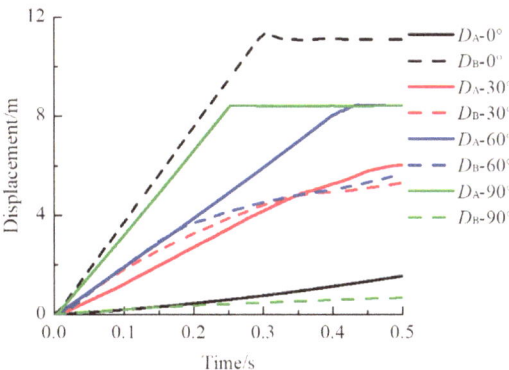

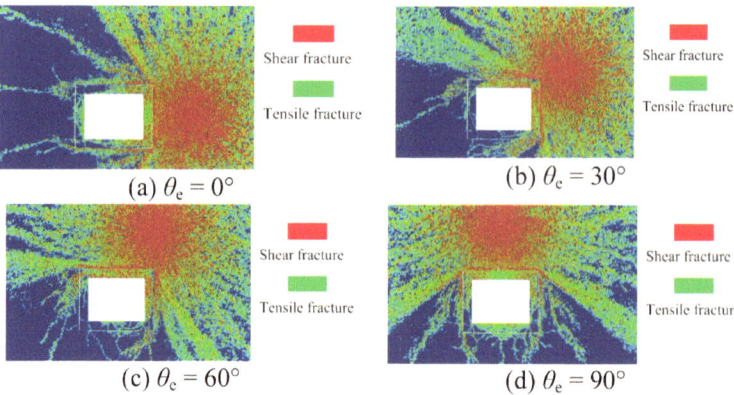

Fig. 4 Initial fracture nephograms of concrete interface

the fracture degree and the spatial distribution of fracture type are different in four cases.

In the case of $\theta_e = 0°$, the concrete at the right boundary is the most severely damaged. The concrete in the middle part mainly undergoes tensile fracture, while the fracture type on both sides includes tensile fracture and shear fracture. For the top boundary and bottom boundary, the concrete on the right part suffers shear fracture and tensile fracture, and the concrete on the left part mainly suffers tensile fracture. The concrete at the left boundary mainly suffers tensile fracture.

In the case of $\theta_e = 30°$, the concrete at the right boundary and top boundary is the most severely damaged, and a little concrete at the bottom boundary and left boundary is damaged. For the right boundary, shear fracture occurs mainly in the middle-upper part, and tensile fracture occurs mainly in the middle-lower part. For the top boundary, the fracture type includes tensile fracture and shear fracture, and the closer is to the explosive, the higher the proportion of concrete is with shear fracture. A little concrete at the bottom boundary and left boundary suffers tensile fracture.

In the case of $\theta_e = 60°$, the concrete at the top boundary is the most severely damaged, followed by the right boundary, and only part of the concrete at the bottom boundary and left boundary is damaged. For the top boundary, the closer is to the explosive, the higher the proportion of concrete is with shear fracture. For the right boundary, the concrete at the middle-upper part mainly undergoes shear fracture, and the concrete at the middle-lower part mainly undergoes tensile fracture.

In the case of $\theta_e = 90°$, the concrete at the top boundary is the most severely damaged. The concrete in the middle part mainly undergoes tensile fracture, while the fracture type on both sides includes tensile fracture and shear fracture. The cracked interface of the left boundary and the right boundary is mainly located in the middle-upper part, which suffers tensile fracture and shear fracture, and the concrete in the middle-lower part mainly undergoes tensile fracture. The concrete at the bottom

Fig. 5 Time-history curves of α_T and α_S

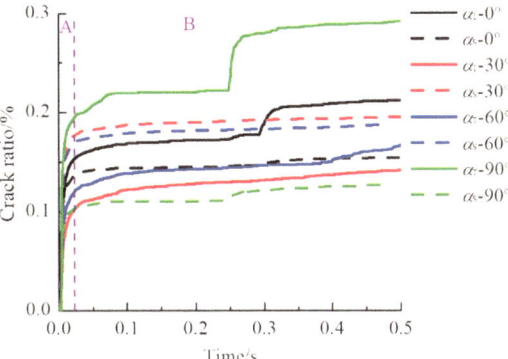

boundary mainly suffers tensile fracture, and the damage at the middle part is the most serious.

To quantitatively study the changing trend of the concrete cracked interface, two dimensionless indices, tensile crack ratio α_T and shear crack ratio α_S are introduced, and the time-history curves are plotted in Fig. 5. It is observed that the time-history curves of α_T and α_S have obvious stage characteristics. In stage A, α_T and α_S increase sharply with time. In stage B, α_T and α_S increase slowly with time. For the tensile crack ratio α_T, when the azimuth angle θ_e increases from $0°$ to $90°$, the value at the same moment decreases first and then increases, and the maximum value is obtained at $\theta_e = 90°$. For the shear crack ratio α_S, when the azimuth angle θ_e increases from $0°$ to $90°$, the value at the same moment increases first and then decreases, and the maximum value is obtained at $\theta_e = 30°$. The overall crack ratio α_M of the underground structure is obtained by accumulating α_T and α_S. The result indicates that the cracked concrete area is the largest at $\theta_e = 90°$, and the cracked concrete area is the smallest at $\theta_e = 30°$.

3 Conclusions

Based on the continuum-discontinuum element method, the difference in the dynamic response of underground structures at different explosion azimuth angles is investigated. The following conclusions can be drawn:

(1) The value and spatial distribution of displacement at the top boundary and right boundary change with the increase of azimuth angle, while the spatial distribution of displacement at the bottom boundary and left boundary remains unchanged. The concrete at the midpoint of the top boundary reaches the bottom boundary earliest at $\theta_e = 90°$. The concrete at the midpoint of the right boundary reaches the left boundary earliest only at $\theta_e = 0°$.

(2) The initial fracture type of concrete interface is composed of tensile fracture and shear fracture, and the spatial distribution of fracture type is closely related to the position of the explosive. The time-history curves of α_T and α_S have obvious stage characteristics, the maximum α_T is obtained at $\theta_e = 90°$, and the maximum α_S is obtained at $\theta_e = 0°$. The total cracked concrete area is the largest at $\theta_e = 90°$, and the total cracked concrete area is the smallest at $\theta_e = 30°$.

References

1. Wei, X., Zhou, H. Z., Cui, H., Zhou, C. Z., Yuan, Q. Q., & Zhang, B. (2020). Quick validation of earth-penetrating nuclear weapon attack on underground hardened targets. *Protective Engineering, 42*, 47–51.
2. Zhang, G. X., Qiang, H. F., Chen, F. Z., & Shi, C. (2018). Research and development of earth projectile penetrating underground fortification. *Aerodynamic Missile Journal, 6*, 34–38.
3. Liu, F., Wang, H. M., Yan, L. H., & Li, H. Q. (2021). Damage effect of shallow buried civil air defense engineering structures under nearby blast loading. *Acta Armamentarii, 42*, 625–632.
4. Zheng, Y. M., Tian, Q., Wang, Q. F., Xiao, L., & Feng, G. X. (2011). The damage influence of explosion location of cylindrical charge on underground structures. *Protective Engineering, 33*, 5–10.
5. Liu, G. K., Liu, R. C., Wang, W., Wang, X., & Zhao, Q. (2021). Blast resistance experiment of underground reinforced concrete arch structure under top explosion. *Chinese Journal of Energetic Materials, 29*, 157–165.
6. Liu, S. L., Sun, H. X., Zhang, Y., Huang, W. W., & Feng, T. (2017). The research on an empirical formula for earth penetrator weapons penetrating rock targets. *Journal of Air Force Engineering University (Natural Science Edition), 18*, 99–103.
7. Cao, A. S., Wang, G. Y., Dun, Z. L., Ren, L. W., & Sun, X. W. (2021). Dynamic responses and cumulative damage of the underground cavern under cyclic explosion. *Chinese Journal of High Pressure Physics, 35*, 156–165.
8. Huo, Q., Wang, Y. P., Liu, G. K., & Wang, W. (2021). Failure mode and influencing factors of underground arched structure subjected to side top blast. *Acta Armamentarii, 42*, 105–116.
9. Keskin, İ, Ahmed, M. Y., Taher, N. R., Gör, M., & Abdulsamad, B. Z. (2022). An evaluation on effects of surface explosion on underground tunnel; availability of ABAQUS finite element method. *Tunnelling and Underground Space Technology, 120*, 104306.
10. Sun, H. X., Lu, F., Chi, W. S., Kang, T., & Liu, Y. F. (2017). Dynamic interaction between surrounding rock and initial supporting structure subjected to explosion shock wave. *Explosion and Shock Waves, 37*, 670–676.
11. Feng, C., Li, S. H., Liu, X. Y., & Zhang, Y. N. (2014). A semi-spring and semi-edge combined contact model in CDEM and its application to the analysis of Jiweishan landslide. *Journal of Rock Mechanics and Geotechnical Engineering, 6*, 26–35.
12. Lin, Q. D., Li, S. H., Gan, Y. D., & Feng, C. (2022). A strain-rate cohesive fracture model of rocks based on Lennard-Jones potential. *Engineering Fracture Mechanics, 259*, 108126.
13. Zhu, X. G., Feng, C., Cheng, P. D., Wang, X. Q., & Li, S. H. (2021). A novel three-dimensional hydraulic fracturing model based on continuum-discontinuum element method. *Computer Methods in Applied Mechanics and Engineering, 383*, 113887.

Study on Progressive Collapse Resistance Performance of Main Building of Conventional Island in Nuclear Power Plant Based on Corner Column Removal Method

Jiaxing Di

Abstract In the event of an accidental collapse of the main plant of the conventional island, it could threaten the safe operation of the nuclear island and even cause a nuclear leakage accident. Therefore, the study of the collapse resistance of the main plant structure of a conventional island nuclear power plant should be of great interest. In this paper, the collapse resistance of conventional island main plant structure is analyzed by Pushdown analysis method, and the damage of beam and column members and the position of plastic hinge when the conventional island main plant structure is subjected to accidental action are simulated by the method of removing members. The analysis results show that the damage of the column increases with the increase of the column span after the removal of the corner columns, and the structural displacement of the upper floors within the beams and failed columns connected to the control nodes will be dispersed with the increase of the vertical load, and the damage will spread to the top of the structure; It can be concluded from the order of the appearance of the plastic hinge: the main plant structural system conforms to the principle of "strong columns and weak beams", so the main nuclear power plant conventional island studied in this paper The progressive collapse resistance of dismantled corner columns of the structural system of plant has certain guiding significance for practical engineering.

Keywords Main plant of the conventional island · Progressive collapse · Removal of components method

J. Di (✉)
College of Civil Engineering and Architecture, Dalian University, Dalian 116620, Liaoning, China
e-mail: 13052670250@163.com

© The Author(s) 2025
D. Li and Y. Zhang (eds.), *Advances in Frontier Research on Engineering Structures II*,
Lecture Notes in Civil Engineering 535, https://doi.org/10.1007/978-981-97-6238-5_33

403

1 Introduction

With the progress of human civilization and the development of engineering technology, the safety requirements of building structures are becoming higher and higher. The service life of a structure is usually decades or even centuries, and it may be subject to different degrees of accidental action during its life cycle, such as earthquake, explosion, impact, etc. Since accidental actions are not thoroughly considered in the structural design, once the structure is subjected to incalculable accidental actions, local instability and failure will occur, which in turn will lead to changes in the internal forces of the entire structural system. Ellingwood [1] defines it as progressive collapse: partial structural damage due to an accidental event that causes a series of chain reactions, which leads to the spread of structural damage and eventually results in the entire structural system to collapse on a large scale.

Since the 1990s, experts and scholars in the field of engineering have done a lot of research on the resistance to progressive collapse. Among them, Xinzheng [2], Qingfeng [3], Pei [4–6] and others have conducted studies in terms of experiments and numerical simulations. Various codes, such as the Unified Facilities Criteria (UFC), recommend that special structural features be considered in the initial steps of structural design to reduce the possibility of progressive collapse of the structure. These codes mainly point out to design stronger structures to reduce the possibility of progressive collapse in the face of uncommon loading situations [7]. In recent years, nuclear power plant accidents caused by earthquakes have occurred frequently in Japan, such as Fukushima, Kashiwazaki-Kariwa, and Aomori nuclear power plants. Therefore, the safety assessment of nuclear power plant main plant structures and elements (SSCS) under over-design basis earthquakes has become an important research task to ensure the safe operation of nuclear power plant main plants under over-design basis earthquakes without progressive collapses [8, 9].

2 Theoretical Basis of Structural Resistance to Progressive Collapse

2.1 Load Combination Resistance Progressive Collapse

There are different national codes regarding the provisions for load resistance to progressive collapse. Some national codes introduce the effect of contingencies in addition to consideration of load combinations [10–12]. The more important code equivalent static load combinations are shown in Table 1.

Table 1 Equivalent static load combination

Specification name	Load combination value	Accidental load
British norm	$(1\pm0.5)\,DL + LL/3 + W/3$	34 kPa
European norm	$2D + 0.25L + 0.2W$	/
ACSE 7-98,02,05	$(0.9\,or\,1.2)DL + (0.5LL\,or\,0.2S) + 0.2W$ (Remove unit)	
	$1.2DL + A_k + (0.5LL\,or\,0.2S)$ (Local reinforcement)	A_k
	$(0.9\,or\,1.2)DL + A_k + 0.2W$ (Local reinforcement)	
DOD2010	$(0.9\,or\,1.2)DL + (0.5LL\,or\,0.2S) + 0.2W$ (Nonlinear dynamic analysis)	
	$2.0[(0.9\,or\,1.2)DL + (0.5LL\,or\,0.2S)] + 0.2W$ (Linear static analysis)	/
	$DL + 0.5\,LL$ (Overall strengthening)	
GSA 2003	$2(DL + 0.25\,LL)$ (Linear static analysis)	/
	$DL + 0.25\,LL$ (Dynamic analysis)	
China's current norms	$S_d = \eta(S_{GK} + \sum \psi_{qi} S_{Qi,k}) + \psi_w S_{wk}$	/

DL, LL, W, S: constant, live, wind and snow loads; Q_{ak}: accident eigenvalues; G_k, Q_k: constant and live load eigenvalues per unit area; ψ_{qi}, ψ_ω: Combination factor of variable and wind loads; A_k: Incidental load, η: Load subfactor, S_d: Design value of load effect combination; $S_{GK}, S_{Qi,k}, S_{\omega k}$: Standard values of permanent, variable, wind load effects

2.2 Pushdown Analysis Method

The pushdown analysis method is to change the horizontal load in the pushdown analysis to vertical load and transform it into vertical pushdown analysis. In the analysis process, the deformation criterion is used, and the vertical displacement of the upper node of the removed corner column is taken as the control option through displacement control, using the static load condition: 2 × (constant load + 0.25 × live load), loaded in 20 steps, with the beam and column hinges selected as P-M-M related hinges and the skeleton type selected as FEMA type as shown in Fig. 1. After the corner column is removed, the vertical load value in the region directly above the failed member is gradually increased until the target displacement or structural collapse is controlled. This approach allows for a more accurate determination of the load-carrying capacity limit for progressive collapse of the structure, and its analytical procedure allows for a more realistic simulation of the development of plastic hinges for each member of the structure.

Fig. 1 FEMA skeleton
curve

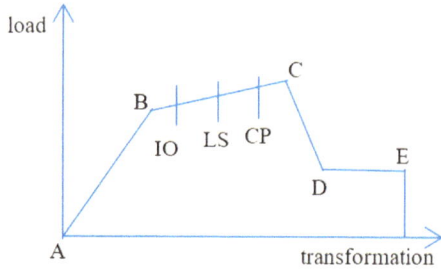

Table 2 Ductile fracture criterion

Component	Ductility	Turning angle (degrees)	Notes
Beam		6	
Tension column		6	
Compression column	1		
Frame		2	Maximum side H/25

The ductility in the table refers to the ratio of the maximum displacement to the elastic limit
displacement, and H represents the height of the structure

2.3 Criterion for Progressive Collapse

The material nonlinearity is taken into account in the static nonlinear analysis method,
so the ductile failure criterion is adopted in the member failure criterion. With reference to GSA2003 and other related data, the ductile failure criterion of concrete
structural members is shown in Table 2.

3 Finite Element Model of Conventional Island Main Powerhouse of Nuclear Power Plant

3.1 Model Size Parameters

The main plant of the plant's conventional island is a semi-basement reinforced
concrete frame structure system consisting of two parts: a steam engine room and an
auxiliary room. Among them, the steam engine room is a three-story frame structure,
with a plan size of 108 m × 44 m, the top elevation of the basement floor is -11.000 m,
the first floor ± 0.000 m, the second floor 8.500 m, the roof is a truss system, the lower
chord elevation is about 31.500 m, the plant is equipped with 2 sets of overhead cranes
and related equipment. The elevation of the floor is: basement roof -11.000 m, first
floor ± 0.000 m, second floor 8.500 m, third floor 11.700 m, fourth floor 20.500 m,
slope roof system, the lowest elevation is about 33.000 m. The main plant's columns

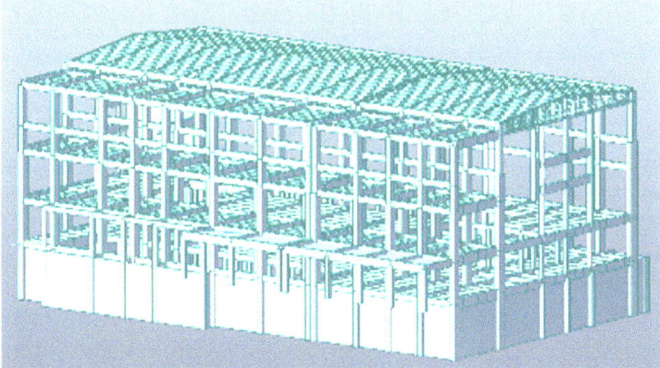

Fig. 2 The main factory building finite element model

are 8.000 m, 12.000 m and 10.500 m apart, and the site has an earthquake intensity of 7 degrees. The design seismic grouping is the third group, with a design fundamental seismic acceleration of 0.10 g and a Class I site category.

3.2 Finite Element Model

The finite element model of the main plant building was constructed by Midas Gen and is shown in Fig. 2. To accurately simulate the structural form of a regular island main factory building, beams and columns were modelled using beam elements; the basement walls and floors of the main factory building were modelled using plate elements; the roof of the main plant building was modelled on a truss element. As the foundation of the main factory building was below zero meters for the reinforced concrete basement, the floor was laid with a raft foundation and the consolidated form was adopted in the model. In the analysis of anti- progressive collapse, members of the upper layers are chosen as objects to study the anti-progressive collapse properties of the entire structure.

3.3 Load Combination

The floor load of the main plant was analyzed for progressive collapse resistance with reference to the US GSA2003 code load combination, and the load of the main plant was reduced to constant load and uniform live load. The structural system studied in this paper is a multi-story structure, and the influence of wind load on the structure is relatively slight, and the load combination does not consider the influence of wind load.

4 Analysis of Structural Resistance to Sustained Collapse After Corner-Post Removal

The corner column of the main plant structure is removed, as shown in Fig. 3, the horizontal beam connected to the corner column is defined as L1, the vertical beam is L2, and the beam I end and J end are defined at the corner column node, then the 252 nodes where the L1 and L2 nodes intersect are the control points for displacement control. Based on the Pushdown analysis step, the curve of the base reaction force versus the displacement of node 252 is shown in Fig. 4, and the development of the plastic hinge is shown in Fig. 5.

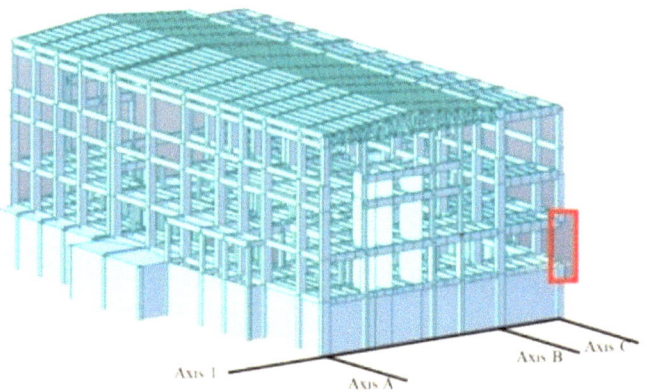

Fig. 3 Removal of corner post models

Fig. 4 Displacement-base reaction curve after the removal of corner 252 node

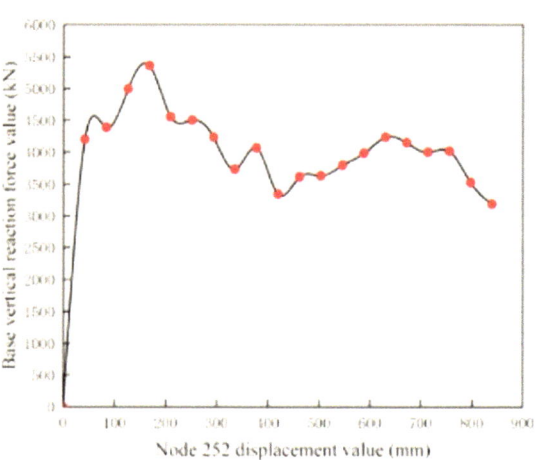

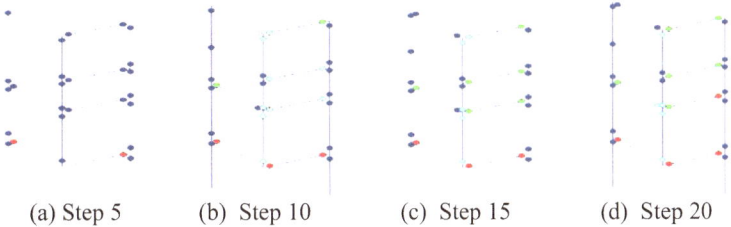

        (a) Step 5   (b) Step 10   (c) Step 15   (d) Step 20

Fig. 5 Pushdown hinge development results the demolition of corner

It can be seen from Fig. 4 that with the gradual increase of vertical load, the structural system undergoes the development process from elastic stage to plastic stage to collapse stage. When the load increases, the value of the base reaction force and the vertical displacement of the control nodes show a curve that rises and then falls. When the fourth step is reached, the vertical reaction force of the base reaches the maximum value, which is 5346 kN. At this time, the vertical load value of the member is the ultimate load carrying capacity of the structural member, and the displacement value of joint 252 is 168 mm. Therefore, the displacement of 168 mm under the fourth step load is the ultimate displacement of the progressive collapse of the structure. The ultimate load corresponding to this displacement is the ultimate load.

After the corner posts are removed, including results for the hinge at steps 5, 10, 15, and 20. By comparing the results of L1 hinges of horizontal beams and L2 hinges of vertical beams, it can be seen that when the vertical load reaches step 5, seven plastic hinges appear in each layer beam L1, and the J end of the first layer L1 has entered the failure stage, the plastic hinges at both ends of the second, third and top floor beams L1 are in the B-IO stage, two plastic hinges appear in each layer beam L2, and the J end of the first layer L2 has entered the failure stage. The plastic hinge at the L2J end of the two-level beam is in the B-IO phase. In step 10, there are eight plastic hinges in each layer beam L1, and the I and J ends of the first layer L1 have entered the failure stage, the two ends of the second and third story beam L1 and the J end of the top beam L1 are in the IO-LS stage, the I ends of the top beam L1 are in the LS-CP stage, and there are a total of three plastic hinges in each layer beam L2, and the J end of the first layer L2 has entered the failure stage, and the second layer I end plastic hinge is in the IO-LS stage. The plastic hinge at the end of the second layer, J, is in the LS-CP phase. In the 15th step, the plastic hinges of the two ends of the second and third story beam L1 and the J end of the top beam L1 are in the LS-CP stage, the plastic hinges of the I and J ends of the top beam L2 are in the IO-LS stage, and the plastic hinges of the J end of the second layer are in the B-IO stage. As a result, it can be seen that the effect of removing the angle C/1 on the C-axis is larger than that on the 1-axis. The reason is that the span of the horizontally oriented beam is smaller than that of the vertically oriented beam, and the load is shifted more along the direction of the smaller span.

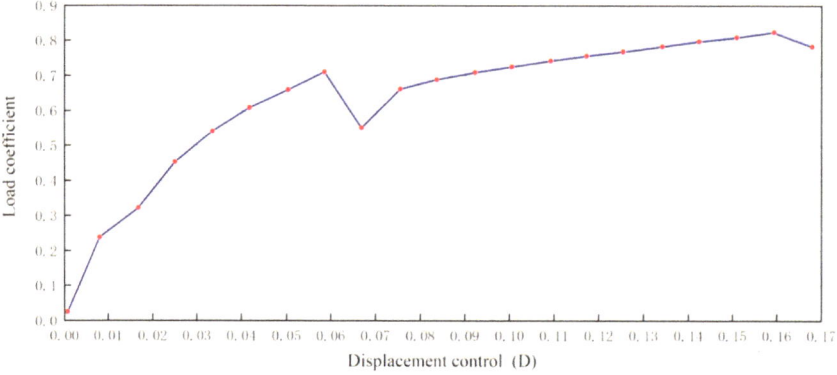

Fig. 6 Displacement—base reaction curve after the demolition of the corner 252 node

When the structural displacement reaches the limit displacement, then the vertical load value is the limit load of the structure to resist progressive collapse. Through the curve of the relationship between the base reaction force and the vertical displacement of the 252nd node in Fig. 4, it is obtained that after the ultimate displacement of the structure is 168 mm, the vertical load of the structure is gradually increased from zero. The curve of the relationship between the vertical load coefficient and the control displacement of the 252nd node when the corner column is removed is shown in Fig. 6.

In Fig. 6, the X-axis represents the vertical displacement of the 252 node, and the Y-axis represents the ratio of the load corresponding to the displacement value of Node 252 to the total vertical load. It can be seen from Fig. 6 that after the corner column is removed, when the vertical load acting on the structure is $78.3\% \times 2(DL + 0.25LL)$, the vertical displacement of the 252nd node reaches the ultimate load, which is 168 mm. Since then, with the increase of the vertical load, the displacement of the superstructure connected to the control node and the failure column will diverge, and the failure range will spread to the top of the structure. The structure cannot bear the vertical load of $2 \times (DL + 0.25LL)$. This shows that in the case of corner column failure, the structure does not meet the requirements of progressive collapse resistance.

5 Conclusion

A nonlinear static analysis of the main plant structure of the conventional island of the nuclear power plant was carried out using the analysis method of Pushdown. The results of plastic hinge and load factor-displacement curves were obtained by removing the corner columns. Finally, the ability of the main plant structure to resist progressive collapse was evaluated. In summary, the following conclusions can be drawn:

(1) The final displacement of the structure against progressive collapse can be determined by a Push-down analysis, and then the final load of the structure against progressive collapse can be determined exactly;

(2) After removal of the corner columns, with the increase of vertical load, the structural displacement of the beams connected to the control nodes and the upper floors within the failed columns will disperse and the damage will spread to the top of the structure;

(3) It can be seen from the order of appearance of plastic hinges that the structural system of the main plant conforms to the principle of "strong columns and weak beams", and the beams of small span are easy to form plastic hinges and may be damaged before the beams of large span; Therefore, during the actual design process, engineers should pay extra attention to frame beams with small span.

Acknowledgements This research was supported by Dalian University Research Platform Project (grant no. 202301ZD01), the National Natural Science Foundation of China (grant no.51878108) and the Department of Science and Technology Guidance Plan Foundation of Liaoning Province (grant no. 2019JH8/10100091).

References

1. Ellingwood, B. R. (2006). Mitigating risk from abnormal loads and progressive collapse. *Journal of Performance of Constructed Facilities, 20*(4), 315–323.
2. Yi, L., & Xinzheng, L., et al. (2007). Design method to resist progressive collapse for a three story RC frame. *Journal of PLA University of Technology (Natural Science Edition), 8*(6), 659–664.
3. Qingfeng, H., & Weijian, Y. (2011). Experimental study of the collapse resistant behavior of RC beam-column sub-structures considering catenary action. *Journal of Civil Engineering, 44*(4), 52–59.
4. Pei, Q., Qi, P., Ma, F., et al. (2023). Resistance of gable structure of nuclear island to progressive collapse in conventional island shield building of nuclear power plants. *Buildings, 13*(1257), 1257.
5. Pei, Q., & Cai, B., et al. (2023). The progressive collapse resistance mechanism of conventional island main buildings in nuclear power plants. *Buildings, 13*(04).
6. Xue, Z., Yan, P. Q. Z., Pei, Q., et al. (2023). Mechanical properties and crack resistance of basalt fiber self-compacting high strength concrete: an experimental study. *Materials, 16*(12), 4374.
7. Jian, Z., Jiachun, C., et al. (2015). Comparative study on design codes and methods for structural resistance of progressive collapse. *Building Structure, 45*(23), 98–105.
8. Liang, P., Boquan, L., et al. (2015). Analysis of progressive collapse resistant behavior of RC frame structure after removal of key column of ground floor. *World Earthquake Engineering, 31*(04), 170–180.
9. Yuzhe, X., & Yi, L., et al. (2019). Concrete beam-column sub-structure dynamic effect of experimental study on the progressive collapse. *Engineering Mechanics, 36*(5), 44–52.

10. Liping, X., & Haifeng, Y. (2018). Progressive collapse analysis of steel frame under bottom implosion. *Structural Engineer, 34*(5), 66–73.
11. Wei, X., Lai, W., et al. (2020). Theoretical and experimental study on progressive collapse resistance of concrete filled steel tubular column-composite beam fram under failure condition of middle column. *Journal of Vibration and Shock, 39*(3), 76–87.
12. Lide, S., Jianshuai, H., et al. (2021). Progressive collapse analysis of frame supported shear wall structure with beam transfer story. *Science Technology and Engineering, 21*(14), 5898–5906.

Investigation on Internal Force Change of Continuous Steel-Box Girder Bridge During Construction Based on Support Position Optimization

Peng Du, Man Liu, Yongqing Ma, Jun Tang, Haonan Jiang, and Qingfei Gao

Abstract During the installation and removal of the support of continuous steel box girder bridge, the internal force of the main girder will change, and the range of internal force change is closely related to the position of the support. However, at present, the position of bridge support is mainly determined by experience, which can not effectively reduce the internal force change of main girder during construction and affect the safety of bridge construction. Based on the matrix displacement method, this paper theoretically analyzes the internal force changes of the main girder with or without support, and puts forward the method to determine the optimal position of the support of the continuous steel box girder bridge. According to the theoretical method, the support position is determined and the finite element model is established. The internal force changes of the key sections of the bridge are calculated by numerical simulation and theoretical analysis respectively. By comparing the calculation results and combining the bending moment diagram of the main girder, it can be concluded that the calculation error of the two methods is less than 6%, which shows that the theoretical method has high accuracy and engineering applicability; The optimal position of the support determined by theoretical method is located near the transition point of positive and negative bending moments. Based on this rule, the position of the support can be preliminarily judged in engineering to improve the efficiency and safety of construction.

Keywords Support · Steel box girder · Continuous beam bridge · Bridge construction

P. Du · M. Liu · Y. Ma · J. Tang
China Railway No.5 Engineering Group Co., Ltd. Urban Rail Transit Engineering Branch, Changsha 410001, Hunan, China

H. Jiang · Q. Gao (✉)
School of Transportation Science and Engineering, Harbin Institute of Technology, Harbin, Heilongjiang 150001, China
e-mail: gaoqingfei@hit.edu.cn

1 Introduction

Currently, common construction methods for steel structure bridges include pushing construction and hoisting construction [1, 2]. Compared to traditional full-span scaffolding, these methods are more efficient and precise. However, this does not imply that construction scaffolding is unnecessary during the construction process. The appropriate form and location of scaffolding can effectively enhance overall construction efficiency, whereas incorrect placement can potentially jeopardize construction safety [3].

When determining the optimal placement of scaffolding, it is crucial to consider the changes in internal forces in the main beams before and after the removal of scaffolding [4, 5]. On one hand, the removal of scaffolding will increase the degrees of freedom of continuous steel box girders [6]. On the other hand, removing scaffolding may lead to a redistribution of internal forces [7]. Therefore, prior to erecting construction scaffolding, it is necessary to identify suitable positions and conduct a detailed structural analysis and assessment of the changes in internal forces before and after scaffolding installation and removal to ensure the bridge can safely withstand the new force distribution [8].

Some scholars have summarized key points in bridge construction scaffolding design and construction processes. Based on the actual engineering context, Guang Ming has provided detailed explanations regarding the installation and removal process of scaffolding for construction with fewer supports [9]. Shang Jian used large-scale finite element software to establish an overall finite element model for a certain steel bridge and conducted a simulated study on the dismantling process of temporary support systems, selecting the optimal removal approach [10]. An Shao Bo conducted theoretical research on the influence of scaffold top modeling elevation and scaffold removal sequence on main beam internal forces based on main beam alignment control [11]. Although the existing research has analyzed the influence of support on the internal force change of main girder, no scholars have studied the influence of support position on the internal force during bridge construction. In engineering practice, the support position is determined according to experience on the principle of facilitating the main girder construction. In this paper, utilizing the matrix displacement method, we establish a mechanical model for a three-span continuous beam, derive the equation for calculating negative bending moments at the top of bridge piers in the absence of scaffolding, and then consider the supporting effect of bridge scaffolding, proposing a method for calculating main beam internal forces under scaffold support conditions. Furthermore, to enhance the applicability of the above method, we conduct a theoretical analysis of the variations in internal forces for continuous beams with unequal spans. Finally, through numerical simulation, we compare the theoretical calculation results with finite element simulation results, validating the accuracy of the theoretical analysis and providing a reference method for determining the optimal placement of scaffolding in practical bridge construction processes. Compared with the empirical method, the support position determination method proposed in this paper has scientific theoretical support, and simplifies the internal force calculation process on the basis of matrix displacement method, which not only guarantees the safety of bridge construction from the perspective of construction convenience, but also has higher reliability and accuracy.

2 Methods

To solve the internal force distribution in continuous beams using the matrix displacement method, it is necessary to first divide the continuous beam into several finite elements by discretizing the beam into a series of small segments or nodes. The stiffness matrices for each element are assembled to form the stiffness matrix of the entire structure, establishing the global stiffness matrix in the global coordinate system. Considering support conditions and boundary conditions, the displacements of each node are determined by solving the linear algebraic system of equations for the displacement matrix. The internal force distribution within each element is then calculated using the elemental displacement-force relationships and the global stiffness matrix.

Taking the continuous beam with equal span and three spans as an example, the specific process of internal force solution is deduced. In Fig. 1, the continuous beam is discretized into three segments by introducing four nodes labelled A, B, C, and D. The global coordinate system aligns with the element coordinate system, with the x-axis oriented to the right along the bridge span as the positive direction and the y-axis oriented downward as the positive direction. Within the global coordinate system, F represents the applied loads at the nodes, and Δ represents the angular displacements at each node. In the element coordinate system, i signifies the axial stiffness of each segment, P denotes the fixed-end forces for each segment, and δ represents the displacement at the segment's end.

In the element coordinate system, the nodal angular displacements are taken as unknown variables, and by considering the physical relationship between fixed-end forces and end displacements, Eqs. (1)–(3) can be formulated.

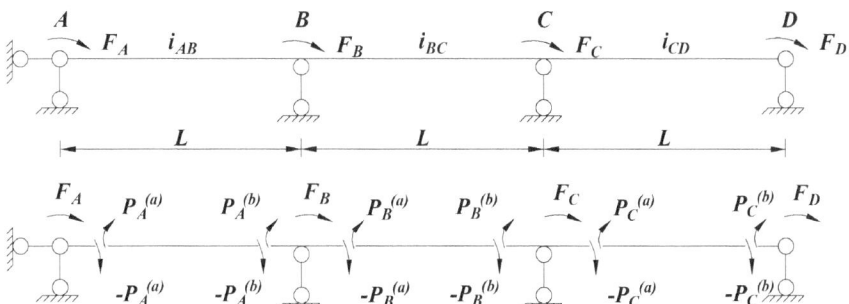

Fig. 1 Simplified mechanical model of three-span continuous beam

$$P_A^{(a)} = 4i_{AB}\delta_A + 2i_{AB}\delta_B$$
$$P_A^{(b)} = 2i_{AB}\delta_A + 4i_{AB}\delta_B$$
(1)

$$P_B^{(a)} = 4i_{BC}\delta_B + 2i_{BC}\delta_C$$
$$P_B^{(b)} = 2i_{BC}\delta_B + 4i_{BC}\delta_C$$
(2)

$$P_C^{(a)} = 4i_{CD}\delta_C + 2i_{CD}\delta_D$$
$$P_C^{(b)} = 2i_{CD}\delta_C + 4i_{CD}\delta_D$$
(3)

In accordance with Eqs. (1)–(3), the elemental stiffness matrices for the AB, BC, and CD beam segments can be derived, as illustrated in Eq. (4).

$$k_e = \begin{bmatrix} \frac{4EI}{L} & \frac{2EI}{L} \\ \frac{2EI}{L} & \frac{4EI}{L} \end{bmatrix}$$
(4)

In Eq. (4), E denotes the elastic modulus of the material, I represents the moment of inertia of the continuous beam's cross-section, and L stands for the length of each span. As for nodes A, B, C, and D, the respective moments at these nodes can be expressed according to Eq. (5), which is derived from the equilibrium conditions at each specific point.

$$F_A = P_A^{(a)} = 4i_{AB}\delta_A + 2i_{AB}\delta_B$$
$$F_B = P_A^{(b)} + P_B^{(a)} = 2i_{AB}\delta_A + 4i_{AB}\delta_B + 4i_{BC}\delta_B + 2i_{BC}\delta_C$$
$$F_C = P_B^{(b)} + P_C^{(a)} = 2i_{BC}\delta_B + 4i_{BC}\delta_C + 4i_{CD}\delta_C + 2i_{CD}\delta_D$$
$$F_D = P_C^{(b)} = 2i_{CD}\delta_C + 4i_{CD}\delta_D$$
(5)

Combined with Eq. (4) and Eq. (5), the total stiffness matrix of three-span continuous beam can be further given, as shown in Eq. (6).

$$K = \begin{bmatrix} k_{11}^1 & k_{12}^1 & 0 & 0 \\ k_{21}^1 & k_{22}^1 + k_{22}^2 & k_{23}^2 & 0 \\ 0 & k_{32}^2 & k_{33}^2 + k_{33}^3 & k_{34}^3 \\ 0 & 0 & k_{43}^3 & k_{44}^3 \end{bmatrix} = \begin{bmatrix} \frac{4EI}{L} & \frac{2EI}{L} & 0 & 0 \\ \frac{2EI}{L} & \frac{8EI}{L} & \frac{2EI}{L} & 0 \\ 0 & \frac{2EI}{L} & \frac{8EI}{L} & \frac{2EI}{L} \\ 0 & 0 & \frac{2EI}{L} & \frac{4EI}{L} \end{bmatrix}$$
(6)

By discretizing each element and solving for the fixed-end forces $P_A^{(a)}$, $P_A^{(b)}$, $P_B^{(a)}$, $P_B^{(b)}$, $P_C^{(a)}$, and $P_C^{(b)}$ in the element coordinate system, combined with Eq. (5), we can determine the loads at various points. These loads can then be expressed in matrix form through Eq. (7).

$$F = \begin{bmatrix} \frac{qL^2}{12} & 0 & 0 & -\frac{qL^2}{12} \end{bmatrix}^T$$
(7)

By establishing the comprehensive stiffness equation for the continuous beam and incorporating the expressions for node loads, we arrive at the nodal displacement matrix, as exemplified in Eq. (8).

$$
\Delta = \begin{pmatrix} \Delta_A \\ \Delta_B \\ \Delta_C \\ \Delta_D \end{pmatrix} = \begin{bmatrix} \frac{4EI}{L} & \frac{2EI}{L} & 0 & 0 \\ \frac{2EI}{L} & \frac{8EI}{L} & \frac{2EI}{L} & 0 \\ 0 & \frac{2EI}{L} & \frac{8EI}{L} & \frac{2EI}{L} \\ 0 & 0 & \frac{2EI}{L} & \frac{4EI}{L} \end{bmatrix}^{-1} \begin{bmatrix} \frac{ql^2}{12} \\ 0 \\ 0 \\ -\frac{ql^2}{12} \end{bmatrix} = \frac{qL^3}{EI} \begin{bmatrix} \frac{1}{40} \\ -\frac{1}{120} \\ \frac{1}{120} \\ -\frac{1}{40} \end{bmatrix}
\tag{8}
$$

Substituting Δ_A, Δ_B, Δ_C, and Δ_D into Eq. (9) enables us to calculate the internal forces within each element. In this equation, the left-hand side represents the element's end forces, while the right-hand side comprises the fixed-end forces and the equivalent loads due to node displacements. The equivalent loads can be regarded as the forces induced at that node due to displacements.

$$
\overline{F}^e = \begin{bmatrix} F \\ F' \end{bmatrix} + k_e = \begin{bmatrix} K_{ii} & K_{i(i+1)} \\ K_{(i+1)i} & K_{\Delta i+1)(i+1)} \end{bmatrix} \begin{bmatrix} \Delta \\ \Delta' \end{bmatrix}
\tag{9}
$$

To simplify calculations, this paper assumes that the impact of supports on the bridge is equivalent to providing vertical support. Consequently, a three-span continuous beam under the influence of support is treated as a four-span continuous beam subjected to uniform distributed loads. This is similar to truncating the central spans of the three-span continuous beam and placing supports accordingly. As a result, the sum of the central two spans, denoted as L_1 and L_2, should be equal to the central span of the three-span continuous beam. In alignment with the characteristics of a four-span continuous beam, and following a similar derivation process as previously described, the total stiffness matrix for the bridge structure is presented in Eq. (10).

$$
K = \begin{bmatrix} k_{11}^1 & k_{12}^1 & 0 & 0 & 0 \\ k_{21}^1 & k_{22}^1+k_{22}^2 & k_{23}^2 & 0 & 0 \\ 0 & k_{32}^2 & k_{33}^2+k_{33}^3 & k_{34}^3 & 0 \\ 0 & 0 & k_{43}^3 & k_{44}^3+k_{44}^4 & k_{45}^4 \\ 0 & 0 & 0 & k_{54}^4 & k_{55}^4 \end{bmatrix} = \begin{bmatrix} \frac{4EI}{L} & \frac{2EI}{L} & 0 & 0 & 0 \\ \frac{2EI}{L} & \frac{4EI}{L}+\frac{4EI}{L_1} & \frac{2EI}{L_1} & 0 & 0 \\ 0 & \frac{2EI}{L_1} & \frac{4EI}{L_1}+\frac{4EI}{L_2} & \frac{2EI}{L_2} & 0 \\ 0 & 0 & \frac{2EI}{L_2} & \frac{4EI}{L_2}+\frac{4EI}{L} & \frac{2EI}{L} \\ 0 & 0 & 0 & \frac{2EI}{L} & \frac{4EI}{L} \end{bmatrix}
\tag{10}
$$

According to the stiffness equations, in order to determine the displacements at each node, and building upon the acquired stiffness coefficients for the entire bridge structure, we must solve for the fixed-end forces of each individual straight member. Drawing parallels with the derivation process conducted without supports, the fixed-end force matrix is expressed as depicted in Eq. (11).

$$
F = \begin{bmatrix} -\frac{qL^2}{12} & \frac{qL^2}{12} & -\frac{qL_1^2}{12} & \frac{qL_1^2}{12} & -\frac{qL_2^2}{12} & \frac{qL_2^2}{12} & -\frac{qL^2}{12} & \frac{qL^2}{12} \end{bmatrix}^T
\tag{11}
$$

After acquiring the stiffness matrix and the vector of fixed-end forces, the displacement values at each node can be determined by solving the structural equations of the bridge through matrix operations. Subsequently, by substituting the obtained values of Δ_A, Δ_B, Δ_C, Δ_D and Δ_E into Eq. (10) for calculating the axial forces at each node, precise numerical values for the axial forces can be ascertained. In the case of non-uniform continuous beams, a similar derivation process to that of uniform continuous beams can be applied, involving the division of the span L into end-span

length $L' \begin{pmatrix} a_{11} & a_{12} & a_{13} \\ a_{21} & a_{22} & a_{23} \\ a_{31} & a_{32} & a_{33} \end{pmatrix}$ and mid-span length L'', and the corresponding substitu-

tion of span parameters to analyse the internal forces within the bridge. Through the integration of numerical analysis methods, multiple interpolation calculations on L_1 can be performed to identify the optimal support locations that minimize variations in internal forces during the bridge construction project.

3 Applications

3.1 Establishment of Finite Element Model

In this study, finite element analysis was conducted using Midas Civil software to create a model of the main girder for a three-span continuous steel box girder bridge. The main girder has a twin-box single-cell design with dimensions of 16.5 m in width, 2.0 m in height, and 2.0 m in flange width. For the uniform-span continuous bridge, the spans are 3×50 m, while for the non-uniform continuous span bridge, the spans are arranged as $45 + 60 + 45$ m. The main girder was divided into a total of 322 elements. Regarding boundary conditions, supports were simulated based on standard continuous girder layout principles, with four general supports placed at the bearing locations and modelled using elastic connections.

3.2 Comparison of Results

Based on theoretical analysis results, the deployment of supports was simulated at the optimal positions, and the values of the negative bending moments at the pier caps were recorded. The boundary conditions at the support locations were then removed to simulate the dismantling process of the supports, while recording the changes in internal forces within the main girder during the removal process. The optimal support positions were determined through theoretical methods, with L' values of 12 m and 9.75 m. The negative bending moments at the pier caps with support conditions were -8758.6 kN·m and -7465.3 kN·m, respectively, while without support conditions, they were $-10,328.4$ kN m and $-11,435.6$ kN m. In

Fig. 2 Bending moment of left pier top under the support of continuous beam with equal span

Fig. 3 Bending moment of left pier top of continuous beam with equal span

Fig. 4 Bending moment of left pier top under the support of continuous beam with unequal span

Fig. 5 Bending moment of left pier top of continuous beam with unequal span

the finite element model, simulations were conducted according to the theoretically determined support positions, and the simulation results are shown in Figs. 2, 3, 4 and 5.

For the uniform continuous beam model, the values of the left pier cap's negative bending moment with and without support conditions were -8468.7 kN m and -9709.0 kN m, respectively. For the non-uniform continuous beam model, the corresponding values were -7284.9 kN m and $-10,780.7$ kN m. Comparatively, the error in results for uniform and non-uniform continuous beams with support conditions was 5.47% and 4.96%, respectively, and without support conditions, it was 6.00% and 5.73%, respectively. All errors were within 6%, indicating that the method for determining the optimal support placement has a high level of accuracy. Further, observing the bending moment diagram of the main girder without supports, we can find that the support positions determined by theoretical methods are all located near the conversion point of positive and negative bending moments.

4 Conclusions

Through a combination of theoretical analysis and finite element verification, this study systematically investigates the variations in internal forces of continuous steel box girder bridges during the support construction process. The following conclusions can be drawn: During the period from the completion of support installation to dismantling, the bridge can be considered to have a constant overall length, an increasing number of spans, and a decreasing negative bending moment at the pier caps. The placement of supports significantly influences the magnitude of internal force variations before and after the bridge construction process. Given the known basic parameters of the bridge, an optimal support position can be determined by employing the matrix displacement method and numerical analysis techniques. The results obtained through this method align closely with the outcomes of numerical simulations. For the three-span continuous beam with equal span and unequal span, the support position determined by theoretical method is near the transformation point of positive and negative bending moments of the main beam.

References

1. Rao, C. H., Huang, M. J., & Zhao, T. F. (2023). Research on decision-making of steel box girder construction scheme based on IAHP and TOPSIS. *Journal of Nonlinear and Convex Analysis., 24*, 1651–1661.
2. Wang, J. F., Wu, T. M., Zhang, J. T., Xiang, H. W., & Xu, R. Q. (2020). Refined analysis and construction parameter calculation for full-span erection of the continuous steel box girder bridge with long cantilevers. *Journal of Zhejiang University-Science A., 21*, 268–279.
3. Ma, F. B., Wang, H. C., Chen, X. X., Feng, D. M., Wu, G., & Hou, S. T. (2015). Experimental investigation and application evaluation case of an adjustable height temporary support for bearing replacement in large-tonnage HSR bridges. *Journal of Bridge Engineering, 44*, 123–125+129.
4. Li, Y. (2022). Jacking and dismantlement techniques for steel pipe column bracket of bridge pier. *Architecture Technology., 53*, 320–323.
5. Chen, Y.Q. (2019) Key technologies for support removal of offshore super-heavy steel box girder. Urban Roads Bridges and Flood Control, 06:205–208+215+23–24.
6. Zhou, W. M. (2019). Construction technology study on lifting and floating removal of large-span bailey beam support. *Railway Construction Technology., 10*, 77–81.
7. Shen, L. M. (2019). Construction technology of overall dismantling the bracket of cast-in-place beams. *Journal of Municipal Technology., 36*, 67–69.
8. Lozano-Galant, J. A., Payá-Zaforteza, I., & Turmo, J. (2015). Effects in service of the staggered construction of cable-stayed bridges built on temporary supports. *Baltic Journal of Road and Bridge Engineering., 10*, 247–254.
9. Guang, M., Yao, F., & Wang, J. H. (2015). Application of scaffolds with less steel tube for cast-in -situ box Girder with high Piers in mountain area. *Construction Technology, 44*, 123–125+129.
10. Shang, J., & Feng, H. B. (2015). Simulation Study of removing temporary support of a steel Bridge. *Steel Construction, 30*, 89–91+79.
11. An, S. B. (2013). Removal of continuous beam support and linear control of main beam based on internal force control of main beam. *Journal of Shijiazhuang Tiedao University (Natural Science), 26*, 1–3.

Modelling the Mechanical Behaviour of the RC Pipe Pile with Construction-Induced Damage and Cracks

Weixin Yu and Xiaohui Wang

Abstract During the construction of the high-piled wharf, cracks may be induced in the top zone of the pre-stressed high-strength concrete pipe pile due to the high-frequency hitting. In this paper, the influence of the construction-induced damage and cracks on the mechanical properties of prestressed reinforced concrete (RC) pipe piles is studied. Finite element analysis is used to simulate the mechanical behaviour of the RC pipe piles with construction-induced damage and the corresponding ultimate bearing capacity of the damaged RC piles is predicted. Influence of different lengths of the construction-induced damage zone and maximum crack widths within the damage zone on the mechanical behaviour of the RC pipe pile is discussed. Results indicate that, the ultimate bearing capacity of the RC pipe piles with construction-induced damage in the top zone of the pile is reduced due to the damage. The failure of the RC pipe piles with construction-induced damage may occur at both the bottom and middle part of the pile for longer damage length and wider maximum crack width.

Keywords RC pipe pile · High-frequency hitting-induced damage and cracks · Finite element method · Mechanical behaviour · Ultimate bearing capacity

1 Introduction

Pre-stressed high-strength concrete pipe pile (PHC pipe piles) are widely used in the high-pile wharf due to their good mechanical properties. The pipe piles are usually constructed by the hammering method. However, this method may cause some damage in the top zone of the pipe piles. In-situ checks on the pipe piles of a general cargo wharf in Suzhou Port showed that, longitudinal cracks on the concrete surface were observed in the top zone of the pipe piles, where the maximum

W. Yu · X. Wang (✉)

College of Ocean Science and Engineering, Shanghai Maritime University, Shanghai 201306, China

e-mail: w_xiaoh@163.com; xiaohwang@shmtu.edu.cn

© The Author(s) 2025

D. Li and Y. Zhang (eds.), *Advances in Frontier Research on Engineering Structures II*, Lecture Notes in Civil Engineering 535, https://doi.org/10.1007/978-981-97-6238-5_35

crack width reached 0.40 mm [1]. How this construction-induced damage affects the mechanical properties of prestressed reinforced concrete (RC) pipe piles?

In the past years, the effect of the early-aged cracks due to hydration heat and drying shrinkage on the chloride penetration of the sound and cracked concrete of wharf structures was studied [2]. For the RC pipe pile with shrinkage-induced cracks, its service life was predicted by proposed equivalent diffusion coefficient and the Fick's second law of diffusion [3, 4]. For the RC pipe pile with original incomplete micro-cracks, theoretical approach for prediction of its service life was also proposed [5]. Compared with the influence of the cracks and crack width on the behaviour of the concrete, the crack depth had a more significant effect on chloride ion permeability [6].

In addition, mechanical performance of the RC pipe piles was also evaluated[7, 8]. The time-dependent lateral bearing behaviour of corrosion-damaged RC pipe piles was investigated by the finite difference method [8] while the fatigue performance of the cylinder full-scale RC pipe pile was studied by Liu et al. [8]. In-situ test and field test are also carried out on the pre-stressed high-strength concrete pipe pile embedded in saturated sandy soil [9] and extra-long high-strength concrete pipe piles [10], respectively, to evaluate their bearing capacity.

It can be seen from the above literatures that no research focuses on effect of the construction-induced damage and cracks on the mechanical properties of prestressed RC pipe piles. In most cases, the load-induced cracks on the behaviour of the RC elements were mainly studied. In this paper, the mechanical behaviour of the RC pipe piles with construction-induced damage is simulated by modeling the degradation of the compressive strength of concrete and the corresponding ultimate bearing capacity is predicted. Influence of different damage lengths and maximum crack widths on the mechanical behaviour and ultimate bearing capacity of the damaged RC pipe pile is discussed.

2 Finite Element Modeling of the RC Pipe Pile with Construction-Induced Damage

2.1 Modelling of the Steel bar

Numerical simulation of RC pipe pile is carried out by using Abaqus software. The stress–strain relationship of the reinforcement under monotonic loading in the Chinses code [11] is adopted.

2.2 Modelling of the Sound Concrete

For concrete materials of the RC pipe pile, the plasticity model in Abaqus is used to describe the compressive and tensile behavior of the concrete [11].

2.3 Modelling the Concrete with Construction-Induced Damage and Cracks

Nakamura et al. [12] and Miura et al. [13] experimentally investigated the influence of cracks on the compressive strength of the concrete. It was concluded that the compressive strength of the longitudinally cracked concrete is linearly related to the crack width [12]. Based on the previous studies, Miura et al. [13] proposed the following equations to consider the effect of crack width on the compressive strength of the longitudinally cracked concrete.

$$\begin{cases} \sigma_c/f_c' = 1 - 0.6 w_{max} & w_{max} < w_{max\,c} \\ \sigma_c/f_c' = 1 - 0.6 w_{max\,c} & w_{max} \geq w_{max\,c} \end{cases} \tag{1}$$

$$w_{max\,c} = G_{max}/40 \tag{2}$$

where $w_{max\,c}$ is the threshold value of maximum average crack width (mm); w_{max} is the maximum average crack width (mm); G_{max} is the maximum coarse aggregate particle size; σ_c is compressive strength of the longitudinally cracked concrete; f_c' is the average compressive strength of sound and uncracked concrete specimens (cylindrical specimens $\Phi 100 \times 200$ mm).

2.4 Equivalent Length of RC Pipe Pile

For an elastic long pile used in the high-pile wharf [14], the depth of the bending embedded point of the pile is determined firstly by the hypothetical embedded point method as follows

$$t = \eta T \tag{3}$$

$$T = \sqrt[5]{\frac{E_p I_p}{m \cdot b_0}} \tag{4}$$

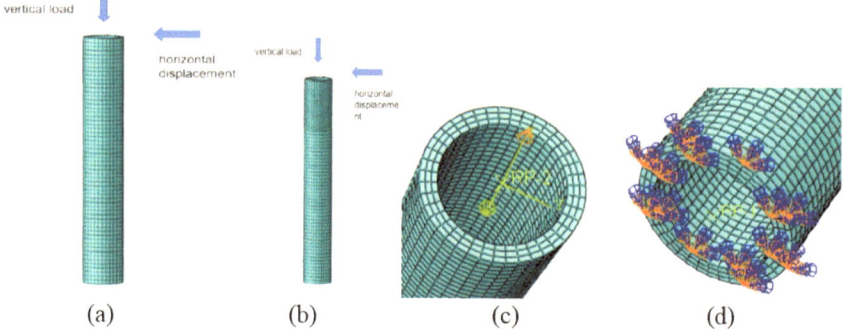

Fig. 1 Schematic diagram of pile boundary conditions and the damaged length of the RC pipe pile: **a** undamaged and sound pile; **b** pile with construction-induced damage and cracks in the top zone; **c** loading conditions; **d** boundary conditions

When the under-mud depth t of the bending embedment point from the mud surface is determined, the equivalent pile length of the pile is the pile length above the mud surface plus t.

2.5 Setting of Loading and Boundary Conditions

End-bearing piles are selected for this modeling without considering the interaction between soil and tubular piles. In the finite element modeling, the RC pipe pile is assumed to be completely fixed in the soil and the top of it is free, see Fig. 1. Applying axial force at the top of the pile and remains constant during the loading process. Then a horizontal displacement load is applied on the top of the pile until failure.

2.6 Model Validation

A sound and uncracked RC pipe pile was designed to validate the finite element model. A C105 concrete pipe pile was selected [15]. The pipe pile is 1 m long with a diameter of 400 mm. The concrete is C105. Good agreement is shown in the calculated and predicted results. As shown in Fig. 2, the trend of the load–displacement curves of the finite element simulation and test basically coincide, and it can be considered that the results of the finite element simulation basically coincide with the actual situation, and the finite element model in this paper is reliable.

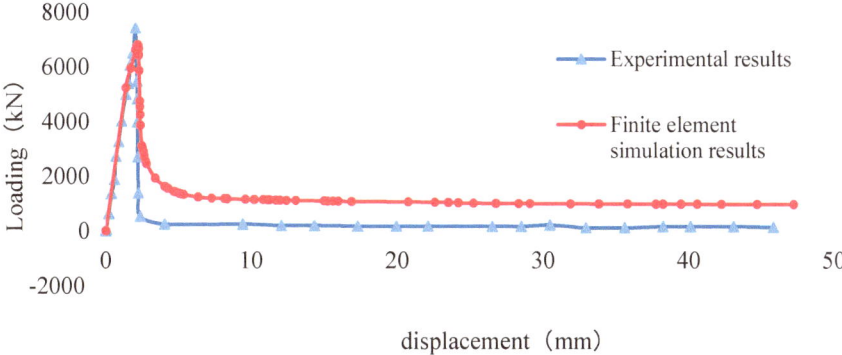

Fig. 2 Comparison of load–displacement curves of the experimental and modelling results

3   Parametric Discussion

3.1   *Geometry Shape and Material Properties of the RC Pipe Piles Model*

The cylindrical RC pipe pile is selected from a wharf in Suzhou port. The length of the pile is 23 m while the outer and inner diameters of the pipe pile are 1000 mm and 800 mm, respectively. The concrete strength grade of the RC pipe pile is C80. 8 pieces of 12 mm diameter HRB400 steel bars were used as longitudinal bars and the HRB400 steel bars with a diameter of 8 mm was selected for stirrups. The clear concrete cover of the stirrups is 46 mm [1]. In the top zone of the RC piles, visible longitudinal cracks along the pile lengths were observed [1], where the crack lengths ranged from 0.40 m to 2.20 m, the maximum crack widths ranged from 0.04 mm to 0.40 mm, and the distances from the top of the pile is 0 ~ 1.82 m.

The total length of the pipe pile is 23 m. Equations (3–4) are used to determine the under-mud depth t of the bending embedment point from the mud surface and the calculated $t = 2.7$ m. The equivalent length of the RC pipe pile is 6 m. The compressive strength of concrete in the length with construction-induced damage and cracks was predicted by Eqs. (1–2) to consider the main effect of cracks on concrete strength. For RC pipe pile with cracks, single maximum crack is assumed in the construction-induced damage length. The material parameters of the model are shown in Table 1. The concrete damage plasticity model (CDP model) is used for this model. The following parameters are used in the CDP model simulation: dilation angle; eccentricity; compressive strength ratio f_{b0}/f_{c0}; ratio K of the second stress invariant on the tensile meridian to that on the compressive meridian for the yield function and viscosity parameter. These parameters are shown in Table 2

Table 1 Material parameters of pipe pile model

Material	Young's modulus(MPa)	Poisson's Ratio	Density(kg/m³)
Concrete (C80)	37,968	0.2	2500
Steel bar (HRB400)	200,000	0.3	7800

Table 2 Material parameters in C80 concrete CDP model

Dilation angle	Eccentricity	f_{b0}/f_{c0}	K	Viscosity parameter
30°	0.1	1.16	0.6667	0.0008

3.2 Influence of the Different Damage Lengths on the Mechanical Behaviour of RC Pipe Pile

To discuss the influence of the different damage lengths on the mechanical behaviour of the RC pipe pile, the maximum crack width in the damage zone is assumed 0.4 mm while the damage lengths from the from the top of the pipe piles are assumed 0 m, 1 m, 2 m and 3 m, respectively. Considering the actual situation, both horizontal displacement and vertical loads are applied on the pipe pile. The load–displacement curves of the four piles are shown in Fig. 3. The ultimate bearing capacity of the undamaged pipe pile is 250.36 kN. For the RC pipe piles with the same maximum crack width in the damage zone, the ultimate bearing capacities corresponding to the 1, 2 and 3 m damage lengths are 241.2 kN, 238.07 kN and 234.07 kN, respectively. Thus, for the RC pipe pile with construction-induced damage, under the constant maximum crack width, the ultimate bearing capacity of the pipe pile decreases with the increased damage lengths and the maximum reduction is 6.5%. The increase of the damaged lengths leads to the decrease of the mechanical properties of the concrete, which in turn leads to the decrease of the ultimate bearing capacity of the pile.

The concrete stress cloud of the pile under different cases are shown in Fig. 4, where the stress is the compressive stress. It can be observed from Fig. 4 that, for both undamaged and damaged piles, the main stress is concentrated at the bottom of the pile, and the stress in the compressed area is greater than the stress in the tensile area. In addition, the difference among the maximum concrete stresses of the undamaged and damaged piles is very little.

Figure 5 show the equivalent plastic concrete strain for RC pipe piles with different damage lengths, where the strain is the compressive strain. For undamaged pile, the equivalent plastic strain is concentrated at the bottom of the pile and decreases gradually from the bottom to the top, while the main equivalent plastic strain occurs in the compressed section, see Fig. 5a. Comparing Fig. 5b–d with Fig. 5a, the equivalent plastic concrete strains become more obvious in the pile with construction-induced damage and cracks. When the damaged length increases, the area of equivalent plastic strain also increases. As shown in Fig. 5d, the significant equivalent plastic strain occurs at both the bottom and middle of the pile when the damage length reaches

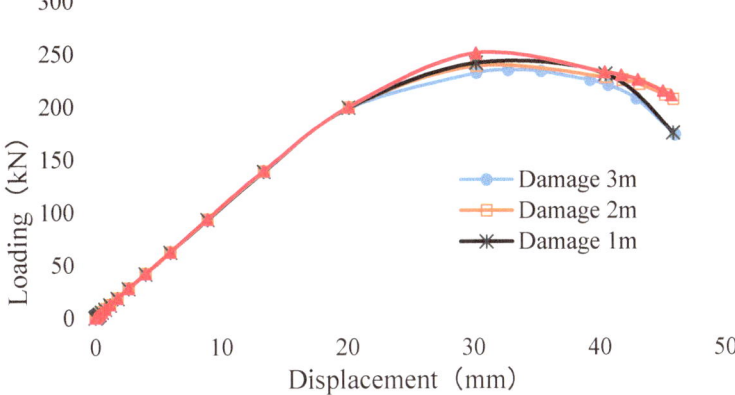

Fig. 3 Load–displacement curves of the top of the pipe piles with different damage lengths

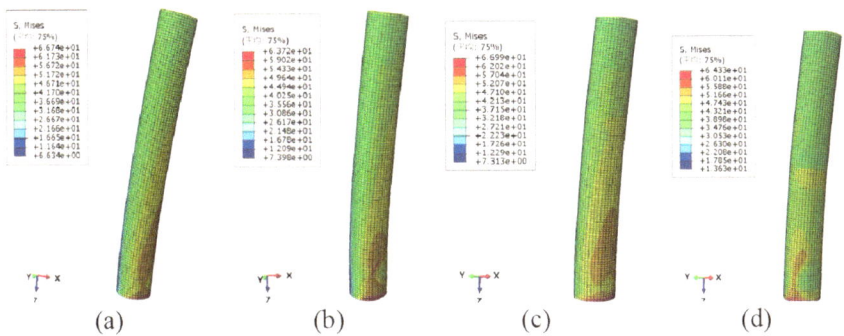

Fig. 4 Stress cloud of the concrete: **a** undamaged pile; **b** pile with 1 m damage length and 0.4 mm crack width; **c** pipe pile with 2 m damage length and 0.4 mm crack width; **d** pipe pile with 3 m damage length and 0.4 mm crack width

3 m. In addition, the main equivalent plastic strain area moves up to the middle part of the pile.

3.3 Influence of the Different Maximum Crack Widths on the Mechanical Behaviour of RC Pipe Pile

To discuss the influence of the different crack widths on the mechanical behaviour of the RC pipe pile, the lengths in the damage zone is assumed 3 m while the maximum crack widths are assumed 0 mm, 0.1 mm, 0.2 mm, 0.3 mm and 0.4 mm, respectively. The load–displacement curves of the five piles are shown in Fig. 6. For the RC pipe pile with construction-induced damage, under the constant damage length, the

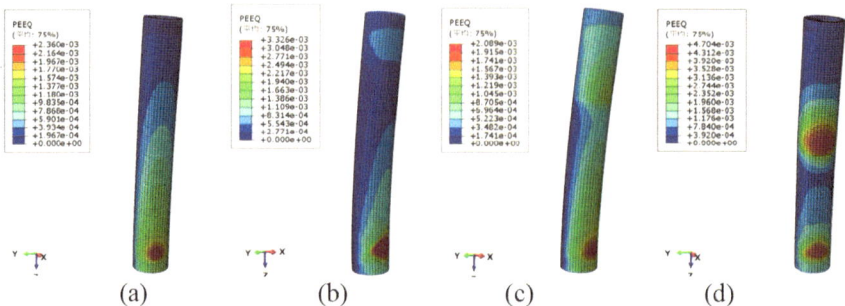

Fig. 5 Equivalent plastic strain cloud of the concrete: **a** undamaged pile; **b** pile with 1 m damage length and 0.4 mm maximum crack width; **c** pipe pile with 2 m damage length and 0.4 mm maximum crack width; **d** pipe pile with 3 m damage length and 0.4 mm maximum crack width

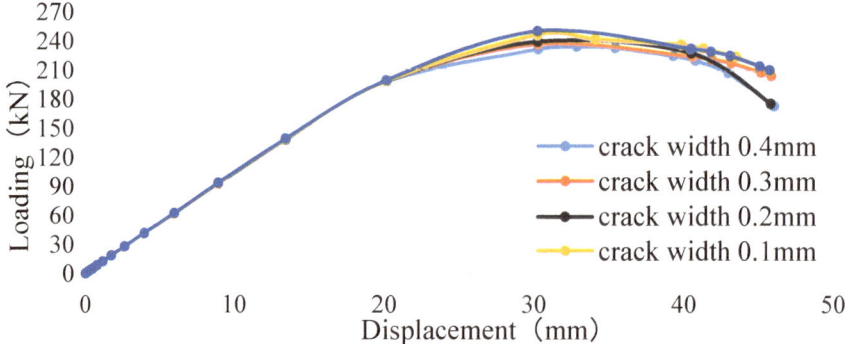

Fig. 6 Load–displacement profile of the top of pipe pile under different maximum crack widths

ultimate bearing capacity of the pipe pile decreases with the increased maximum crack widths in the damage zone.

Figure 7 shows that the equivalent plastic concrete strain of the RC pipe piles, where the strain is the compressive strain. The equivalent plastic strain of the undamaged pile is concentrated at the bottom of the pile and decreases gradually from the bottom to the top. The equivalent plastic strain in the damaged zone of the damaged pile is higher than that in the undamaged pile, see Fig. 7b–d. For damaged pile, when the maximum crack width is less than 0.2 mm (see Fig. 7b and c), the equivalent plastic strain is also concentrated at the bottom of the pile while the area of the main equivalent plastic strain increases with the increased maximum crack widths. When the maximum crack width reaches 0.3 mm, a comparatively lower equivalent plastic strain is observed in the middle part of the pipe pile, see Fig. 7d. With the increase of the maximum crack width in the damaged zone, main equivalent plastic strain is observed in two areas, see Fig. 7e, and one larger main equivalent plastic strain area rises to the middle part of the pile, indicating the failure of the damaged RC pipe pile transferring from the bottom zone to the middle zone of the pile. It seems that,

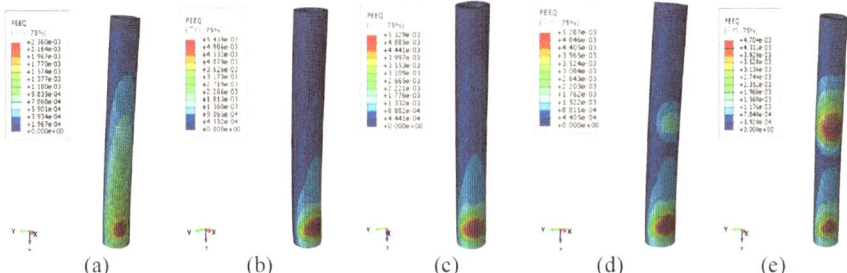

(a)   (b)   (c)   (d)   (e)

Fig. 7 Equivalent plastic strain cloud of the concrete: **a** undamaged pipe pile; **b** pipe pile with 3 m damaged length and 0.1 mm maximum crack width; **c** pipe pile with 3 m damaged length and 0.2 mm maximum crack width; **d** pipe pile with 3 m damaged length and 0.3 mm maximum crack width; **e** pipe pile with 3 m damaged length and 0.4 mm crack width

when the maximum width of the crack increases, the degree of degradation of the mechanical properties of the concrete of the pipe pile gradually increases, leading to the increased main plastic strain region of the pipe pile.

4   Conclusion

In this paper, the mechanical behaviour of the RC pipe piles with construction-induced damage is simulated by finite element model and the corresponding ultimate bearing capacity is predicted. The construction-induced damage in the concrete is considered by the reduced compressive strength of the concrete in the damaged zone of the piles. The following conclusions can be drawn:

1. The ultimate bearing capacity of the RC pipe piles with construction-induced damage in the top zone of the pile is reduced due to the damage. Both the increased damage length and crack width cause the reduction of the ultimate bearing capacity of the RC pipe pile. Considering the worst case, i.e. 0.4 mm maximum crack width in the damage zone and 3 m damage length, the maximum reduction of the ultimate bearing capacity of the RC pipe pile is 6.5%.
2. Under horizontal displacement and vertical loads, the failure of the undamaged pile occurs at the bottom of the pile. While for the pile with construction-induced damage and cracks, with the increase of the damaged length from top to the middle zone of the pile and the increased maximum crack width in the damage zone, the failure of the pile may occur at both the bottom of the pile and the middle of the pile. The construction-induced damage and cracks of the pile result in a shift of the plastic strain region in the failure state.

In this study, it is assumed that the crack in the cracked concrete is a single crack and the steel bars are not damaged. In practical engineering, the number of cracks in the damaged concrete will be more complicated, and the cracked concrete will influence the corrosion rate and mechanical properties of the reinforcement. The conclusions obtained in this study may not be applicable to the situations different from this study.

Acknowledgements This research was financially supported by the Shanghai Natural Science Foundation (No. 21ZR1426800).

References

1. Huang, W., Chen, C., Lin, S., & Chu, N. (2013). Crack reinforcement design of PHC pile foundation for a general cargo terminal project in Suzhou port. *China Jiangsu Constr, 06*, 77–80.
2. Kwon, S. J., Na, U. J., Sang, S. P., & Sang, H. J. (2009). Service life prediction of concrete wharves with early-aged crack: Probabilistic approach for chloride diffusion. *Structural Safety, 31*(1), 75–83.
3. Shao, W., & Li, J. (2014). Service life prediction of cracked RC pipe piles exposed to marine environments. *Construction and Building Materials, 64*, 301–307.
4. Li, J., & Shao, W. (2014). The effect of chloride binding on the predicted service life of RC pipe piles exposed to marine environments. *Ocean Engineering, 88*, 55–62.
5. Li, L., Li, J., & Yang, C. (2019). Theoretical approach for prediction of service life of RC pipe piles with original incomplete cracks in chloride-contaminated soils. *Construction and Building Materials, 228*, 116717.
6. Yang, Z., Tang, M., & Zhang, J. (2018). The influences of chloride ion transmission on cracking and chemical corrosion resistance of PHC pipe piles. *Chemical Engineering Transactions, 71*, 1183–1188.
7. Shao, W., Shi, D., Jiang, J., & Chen, Y. (2017). Time-dependent lateral bearing behaviour of corrosion-damaged RC pipe piles in marine environments. *Construction and Building Materials, 157*, 676–684.
8. Liu, R., Su, J., Wu, F., Lv, Y., Guo, Y., Bao, J., et al. (2022). Full-scale experimental research on the bending fatigue performance of post-tensioned prestressed concrete pipe piles. *Ocean Engineering, 260*, 112025.
9. Wei, Y., Wang, D., Li, J., Jie, Y., Ke, Z., Li, J., et al. (2020). Evaluation of ultimate bearing capacity of pre-stressed high-strength concrete pipe pile embedded in saturated sandy soil based on in-situ test. *Applied Sciences, 10*, 6269.
10. Xiao, Y., Liu, X., Zhou, J., & Song, L. (2023). Field test study on the bearing capacity of extra-long PHC pipe piles under dynamic and static loads. *Sustainability, 15*(6), 5161.
11. Standard, C. (2015). *Code for the Design of Concrete Structures GB50010-2015*. China Construction Industry Press.
12. Nakamura, H., Nanri, T., Miura, T., & Roy, S. (2018). Experimental investigation of compressive strength and compressive fracture energy of longitudinally cracked concrete. *Cement and Concrete Composites, 93*, 1–18.
13. Miura, T., Sato, K., & Nakamura, H. (2021). Influence of primary cracks on static and fatigue compressive behavior of concrete under water. *Construction and Building Materials, 305*, 124755.
14. Standard, C. (2012). *Code for Pile Foundation in Port Engineering JTS 167-4-2012*. China communication press.
15. Yin, X. X., Du, Q. Q., & Huang, C. H. (2017). Reliability analysis of axial compressive bearing capacity of C105 concrete pipe piles. *Concrete, 03*, 56–58. (In Chinese).

3D Printed Concrete and Visualization Monitoring

Research and Application of Simple Concrete 3D Printer

Yilong Li, Xiaoyu Chen, Yang Wei, Di Zhu, Wenhe Zhang, and Xin Li

Abstract In response to the call for the national concept of carbon neutrality, this paper made a simple concrete 3D printer by utilizing its principle, studied the mix ratio of printing materials, and applied the research results to the restoration and reconstruction of public service facilities in and around the campus to promote the green transformation of urban development.

Keyword Concrete; 3D printing; Carbon neutrality

1 Introduction

In 2020, China put forward the dual-carbon goal of "carbon peaking" and "carbon neutrality", which elevated China's green development road to a new height and greatly boosted the comprehensive intelligent transformation and green transformation of China's construction industry [1–3]. As an efficient means to realize digitalization, intelligence, and automation of construction, concrete 3D printing technology is an important means to transform and upgrade the traditional construction industry [4]. Concrete 3D printing technology boasts the advantages of fast construction speed, low cost, energy conservation, and environmental protection. It can eliminate the need for formwork construction by utilizing local materials and waste and can also produce special-shaped buildings and special-shaped components, which outperforms traditional concrete technology.

In this paper, by exploring the principle of concrete 3D printing technology, a simple concrete 3D printer is fabricated. The research results of this experiment are used to repair the public service facilities in and around the campus, which not only ensures the safe use of campus facilities and beautifies the campus environment, but also makes the most direct response to the green construction and dual-carbon policy.

Y. Li · X. Chen · Y. Wei · D. Zhu · W. Zhang · X. Li (✉)
School of Architecture and Engineering, City College, Dalian University of Technology, Dalian, Liaoning, China
e-mail: 86843336@qq.com

© The Author(s) 2025   437
D. Li and Y. Zhang (eds.), *Advances in Frontier Research on Engineering Structures II*,
Lecture Notes in Civil Engineering 535, https://doi.org/10.1007/978-981-97-6238-5_36

2   Design and Manufacture of Simple Concrete 3D Printer

2.1   Research Status and Research Background of Concrete 3D Printing Technology

With the switch development of the global intelligent industry, the construction industry has begun to transform into intelligence. The advantages of concrete 3D printing technology, such as high efficiency and intelligence, endow it with great prospects for development in building construction. At present, concrete 3D printing technology is mainly used in low-rise houses, simple bridges, architectural decoration, and simple reconstruction and reinforcement, but it is not mature in high-rise buildings [5]. Because of the high labor cost in foreign countries, there are many houses built with concrete 3D printing technology, and the built houses can meet the requirements of safety and normal use after being renovated and put into use. However, due to various technical reasons, the concrete 3D printing technology has not witnessed sufficient popularization and application in China, featuring immature development mode and expensive prices. Therefore, this paper makes a simple concrete 3D printer according to its principle, uses waste materials as raw materials for concrete 3D printing, determines the appropriate mix ratio, makes concrete 3D printing specimens, tests whether the strength meets the requirements, compares with traditional concrete specimens, and then further applies the research results to house repair and transformation, contributing to the response to carbon peaking and carbon neutrality.

2.2   Design Principle and Structure of the Simple Concrete 3D Printer

Concrete 3D printer involves extruding the configured concrete slurry through the extrusion device, which is controlled by the 3D software and printed by nozzle extrusion according to the pre-set printing program to finally obtain the designed concrete components. In the printing process, concrete 3D printing technology dispenses with the formwork support process of traditional concrete forming, which is the latest concrete mold-less molding technology. However, the traditional concrete 3D printer is characterized by large size, heavy weight, difficulty in transportation, and high cost. Therefore, this paper designs a simple and lightweight concrete 3D printer, which is simple in structure, lightweight in model, and transportable at will. In addition, the printing material is unfettered and available locally, providing ease of use, simplicity, efficiency, and proper accuracy.

Figure 1 is a three-dimensional design drawing of a simple concrete 3D printer. The whole printer consists of a cylindrical cylinder, an air inlet, an outlet, a cylinder support component, two sets of guide rails perpendicular to each other, four X-axis running wheels, and four Y-axis running wheels. The upper cover plate is tightly connected to the cylinder through flanges. The upper air inlet can be connected with an air pump to apply the gas load, that is, pressure, to the materials in the cylinder. Materials are extruded from the lower outlet under the action of gas pressure. With the aid of rails and wheels, the cylinder is manually pushed to move freely in X and Y directions, so that long strip materials in any direction can be printed. Also, hoses with different diameters can be connected at the outlet, and components with different shapes can be printed by manually moving the hoses. It is available to send the plane, elevation, sectional view, and detailed structure drawing of the printer to the manufacturer, and determine the processing materials and specific dimensions and structures of each part of the printer through detailed communication with the manufacturer. The printer after processing and molding is shown in Fig. 2.

Fig. 1 3D design drawing of simple concrete 3D printer

Fig. 2 Physical diagram of simple concrete 3D printer

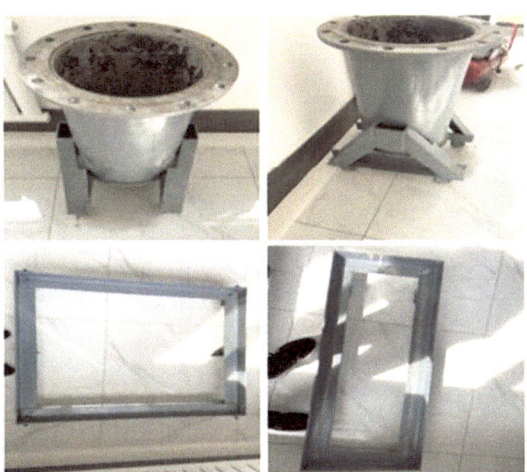

3   Research on Materials Used in 3D Printing of Concrete

3.1   Study on Mix Ratio of Materials

As opposed to traditional concrete construction, 3D printed concrete construction eliminates the need for formwork, and 3D printed concrete needs to be set quickly in a short time, so the fluidity of concrete should not be too large. Otherwise, it will make the structure difficult to form. However, if the concrete flow is too small, the nozzle will be susceptible to blockage, so it is a key issue that we need to solve to ensure that the concrete material used for 3D printing has the right flow and that the concrete can set quickly after being ejected from the nozzle.

Printing concrete materials should meet the requirements of fluidity, extrudability, buildability, setting time, and mechanical properties [6]. Extrudability and fluidity are the direct validity states for testing various indexes of 3D printed concrete, the key factor affecting which is water content. If the water content is too small, the printing materials cannot be extruded smoothly from the equipment, and even if they can be extruded, the adhesion between layers cannot be well achieved, resulting in a serious reduction of structural stability; if the water content is too large, the printed samples will be difficult to form and easy to collapse. Adding water-reducing agents can effectively control the water content of printed materials without affecting the fluidity of materials. The research found that concrete has good cohesion and fluidity after adding calcium carbonate powder, and calcium carbonate powder has a certain water-reducing effect [7, 8].

In comparison with the traditional concrete formwork forming technology, 3D printed concrete needs to be piled up and printed layer by layer, and the specimen cannot be vibrated, so it is difficult to guarantee the good internal compactness of the material. The printed specimen should not only support its weight but also consider the pressure from the concrete load attached to it. Hence, the process of stacking layer by layer will lead to gaps at the interlayer interface. To this end, interlayer adhesion is a necessary condition to allow the specimen to present good integrity and structural stability [9, 10].

According to the above evaluation of the performance index of 3D printed concrete, the experimental scheme is determined. The traditional concrete materials are composed of water, sand, gravel, and cement. In this experiment, based on the traditional material, cement (composite Portland cement P.C 42.5 grade), sand (natural river sand), and water are used as basic mixture materials, into which a small amount of short fiber and calcium carbonate are added. Short fiber can delay the appearance of cracks in concrete and improve the tensile strength of concrete, while calcium carbonate can improve the fluidity and continuity of concrete materials. The results show that the greater the water-cement ratio is, the harder it is to form the printed concrete material, but a lower water-cement ratio will affect the fluidity of the printed material. In this regard, it is necessary to take an appropriate mixture ratio between them, and then increase the amount of calcium carbonate in an appropriate amount, so that the printed concrete material under this mixture ratio

Table 1 Experimental results of mix design of 3D printed concrete materials

Level	Cement (g)	Sand (g)	Water (ml)	Short fiber (g)	Calcium carbonate (g)	Cement-sand ratio	Water-cement ratio
The first time	25	72	26	3	0	0.38	0.92
The second time	100	245	62	3	3	0.43	0.56
The third time	160	267	120	3	5	0.63	0.70

can meet the fluidity and the good plasticity of concrete. Therefore, many experiments have been carried out, and the third experiment result is finally taken as the best mixture ratio. The mixture ratio data of the materials used in the experiment are shown in Table 1.

3.2 Study on Strength of 3D Printed Concrete Specimens

The concrete compressive strength can be calculated by the following formula according to the current codes of China:

$$f_{cc} = \frac{F}{A}$$

in the formula:

f_{cc}—the compressive strength of concrete cube specimen (Mpa).

F—the failure load of the specimen (N).

A—the compression area of the specimen (mm^2).

In order to compare the strength of concrete 3D printing materials with traditional concrete materials, traditional concrete materials are made with the same kind of cement, water, sand, and other materials according to the common mix ratio of P.C 42.5 grade cement in the project, that is, the masses of cement, sand, and water per square meter of concrete are 220, 1560 and 250 kg respectively. Single-layer and multi-layer specimens were made of traditional concrete materials and 3D printed concrete materials with the best mix ratio, then vertical loading tests were carried out on all specimens after curing for 14 days, and finally, the strength was compared.

First of all, the vertical loading test was carried out on the traditional single-layer concrete specimens. When loaded with a loading rate of 5.93 KN/S at 55.93 KN, the specimen showed tiny cracks, and the compressive strength was 2.51 MPa at this time. When the specimen was loaded with a loading rate of 8.02 KN/S at 72.56 KN, cracks appeared all over the specimen, and the compressive strength was 3.56 MPa

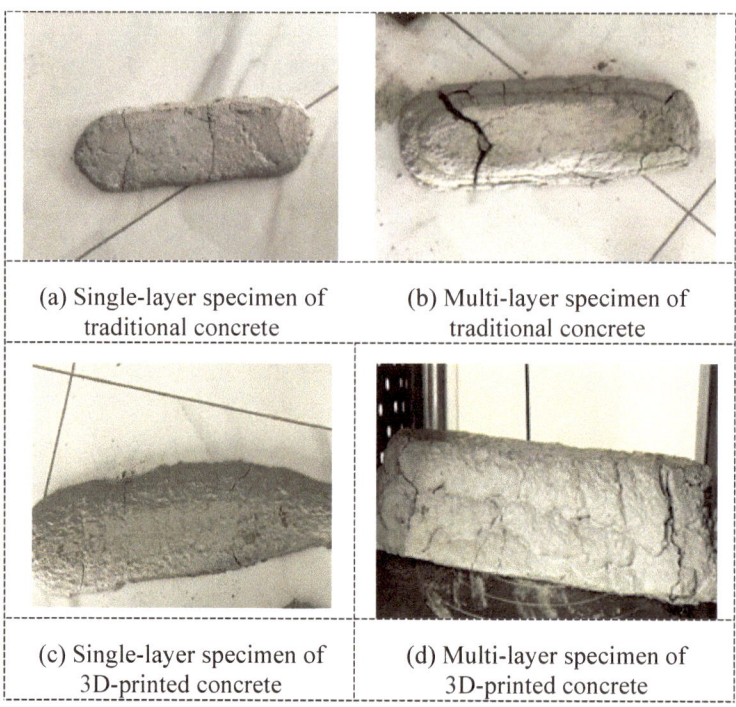

| (a) Single-layer specimen of traditional concrete | (b) Multi-layer specimen of traditional concrete |
| (c) Single-layer specimen of 3D-printed concrete | (d) Multi-layer specimen of 3D-printed concrete |

Fig. 3 Failure mode of specimens for compressive strength test

at this moment. Finally, when loaded at 142.52 KN at a loading rate of 3.47 KN/ S, the specimen was fractured directly, and the compressive strength at this time is 6.37 MPa. Figure 3a depicts the crack on the specimen.

In the vertical loading test of traditional multi-layer concrete specimens, when the specimen was loaded with the loading rate of 9.46 KN/S at 88.85 KN, concrete fragments began to appear at the edge of the specimens, but no obvious cracks were found; at this time, the compressive strength was 4.25 MPa. When 172.32 KN is loaded at a loading rate of 21.77 KN/S, a large number of cracks appear on both sides and a small number of fine cracks appear in the middle, with a compressive strength of 7.89 MPa. After that, cracks occurred continuously in the specimen. Finally, when loaded at 235.86 KN with a loading rate of 7.33 KN/S, the specimen was directly crushed and separated, and the compressive strength was 10.48 MPa. Figure 3b shows the cracked specimen.

The vertical loading test for the single-layer specimen of the 3D-printed concrete under the best mix ratio was carried out. When the specimen was loaded at 86.38 KN with a loading rate of 22.50 KN/S, cracks appeared in the specimen, with predominantly oblique cracks, and the compressive strength was 4.25 MPa. When the specimen was loaded at 178.05 KN with a rate of 15.50 KN/S, the oblique cracks first appeared at the bottom of the specimen and then developed obliquely to the middle of the side. When the cracks developed to the top surface, the horizontal section cracks appeared, the horizontal length of the cracks was about 3–5 cm, and the compressive strength was 8.01 MPa. Until the specimen was loaded at 257.56 KN with a loading rate of 1.25 KN/S, obvious cracks appeared in the middle and even penetrated in some places, resulting in direct separation. The phenomenon that the fiber is broken can be distinctly seen between the cracks. At this moment, the compressive strength was 11.22 MPa, and the cracked specimen was shown in Fig. 3 (c).

The vertical loading test for the multi-layer specimen of the 3D-printed concrete under the optimum mix ratio was carried out. When the load was 124.85 KN with a loading rate of 6.55 KN/S, cracks appeared in the specimen, and the compressive strength was 5.65 MPa. When the loading rate was 13.25 KN/S and the load was 298.13 KN, substantial cracks began to appear on the vertical surface of the specimen, dominated by oblique cracks. At the same time, there were a few cracks in the horizontal section, the length of which was about 1–3 cm, and the compressive strength was 13.25 MPa. Finally, when the loading rate was 10.78 KN/S and the load was 410.53 KN, plenty of cracks occurred in the whole specimen, mainly at both ends of the specimen. Although there was no penetration separation phenomenon in all cracks, the surface concrete broke away. At this time, the compressive strength was 18.89 MPa. Figure 3d describes the cracked specimens. The compressive strength and crack development of different specimens during failure are shown in Table 2.

The experimental results show that the compressive strength of 3D printed concrete specimens made of the best mix ratio is significantly higher than that of traditional concrete specimens, and the strength is close to that of C20 concrete with fewer cracks. Figure 4 shows the curves of load changes with time for four kinds of specimens.

Table 2 Vertical loading test results of different types of specimens

Specimen type	Traditional concrete single-layer specimen	Traditional concrete multilayer specimen	3D printed concrete single-layer specimen under the best mix ratio	3D printed concrete multilayer specimen under the best mix ratio
Compressive strength (MPa)	6.37	10.48	11.22	18.89
Crack development	Crack penetration and separation	Crack penetration and separation	A few cracks, excluding penetration and separation	A few cracks, excluding penetration and separation

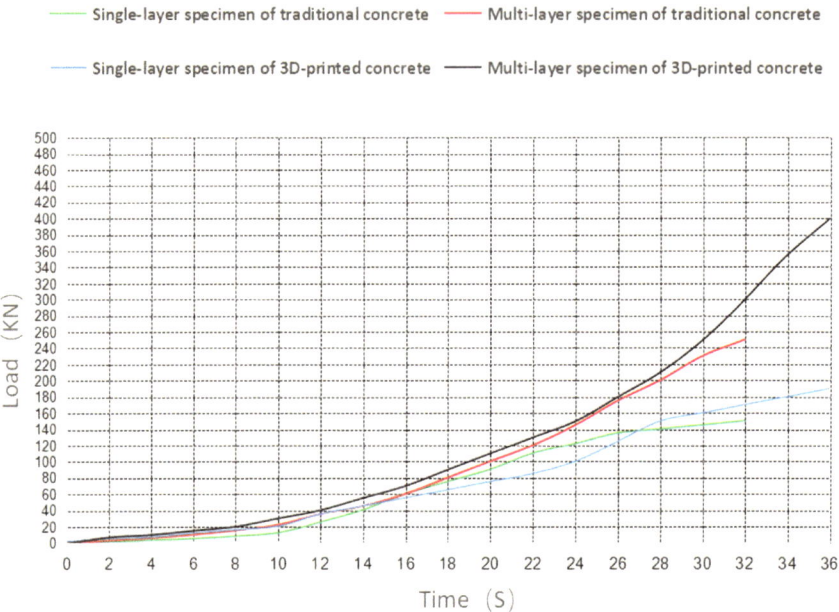

Fig. 4 Comparison of load-time curves of different specimens

4   Research and Application of Experimental Results

4.1   *Practical Application of the Simple Concrete 3D Printer*

Through the above experiments, the best mix ratio of materials used in the concrete 3D printer was determined, and the experimental results were applied to the reconstruction and repair of actual houses. Through investigation, it was found that there were large cracks in some pavements on the campus, and the scattered water of the teaching building had a large tension crack with the main structure due to uneven settlement of the foundation. Some old houses around the campus were seriously damaged. The above-damaged places were partially repaired by using the self-made concrete 3D printer and experimental research results. Figures 5 and 6 are comparative pictures before and after repair. If the traditional construction method is used for repair, the construction is difficult, the cost is high and the efficiency is low. While the self-made simple concrete 3D printer can effectively repair damaged components and houses without formwork, which not only reduces the construction difficulty but also improves the repair efficiency and quality. Moreover, the construction method is energy-saving and environmentally friendly.

Fig. 5 Comparative diagram before and after the repair of the teaching building

Fig. 6 Comparison of
shoulder and pavement crack
before and after repair

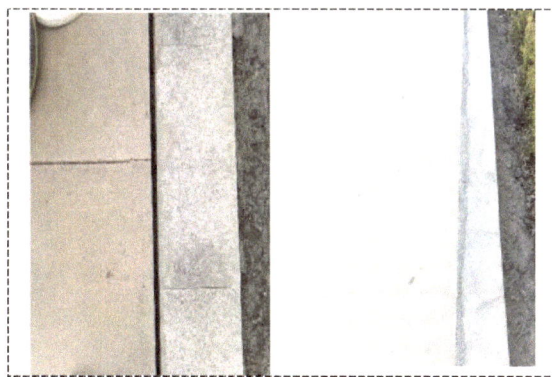

4.2 Problems and Solutions in Experimental Research

The simple concrete 3D printer made in this paper is a cylindrical cylinder. Under the action of gas load, the concentrated pressure is not uniform enough, which leads to the fact that when there are few materials in the later period, the materials cannot be discharged at the edge of the cylinder wall near the discharge port, and the gas pressure is not uniform enough, which leads to the uneven delivery of concrete materials. To cope with this problem, the cylindrical cylinder can be changed into a cone, and a gas distributor is set at the top of the cylinder, so that the concrete materials in the cylinder can be delivered evenly. The schematic diagram of the improved concrete 3D printer is shown in Fig. 7. In addition, since only the compressive strength of printed concrete specimens was studied, the research on tensile strength needs further discussion.

Fig. 7 Schematic diagram of the appearance of the improved concrete 3D printer

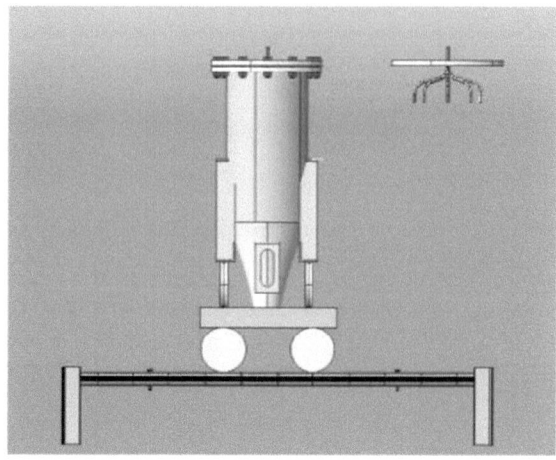

5 Conclusion

This paper actively responded to the green development strategy of "carbon neutrality" put forward by our country, took concrete 3D printing technology as the background, made a simple concrete 3D printer, studied the materials and mix ratio used for printing, made specimens and tested the strength of specimens, then compared the strength with traditional concrete specimens, and applied the research results to the repair and transformation of public facilities inside and outside the campus. The research results show that the simple concrete 3D printer can realize its basic functions, print special-shaped components without templates, save energy, protect the environment, and have high efficiency. It can be used for the repair and reconstruction of small house components and public service facilities, thus contributing to ensuring living safety, green construction, energy conservation, and emission reduction.

References

1. Sun, X., Wang, Q., Wang, H., et al. (2020). Influence of multi-walled nanotubes on the fresh and hardened properties of a 3D printing PVA mortar ink. *Construction and Building Materials, 247*(4), 118590.
2. Zhang, Y., Zhu, Y. M., R. Q., et al. (2021). Progress on 3D printing construction technology and its cement-based materials. *Bulletin of the Chinese Ceramic Society, 40* 6), 1796–1807.
3. Zhao, X. G., Jiang, G. W., Li, A., et al. (2016). Technology, cost, a performance of waste-to-energy incineration industry in China. *Renewable and sustainable energy reviews, 55,* 115–130.
4. Ding, L. Y., Xu, J., & Qin, Y. W. (2015). Summary of research and application of 3D printing digital construction technology. *Journal of Civil Engineering and Management, 32*(3), 1–10.
5. Cai, J. G., Zhang, Q., Du, C. X., et al. (2021). Research status and development trend of 3D printing concrete technology. *Industrial Construction, 51*(6), 1–11.
6. Zhang, Y., Zhang, Y. S., She, W., et al. (2019). Rheological and hardening properties of the high-thixotropy 3D printing concrete. *Construction and Building Materials, 201,* 278–285.

7. Nan, F., Jin, R. L., Liu, X. W., et al. (2012). Effect of limestone powder on the performance of self-compacting pumping concrete. *Concrete, 11*, 91–93.
8. Zhou, Y. X., Wang, Y. H., Wang, S. Y., et al. (2014). Characteristics of limestone powder and its influence on concrete performance. *Construction Technology, 43*(09), 23–27.
9. Jay G., Sanjayan., Behzad N., et al. (2018). Effect of surface appearance on inter-layer strength of 3D printed concrete. *Construction and Building Materials, 172*, 468–475.
10. Lei, B., Ma, Y., Xiong, Y., et al. (2017). Evaluation method of shaping performance of 3D printing concrete material. *Bulletin of the Chinese ceramic society, 10*.

Experimental Study on the Durability of High-performance Concrete Against Chloride Ions Penetrations

Tao Ge and Chenxi Liu

Abstract To study the durability of concrete structures against chloride ions penetrations under subway environmental conditions, permeability performance tests were conducted on high-performance concrete using the conductivity method and corrosions method, respectively. The experimental results show that the additions of fly ash, slag, and other admixtures in high-performance concrete have a significant effect on resisting the infiltrations and diffusions of chloride ions into the concrete. Considering the long-term durability of chloride ions penetrations, high-performance concrete with dual additions of fly ash and slag has better durability improvement than single additions. The durability life of high-performance concrete in chloride ions resistance test is 3–4 times longer than that of ordinary concrete.

Keywords High performance concrete · Chloride ions · Permeability · Durability

1 Introductions

The main factors affecting the durability of concrete structures under subway environmental conditions are the erosions of harmful media and the corrosion of stray currents in the underground environment [1–4]. Therefore, chloride ions are generally considered to be one of the main harmful media affecting the durability of concrete structures. Therefore, the resistance of concrete to chloride ions penetrations is an important indicator for measuring the 100 years of design service life of subway engineering concrete and reinforced concrete structures. The High-Performance Concrete (HPC) designed for the Nanjing Metro project is a C30 concrete with separate or simultaneous additions of fly ash and slag powder. This article conducts experimental research on its resistance to chloride ions penetrations and analyzes and demonstrates whether its durability against chloride ions penetrations meets the 100 years of design service life.

T. Ge (✉) · C. Liu
Aeronautics Engineering College, Air Force Engineering University, Xi'an, China
e-mail: getaoge@163.com

D. Li and Y. Zhang (eds.), *Advances in Frontier Research on Engineering Structures II*,
Lecture Notes in Civil Engineering 535, https://doi.org/10.1007/978-981-97-6238-5_37

Table 1 Chemical compositions of raw materials (ω/%)

Raw material	Sio_2	Al_2O_3	Cao
Cement	25.63	8.27	53.36
Fly ash	50.24	33.12	5.18
Slag powder	32.05	15.59	37.28
Raw material	Mgo	Fe_2O_3	Na_2O
Cement	2.73	3.22	–
Fly ash	0.92	4.01	–
Slag powder	10.18	0.84	–
Raw material	K_2O	SO_3	Loss on ignitions
Cement	0.62	1.69	3.62
Fly ash	–	0.46	1.45
Slag powder	–	0.89	1.88

2 Experimental Design

2.1 Raw Material

The specimen is C30 high-performance concrete. The raw materials mainly include ordinary Portland 32.5 cement, fly ash, slag powder, sand, crushed stone, and anti-crack and anti-seepage agent. The HLC additive produced by a certain company is used as the anti-crack and anti-seepage agent.

The chemical compositions of the test raw materials are shown in Table 1.

2.2 Mix Ratio Design

The design of the experimental concrete mix Ratio is shown in Table 2.

In the mix proportions specimens shown above, the additive HLC is all internally added. C310 and C320 are reference concrete without additives, with HLC content of 29 kg m^{-3} and 33 kg m^{-3}, and water-cement ratio of 0.43 and 0.38, respectively; C312–C336 are concrete with separate or simultaneous additions of fly ash and slag, with an HLC content of 33 kg/m^3 and a water-cement ratio of 0.38.

According to the designed mix ratio, the molding dimensions are φ 100 mm × 60 mm and size 100 mm × 100 mm × 100 mm. Two types of concrete specimens with a thickness of 100 mm are embedded with steel bars in the cube specimens.

Table 2 Test concrete mix ratio

Amount of raw materials used per cubic meter of concrete (kg m^{-3})

	Cement	Fly ash	Slag	Sand
C310	334	–	–	747
C320	377	–	–	723
C312	229	152	–	700
C321	214	–	164	713
C322	132	–	246	708
C332	190	76	114	708
C333	154	115	115	714
C335	116	116	154	706
C336	116	77	193	713
	Fine crushed stone	Small gravel	Water	HLC
C310	591	484	156	29
C320	596	488	156	33
C312	579	473	157	33
C321	588	482	156	33
C322	584	480	156	33
C332	584	478	156	33
C333	590	482	158	33
C335	583	477	159	33
C336	588	482	159	33

2.3   Test Method

This article conducts tests on the resistance of concrete to chloride ions penetrations based on the industry standards of the People's Republic of China [5, 6].

Conductivity Method

This method involves applying an electric field to the concrete to test the difficulty of chloride ions passing through the concrete specimen under the actions of the applied electric field, in order to compare the resistance of concrete to chloride ions penetrations. This method measures the relative chloride ions diffusion coefficient of concrete and cannot be directly used to calculate the durability life of concrete.

Etching Method

This method simulates the corrosive environment of steel bars in concrete in the splash zone and water level fluctuations zone of marine engineering, which involves alternating dry and wet seawater. The concrete specimens are immersed in a 3.5% concentration of sodium chloride solutions for 1 day, then dried at 60 °C for 13 days, followed by soaking and drying until multiple cycles. Then drill samples of concrete mortar powder at different depths to determine the water-soluble and acid-soluble chloride ions content in the concrete mortar. Based on this, calculate the effective diffusions coefficient of chloride ions and the durability of concrete against chloride ions penetrations under these conditions.

100 mm is to be formed × 100 mm × after standard curing for 60 days, and the 100 mm cube specimen is taken out. Except for the two symmetrical planes parallel to the steel bar for external chloride ions penetrations, the other four surfaces are sealed with wax and then the soaking and drying cycle begins.

3 Testing Results and Analysis

3.1 Conductivity Method

To be formed φ 100 mm × The 60 mm cylindrical specimens were cured in water at 20 ± 3 °C for 28 days and 90 days respectively, and their electrical conductivity was measured. The relative chloride ions diffusion coefficient of the concrete was calculated based on the Nernst Plank equations [7], and the results are shown in Table 3.

From Table 3, we can see that whether one admixture is added alone or two admixtures are added simultaneously, the additions of admixtures can greatly reduce the relative chloride ions diffusions coefficient of concrete, and with the increase of the dosage, the relative chloride ions diffusions coefficient decreases. Under the same dosage conditions, slag has a better effect on reducing the relative chloride ions diffusions coefficient of concrete than fly ash, especially in the early stage; At 28 days, the relative chloride ions permeability coefficient of C322 concrete specimens with 60% slag added separately is the smallest, which is 20% of the reference concrete. However, in the later stage, due to the strengthening trend of fly ash in the later stage, under the same dosage conditions, adding slag at the same time has a better effect on reducing the relative chloride ions diffusions coefficient of concrete than adding slag alone, especially in high dosage situations; At 90 days, the relative chloride ions permeability coefficient of C336 concrete specimens with the additions of 18% fly ash and 46% slag is the smallest, which is 20% of the reference concrete. Therefore, considering the long-term durability of chloride ions penetrations resistance, high-performance concrete with the additions of fly ash and slag at the same time has a better effect on durability than adding them alone.

3.2 Etching Method

After 8 cycles of etching and baking, concrete specimens were drilled to obtain powder samples of concrete mortar at different depths, and the water-soluble and

Table 3 Testing the resistance of concrete to chloride ions penetrations using the conductivity method

Sample number	Conductivity C_i ($\times 10^{-4}$S)	
	28d	90d
C310	8.861	4.244
C312	4.814	2.044
C321	2.362	1.440
C322	1.777	1.416
C331	4.417	1.875
C332	2.590	1.158
C333	2.428	0.935
C334	2.309	1.527
C335	1.962	0.870
C336	1.924	0.846
Sample number	Relative chloride ions diffusion coefficient D_i ($\times 10^{-8}$cm^2 s^{-1})	
	28 d	90 d
C310	2.082	0.997
C312	1.131	0.480
C321	0.555	0.338
C322	0.418	0.333
C331	1.038	0.441
C332	0.609	0.272
C333	0.571	0.220
C334	0.543	0.359
C335	0.461	0.204
C336	0.452	0.199

acid-soluble chloride ions content in the concrete mortar was measured to obtain Cl in the concrete. The variations curve of content with diffusions depth is shown in Figs. 1, 2, 3 and 4.

The experimental results show that the water-soluble chloride ions content is approximately equal to the free chloride ions content in the pores of concrete; The content of acid-soluble chloride ions is equal to the total content of chloride ions in concrete; The difference between the two in the same layer is the combined chloride ions content in the concrete of that layer. From the table, it can be seen that the percentage of chloride ions bound in concrete specimens with admixtures increases significantly from the third layer compared to the reference concrete, indicating that the ability of chloride ions bound inside high-performance concrete with admixtures is stronger than that of the reference concrete. In fact, the damage to the steel bars in concrete is not caused by the total chloride ions content, but by the free chloride ions content in the pores of the concrete. When the total content of chloride ions is fixed, if the binding capacity of concrete to chloride ions is strong, the content of

Fig. 1 Water-soluble Cl in concrete when one type of admixture is added separately-Variations curve of content with diffusions depth

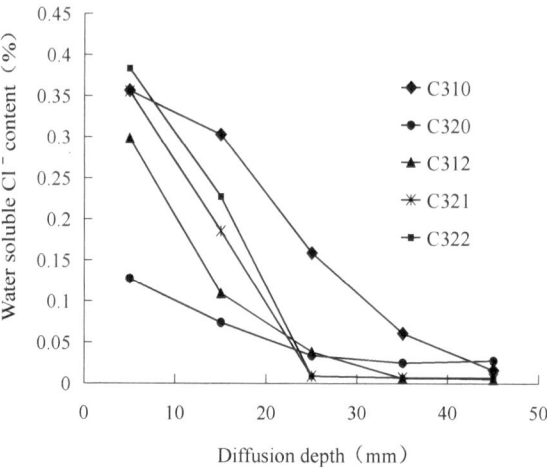

Fig. 2 Water-soluble Cl in concrete when two types of admixtures are added simultaneously-Variations curve of content with diffusions depth

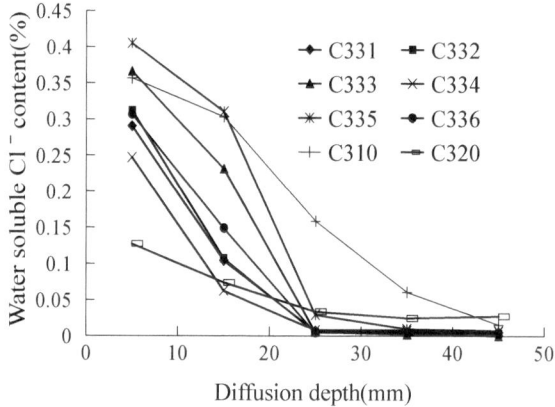

Fig. 3 Acid-soluble Cl in concrete when adding one type of admixture separately-Variations curve of content with diffusions depth

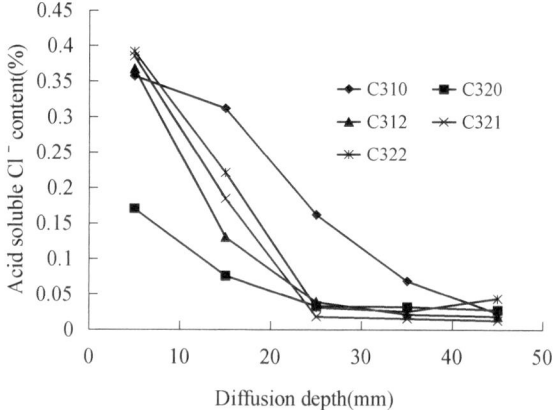

Fig. 4 Acid soluble Cl in concrete when two types of admixtures are added simultaneously -Variations curve of content with diffusions depth

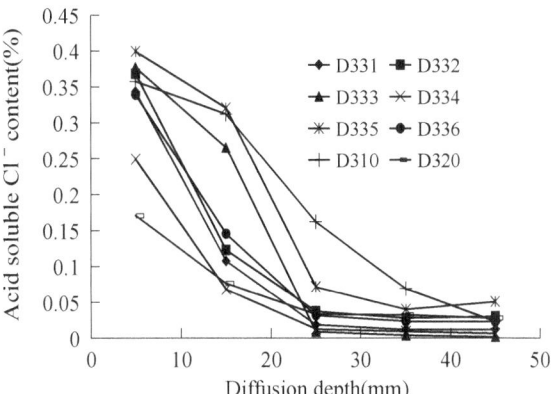

free chloride ions in the pores will be reduced, and the ability of concrete to resist chloride ions corrosion will be improved. Therefore, fly ash and slag contribute to the improvement of concrete's resistance to chloride ions corrosions.

From Figs. 1, 2, 3 and 4, it can also be seen that the water-soluble and acid-soluble Cl in concrete. The content decreases with the increase of diffusions depth. The variations curves in benchmark concrete C310 and C320 are relatively flat, while in concrete with admixtures, Cl⁻. The content has sharply decreased in the third layer. Starting from the third layer, the content of chloride ions changes very little and only in trace amounts. This indicates that the additions of fly ash, slag, and other admixtures in concrete have a significant effect on resisting the diffusions of chloride ions into the interior of the concrete.

Due to the fact that the concentrations of chloride ions in concrete vary with depth and time, the diffusions of chloride ions in concrete in this experiment follow Fick law, namely:

$$c(x, t) = c_0 \left(1 - \operatorname{erf} \frac{x}{2\sqrt{Dt}} \right) \tag{1}$$

$c(x,t)$—After time t, the chloride ions content at depth x in the concrete;

   c_0—Chloride ions content on the surface of concrete;

   D—Diffusions coefficient of chloride ions in concrete;

   erf—Error functions.

Because the chloride ions content in the concrete measured in this experiment is the weight percentage c (x, t) of chloride ions in the unit mortar, while the chloride ions content in the sodium chloride solutions soaked in the concrete specimen is the weight percentage of chloride ions in the unit water, it cannot be directly compared. Therefore, c_0 is determined based on empirical methods as the concentrations of chloride ions in the surface layer of concrete [8–10]. We calculate the diffusion coefficients of water-soluble and acid-soluble chloride ions using the above formulas, as shown in Table 4.

Table 4 Corrosions method for testing the resistance of concrete to chloride ions penetrations

Sample number	Diffusions coefficient of water-soluble chloride ions D_i ($\times 10^{-8} cm^2\ s^{-1}$)
C310	19.62/166
C320	11.84/100
C312	6.34/54
C321	5.51/47
C322	5.00/42
C331	5.14/43
C332	4.33/37
C333	4.46/38
C334	4.07/34
C335	3.36/28
C336	3.02/26
Sample number	Acid soluble chloride ions diffusions coefficient D_i ($\times 10^{-8} cm^2\ s^{-1}$)
C310	23.93/162
C320	14.76/100
C312	6.78/46
C321	5.65/38
C322	5.19/35
C331	5.35/36
C332	4.66/32
C333	5.28/36
C334	4.79/32
Sample number	Acid soluble chloride ions diffusion coefficient D_i ($\times 10^{-8} cm^2\ s^{-1}$)
C335	4.59/31
C336	4.26/29

From the results in Table 4, it can be seen that the water-cement ratio has a significant impact on the diffusion coefficient of chloride ions. In the two benchmark concrete groups, the water solubility and acid solubility chloride ions diffusions coefficients of C310 specimens with a water-cement ratio of 0.43 are 66% and 62% higher than those of C320 specimens with a water-cement ratio of 0.38, respectively. That is to say, a low water-cement ratio is beneficial for improving the chloride ions penetrations resistance of concrete. Under the same water-cement ratio (0.38), the addition of fly ash and slag significantly reduces the diffusion coefficient of chloride ions in concrete, whether water-soluble or acid-soluble. The C336 concrete specimen has the best resistance to chloride ions penetrations, with water solubility and acid solubility chloride ions diffusion coefficients only 26% and 29% of the reference concrete C320, respectively, which is consistent with the conclusions of the conductivity method.

Due to the corrosions caused by free chloride ions in concrete, the service life of concrete can also be calculated using Fick's law based on the diffusion coefficient of water-soluble chloride ions in concrete. Under marine environmental conditions, the chloride ions content C0 on the surface of concrete can be taken as 2.4% (by weight of cementitious materials), and the critical chloride ions content c (x, t) at the beginning of steel corrosions is taken as 0.5% (by weight of cementitious materials) [11]. The Fick formula becomes:

$$x^2 = 3.17Dt \tag{2}$$

$$t = 0.32\frac{x^2}{D} \tag{3}$$

x—Concrete cover (cm);

 t—Start time of steel corrosions (s);

 D—Diffusion coefficient of water-soluble chloride ions in concrete ($cm^2\ s^{-1}$).

If the thickness of the concrete protective layer is 6.5 cm, the results of the initiations age of steel corrosion are shown in Table 5.

From Table 5, it can be seen that in the marine environment (with chloride ions concentrations of 3.5%, calculated as a percentage of solutions concentrations), in the water level fluctuations zone and splash zone (alternating dry and wet), and under harsh test conditions where the temperature of the concrete reaches 60 °C during drying, the durability of high-performance concrete against chloride ions is significantly improved compared to ordinary concrete. Under the same water-cement ratio, the durability time of ordinary concrete against chloride ions is only 3–4 years, while the durability time of high-performance concrete with fly ash and slag added can be as high as 10–15 years, increasing the durability life by 3–4 times.

Because the subway is located inland, the chloride ions content in groundwater is very low (<0.1%), and the environment in which the subway is located is relatively closed, with little change in external temperature. The chloride ions content on the surface of the concrete will not exceed 0.5% (by weight of cementitious materials). If the critical chloride ions amount c (x, t) at the beginning of steel reinforcement corrosions is 0.5% (by weight of cementitious materials), according to Formula (3), The subway concrete will not be affected by chloride ions and can fully meet the durability life of 100 years against chloride ions.

4 Conclusions

(1) Reducing the water-cement ratio of concrete and improving its compactness are beneficial for improving its resistance to chloride ions penetrations. Under the same water-cement ratio, the additions of fly ash and slag micro powder, whether alone or simultaneously, can greatly reduce the diffusion coefficient

Table 5 The durability of concrete against chloride ions penetrations

Sample number	Diffusions coefficient of water-soluble chloride ions D_i ($\times 10^{-8}$cm^2 s^{-1})
C310	19.62
C320	11.84
C312	6.34
C321	5.51
C322	5.00
C331	5.14
Sample number	Diffusions coefficient of water-soluble chloride ions D_i ($\times 10^{-8}$cm^2 s^{-1})
C332	4.33
C333	4.46
C334	4.07
C335	3.36
C336	3.02
Sample number	Starting years of steel corrosion in concrete (y)
C310	2.2
C320	3.6
C312	6.8
C321	7.8
C322	8.6
C331	8.3
C332	9.9
C333	9.6
C334	10.5
C335	12.8
C336	14.2

of chloride ions in concrete. Moreover, with the increase in the dosage, the diffusion coefficient of chloride ions decreases. The high-performance concrete with the additions of fly ash and slag micro powder simultaneously has a better effect on improving the durability against chloride ions than the single additions.

(2) The durability life of high-performance concrete in chloride ions resistance test is 3–4 times longer than that of ordinary concrete, and the inland subway can fully meet the 100 years chloride ions resistance durability life.

References

1. Huaxia, Z., Yuebo, C., Xunjie, C., etc. (2019). Review of stray current corrosions on reinforced concrete. *Concret, 6* 31–36.

2. Kai, H., Feng, X., Shuguang, W., et al. (2016). Influence of the coupled effect of stray current and sulfate upon chloride diffusing into concrete of underground structures. *Concret, 5*, 45–48.
3. Zhang Xueqin, X., Feng, W. S., et al. (2016). Influence of stray current on chloride ions transport mechanism in underground structure concrete. *Concret., 5*, 45–48.
4. Kunlin, M, Youjun, X., Can, L., et al. (2004). *Experimental study on the influence of concrete resistance to chloride ingress. Concret, 6*, 20–21, 32.
5. Ministry of Housing and Urban Rural Development of the People's Republic of China. (1985). Long term performance and durability test method for ordinary concrete: GBJ8..
6. Ministry of Transport of the People's Republic of China. (1998). Concrete Testing Regulations for Water Transport Engineering: JTJ270.
7. Renmin, L., & Songyu, L. (2007). Numerical analysis of migrations test based on multi-species model. *Concret, 8*, 34–36,43.
8. Zhinong, H. (1989). Study on chloride ions diffusions of fly ash cement mortar and concrete. Nanjing City: Nanjing Hydraulic Research Institute, Master's Thesis, pp. 32–38.
9. Bilderbeek, D. W. (1986). Durability of structures in marine environment recent Dutch research. In: *Collected Essays on the 26th International Shipping Conference*, pp. 56–62.
10. Dhir, R. K., Jones, M.R., & Ng, S. L. D. (1998). Predictions of total chloride content profile and concentrations/time-dependent diffusions coefficient for concrete. *Magazine of Concrete Research*, 39–47.
11. Zhongwei, W., & Huizhen, L. (1999). *High performance concrete* (pp. 1–5). China Railway Press.

Research on Point Cloud Reconstruction of Pumping and Storage Construction Site Under Ground 3D Laser Scanning Technology

Feng Cao, Pengtao Ying, Jianxu Zhong, and Jishuang Han

Abstract The current point cloud reconstruction matrix of the pumping and storage construction site is generally set to be unidirectional, and the efficiency of point cloud reconstruction is low, which leads to the prolongation of point cloud reconstruction time. Therefore, the design and verification of point cloud reconstruction method for pumping and storage construction site underground 3D laser scanning technology are proposed. According to the actual point cloud reconstruction requirements and standards, the 3D point cloud primitive library is constructed first, and the multi-step method is adopted to improve the efficiency of point cloud reconstruction. The multi-step point cloud reconstruction matrix is designed. Based on this, the point cloud reconstruction model of the ground 3D laser scanning and storage construction site is established, and the point cloud reconstruction processing is realized by building topology matching. The test results show that after the measurement and analysis of five areas, the final point cloud reconstruction time of the unit project is well controlled below 0.25 s, which indicates that, with the assistance and support of the ground 3D laser scanning technology, the current point cloud reconstruction for the G pumping construction project is more effective, more targeted and it has a larger coverage of point cloud reconstruction, which has practical application value.

Keywords Ground 3D laser · Laser scanning technology · Pumping and storage construction · On-site point cloud · Point cloud reconstruction · Reconstruction design

F. Cao (✉) · P. Ying · J. Han
Engineering Construction Management Branch of China Southern Power Grid Peak Shaving Frequency Modulation Power Generation Co., Ltd., Guangzhou 511402, Guangdong, China
e-mail: 22823401@qq.com

J. Zhong
China Southern Power Grid Energy Storage Co., Ltd., Information and Communication Branch, Guangzhou 511400, Guangdong, China

© The Author(s) 2025
D. Li and Y. Zhang (eds.), *Advances in Frontier Research on Engineering Structures II*,
Lecture Notes in Civil Engineering 535, https://doi.org/10.1007/978-981-97-6238-5_38

1 Introduction

Point cloud reconstruction, also known as 3D point cloud reconstruction, refers to reconstructing the surface and contour of buildings or objects according to the initial reconstruction target set by the 3D point cloud, restoring the external shape and contour of recognized objects to facilitate subsequent structural adjustment and construction reconstruction [1]. The 3D object surface reconstructed from the point cloud is generally composed of planes, a continuous state [2] is formed at the surface by a specific format. Point cloud reconstruction is a practical construction procedure commonly used at the pumping and storage construction site, which has achieved relatively good results in the initial application [3]. In general, new construction, capping and demolition are common processing links and tasks of pumping and storage construction, which are more related to construction planning, municipal roads, traffic routes, 3D maps, etc., facilitating the extraction of point cloud data and information [4]. In addition, at the pumping and storage construction site, texture confusion between ground objects, shading, mixed pixels, etc. are also key issues that affect the final point cloud reconstruction and even seriously hinder the implementation of follow-up tasks at the pumping and storage construction site [5]. To alleviate the above problems, the relevant construction personnel designed the point cloud reconstruction method of pumping and storing the construction site. The traditional reconstruction method is mostly a single-structure form. Although it can achieve the expected reconstruction processing goal, it lacks pertinence and stability. In addition to the impact of the external environment and specific factors, the final point cloud reconstruction result is difficult to achieve the expected effect [6]. Moreover, the efficiency of the single-point cloud reconstruction method of pumping and storing the construction site is too low, and the reconstruction link is also very complex and tedious, which makes it difficult to control the construction stability [7]. Therefore, the design and verification of point cloud reconstruction methods for pumping and storage construction site underground 3D laser scanning technology are proposed. The so-called Terrestrial Laser Scanning (TLS) mainly refers to a new surveying and mapping technology [8] integrating a variety of high-tech. The orientation of the scanning control module is adjusted through the scanning instrument. At the same time, the real-time angle of each pulse laser is measured and calculated in different environments, and the corresponding ground scanning point [9] is defined. The integration of this technology with the point cloud reconstruction work at the pumping and storage construction site can further expand the current point cloud reconstruction scope to a certain extent and gradually build a more flexible and changeable point cloud reconstruction structure [10]. In the actual application construction, with the assistance and support of the ground 3D laser scanning technology, the reconstruction of 3D point clouds can also be directly carried out using image-intensive matching, the structured point cloud location scheduling can be carried out, and the reconstruction prior processing can be realized through feature selection and equivalent balance design, providing convenient conditions for large-scale dense point cloud processing at the later stage of pumping and storage construction site [11].

In addition, the extraction and analysis of abstract point cloud features can also provide more assistance for power reconstruction. Reconstruction links with high initial complexity are simplified, specific and explicit, which greatly improves the recognition and measurement accuracy of point cloud reconstruction [12]. In [13], problems such as large amounts of point cloud data, unstructured spatial distribution and uneven point transport density are solved, and the point cloud reconstruction technology of the pumping and storage construction site is promoted to a new level of development [13].

2 Reconstruction Method of 3D Laser Scanning Point Cloud on the Ground of Design Pumping and Storage Construction Site

2.1 Construction of 3D Point Cloud Primitive Library

The so-called point cloud primitive library mainly refers to the basic matching types of point cloud reconstruction. The description forms mainly include plane, box, matrix, cone and other geometric models with single lines. In addition, the point cloud can be set and reconstructed in a combined way to form a diversified building shape [14]. In combination with the current measurement and reconstruction requirements, to build a targeted primitive library, two types of primitives need to be set, namely, basic primitives and hierarchical combination primitives [15].

Generally speaking, simple buildings mainly include a flat roof layer, a double slope layer, four slope layers and four corner top layers. The complex structure mainly consists of a double slope layer plus four slope layers, which present L type and T type. Combined with the ground 3D laser technology, a closed 3D solid structure is built to form a basic 3D point cloud primitive library, and the storage model of the primitive library is built, as shown in Fig. 1.

According to Fig. 1, the presentation of the storage model of the 3D point cloud primitive library and the directional design of the primitive library are completed. Next, with the assistance and support of ground 3D laser technology, the current depositor range and data have been further increased and improved, forming a complete and specific 3D point cloud primitive repository, which provides a reference basis for subsequent point cloud reconstruction and construction conversion. However, it should be noted that the coverage of the currently designed point cloud

Flat floor   Double slope L-shaped   pyramid roof

Fig. 1 Representation of 3D point cloud primitive database storage model

primitive database is not fixed, but it is adjusted in real-time with the change of building coverage and directional fluctuations. And then it completes the corresponding point cloud reconstruction goals and tasks with the flexibility and stability of the party.

2.2 Design of Reconstruction Matrix of Multi-Level Point Cloud

Compared with the traditional one-way point cloud reconstruction form, the multi-step point cloud reconstruction form designed this time is more flexible and refined and has certain point cloud change characteristics, which can further expand the coverage of the matrix and form an integrated point cloud reconstruction basic standard. First, the corresponding reconstruction standard is set and the matrix basic control indicators and values are set in combination with the current point cloud reconstruction target, as shown in Table 1.

According to Table 1, the setting and analysis of control indicators and values are completed for the reconstruction matrix of multi-level point clouds. Then, based on this, the recognition range of the current point cloud reconstruction matrix is adjusted, the basic reconstruction area is calibrated which can be covered, and the difference existing in the point cloud reconstruction process is strengthened through a multi-level way to lay a solid foundation.

Table 1 Control indicators and numerical settings of multi-order point cloud reconstruction matrix

Control index for multi-order point cloud reconstruction matrix	Directional parameter reference value	Measured parameter reference value	Controllable edge range
3D recognition difference	2.01	1.72	1.01–5.73
Mean value of point cloud primitives	10.71	15.72	10–19.28
Directional conversion ratio	3.2	2.1	1.5–5.3
Variable characteristic value	11.72	15.62	8.16–19.34

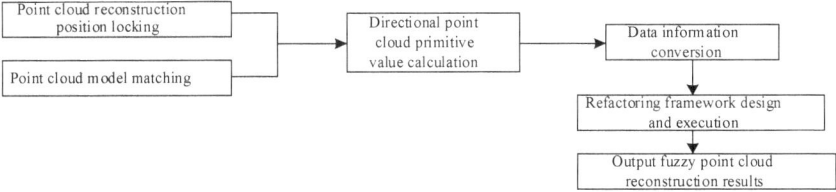

Fig. 2 Structure diagram of point cloud reconstruction model of ground 3D laser scanning pumping construction site

2.3 Establishment of Point Cloud Reconstruction Model of Ground 3D Laser Scanning Pumping Construction Site

After completing the design of the multi-step point cloud reconstruction matrix, the next step is to design the point cloud reconstruction model of the pumping construction site based on the ground 3D laser scanning technology. By carrying out point calibration relative to the location to be constructed, unidirectional or combined point cloud reconstruction models in the 3D primitive library are selected, matching with the current building and building the corresponding reconstruction framework, as shown in Fig. 2.

According to Fig. 2, the design and verification analysis of the point cloud reconstruction model structure of the ground 3D laser scanning and storage construction site were completed. Then, based on this, the three-dimensional scanning of buildings was carried out through the ground 3D laser scanning technology. At the same time, the data obtained were compared and adjusted with the data in the primitive database to ensure the authenticity and reliability of the subsequent point cloud reconstruction results.

2.4 Building Topology Matching for Point Cloud Reconstruction

The so-called building topology matching mainly refers to matching the results obtained from the reconstruction model of the ground 3D laser scanning and pumping construction site point cloud designed above with the model of the primitive library. After the difference is obtained, the point cloud is reconstructed through the building topology matching correction program, and the correction limit is calculated, as shown in Formula 1.

$$W = (e - v)^2 \times \tau h \qquad (1)$$

In Formula 1: W is the correction limit, e represents the conversion ratio, v represents the directional topological ratio, τ represents the corrected fixed value, h indicates the three-bit scanning range. Combined with the current measurement, according to the correction limit obtained, multidimensional correction is made to the results obtained from the model to ensure the authenticity and reliability of the final construction building point cloud reconstruction results.

3   Method Test

It is mainly to analyze and verify the actual application effect of the point cloud reconstruction method of the pumping and storage construction site under the ground 3D laser scanning technology. Considering the authenticity and reliability of the final test results, the analysis is carried out through comparison. The G pumping and storage construction project is selected as the main target of the test, and professional equipment is used to collect the current construction data and information. After the summary and integration, it is ready for subsequent use. Next, the initial test environment is built based on the ground 3D laser scanning technology.

3.1   Test Preparation

Combined with the 3D laser scanning technology on the ground, the test environment for the point cloud reconstruction method of the G pumping and storage construction project site was set up. The corresponding point cloud public data set was designed, accessing 10 airborne lidars in the measured point cloud reconstruction program. The program used was connected to the Leica ALS50 system by measuring the current point cloud average density between 4.25 and 65.67 points/m^2. CloudCompare software was used to adjust and process the basic point cloud data source, measure and study 5 selected areas in the project, form unit-building point cloud blocks, and filter the point cloud noise to ensure the authenticity and stability of the subsequent point cloud data practical application. In the current measurement environment, the basic point cloud reconstruction monitoring nodes are set for each block. Generally, the nodes are set independently, but once the point cloud reconstruction is performed, the nodes within the range need to be associated to form a circular point cloud reconstruction program. In combination with the RTG adjacency matrix of 3D primitives, first, the elevation value is filtered in the point cloud reconstruction process, the top point in the building is set as the candidate point for point cloud reconstruction, the basic point cloud reconstruction structure is designed after optimization, and the point cloud reconstruction boundary value is calculated, as shown in Formula 2.

$$D = (1 + l)^2 \times \eta - \sum_{i=1} \varepsilon i + \Re \qquad (2)$$

Table 2 Initial point cloud reconstruction indicators and parameter settings

Initial point cloud reconstruction indicator name	Standard values of basic controllable parameters	Standard values of measured controllable parameters	Edge controllable range
Fitted mean	16.35	18.11	10.25–21.05
Number of boundary candidate points/piece	12	18	10–22
Point cloud reconstruction phase	Data collection stage + point cloud combination stage + point cloud reconstruction stage	Data collection stage + point cloud combination stage + data adjustment stage + point cloud reconstruction stage	Basic construction and reconstruction area within the coverage area
Constraint condition	Data information, construction environment	Data information, construction environment, pumping range	Basic conditions
Objective function	10.25	16.34	10–25.64
Reconstruct controllable difference	1.16	0.96	0.25–1.35

In Formula 2, D represents the boundary value of point cloud reconstruction, l represents the matching value, η represents a reconstruction primitive, ε represents denoising value, i represents the reconstruction range, $\mathfrak{R}$ represents an objective function. Combined with the current measurement requirements and standards, by completing the calculation of the point cloud reconstruction boundary value, it is set as the basic point cloud reconstruction limit value and adjusted to the current reconstruction standards and constraints. Next, the initial point cloud reconstruction indicators and parameters are set, as shown in Table 2.

The initial point cloud reconstruction indicators and parameters are set according to Table 2. Then, based on this, the construction area of the project will be covered by 3D laser scanning technology, and the basic data and information will be collected and summarized for future use. So far, the analysis of the test results has been completed. Next, specific measurements and verification are carried out.

3.2 Analysis of Test Process Results

In the above-built test environment, combined with the ground 3D laser scanning technology, the point cloud reconstruction method of the G pumping construction site was measured. First, the nodes deployed above were used to collect the basic data

Fig. 3 Structure diagram of fuzzy point cloud identification at G pumping and storage construction location

and information about the building, and fuzzy point cloud recognition was conducted for the pumping construction location, as shown in Fig. 3.

According to Fig. 3, the design and practical analysis of the fuzzy point cloud recognition structure for the G pumping and storage construction location were completed. Then, the current point cloud reconstruction system was integrated, and the position of each link of the project and buildings were scanned directionally through the ground 3D laser scanning technology to form a basic scanning framework, and the double-layer perception point cloud reconstruction field value was calculated, as shown in Formula 3 below:

$$D = (1 - \alpha)^2 \times \aleph \sum_{n=1} \chi n + \delta \tag{3}$$

In Formula 3, D represents the reconstruction domain value of the double-layer perception point cloud, α indicates the controllable identification range, $\aleph$ indicates a directional reconstruction area, χ represents the point cloud shift value, n represents the number of point clouds, δ represents the stacking range, and completes the calculation of the reconstruction domain value of the double-layer perception point cloud in combination with the current measurement requirements. Then, based on this, the current building's point cloud reconstruction needs were identified, and the construction coverage of point cloud pumping and storage was included. Through the ground 3D laser scanning technology, combined with the current calibration needs and adjustment standards, the point cloud position was reconstructed. The construction location of the project is divided into five areas. The point cloud reconstruction task is performed according to the above method. The time consumption for point cloud reconstruction of the unit project is calculated, as shown in Formula 4.

$$A = \sum_{t=1} (\theta t - b)^2 \times \rho + \frac{1}{c} \tag{4}$$

In Formula 4, A indicates that the point cloud reconstruction of the unit project is time-consuming, θ represents the coverage of point cloud reconstruction, t represents the reconstruction frequency, b represents the field reference value, ρ represents the point cloud conversion ratio, c represents the vector mean. Combined with the current measurement, the calculation of the time consumption is completed for the point cloud reconstruction of the unit project. According to the current measurement requirements and standards, the analysis of the test results is completed, as shown in Table 3.

Table 3 Comparison and Analysis of Test Results

Basic testing area	Point cloud grid unit value	Point cloud reconstruction coverage ratio	Unit project point cloud reconstruction time consuming/s
Region1	10.71	3.81	0.21
Region2	11.82	4.27	0.19
Region3	14.28	3.02	0.13
Region4	12.82	4.21	0.23
Region5	13.22	3.72	0.24

According to Table 3, the analysis of the test results is completed. After the measurement and analysis of five areas, the final point cloud reconstruction time of the unit project is better controlled below 0.25 s, which indicates that, with the assistance and support of the ground 3D laser scanning technology, the current point cloud reconstruction for the G pumping and storage construction project is more effective and targeted, and the coverage of point cloud reconstruction is larger. It has practical application value.

4 Conclusion

To sum up, this is the design and verification research on the point cloud reconstruction method of the pumping and storage construction site under the ground 3D laser scanning technology. Different from the initial point cloud reconstruction method, the current combination of the ground 3D laser scanning technology makes the designed point cloud reconstruction form of the pumping and storage construction site more flexible and changeable with stronger pertinence and stability. It can enhance the effect and coverage of point cloud reconstruction in complex background environments. In addition, combined with UAV tilt photogrammetry technology, 3D laser scanning technology can also be used to determine the actual reconstruction range and obtain the point cloud data and information of each link. After preprocessing, a filtering algorithm or depth learning method is used to conduct combined summary processing on the data to achieve rapid and intelligent classification of building point clouds. In the process, it is necessary to classify and eliminate the abnormal point cloud data to ensure the subsequent reconstruction effect and set the reconstruction space with the point cloud. To some extent, it can alleviate the noise, complex boundaries and other problems in point cloud reconstruction, improve the effect of point cloud reconstruction, and lay a solid foundation for the development and improvement of subsequent related technologies.

References

1. Yu, Q., Yang, C., & Wei, H. (2022). Part-wise AtlasNet for 3D point cloud reconstruction from a single image. *Knowledge-Based Systems, 242*, 108395.
2. Yang, Y., Zhang, J., Wu, K., Zhang, X., Sun, J., Peng, S., Li, J., & Wang, M. (2021). 3D point cloud on semantic information for wheat reconstruction. *Agriculture, 11*(5), 450.
3. Li, G., Zhu, W. D., Dong, H., & Ke, Y. (2022). Error compensation based on surface reconstruction for the industrial robot on the two-dimensional manifold. *Industrial Robot: The International Journal of Robotics Research and Application, 49*(4), 735–744.
4. Huang, J., & Huang, L. (2021). Research on 3D modeling of Wupaolong based on sparse point cloud reconstruction of SFM algorithm. *Journal of Imaging Science and Technology.*
5. Wang, P., Liu, L., Zhang, H., & Wang, T. (2021). CGNet: A cascaded generative network for dense point cloud reconstruction from a single image. *Knowledge-Based Systems, 223*, 107057.
6. Puliti, M., Montaggioli, G., & Sabato, A. (2021). Automated subsurface defects' detection using point cloud reconstruction from infrared images. *Automation in Construction, 129*, 103829.
7. Pan, Z., Hou, J., & Yu, L. (2022). Optimization algorithm for high precision RGB-D dense point cloud 3D reconstruction in indoor unbounded extension area. *Measurement Science and Technology, 33*(5), 055402.
8. Liu, X., Zheng, W., Mou, Y., Li, Y., & Yin, L. (2021). Microscopic 3D reconstruction based on point cloud data generated using defocused images. *Measurement and Control, 54*(9–10), 1309–1318.
9. Javadnejad, F., Slocum, R. K., Gillins, D. T., Olsen, M. J., & Parrish, C. E. (2021). Dense point cloud quality factor as a proxy for accuracy assessment of image-based 3D reconstruction. *Journal of Surveying Engineering, 147*(1), 04020021.
10. Gao, H., Yu, J., Sun, J., Yang, W., Jiang, Y., Zhu, L., & Ju, Z. (2021). Robust 3D model reconstruction based on continuous point cloud for autonomous vehicles. *Sensors & Materials, 33.*
11. Li, Y., Gao, J., Wang, X., Chen, Y., & He, Y. (2022). Depth camera-based remote three-dimensional reconstruction using incremental point cloud compression. *Computers and Electrical Engineering, 99*, 107767.
12. Wen, S., Liu, X., Zhang, H., Sun, F., Sheng, M., & Fan, S. (2021). Dense point cloud map construction based on stereo VINS for mobile vehicles. *ISPRS Journal of Photogrammetry and Remote Sensing, 178*, 328–344.
13. Rao, J., Wang, J., Kollmannsberger, S., Shi, J., Fu, H., & Rank, E. (2022). Point cloud-based elastic reverse time migration for ultrasonic imaging of components with vertical surfaces. *Mechanical Systems and Signal Processing, 163*, 108144.
14. Yuniarti, A., Arifin, A. Z., & Suciati, N. (2021). A 3D template-based point generation network for 3D reconstruction from single images. *Applied Soft Computing, 111*, 107749.
15. Xi, L., Zhao, Y., Chen, L., Gao, Q. H., Tang, W., Wan, T. R., & Xue, T. (2021). Recovering dense 3D point clouds from a single endoscopic image. *Computer Methods and Programs in Biomedicine, 205*, 106077.

Research on 3D Visualization Modeling Method of Pumped Storage Power Plant

Zengtao Zhao, Feng Cao, Li Chen, and Fanqi Huang

Abstract Pumped storage plants usually occupy a large area and contain multiple components and facilities. The details and actual geomorphological features of the entire area are taken into account when modeling to provide comprehensive visualizations. However, dealing with large-scale scenarios increases the amount of data and computational complexity, further increasing the difficulty. Therefore, a new 3D visualization modeling method of pumped storage power plant is designed. BRCM software is selected as the 3D visualization dynamics modeling and analysis software to form an effective 3D visualization modeling design flow, transient analysis of 3D nonlinear dynamics, optimization of the accuracy of real scene 3D modeling, and finally the realization of 3D visualization modeling. The result of case analysis shows that the designed 3D visualization modeling method of pumped storage power plant has high modeling precision, good modeling effect, reliability, and certain application value, and has made a certain contribution to improving the operation safety of pumped storage power plant.

Keywords Pumped storage · Power plant · Real scene · Three dimensions · Visualization · Modeling · Way

1 Introduction

Under the background of computer development and information technology progress, various digital technologies have emerged and been applied in various fields such as engineering construction. Pumped storage power plant is a common hydropower station, which can not only realize peak regulation and frequency modulation of power grid, but also reduce the operating load [1–3] of power grid. Pumped storage power stations can use the peak-valley relationship of power supply and distribution load to realize electric energy conversion, improve the operation efficiency of the power grid and keep the power grid in a stable state all the time. Therefore,

Z. Zhao (✉) · F. Cao · L. Chen · F. Huang
China Southern Power Grid Energy Storage Co. Ltd., Guangzhou 511400, Guangdong, China
e-mail: xyzztpro@qq.com

© The Author(s) 2025
D. Li and Y. Zhang (eds.), *Advances in Frontier Research on Engineering Structures II*,
Lecture Notes in Civil Engineering 535, https://doi.org/10.1007/978-981-97-6238-5_39

pumped storage power stations [4–6] of different scales have been built in China in recent years. Under the background of new energy development, the configuration of pumped storage power stations in China has changed to varying degrees, and gradually changed from the original single power supply and distribution load center to a diversified power supply and distribution center. However, the pumped storage power station is always in a state of high load and long-term operation, so it is very prone to operation failure [7–10], which causes abnormal power supply and distribution of the whole power grid. In order to improve the operation reliability of the pumped storage power station, it is necessary to build a visual three-dimensional model.

In fact, in recent years, China's power demand is increasing day by day, and it is in the era of rapid development of power construction. Therefore, there are often problems of tight power supply and distribution and insufficient peak reserve of power supply and distribution, which is difficult to meet the requirements of large industrial enterprises. In addition, users are increasingly demanding the power quality of the load side [11, 12], and it is difficult to meet the peak shaving requirements if only distributed energy storage is carried out. Pumped storage is a reliable energy storage technology, which needs to rely on high-performance energy storage devices. The research shows that the existing pumped storage power stations in China are mainly divided into several different types. The first type is pure pumped storage power station, which mainly completes the reciprocating utilization of reservoirs in a specified period and undertakes the task of integrated power generation; The second type is a hybrid pumped storage power station. This type of pumped storage power station is mainly composed of generator sets and hydraulic turbines, which has high complexity and is prone to failure [13–16], but its integration effect is relatively good and its functions are rich. According to the operation characteristics of pumped storage power plants, some conventional three-dimensional visualization modeling methods are designed. The first one is the three-dimensional visualization modeling method of pumped storage power plants based on GIS, and the second one is the three-dimensional visualization modeling method of pumped storage power plants considering the complex combination relationship. However, the conventional three-dimensional visualization modeling method mainly uses BRCM three-dimensional laying technology to complete three-dimensional layout, which is easily influenced by local fuzzy optimization, resulting in low modeling precision and does not meet the current requirements of three-dimensional visualization modeling. Therefore, this paper designs a new three-dimensional visualization modeling method for pumped storage power plants.

2 Design of 3D Visualization Modeling Method for Pumped Storage Power Plant

2.1 Transient Analysis of 3D Nonlinear Dynamics Modeling

There is hydraulic coupling problem in pumped storage power plant. Therefore, in order to reduce the influence of hydraulic power on three-dimensional visual modeling, the method designed in this paper first carries out transient analysis of three-dimensional nonlinear dynamic modeling. Firstly, the flow rate is decomposed according to the coupling effect of each pipe section of the pumped storage power plant [17–19], and the excessive calculation is carried out according to the dynamic characteristics of the hydraulic turbine. At this time, the schematic layout of the pipe sections is shown in Fig. 1.

It can be seen from Fig. 1 that whether there is an elastic water hammer relationship in the pipeline can be determined by combining the above schematic layout of the pipeline section, so as to judge the hydraulic difference of the pumped storage power station. In order to improve the dynamic response function of the three-dimensional visual model, PID controller can be used to optimize the control parameters, assuming the basic operating parameters of the power plant [20–22], and numerical simulation can be carried out.

In this paper, BRCM software is selected as the 3D visual dynamic modeling and analysis software, which can be linked with Substation to define parameters and generate an effective 3D visual modeling and design process, as shown in Fig. 2.

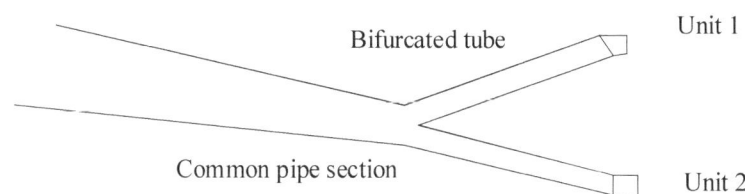

Fig. 1 Schematic diagram of pipe section layout

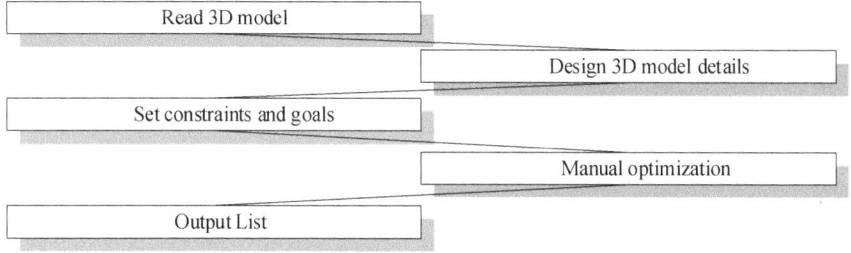

Fig. 2 3D visual modeling design process

As can be seen from Fig. 2, in the process of three-dimensional visual modeling, the first step is to read the three-dimensional model and match it according to the relevant parameters of the pumped storage power plant. The second step is to judge the conflict relationship of modeling according to the corresponding boundary conditions and make optimization and adjustment. Step 3, set effective constraints and optimization objectives to avoid model conflicts; Step 4, optimize the model and manually adjust the model optimization path; The fifth step is to output the modeling list and number different modeling points to improve the reliability of modeling.

The operation parameters of pumped storage power plant are different under different working conditions, and limit parameters may appear. Therefore, in the process of transient analysis, it is necessary to judge the conditions of extreme value according to the operating state of each working condition and over-calculate the working conditions to ensure the stability of three-dimensional visual modeling. Under the normal operation conditions of transmission stations, there is the problem of calculating the parameters of hydraulic over-boundary. Therefore, the induced factors of modeling can be judged by considering the load increase and decrease of hydropower stations and the relationship between operation and mining, and the hydraulic over-condition can be checked while ensuring the reliability of unit operation, so as to minimize the destructive impact of over-construction on model construction and improve the accuracy of the model.

2.2 Generate a Three-Dimensional Visualization Model of Pumped Storage Power Plant

After the transient analysis of the three-dimensional nonlinear dynamics modeling, the modeling data can be predicted visually. In this paper, the fault tree model is used to judge the logical relationship between parameters, and the fault diagnosis knowledge center of pumped storage hydropower station is generated. Then, according to the mapping relationship of faults, the automatic judgment rule is obtained, and the change symptoms of fault data are analyzed to support the subsequent model optimization construction. The three-dimensional visualization model of pumped storage power plant includes a lot of information, so the sensitivity index of the main dam and the sensitivity index of the auxiliary dam can be selected as the model construction index, and the model construction parameters can be selected. At this time, the three-dimensional visualization model of pumped storage power plant is shown in Fig. 3.

As can be seen from Fig. 3, statistics can be carried out according to the above-mentioned three-dimensional visualization model to obtain statistical displacement parameters as shown in the following (1).

$$\sigma = \sigma_a + \sigma_s \tag{1}$$

Fig. 3 Three-dimensional visualization model of pumped storage power plant

In the formula (1), σ_a represents the displacement water pressure of pumped storage power plant, σ_s represents the aging component. At this time, a statistical model of seepage pressure can be generated by combining the aging component parameters of pumped storage power plants H, , as shown in the following (2).

$$H = \sum_{i=1}^{n} a_i H_I \tag{2}$$

In the formula (2), a_i represents the water pressure component of the power plant, H_I represents the periodic harmonic factor, the periodicity of pumped storage power station will change under unstable seepage pressure conditions. Therefore, the modeling method designed in this paper introduces the periodic influence factor to optimize the model parameters p, , as shown in the following (3).

$$p = f(h) + f(H) \tag{3}$$

In the formula (3), $f(h)$ represents modeling time parameters, $f(H)$ represents the modeling data parameters. After the parameter optimization is completed, the relationship between periodic harmonics and temperature change needs to be considered, and the comprehensive modeling parameters C is obtained at this time as shown in the following (4).

$$C = B \cdot \sum_{i=1}^{n} a_i + \tau \tag{4}$$

In the formula (4), B represents a nonlinear support vector, τ represents the three-dimensional displacement index, the three-dimensional visual modeling method of pumped storage power plant can effectively adjust the output results of the model and improve the fitting of the model to the greatest extent.

3   Example Analysis

3.1   Overview and Preparation

For example, a pumped storage power station has a total installed capacity of 1200 megawatts (MW) and a total of six generating units. Each generating unit has an installed capacity of 200 MW. The pumped storage power station has two reservoirs, an upstream reservoir and a downstream generator cell. As needed, water can be pumped from the upstream reservoir to the downstream generator cell, using excess power to pump water from the downstream back upstream, forming a continuous cycle. Based on this example, according to the analysis requirements of point-to-point visual modeling in pumped storage power plant, this paper selects Renderbus cloud rendering platform as the experimental platform. The real-life modeling of pumped storage power plant includes polygon modeling, satellite data gridding and other steps. Based on this, this paper integrates information in the process of case analysis and sets up a unified modeling code, as shown in Table 1.

From Table 1, it can be seen that the model can be further lightened according to the above model coding, and the model accuracy level can be set according to the requirements of case analysis and the required dimensions, and then the model volume can be reduced by mosaic subdivision method, and BIM data can be imported. In order to reduce the difficulty of case analysis, this paper uses Revit as the design parameter processing software to render, which improves the reliability of the experimental model to the greatest extent. The number of segments of the control surface determines the final performance of the model. In this paper, Pixyz Stidio software is used to convert polygons, and then Decimate tool is used to reduce the amount of meshes, so as to avoid the damage of the model and realize the lightweight of the model. The infrastructure of the experimental platform generated at this time is shown in Fig. 4.

The 3D visualization effect of pumped storage power plant is shown in Fig. 5.

The scene of pumped storage power plant presented by three-dimensional visualization technology can clearly show the structure and layout of pumped storage power plant, so that viewers can more accurately understand its structure and operation mode. The 3D visualization technology can show the details and operation process of the pumped storage power plant, so that people can more intuitively understand its working principle, and enhance the viewer's cognition of the facility.

Table 1 Three-dimensional visual modeling coding of pumped storage power plant

Model number	Model name	Model encoding	Model program coding
1	Pumped storage power station	JHSDZ	DZQ
2	Upper and lower reservoirs	JHSDZ_SD	DZQ_ ShuiKu
3	Upper reservoir	JHSDZ_SSD	DZQ_ ShangSK
4	Concrete face rockfill dam	JHSDZ_MBDSB	MianBanDuiShiBa
5	Inlet and outlet	JHSDZ_JCSK_S	JinChuShuiKou_ S
6	Emergency gate shaft	JHSDZ_SGZMJ	ShiGuZhaMenJing
7	Concrete face	DZQ_XK	DZQ_ XiaXK
8	Inlet and outlet	JHSDZ_HNTMBB	HunNingTuMianBanB
9	Maintenance gate shaft	JHSDZ_JCSK_X	JinChuShuiKou_X
10	Underground powerhouse	JHSDZ_JXZMJ	JianXiuZhaMenJing
11	Main power house	JHSDZ_DXCF	DZQ_ DiXiaChangFang
12	Main power house	JHSDZ_ZCF	ZhuChangFang
13	Access tunnel	JHSDZ_TFD	TongFengDong
14	Drainage gallery	JHSDZ_JTD	JiaoTongDong
15	Drainage gallery	JHSDZ_PSLD	MuXianDong
16	Main Electrical Wire Hall	JHSDZ_JTDLD	JiaoTongDianLanDong
17	Tailgate chamber cable hole	JHSDZ_WSDLD	WeiZhaShiDianLanDong

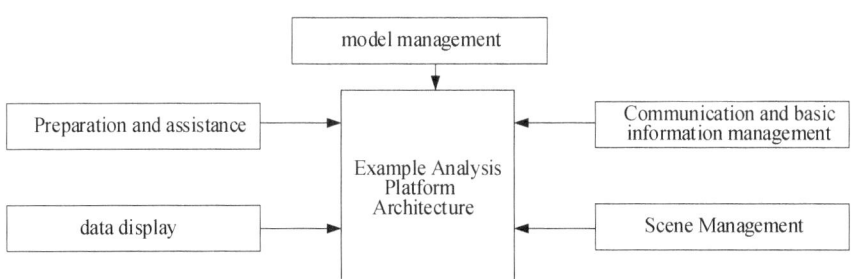

Fig. 4 Experimental platform infrastructure

3.2 Index Verification

Combined with the above general situation and preparation, we can analyze the three-dimensional visual modeling examples of pumped storage power plant, that is, we

(a) Pumped storage power plant reality   (b) External reality   (c) Interior reality

Fig. 5 3D visual rendering of pumped storage power plant

use the three-dimensional visual modeling methods of pumped storage power plant designed in this paper, the three-dimensional visual modeling methods of pumped storage power plant based on GIS and the three-dimensional visual modeling methods of pumped storage power plant based on complex combination, and use formula (5) to calculate the modeling accuracy of the three methods.

This paper selects the modeling fineness S_A as an example analysis index, the calculation formula of this index is shown in the following (5).

$$S_A = \left(\frac{D_R + C_V + Q_C}{F_A} \right) \times 100\% \tag{5}$$

In the formula (5), D_R represents the model accuracy parameters, C_V represents the model texture clarity parameter, Q_C representative model detail parameters, F_A represents the comprehensive parameters of the model rendering appearance, the higher the modeling fineness, the better the modeling effect of the real-life 3D visualization modeling method; conversely, the lower the modeling fineness, the worse the modeling effect of the real-life 3D visualization modeling method; and effective case analysis results can be obtained by calculation according to the selected case analysis indicators.

The results of the example analysis are shown in Table 2.

As can be seen from Table 2, the 3D visualization modeling method of pumped storage power plant designed in this paper has higher modeling precision under different types of models, while the 3D visualization modeling method of pumped storage power plant using GIS and the 3D visualization modeling method of pumped storage power plant considering complex combination relationship have lower modeling precision under different types of models. The analysis results of the above examples prove that the modeling effect of the three-dimensional visualization modeling method of pumped storage power plant designed in this paper is good, reliable and has certain application value.

Table 2 Example analysis results

Model Naming Number	The modeling precision of the 3D visualization modeling method for pumped storage power plants designed in this article (%)		Modeling Precision of 3D Visualization Modeling Method for Pumped Storage Power Plants Using GIS (%)		Modeling precision of 3D visualization modeling method for pumped storage power plants considering complex combination relationships (%)	
	Models with relatively large dimensions	Models with relatively small dimensions	Models with relatively large dimensions	Models with relatively small dimensions	Models with relatively large dimensions	Models with relatively small dimensions
001-1.CATPart	95.5453	99.4567	65.4512	65.8443	65.8745	65.8952
002-2.CATPart	92.5235	95.5525	62.9625	61.4251	51.1142	61.4473
003-1.CATPart	96.7452	94.4556	66.2658	72.3659	62.3958	62.8854
004-2.CATPart	95.8446	95.7284	63.2745	73.8741	73.4755	73.2784
005-1.CATPart	94.8441	96.5414	75.4168	66.3692	56.2566	66.6539
006-2.CATPart	95.4823	98.7455	69.2856	59.4452	69.8545	79.8554
007-1.CATPart	98.8541	94.1456	55.8745	78.3874	75.8586	65.2865
008-1.CATPart	94.2775	95.8122	64.8463	54.5523	74.2545	54.1472
009-1.CATPart	95.3856	96.8995	55.8472	65.9985	75.4848	65.3689
010-1.CATPart	96.4523	93.6858	65.3965	52.3648	68.5418	58.6852
011-2.CATPart	99.9854	96.2368	61.8545	65.5523	64.4462	74.4525
012-2.CATPart	98.2863	95.5442	62.6212	74.6954	75.4723	65.3259
013-1.CATPart	94.8541	95.4139	63.9853	62.2545	52.4556	62.8745
014-2.CATPart	95.4785	94.8254	76.8463	53.3884	73.8522	66.4536
015-1.CATPart	96.3985	98.3255	78.8415	66.2653	76.1565	68.5258
016-2.CATPart	99.4235	95.4723	78.8456	58.8956	69.8523	74.4698
017-1.CATPart	98.9678	98.8547	65.2538	64.2659	75.4965	53.7414
018-2.CATPart	94.7456	94.2356	66.8494	58.7474	64.8414	76.6397

4 Conclusion

Pumped storage power plant is a common place for power conversion and peak-shaving distribution, which is of great significance for reducing power supply and distribution load and improving power supply and distribution stability. In the era of economic development and information progress, the demand for electricity in China is increasing day by day, and the composition of pumped storage power plants is becoming more and more complicated, which is difficult to manage. Three-dimensional visualization technology can build a three-dimensional model according to the actual operation state and operation components of pumped storage power plant, manage it online and realize real-time monitoring. Therefore, it is necessary to design an effective three-dimensional visualization modeling method for pumped storage power plant. Conventional three-dimensional visualization modeling methods often ignore the modeling details, and the modeling accuracy is low, which does not meet the requirements of intelligent supervision of pumped storage power plants. Therefore, this paper designs a brand-new three-dimensional visualization modeling method for pumped storage power plants. The example analysis shows that the three-dimensional visualization modeling method of pumped storage power plant has good modeling effect, accuracy and certain application value, and has made certain contributions to promoting the development of intelligent management of pumped storage power plant.

References

1. Li, K., & Zhang, X. (2022). Research on key feature point recalibration methods for ecological landscape 3D models. *Computer Simulation, 11*, 366–370.
2. Guo, Z., Li, C., Cao, Y., Jiang, L., Zhang, Y., Li, P., & Lu, H. (2022). 3D visualization and morphometric analysis of spinal motion segments and vascular networks: A synchrotron radiation-based micro-CT study in mice. *Journal of Anatomy, 240*(2), 268–278.
3. Zhang, X., Hu, B., & Zhang, Z. (2023). A three-dimensional angiogenesis model with time-delay. *Discrete & Continuous Dynamical Systems-Series B, 28*(3).
4. Li, X., Zhou, H., Liu, H., & Chen, Z. (2021). Three-dimensional analytical continuum model for axially loaded noncircular piles in multilayered elastic soil. *International Journal for Numerical and Analytical Methods in Geomechanics, 45*(18), 2654–2681.
5. Wang, J., Shi, Q., Akankwasa, N. T., Zhang, Y., & Wang, J. (2022). Establishment of the three-dimensional model of the nonwoven structure with fiber scale based on the GAN algorithm. *Textile Research Journal, 92*(9–10), 1656–1665.
6. Brzozka, Z., Sokolowska, P., Zukowski, K., Janikiewicz, J., Jastrzebska, E., & Dobrzyn, A. (2020, May). Lab-on-a-chip system for developing and fluorescence imaging a three-dimensional model of pancreatic islets under flow conditions. In *Electrochemical Society Meeting Abstracts 237* (No. 27, pp. 1984–1984). The Electrochemical Society, Inc.
7. L'Hostis, A., & Abdou, F. (2021). What is the shape of geographical time-space? A three-dimensional model made of curves and cones. *ISPRS International Journal of Geo-Information, 10*(5), 340.
8. He, Y. (2021, February). On the research of three-dimensional model supported by computer data collection in the teaching of highway route selection. In *Journal of Physics: Conference Series* (Vol. 1744, No. 4, p. 042085). IOP Publishing.
9. Wiyanto, W., & Hidayah, I. (2021, June). Review of a scientific creativity test of the three-dimensional model. In *Journal of Physics: Conference Series* (Vol. 1918, No. 5, p. 052088). IOP Publishing.
10. Wang, K., Zhao, N., He, Q., & Xu, J. (2021). A quality analysis method for three-dimensional model in aircraft structural parts design. *Advances in Mechanical Engineering, 13*(3), 16878140211008044.
11. Łukasik, A., Szuszkiewicz, M., Wanic, T., & Gruba, P. (2021). Three-dimensional model of magnetic susceptibility in forest topsoil: An indirect method to discriminate contaminant migration. *Environmental Pollution, 273*, 116491.
12. Zhao, X., & Xiao, S. (2021). A three-dimensional model of gas foil bearings and the effect of misalignment on the static performance of the first and second generation foil bearings. *Tribology International, 156*, 106821.

13. Cheng, W., Chen, Z., Zeng, L., Yang, X., Huang, D., Zhai, Y., Lai, M., Zheng, L., Thomashow, L. S., Weller, D. M., Yu, Z., & Zhang, J. (2021). Control of Meloidogyne incognita in three-dimensional model systems and pot experiments by the attract-and-kill effect of furfural acetone. *Plant Disease, 105*(8), 2169–2176.
14. Mustafa, S., Bahar, A., Abidin, A. R. Z., Aziz, Z. A., & Darwish, M. (2021). Three dimensional model for solute transport induced by groundwater abstraction in river-aquifer systems. *Alexandria Engineering Journal, 60*(2), 2573–2582.
15. Ishioka, K., Yamamoto, N., & Fujita, M. (2022). A formulation of a three-dimensional spectral model for the primitive equations. *Journal of the Meteorological Society of Japan. Ser. II, 100*(2), 445–469.
16. Kolozvari, K., Kalbasi, K., Orakcal, K., & Wallace, J. (2021). Three-dimensional shear-flexure interaction model for analysis of non-planar reinforced concrete walls. *Journal of Building Engineering, 44*, 102946.
17. Conigliaro, E., Monti, P., Leuzzi, G., & Cantelli, A. (2021). A three-dimensional urban canopy model for mesoscale atmospheric simulations and its comparison with a two-dimensional urban canopy model in an idealized case. *Urban Climate, 37*, 100831.
18. Chen, W., Wang, L., Yan, Z., & Luo, B. (2021). Three-dimensional large-deformation model of hard-magnetic soft beams. *Composite Structures, 266*, 113822.
19. Wang, J., Schweizer, D., Liu, Q., Su, A., Hu, X., & Blum, P. (2021). Three-dimensional landslide evolution model at the Yangtze River. *Engineering Geology, 292*, 106275.
20. Farokhi, H., & Erturk, A. (2021). Three-dimensional nonlinear extreme vibrations of cantilevers based on a geometrically exact model. *Journal of Sound and Vibration, 510*, 116295.
21. Wang, H. (2021). Construction and discrete processing of screw extruder based on three-dimensional spiral model. *Microprocessors and Microsystems, 82*, 103946.
22. Murphy, R., & Wong, I. (2021). Creation of a three-dimensional printed model for the preoperative planning of hip arthroscopy for femoral acetabular impingement. *Arthroscopy Techniques, 10*(4), e1143–e1147.

Research of Monitor System for Water Lock Gate Based on Digital Twin Technology

Qiyong Liu and Ming Li

Abstract As climate change becomes more severe, droughts and floods are causing increasing losses to people's lives and property. How to fully utilize modern computer technology, communication technology, artificial intelligence, and 3D simulation technology to build a comprehensive control system for water lock gates of a whole river has become a key research focus. The ultimate goal is to achieve remote real-time monitoring of changes in the hydrological conditions of the river basin and precise control of water gates to regulate water resource utilization and reduce disasters. This article introduces a digital twin system for comprehensive control of water gates. It will model buildings and equipment into 3D models, and convert measurement data obtained from sensors into real-time measurement data into 3D numerical model mapping data, reflecting the real-time status of buildings and equipment. Based on real-time data and HYDRONETS models, disaster prediction will be carried out to achieve precise regulation of water resources and reduce disaster losses.

Keywords Digital Twin · Monitoring sensor disaster prediction · LSTM

1 Introduction

With the intensification of the Earth's greenhouse effect in recent years and the increasing frequency of abnormal weather, the resulting flood disasters have caused significant losses to people's lives, property, and social production [1]. Therefore, Our country has increased efforts to invest money and resources in building water conservancy facilities [2]. A large number of culvert gates have been built especially on the main area of the rivers and lakes. At the same time, with the development of computer

Q. Liu (✉)
Water Supply Bureau of Binzhou Yellow River Bureau, Binzhou 256600, Shandong, China
e-mail: 183420399@qq.com

M. Li
Zibo Yellow River Bureau, Zibo 255000, Shandong, China

D. Li and Y. Zhang (eds.), *Advances in Frontier Research on Engineering Structures II*,
Lecture Notes in Civil Engineering 535, https://doi.org/10.1007/978-981-97-6238-5_40

information technology, internet technology, cloud technology, and digital twin technology, those new technologies have also been widely applied in building remote monitoring systems of water conservancy facilities, providing technical support for efficient operation of these water conservancy facilities to serve disaster prevention and comprehensive utilization of water resources [3]. This article introduces the research on water gate monitoring system based on digital twin technology, which applies internet technology, cloud computing technology, digital twin technology and LSTM [4] model to the comprehensive control system of water gates and implements functions such as remote monitoring, unmanned operation, virtual simulation, and disaster analysis [5].

2   Overall Framework of Water Gate Monitoring System

This system is mainly designed according to the architecture of the Internet of Things system, including five layers: Sensor layer, Transport layer, Data layer, Business layer, and Application layer. The whole system framework is showing as Fig. 1.

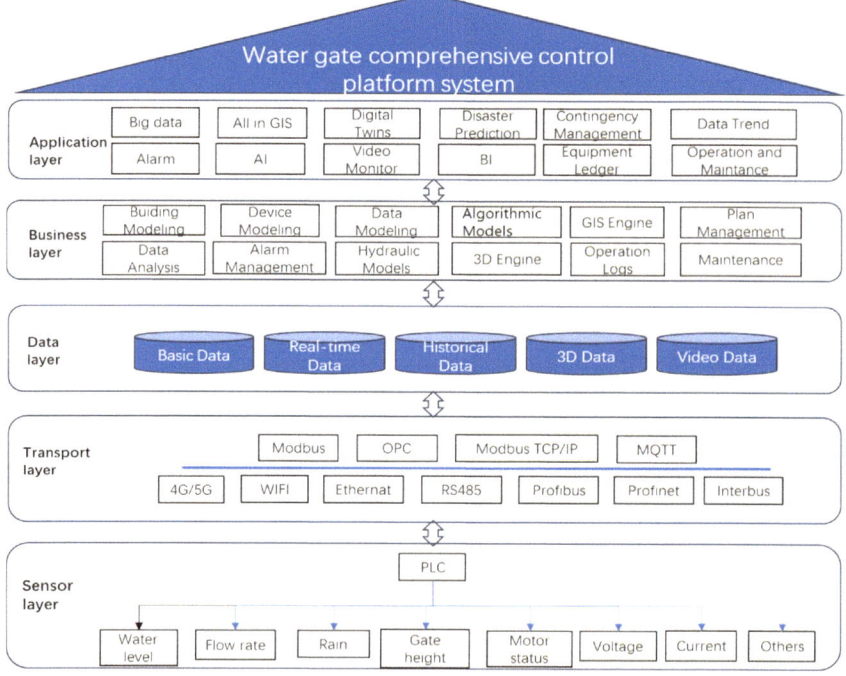

Fig. 1 Overall framework of this system

2.1 Sensor Layer

The Sensor Layer collects real-time data on various states of the water gate, including water levels before and after the gate, flow rate, rainfall data from weather forecast, gate lifting height, lifting motor working status, electric current, electric voltage, and other data. After those sensors complete data collection, those data are transmitted to digital data into the PLC, which packages the data through a remote data interface and uploads them to the server.

2.2 Transport Layer

The transport layer is the medium and protocol for transmitting data between the underlying device and the server. The medium supports networks such as RS485, Ethernet, fiber optic network, 4G/5G, WiFi, etc. The data protocols include OPC, Modbus, MQTT, etc.

2.3 Data Layer

The data layer completes real-time data processing and storage, forming alarm data, refreshing real-time data, recording historical data, this layer also includes basic data for water gate equipment and facilities. The database adopts MySQL or other Database operation system.

2.4 Business Layer

The business logic module for implementing comprehensive control of water gates in the business layer, including real-time data display, alarm management, data analysis, 3D engine, GIS engine, AI recognition, hydraulic model, algorithm model, etc., provides support for the application layer.

2.5 Application Layer

The application layer provides specific business management functions for users, all data in a map, big data display, digital twins, AI recognition, alarm management, operation and maintenance management, emergency management, flood analysis, etc.

3   The Main Modules of This Digital Twins System

This digital twin system is composed by 9 parts including Physics Data, Mathematic Models, GIS Data, Sensor Data, Digital Twin, Rain Condition Data, Remote Monitor, Maintenance and Simulation as showing in Fig. 2.

3.1   Physics Data

The Physics Data includes the real physics size and property data of water lock gate's facilities. It can be obtained from design and construction drawings and documents. Some data can be measured by manual measurement or photographic measurement.

3.2   Mathematic Models

Mathematic Models are collections of useful mathematic functions which implement transmission of real data of sensor to virtual change of 3D models and predicting the trend of development of flood by inputting rainfall condition data and other environment parameters.

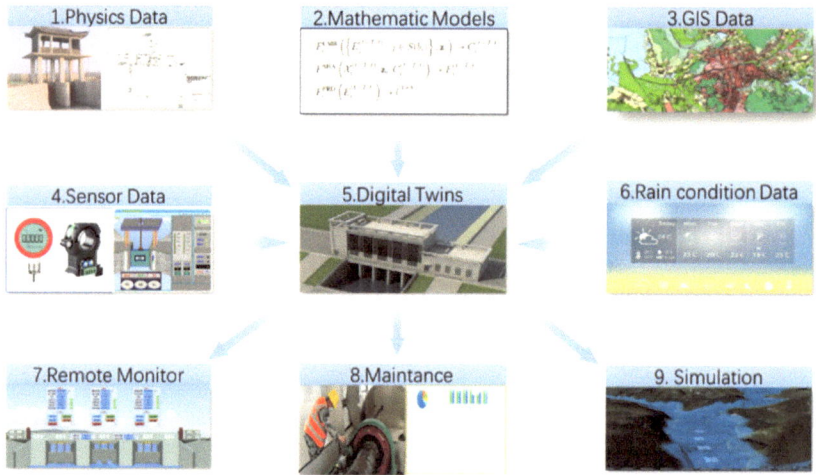

Fig. 2   Data Twin of this system

3.3 GIS Data

In order to visually display the water gate's position and device's status, a three-dimensional GIS was used. The 3D water gate model is put on the 3D GIS like Cesium to more effectively display the impact of water control on the environment.

3.4 Sensor Data

Sensor Data is as one of system inputs for this digital twin system which will display correctly in 3d models.

3.5 Digital Twin

This part is the key component of the whole system which is built by 3D Max and driven by Unity3D engine. 3D models include building model, equipment's models and functions of actions and events.

3.6 Rain Condition Data

The rain condition data is from the weather forecast via extracting data on internet as an input parameter to predict the future flow rate of the target river.

3.7 Remote Monitor

This module is very important for the operator to monitor and control the water gate's devices anytime and anywhere using internet through remote display interface.

3.8 Maintenance

By recording the operation and maintenance records of the water gate, operators can maintenance in time and keep the water gate facilities and equipment in good status to ensure the normal operation of the system.

4 Build Models

The digital twin of water gate is built by steps as showing in Fig. 3. Firstly, CAD drawings of water gate are imported into 3D Max to generate 3D models. Secondly, the digital twin models are regenerated in Unity3D by importing 3D models and calibrated by actual measurement data or drone photography. Then relation models including real data transmission function and HYDRONETS mathematic model are programmed as inner functions to control and display digital twin modes by C#. Finally, the digital twin models are put on the 3D GIS and published in website which implement virtual building, virtual devices, simulation flood and simulation building.

4.1 Simulation Flood

In order to more effectively utilize water resources and reduce flood disasters, flood trend analysis is necessary. Figure 4 shows the digital twin's working principle with input and output. Environment data including rainfall and upstream inflow is input into modes working with the basic information such as current water level and capacity of this water gate as parameters for prediction algorithm. After combining the prediction algorithm model with constraint conditions calculation, on the one hand, sensor simulation is achieved, and on the other hand, flood inundation analysis is achieved.

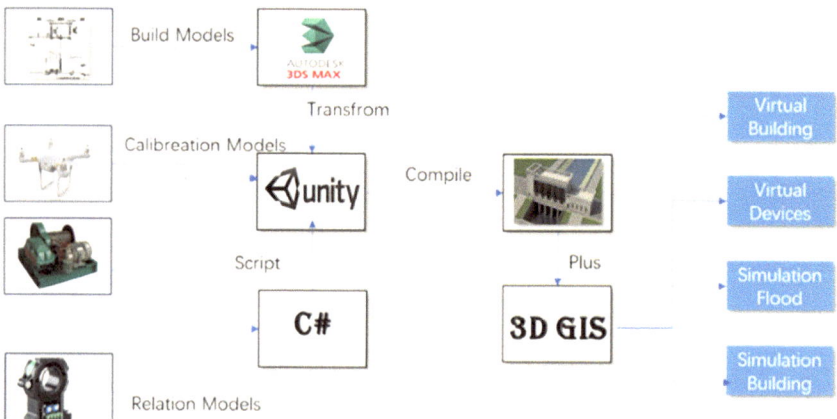

Fig. 3 Building models

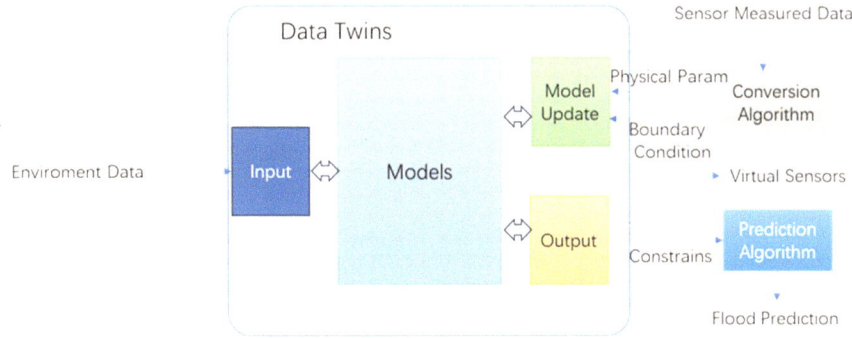

Fig. 4 Working principle of digital twin

4.2 *LSTM*

Time series modeling of control targets using long short term memory (LSTM) is a neural network structure dedicated to time series, for modeling control targets whose output changes according to the history of time series input data [6].

The technical flow chart is showing as Fig. 5 which starts from training LSTM by historical hydrology data of the river including CHENSHAN, QINGHE, PANGKOU, XIAOKAI from August 2021 to August 2023. The inundation analysis will be done by inputting future time to predict the depth of flood while the accuracy of predictive data meets the requirement. The predictive depth of every main point of this river is put into the 3DMap like Cesium with DEM data and render inundation process.

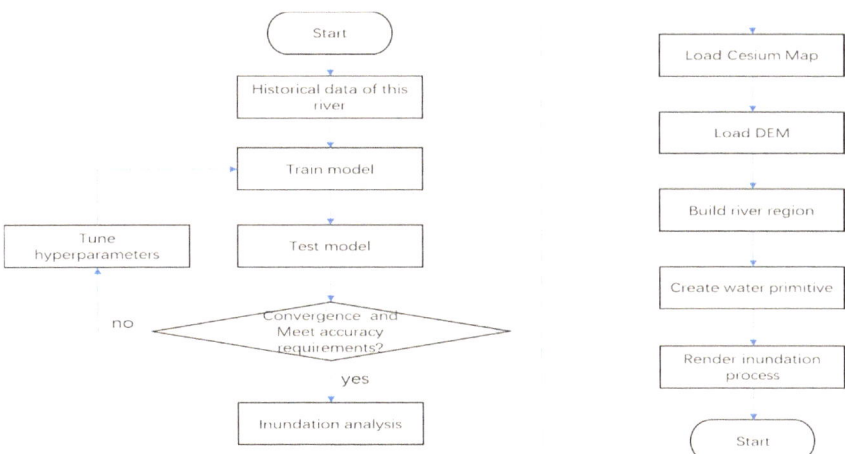

Fig. 5 Technical flow chart

Table 1 The historical hydrology data

ID	Time	CHENSHAN	QINGHE	PANGKOU	XIAOKAI
1	2021-8-1 08:00	40.34	40.33	39.45	14.79
2	2021-8-1 09:13	40.31	40.31	39.41	14.75
3	2021-8-2 08:00	40.28	40.28	39.41	14.76
4	2021-8-3 08:00	41.29	40.27	39.41	14.75
5	2021-9-2 17:03	41.08	40.85	39.6	14.9
…	…	…	…	…	…
506	2023-8-14 08:00	40.42	40.26	39.15	14.5

Table 2 Comparison of measured and numerically modelled ponding depths at inundated water points for 12 August 2022

Location	Measured water level (m)	Simulated water depth (m)	Difference (m)
CHENSHAN	40.32	40.3	0.02
QINGHE	40.63	40.68	−0.05
PANGKOU	30.41	30.81	−0.4
XIAOKAI	2.39	2.86	−0.47

4.3   Flooding Simulation Effect

The historical hydrology data was about 500 rows from August 2021 to August 2023 as showing in Table 1 and the test result was showing Table 2. For training, 80% of the samples were selected for training and 20% for testing.

The measured water level data were compared with water level simulated by the numerical model, as shown in Table 2.

The flood simulation model was tested using Cecium high-precision 3D maps and DEM. The following Fig. 6 shows the prediction and simulation of downstream inundation under different water levels. Firstly, the model is trained using a large amount of historical data on the water level along the river. Then, the predetermined water level and the current downstream water level are input, and combined with the map and elevation data model, a real-time downstream inundation map is derived and displayed on a 3D map.

5   Conclusion and Perspectives

To provide information on the status of every part of water gate, A digital twin system was established based on actual physical dimensions, mathematics, and real-time sensor data, which built a comprehensive control system for the water gate showing high performance and operability. Using algorithms and 3D maps to achieve flood

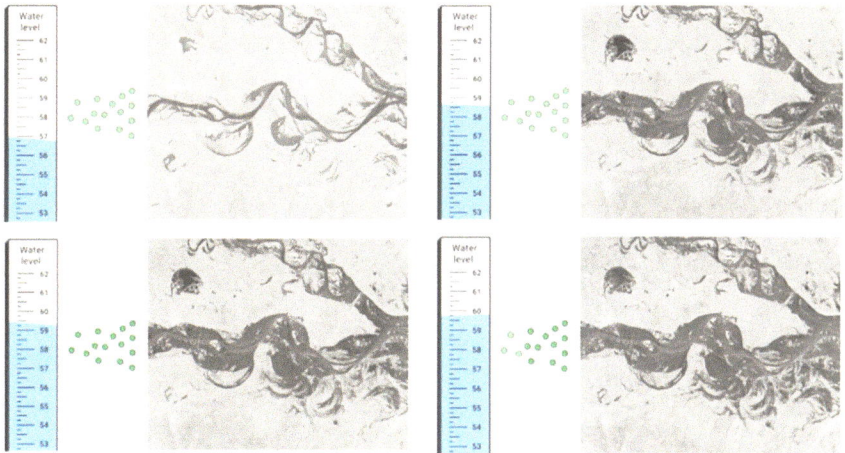

Fig. 6 The result of flood simulation

trend analysis and water regime prediction functions. At present, the system has been constructed and tested in a river sluices in our province, and has achieved good results. With the same tools, we can also provide design validations on new structures or virtual sensors to monitor inaccessible areas of the structure or physical phenomena difficult to measure. This approach is currently validating on other structures such as damp, large size lake or water plant.

Acknowledgements Authors wishing to acknowledge assistance or encouragement from colleagues, special work by technical staff or financial support from organizations should do so in an unnumbered Acknowledgments section immediately following the last numbered section of the paper.

References

1. Liu, Q., & Qi, S. (2023). Quantitative analysis of research literature on urban flood disasters from 1998 to 2022 based on WoS and CNKI citation database. *Cities and Disaster Reduction, 4*, 20–26.
2. Li, Z., & Hang, Z. (2023). Research and application of key technologies for 3D simulation of water conservancy digital twin platform. *People's Yangtze River, 54*(8), 9–18.
3. Fan, Z. L., Jiang, Y. T., Zhang, Y., et al. (2021). Research and application of 3D dynamic visualization method for flood evolution. *China Rural Water and Hydropower, 9*, 91–95.
4. Majumdar, S., Smith, R., Conway, B. D., Butler, J. J., Lakshmi, V., & Dagli, C. H. (2021). Estimating local-scale groundwater withdrawals using integrated remote sensing products and deep learning. In *2021 IEEE International geoscience and remote sensing symposium IGARSS* (pp. 4304–4307).
5. Artificial Intelligence Research. (2021). *Using machine learning for flood prediction and water level prediction.* https://zhuanlan.zhihu.com/p/376347720
6. Chuzo, N. (2022). LSTM AI modeling. *AI Time Series Control System Modelling*, 60–90.

Robustness Study and Deformation Pattern Analysis of Multiple Prediction Regression Models for a Panel Rockfill Dam

Xinyuan Zhu, Zheng Zhou, and Jinzhang Tian

Abstract Panel rockfill dams, due to their cost-effectiveness, short construction periods, high safety standards, and strong adaptability to foundations, have gained rapid development and considerable attention in recent years. However, the inherent material properties and structural behavior of panel rockfill dams change continuously during operation. Particularly, long-term operation results in coordination issues between the panel concrete and the rockfill material, leading to deformations. Regular monitoring and prediction of horizontal displacements in panel rockfill dams are crucial to ensure the dam's safety. This study focuses on the robustness of two prediction regression models, namely, displacement statistical regression models and grey prediction monitoring models, applied to a panel rockfill dam in the Lower Elesai Hydropower Project in Cambodia. The research explores the accuracy, applicability, strengths, and weaknesses of these two commonly used regression models for fitting and predicting deformations, such as displacements, in panel rockfill dams. Furthermore, both prediction regression models are employed to analyze horizontal displacement monitoring data of the panel dam, conducting an in-depth study of deformation patterns. This analysis aims to determine the initial reservoir period's coordination patterns of deformation and provide recommendations based on the findings.

Keywords Panel rockfill dam · Robustness study · Deformation analysis · Multiple prediction regression

X. Zhu
Stung Russei Chrum Department, China Huadian Corporation, Beijing 100031, China

Z. Zhou (✉) · J. Tian
Changjiang Institute of Survey, Planning, Design and Research Co., Ltd., Wuhan 430010, Hubei, China
e-mail: zhouzheng@cjwsjy.com.cn

National Dam Safety Research Center, Wuhan 430010, Hubei, China

© The Author(s) 2025
D. Li and Y. Zhang (eds.), *Advances in Frontier Research on Engineering Structures II*,
Lecture Notes in Civil Engineering 535, https://doi.org/10.1007/978-981-97-6238-5_41

1 Introduction

The concrete panel rockfill dam is a dam type that emerged after the 1960s. Compared to other dam types, it exhibits excellent stability of dam slopes, strong resistance to water pressure in anti-seepage panels, high permeability of dam bodies with minimal impact from seepage, strong adaptability to foundations, and excellent seismic resistance. Additionally, it offers short construction periods, low difficulty in construction diversion and flood discharge, and is less affected by weather conditions. Moreover, it is cost-effective as dam construction materials can be sourced locally. Therefore, it has gained significant favor in China's hydraulic engineering community.

In recent years, there has been rapid advancement in dam construction technology, leading to a rapid increase in the number and height of dams, indicating a promising future for this type of dam. Presently, the world's tallest concrete panel rockfill dam is the Shuibuya Hydropower Station, with a height of 233 m. The Jawa Hydropower Station, with a dam height of 239 m, is currently under construction. However, during the operation of concrete panel rockfill dams, the material properties and structural behaviors undergo continuous changes. Particularly, the long-term operation of the two materials, panel concrete and rockfill, can lead to deformation coordination issues. Therefore, it is essential to focus on monitoring panel deformation and stress changes. Monitoring and analyzing the displacement of concrete panel rockfill dams are of significant importance for the stable operation of dams.

Currently, the development of prediction regression models for concrete dams is well-established and abundant, as evidenced by various studies [1–3]. Examples include the HST regression model (hydrostatic-seasonal-time) and HTT (hydrostatic-thermal-time) regression model, as well as the HSS (hydrostatic-seasonal-state) model that describes aging effects as a state vector [4]. In contrast, the development of prediction regression models for concrete panel rockfill dams lags behind that of concrete dams. Many scholars directly apply prediction regression models developed for concrete dams, such as the HST or HTT models, to analyze horizontal and vertical displacements in panel rockfill dams [5]. For instance, Gamse et al. [6] applied the HTT model to monitor displacements and settlements in the upstream and downstream directions along the dam axis of a rockfill dam. Shen Yang et al. [5] considered pressure components, temperature components, and aging components in their prediction model for horizontal displacements, tailored to the characteristics of the dam section and material properties of panel rockfill dams. However, due to the distinct mechanical properties of panel rockfill dams compared to monolithic concrete dams [7], their structural mechanisms are not as clear, material zoning is evident, and mechanical properties are variable [8]. The deformations of both the rockfill body and the panels need to be coordinated. Therefore, the applicability of regression prediction models for panel rockfill dams still requires further research.

In summary, this study applied two prediction models, displacement statistical regression models, and grey prediction monitoring models, to analyze horizontal displacement monitoring data of the panel rockfill dam in the Lower Elesai Hydropower Project in Cambodia. The research investigated the goodness of fit of these two regression models concerning various monitoring parameters of the panel rockfill dam. Additionally, based on these prediction models, the study conducted an in-depth analysis and discussion of the monitoring data and displacement characteristics of the project.

2 Engineering Background

The Upper Dam of the Lower Elesai Hydropower Project is located downstream of the Sesan River in the northern part of Koh Kong Province, Cambodia. It is classified as a Grade II large-scale (Type 2) project, with its primary purpose being electricity generation [9, 10]. The key components of the hydropower station complex include the main dam, right-bank open spillway, right-bank flood discharge tunnel, left-bank water intake, and power generation system. The main dam of the hydropower station is a concrete panel rockfill dam with a crest elevation of 266.0 m, a dam axis length of 428.8 m, a maximum dam height of 125 m, a crest width of 8.7 m, and a maximum bottom width of 361.5 m in the riverbed cross-section. The dam filling materials from upstream to downstream consist of rockfill compaction zone (1B), clay cover zone (1A), special cushion layer zone (2B), cushion layer zone (2A), transition zone (3A), main rockfill zone (3B), downstream rockfill zone (3C), and large stone and dry rubble stone revetment slope. The upstream slope ratio of the dam is 1:1.4, and the panels are made of reinforced concrete, with a total of 35 blocks which are mainly 10, 12 and 16 m widths. The thickness of the panels gradually becomes thinner from bottom to top, with the top thickness of 30 cm and the maximum bottom thickness of the 65.97 cm. The panel concrete is made of secondary concrete.

The exposed geological strata at the dam site primarily consist of Middle Jurassic quartz sandstone, mudstone, sandy mudstone, and silty mudstone. The Quaternary loose deposits exposed in the dam area mainly include river alluvial deposits, slope debris deposits, and slope residual deposits. There are no major faults in the dam area, and the main geological features are fractures. The physical geological phenomena at the dam site mainly manifest as weathering and unloading, with intercalated weathering layers observed in the bedrock. The bedrock of the riverbank slopes on both sides of the dam site is exposed, composed of sandstone and mudstone, forming a lithological alternation of soft and hard rock masses. The riverbed of the Sesan River and the steep cliffs on the riverbanks have various-sized cavities, which are the result of the weathering and softening of gravel in sandstone, subsequently eroded by surface water flow.

The Elesai dam started construction in December 2010, the reservoir were successfully stored in May 2013, and the power generation was put into operation in September 2013. It's crucial to monitor dam horizontal displacements state during operation to ensure the dam's safety under this geological conditions. Figure 1. Is the upper dam of the lower Elesai hydropower project.

Fig. 1 The photograph of upper dam of the lower elesai hydropower project

3 Predictive Regression Models

In this section, two predictive regression models commonly used for concrete dams are introduced. These models are then applied to analyze the deformation patterns in horizontal displacements of panel rockfill dams. The study explores the accuracy, applicability, as well as the strengths and weaknesses of these models in the context of panel rockfill dams.

(a) Displacement Statistical Regression Model

The displacement statistical regression model utilizes environmental factors as independent variables and monitoring values (horizontal displacement, settlement, stress–strain) as response variables. It is established using mathematical and statistical analysis methods and real monitoring data as training samples. This model is an empirical model derived from the analysis of historical monitoring data.

Based on the research experiences of previous scholars, it is evident that the displacement at any point of the dam at time t is primarily influenced by factors such as upstream and downstream water levels (water pressure), temperature, and time effects. Thus, the displacement statistical model for any point of the dam at time t mainly consists of water pressure components, temperature components, and time-effect components [11, 12]. It can be expressed as follows:

$$S = y_H(t) + y_T(t) + y_\theta(t) \tag{1}$$

where $y_H(t)$ represents the water pressure displacement component at time t. It can be expressed as a function of reservoir water level H, denoted as $y_H(t) = a + bH$; $y_T(t)$ represents the temperature displacement component at time t. $y_\theta(t)$ represents the time effect displacement component, which is an irreversible component evolving in a specific direction with time. This component is influenced by various factors such as dam material creep, dam foundation rock creep, rock mass joints and fractures, as well as compressive and plastic deformation of weak structures under the action of water pressure and other loads. The origin of the time effect displacement component is complex.

Considering the varying strengths of water pressure, temperature, and time effect components in different types of projects, the displacement statistical model can be simplified accordingly.

(b) Grey Prediction Monitoring Model

In the objective world, there is a vast amount of known and unknown information. Assuming unknown information is represented by black, the system containing unknown information is referred to as a black system; known information is represented by white, and the system containing known information is a white system. A system containing both known and unknown information is called a grey system, and this is the most common type of system.

Dams are typical grey systems. Monitoring data from dams are treated as grey quantities that change within a certain range. Monitoring data are seen as grey processes changing over time. Grey models (GM) are established based on this principle [13–15]. Calculated values derived from the grey models are then restored to the original data, allowing for the analysis of the change patterns and evolution trends in dam monitoring data. Grey prediction models are suitable for regression and prediction with limited monitoring data, low data completeness, and reliability. These models require minimal information for modeling, provide high accuracy, have simple computations, and are easy to validate. The commonly used GM (1, 1) model steps and principles are as follows:

- Set the original data as:

$$x^{(0)}(k) = \{x^{(0)}(1), x^{(0)}(2), \ldots, x^{(0)}(n)\} \tag{2}$$

- Add $x^{(0)}(k)$ once to generate a sequence of numbers:

$$x^{(1)}(k) = \{x^{(1)}(1), x^{(1)}(2), \ldots, x^{(1)}(n)\} \tag{3}$$

where $x^{(1)}(k) = \sum_{i=1}^{k} x^0(i), k = 1, 2, \ldots n.$

The resulting sequence $x^{(1)}(k)$ satisfies the following differential equation:

$$\frac{dx^{(1)}(t)}{dt} + ax^{(1)}(t) = u, t \in [0, \infty] \tag{4}$$

where a, u is the undetermined parameter.

- According to the least square method, the prediction function of the original data can be solved as follows:

$$\hat{x}^{(0)}(1) = x^{(0)}(1)$$

$$\hat{x}^{(0)}(k = 1) = \hat{x}^{(1)}(k + 1) - \hat{x}^{(0)}(k)$$
$$= (1 - e^a)\left(x^0(1) - \frac{u}{a}\right)e^{-ak}, \quad k = 1, 2, \ldots, n - 1 \qquad (5)$$

4 Analysis of Monitoring Data Based on Prediction Models

For horizontal displacement monitoring, the Upper Dam of the Lower Sesan II Hydropower Project utilizes tensioned wire-type displacement meters. The monitoring section is at the $0 + 196.00$ m section on the left side of the dam. Along this section, three tensioned wire-type horizontal displacement meters are installed from bottom to top at elevations of 175 m, 205 m, and 235 m. There are 5, 4, and 2 horizontal displacement meters installed at elevations of 175 m, 205 m, and 235 m, respectively, making a total of 11 horizontal displacement measurement points.

(a) **Regression Analysis of Dam Horizontal Displacement**

Based on the observed dam horizontal displacement data during the initial reservoir period from 2014 to 2017, statistical regression analysis is conducted on the monitoring results of dam horizontal displacement points. The statistical model takes into account factors such as upstream water pressure and time effects. Equation (1) can be rewritten as follows:

$$S = S_0 + ah^\alpha(1 - e^{-\beta t}) + bH^\delta(1 - e^{-\eta t}) \qquad (6)$$

where h represents the height of the fill, which is a constant during the impoundment period. H represents the reservoir water level. t represents the time from the initial measurement day to the observation day, increasing by 0.01 each day. a, β, δ and η are trial parameters. S_0, a and b are regression parameters.

After conducting statistical regression analysis on dam horizontal displacement measurements, the results of the statistical regression models for each measurement point are shown in Table 1. The results indicate that, except for points ID1-2 and ID3-1, the multiple correlation coefficients for each measurement point range from 0.74 to 0.96, and the standard deviations are in the centimeter range. The F-test values are mostly greater than 200, indicating that the overall regression performance of the models is good, and the accuracy of the models is high.

(b) **Dam Horizontal Displacement Grey Prediction Monitoring Model**

Utilizing the GM (1,1) model in equal 5, based on the observed data of the dam's horizontal displacement during the initial reservoir filling period from 2014 to 2017, grey

Table 1 Results of statistical regression analysis of dam horizontal displacement

Measuring point number	Measuring point elevation (m)	Regression equation	Multiple correlation coefficient R	Standard deviation S (cm)	Test value F
ID1-1	235	$S = 0.0515 - 14.6529h^{0.6}(1 - e^{-0.95t})$ $+262.0251(1 - e^{-t})$	0.7363	4.03	66.33
ID1-2	235	$S = -8.6498 - 11.6584h^{0.4}(1 - e^{-0.5t})$ $+50.8306H^{0.1}(1 - e^{-0.6t})$	0.4788	6.79	16.66
ID2-1	205	$S = 1.2755 - 25.4082h^{0.3}(1 - e^{-t})$ $+98.2404(1 - e^{-0.95t})$	0.8755	1.63	183.89
ID2-2	205	$S = -6.8010 + 109.9825h^{0.1}(1 - e^{-0.7t})$ $-161.4391(1 - e^{-0.65t})$	0.9018	2.79	243.91
ID2-3	205	$S = -21.6106 - 63.2550h^{0.2}(1 - e^{-0.65t})$ $+143.9287(1 - e^{-0.6t})$	0.8728	4.63	179.08
ID2-4	205	$S = -1.1508 - 7.6940h^{0.3}(1 - e^{-t})$ $+37.6133(1 - e^{-0.95t})$	0.9488	0.71	505.50
ID3-1	175	$S = 0.2573 - 0.8717h^{0.6}(1 - e^{-t})$ $+8.2636H^{0.1}(1 - e^{-0.95t})$	0.2685	1.56	4.35
ID3-3	175	$S = 57.6219 + 187.3165h^{0.2}(1 - e^{-0.45t})$ $-666.5669(1 - e^{-0.55t})$	0.8988	34.90	235.50
ID3-4	175	$S = -0.2508 - 2.6567h^{0.1}(1 - e^{-0.75t})$ $+5.3135H^{0.1}(1 - e^{-t})$	0.9597	0.48	653.90
ID3-5	175	$S = -6.8533 + 37.3586(1 - e^{-0.05t})$ $+0.8919H^{0.5}(1 - e^{-t})$	0.9064	6.56	257.93

prediction monitoring models were established for two typical monitoring points, ID1-2 and ID2-3, as shown in Eqs. (7) and (8):

$$\hat{x}^{(0)}(1) = x^{(0)}(1)$$

$$\hat{x}^{(0)}(k + 1) = \hat{x}^{(1)}(k + 1) - \hat{x}^{(1)}(k)$$
$$= (1 - e^{-0.0955})(x^0(1) - 51.4679)e^{0.0955k}, \quad k = 1, 2, \ldots, n - 1 \quad (7)$$

$$\hat{x}^{(0)}(1) = x^{(0)}(1)$$

$$\hat{x}^{(0)}(k+1) = \hat{x}^{(1)}(k+1) - \hat{x}^{(1)}(k)$$
$$= (1 - e^{0.0003})(x^0(1) + 64985)e^{-0.0003k}, \quad k = 1, 2, \ldots, n-1 \quad (8)$$

The measured values and the predicted values of the grey prediction models for monitoring points ID1-2 and ID2-3 are compared in Tables 2 and 3, respectively. The fitting results indicate that, especially in the short term and particularly in cases with prominent nonlinearity and abrupt displacement changes, the grey prediction monitoring model fits the dam's horizontal displacement characteristics better than the displacement statistical regression model. It exhibits higher accuracy; for instance, the standard deviation for ID1-2 is 2.74, which is less than the 6.79 of the horizontal displacement statistical regression model, and the standard deviation for ID2-3 is 1.30, smaller than the 4.63 of the horizontal displacement statistical regression model. However, in the case of long time series, the error between the measured values and predicted values of the grey prediction model rapidly increases. For example, the errors for sequences 8 and 9 increase to 23.13 and 15.16, respectively.

(c) **Results Analysis**

Based on the predictive regression model, the maximum horizontal displacement of the panel dam in the upper dam of the Elesai downstream hydroelectric project in 2022 was predicted. The ratio of the maximum horizontal displacement to the maximum dam height, commonly used internationally, was used as an evaluation indicator to assess the horizontal displacement of the dam. The calculated horizontal displacement indicators based on the prediction model for this project were compared with similar projects, and the results are shown in Table 4. From the table, it can be observed that, compared to similar concrete panel dams, the characteristic horizontal displacement value of the upper dam of the Elesai downstream hydroelectric project is slightly larger than that of other similar high dams (such as Shanxi and Gongboxia),

Table 2 ID1-2 Grey prediction results of horizontal displacement

Sequence number	Measured value (cm)	Predicted value (cm)	Error (%)
1	−2.902	−2.902	\
2	−5.758	−5.451	5.33
3	−5.745	−5.997	4.39
4	−6.68	−6.599	1.21
5	−6.68	−7.260	8.68
6	−8.64	−7.988	7.55
7	−8.64	−8.789	1.73
8	−12.58	−9.670	23.13
9	−12.54	−10.639	15.16
Standard deviation	2.74		

Sequences 1–7 are the data from the initial reservoir filling period, used as the initial sequences for the grey prediction model

Table 3 ID2-3 Grey prediction results of horizontal displacement

Sequence number	Measured value (cm)	Predicted value (cm)	Error (%)
1	−19.349	−19.349	\
2	−19.385	−19.371	0.072
3	−19.351	−19.365	0.072
4	−19.353	−19.360	0.036
5	−19.353	−19.354	0.005
6	−19.357	−19.348	0.046
7	−19.341	−19.342	0.005
8	−22.733	−19.337	14.94
9	−23.897	−19.331	19.11
Standard deviation	1.30		

Sequences 1–7 are the data from the initial reservoir filling period, used as the initial sequences for the grey prediction model

Table 4 Comparison of horizontal displacement index of face rockfill dam

Dam	Maximum dam height H_{max}/m	Horizontal displacement of dam body after impoundment/ cm		Horizontal displacement of dam body after impoundment	
		Upstream D_u	Downstream D_d	$D_u/H_{max}(10^{-4})$	$D_d/H_{max}(10^{-4})$
Dong Jing	149.5	−15.7	16.8	11.3	12.1
Sangye	132.5	−8.0	9.1	6.0	6.9
Gongbo Gorge	132.2	−1.5	13.4	1.1	10.1
Elesay	126	−12.186	1.65	9.67	1.31

but still lower than the values observed in the Dongjing Dam. Furthermore, the horizontal deformations downstream are smaller than other dams, being less than 1‰. This indicates that the horizontal displacement values of the panel dam are within a reasonable range. However, attention should be focused on the horizontal displacement values upstream in the future.

5 Conclusion

This study utilized displacement statistical regression models and grey prediction monitoring models to analyze the horizontal displacement monitoring data of the panel rockfill dam in the Lower Elesai Hydropower Project. The research explored the accuracy, applicability, strengths, and weaknesses of these two regression models.

Based on these models, the study analyzed the displacement characteristics of the panel rockfill dam during the initial reservoir period, and the conclusions are as follows:

- Both the statistical regression models and the grey prediction monitoring models built on the horizontal displacement monitoring data fit the actual measured data reasonably well. However, nonlinear changes in the data direction in 2014 led to reduced correlation coefficients for certain points (ID1-2, ID3-1). Despite this, the regression models still effectively captured the long-term trends in horizontal displacement of the panel rockfill dam. The grey prediction monitoring model exhibited lower errors, but its accuracy decreased as the time series grew. Horizontal deformations caused by the initial impoundment in 2014 differed significantly from those after 2015, indicating distinct horizontal deformations during the first impoundment, followed by relatively stable strains in the downstream direction during operation. The characteristic values of horizontal displacements of the panel dam remained within a reasonable range.
- Both displacement statistical regression models and grey prediction monitoring models demonstrated good applicability in the displacement monitoring and analysis of panel rockfill dams. Unlike monolithic concrete dams with clear structural mechanisms, panel rockfill dams exhibit strong nonlinearity in horizontal displacement, including certain abrupt changes before and after the initial impoundment, leading to reduced fitting accuracy for both grey models and displacement statistical regression models. Under nonlinear conditions, the grey prediction monitoring model showed higher accuracy, especially for short time series data. However, its accuracy decreased over longer time series. Although statistical regression models exhibited lower overall multiple correlation accuracy under abrupt displacement conditions, the ultimate trend remained consistent with actual situations, making them suitable for long-term displacement prediction.

In summary, displacement statistical regression models are suitable for regression fitting and analysis of displacement monitoring data over long time series, while grey prediction monitoring models are suitable for regression fitting and analysis of displacement monitoring data with short time series, low data completeness, low reliability, and high nonlinearity.

Acknowledgements This study was supported by the National Key R&D Program of China (No. 2022YFC3005404), the National Natural Science Foundation of China (52309152), the Fundamental Research Funds for the Central Universities(B230201013), the Natural Science Foundation of Jiangsu Province (BK20220978), State Key Laboratory of High Performance Civil Engineering Materials (No. 2022CEM014), Fund of National Dam Safety Research Center (Grant No. CX2023B03).

References

1. Hu, J., & Ma, F. (2019). Zoned safety monitoring model for uplift pressures of concrete dams. *Transactions of the Institute of Measurement and Control, 41*(1), 3952–3969.
2. Zhao, Y., Li, T., Cheng, J., et al. (2015). Inversion of deformation modulus for gravity dam based on statistical models and finite element method. *Hydroelectric Energy Science, 33*(12), 96–100.
3. Zhu, Y., Xie, M., Zhang, K., et al. (2010). A dam deformation residual correction method for high arch dams using phase space reconstruction and an optimized long short-term memory network. *Mathematics, 2023*, 11.
4. Li, F., Wang, Z., Liu, G., et al. (2015). Hydrostatic seasonal state model for monitoring data analysis of concrete dams. *Structure and Infrastructure Engineering, 11*(12), 1616–1631.
5. Xu, M., Pang, R., Zhou, Y., et al. (2023). Seepage safety evaluation of high earth-rockfill dams considering spatial variability of hydraulic parameters via subset simulation. *Journal of Hydrology, 626*(A), 130261.
6. Gamse, S., Zhou, W., Tan, F., et al. (2018). Hydrostatic-season-time model updating using Bayesian model class selection. *Reliability Engineering and System Safety, 169*, 40–50.
7. Zhu, A., Zhang, Y., Liao, J., et al. (2020). Stress and deformation characteristics of large-angle polygonal high-panel block stone dam body and panels. *Advances in Science and Technology of Water Resources, 40*(1), 48–55.
8. Lu, J., & Gao, Z. (2016). Influence of dam material zoning on stress and deformation of high-panel block stone dam. *People's Yellow River, 38*(2), 90–94.
9. Zhu, X., & Chen, Z. (2015). Construction dynamic management practice of panel block stone dam in Elesai Hydropower Station. *Cambodia. Yangtze River, 46*(07), 34–35.
10. Le, J. (2014). Treatment of climate and geological characteristics in the construction area of Elesai Project in Cambodia and important issues in the construction of panel block stone dam. In: *Proceedings of Annual Conference of China Society for Hydropower Engineering.* Yellow River Water Conservancy Press, 177–183.
11. Zhu, Y., Zhang, Z., Gu, C., et al. (2023). A coupled model for dam foundation seepage behavior monitoring and forecasting based on variational mode decomposition and improved temporal convolutional network. *Structural Control and Health Monitoring*, 3879096
12. Zhu, Y., & Tang, H. (2023). Automatic damage detection and diagnosis for hydraulic structures using drones and artificial intelligence techniques. *Remote Sensing, 15*, 615.
13. Chen, X., & Ma, Z. (2006). Application of G (1,1) grey model in horizontal displacement prediction of a certain dam. *Journal of Water Resources and Architectural Engineering, 4*(3), 22–24.
14. Saeed, B., & Hossein, B. (2021). Improving grey prediction model and its application in predicting the number of users of a public road transportation system. *Journal of Intelligent Systems, 30*(1), 104–114.
15. Fu, H., Yang, B., Hu, D., et al. (2018). Improved Non-Equidistance GM (1,1) forecasting model of dam displacement. Yellow River, 127–130+144.

Analytical Investigation of Temperature Field in Rock Surrounding Shallow Buried Circular Tunnel Considering Insulation Layer

Tianyu Zhang, Yan Yu, Danqi Ge, and Dongsheng Xia

Abstract The temperature field in the rock surrounding a shallow-buried circular tunnel by considering the influence of the insulation layer is analytically investigated in this paper. The conformal mapping method is used to transform the heat conduction equation of the tunnel surrounding rock in Cartesian coordinates to one in the polar coordinate system, obtaining the expression of the heat conduction equation in polar coordinates. By applying boundary conditions and continuity conditions, the analytical solution for the steady-state temperature field in the rock surrounding a shallow-buried circular tunnel is obtained. Meanwhile, a finite element model is established to compare the analytical solution obtained in this study with the numerical solution. The consistency between the analytical solution and the numerical solution verifies the accuracy of the analytical solution proposed in this paper. The influences of tunnel radius, tunnel burial depth, insulation layer thickness, and thermal conductivity of the insulation layer on the steady-state temperature field in the rock surrounding the tunnel are analyzed. The research results are significant for the guidance of actual tunnel engineering projects.

Keywords Shallow buried circular tunnel · Steady-state temperature field · Conformal mapping · Analytical solution

T. Zhang · Y. Yu (✉) · D. Xia
Dalian Maritime University, Dalian 116026, China
e-mail: yyan@dlmu.edu.cn

D. Xia
e-mail: xia@dlmu.edu.cn

D. Ge
Aerospace Science and Industry Shenzhen (Group) Co., Ltd., Shenzhen 518000, China

D. Li and Y. Zhang (eds.), *Advances in Frontier Research on Engineering Structures II*,
Lecture Notes in Civil Engineering 535, https://doi.org/10.1007/978-981-97-6238-5_42

507

1 Introduction

In recent years, with the continuous development and application of tunnel engineering, the study of the temperature field in tunnels surrounding rock has attracted widespread attention. The temperature field in the tunnel surrounding rock has a significant impact on the stability of tunnel structures, the safety of underground engineering, and fire prevention and control measures [1]. Furthermore, the temperature field in tunnel surrounding rock is coupled with the seepage field and stress field, thereby influencing the stress distribution in tunnel linings [2]. In practical engineering, the temperature field in a tunnel surrounding rock is influenced by various factors such as insulation layers, flow fields within the tunnel, and ventilation velocity [3], with the insulation layer having a significant impact on the temperature field in the tunnel surrounding rock. Therefore, studying the temperature field in tunnel surrounding rock considering the insulation layer is of great importance for the entire operational lifecycle of tunnels [4].

Currently, research on the temperature field in tunnels surrounding rock mainly focuses on analytical methods, numerical simulations, and model experiments. Among them, the use of model experiments and numerical simulations to study the temperature field in tunnel surrounding rock has become more mature. Most researchers employ finite difference or finite element methods to obtain accurate solutions for the temperature field in tunnel surrounding rock under various operating conditions [5–7]. Although model experiments can yield results similar to the actual situation, they are challenging to implement. On the other hand, analytical methods provide more intuitive solutions through rigorous formula derivation. They facilitate a deeper understanding of the relationship between the temperature field and various influencing factors, enabling easy comparison and evaluation of analytical results. This approach is conducive to further in-depth research.

Zhang et al. [8] obtained the analytical solution for the temperature field of a four-layer structure in a cold region tunnel with insulation layers using the Bessel function orthogonal expansion theorem and differential equation solving method. Xia et al. [9] utilized Laplace transforms and the method of separation of variables to derive the analytical solution for the temperature field in a cold region tunnel with insulation layers. Lai et al. [10] conducted an analytical study on the temperature field of circular cross-section tunnels in cold regions using perturbation techniques and other methods, thereby obtaining an approximate analytical solution for circular tunnels during the freezing process. Hong et al. [11] developed an analytical solution for the steady-state temperature field using the boundary separation method based on a freezing model of the pipe-in-pipe system. Cheng et al. [12] obtained the desired analytical functions using the power series method based on the temperature field distribution derived from Laplace transforms. Liu et al. [13] conducted an analytical analysis of insulation layers in cold region tunnels by combining Laplace transforms with Fourier integral transforms on a three-dimensional model.

Literature studies indicate that there is a significant amount of analytical research on the temperature field in deeply buried tunnels surrounding rock, while there is a lack of analytical research on the temperature field in shallowly buried tunnels surrounding rock, with little consideration given to the influence of insulation layers on the thermal properties of tunnel surrounding rock. This paper presents an analytical study on the temperature field in tunnel surrounding rock considering the presence of insulation layers. By employing conformal transformation, the heat conduction equation of the tunnel surrounding rock in Cartesian coordinates is transformed into a circular annular region for solution. By combining the boundary conditions of tunnel linings and surface temperatures, and considering the continuity of temperature and heat flux functions at the interface between the tunnel insulation layer and the surrounding rock, the coefficients in the series solution of the thermal Laplace equation for the circular annular region are determined, leading to the derivation of an analytical solution for the steady-state temperature field in shallowly buried circular tunnels. A comparison between the analytical solution and the simulation results obtained from finite element software demonstrates good agreement, validating the accuracy of the analytical results in this study. Furthermore, this study investigates the effects of tunnel radius, burial depth, insulation layer thickness, and thermal conductivity of the insulation layer on the temperature field in the tunnel surrounding rock. The research findings have practical implications for guiding actual tunnel engineering projects.

2 Calculation Equation

2.1 The Heat Conduction Equation

Since the longitudinal dimension of the tunnel is much larger than the transverse dimension, and the tunnel cross-section remains unchanged along the length, the temperature field of the surrounding rock in a three-dimensional shallow-buried tunnel can be approximated as a two-dimensional problem of the temperature field in the tunnel cross-section [14]. In the Cartesian coordinate system, a thermal conduction model for the cross-section of a shallow-buried circular tunnel, considering the insulation layer, is established as shown in Fig. 1.

In Fig. 1, h represents the distance from the center of the tunnel cross-section to the ground surface, while r denotes the radius of the tunnel. Points A, B, C, D, and E are control points that facilitate the depiction of the tunnel model after conformal mapping.

According to the principles of heat transfer [15], the temperature field in the surrounding rock of the tunnel satisfies the following equation

$$\frac{\partial^2 T}{\partial x^2} + \frac{\partial^2 T}{\partial y^2} = \frac{1}{\alpha} \frac{\partial T}{\partial t} \tag{1}$$

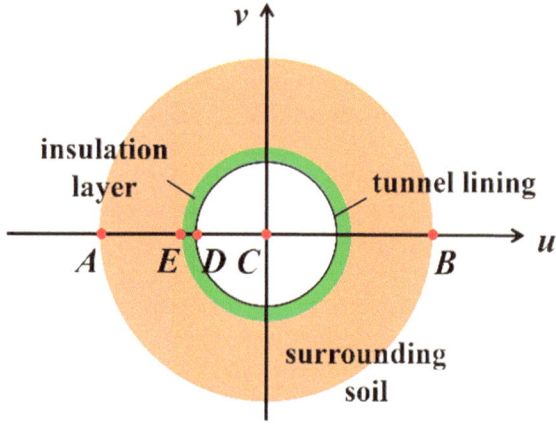

Fig. 1 Geometric model of tunnel after conformal mapping

In the equation, $\alpha = \lambda/c\rho$ represents the thermal diffusivity, where λ is the thermal conductivity, c is the specific heat capacity, and ρ is the density.

2.2 Conformal Mapping

By applying the following conformal transformation equations, the semi-infinite region depicted in Fig. 1 can be transformed into the annular region shown in Fig. 2 [16].

$$z = x + iy = -iA\frac{1+\omega}{1-\omega}, \quad \omega = u + iv = \frac{z + iA}{z - iA} \tag{2}$$

$$A = \frac{1 - \beta_1^2}{1 + \beta_1^2}, \quad \beta_1 = \frac{h}{r} - \sqrt{\left(\frac{h}{r}\right)^2 - 1}, \quad \beta_2 = \frac{h}{r + \Delta r} - \sqrt{\left(\frac{h}{r + \Delta r}\right)^2 - 1} \tag{3}$$

In Eq. (2), z and ω represent the complex numbers corresponding to any point in the semi-infinite circular cavity region and the annular region, respectively. u and v represent the positional coordinates of any point in Fig. 2's annular region. In Eq. (3), A, β_1, and β_2 are the parameters involved in the conformal transformation process.

Table 1 displays the coordinates of the control points before and after the conformal transformation.

By employing the conformal transformation method described by Eq. (2), the expression of Eq. (1) in the annular region of Fig. 2 can be obtained [17]

$$\frac{(1 + \rho^2 - 2\rho\cos\theta)^2}{4a^2}\left(\frac{\partial^2 T}{\partial\rho^2} + \frac{1}{\rho}\frac{\partial T}{\partial\rho} + \frac{1}{\rho^2}\frac{\partial^2 T}{\partial\theta^2}\right) = \frac{1}{\alpha}\frac{\partial T}{\partial t} \tag{4}$$

Fig. 2 Geometric model of shallow buried circular tunnel considering insulation layer

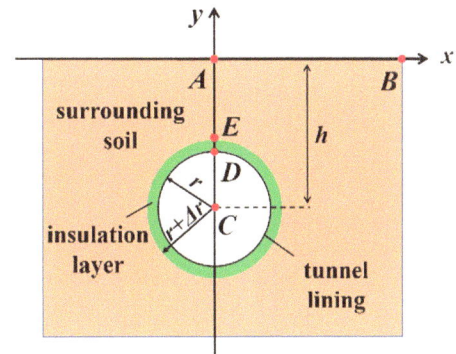

Table 1 Coordinates of control points before and after conformal mapping

Control points	Before conformal transformation	After conformal transformation
A	$(0, 0)$	$(-1, 0)$
B	$(+\infty, 0)$	$(1, 0)$
C	$(0, -h)$	$(0, 0)$
D	$(0, -h + r)$	$(-\beta_1, 0)$
E	$(0, -h + r + \Delta r)$	$(-\beta_2, 0)$

In the above equation, the parameter $a = \sqrt{h^2 - r^2}$, ρ represents the polar radius, and θ represents the polar angle.

3 Analytical Investigation

In the case of steady-state conditions, the temperature fields of the tunnel insulation layer and the soil must satisfy the following equations:

$$\frac{\partial^2 T_i}{\partial \rho^2} + \frac{1}{\rho}\frac{\partial T_i}{\partial \rho} + \frac{1}{\rho^2}\frac{\partial^2 T_i}{\partial \theta^2} = 0 \tag{5}$$

In the given equation, the subscript i takes values of 1 or 2, where 1 represents the tunnel insulation layer, and 2 represents the soil.

Assuming that the temperature of the tunnel lining is constant and denoted as $T1$, and the temperature of the ground surface is constant and denoted as Tg, the boundary conditions can be stated as follows.

$$\begin{cases} T_1|_{\rho=\beta_1} = T_1 \\ T_2|_{\rho=1} = T_g \end{cases} \tag{6}$$

At the interface between the tunnel and the insulation layer, the following continuity conditions should be satisfied for temperature T and heat flux q

$$\begin{cases} T_1|_{\rho=\beta_2} = T_2|_{\rho=\beta_2} \\ \lambda_1 \dfrac{\partial T_1|_{\rho=\beta_2}}{\partial r} = \lambda_2 \dfrac{\partial T_2|_{\rho=\beta_2}}{\partial r} \end{cases} \tag{7}$$

The general solution for the temperature field between the tunnel insulation layer and the surrounding soil can be expressed as follows

$$\begin{cases} T_1(\rho, \theta) = \dfrac{a_1 + b_1 \ln \rho}{2} + \sum_{n=1}^{\infty} \{(a_{n1}\rho^n + c_{n1}\rho^{-n}) \cos n\theta \\ \qquad + (b_{n1}\rho^n + d_{n1}\rho^{-n}) \sin n\theta\}, \ \beta_1 < \rho < \beta_2 \\ T_2(\rho, \theta) = \dfrac{a_2 + b_2 \ln \rho}{2} + \sum_{n=1}^{\infty} \{(a_{n2}\rho^n + c_{n2}\rho^{-n}) \cos n\theta \\ \qquad + (b_{n2}\rho^n + d_{n2}\rho^{-n}) \sin n\theta\}, \ \beta_2 < \rho < 1 \end{cases} \tag{8}$$

Substituting the boundary condition from Eq. (6) into Eq. (8), we obtain

$$\begin{cases} \dfrac{a_1 + b_1 \ln \beta_1}{2} + \sum_{n=1}^{\infty} \{(a_{n1}\beta_1^n + c_{n1}\beta_1^{-n}) \cos n\theta + (b_{n1}\beta_1^n + d_{n1}\beta_1^{-n}) \sin n\theta\} = T_l \\ \dfrac{a_2}{2} + \sum_{n=1}^{\infty} \{(a_{n2} + c_{n2}) \cos n\theta + (b_{n2} + d_{n2}) \sin n\theta\} = T_g \end{cases} \tag{9}$$

Substituting the continuity condition from Eq. (7) into Eq. (9), we obtain

$$\begin{cases} \dfrac{a_1 + b_1 \ln \beta_2}{2} + \sum_{n=1}^{\infty} \{(a_{n1}\beta_2^n + c_{n1}\beta_2^{-n}) \cos n\theta + (b_{n1}\beta_2^n + d_{n1}\beta_2^{-n}) \sin n\theta\} \\ = \dfrac{a_2 + b_2 \ln \beta_2}{2} + \sum_{n=1}^{\infty} \{(a_{n2}\beta_2^n + c_{n2}\beta_2^{-n}) \cos n\theta + (b_{n2}\beta_2^n + d_{n2}\beta_2^{-n}) \sin n\theta\} \\ \lambda_1 \{\dfrac{b_1}{\beta_1} + \sum_{n=1}^{\infty} n[(a_{n1}\beta_2^{n-1} - c_{n1}\beta_2^{-n-1}) \cos n\theta + (b_{n1}\beta_2^{n-1} - d_{n1}\beta_2^{-n-1}) \sin n\theta]\} \\ = \lambda_2 \{\dfrac{b_2}{\beta_2} + \sum_{n=1}^{\infty} n[(a_{n2}\beta_2^{n-1} - c_{n2}\beta_2^{-n-1}) \cos n\theta + (b_{n2}\beta_2^{n-1} - d_{n2}\beta_2^{-n-1}) \sin n\theta]\} \end{cases} \tag{10}$$

Combining the Gaussian elimination method and Fourier series, the coefficients can be obtained as follows

$$
\begin{cases}
a_1 = -2(T_g - T_l)\dfrac{\lambda_2 \ln \beta_1}{\lambda_2 \ln \frac{\beta_2}{\beta_1} - \lambda_1 \ln \beta_2}, \\[3mm]
a_2 = 2(T_g - T_l), \quad b_1 = 2(T_g - T_l)\dfrac{\lambda_2}{\lambda_2 \ln \frac{\beta_2}{\beta_1} - \lambda_1 \ln \beta_2}, \\[3mm]
b_2 = 2(T_g - T_l)\dfrac{\lambda_1}{\lambda_2 \ln \frac{\beta_2}{\beta_1} - \lambda_1 \ln \beta_2} \\[3mm]
a_{n1} = a_{n2} = 0, \ b_{n1} = b_{n2} = 0, \ c_{n1} = c_{n2} = 0, \ d_{n1} = d_{n2} = 0
\end{cases}
\tag{11}
$$

The solution to Eq. (5) can be obtained from Eq. (11)

$$
\begin{cases}
T_1(\rho, \theta) = (T_g - T_l)\dfrac{\lambda_2 \ln \frac{\rho}{\beta_1}}{\lambda_2 \ln \frac{\alpha_2}{\beta_1} - \lambda_1 \ln \beta_2}, & \beta_1 < \rho < \beta_2 \\[4mm]
T_2(\rho, \theta) = (T_g - T_l)\dfrac{\lambda_2 \ln \frac{\beta_2}{\beta_1} + \lambda_1 \ln \frac{\rho}{\beta_2}}{\lambda_2 \ln \frac{\beta_2}{\beta_1} - \lambda_1 \ln \beta_2}, & \beta_2 < \rho < 1
\end{cases}
\tag{12}
$$

4 Verification by Finite Element Method

To validate the correctness and accuracy of the analytical solution given in Eq. (12) in this study, a finite element model was established to compare the numerical analysis results with the analytical solution.

The geometric parameters of the selected tunnel model in this study are as follows: $h = 20$ m, $r = 4$ m, $\Delta r = 0.5$ m the boundary conditions are set as follows: $T_1 = 25\,°C$, $T_g = 0\,°C$. The thermal conductivity parameters of the soil are presented in Table 2.

A two-dimensional thermal conduction model of a shallow-buried circular tunnel was created using finite element software, as illustrated in Fig. 3.

The established finite element model was locally refined using 15-node plane polygonal elements.

Table 2 Values of thermal parameters for tunnel lining and surrounding soil

	ρ (kg/m^3)	c (J/kg·°C)	λ (W/(m °C))	α (m^2/s)
Tunnel lining	1800	1200	0.04	1.85×10^{-8}
Surrounding soil	2000	1500	0.06	2.0×10^{-8}

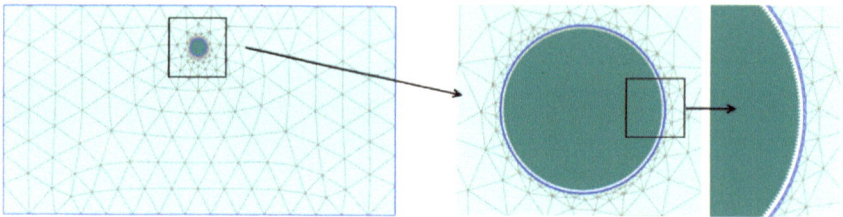

Fig. 3 Finite element model

After calculating the analytical expression (12) using MATLAB, the results were transformed back to the original Cartesian coordinate system using the conformal transformation Eq. (2). The comparison between the analytical results and the numerical simulation results is presented in Fig. 4.

Figure 4 illustrates that the temperature gradient increases as the position gets closer to the crown of the tunnel. Furthermore, the analytical solution calculations are in close agreement with the numerical solution calculations.

The isotherms near the insulation layer are depicted in Fig. 5.

Fig. 4 Comparison of analytical solution and numerical solution for temperature

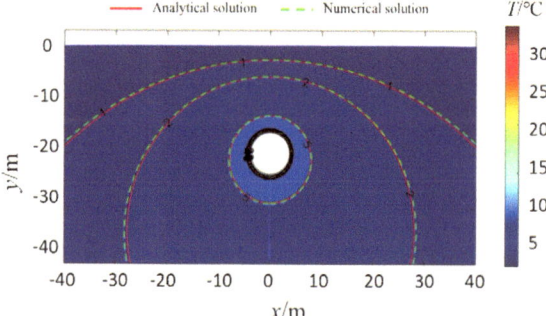

Fig. 5 Comparison of analytical solution and numerical solution for temperature near the tunnel insulation layer

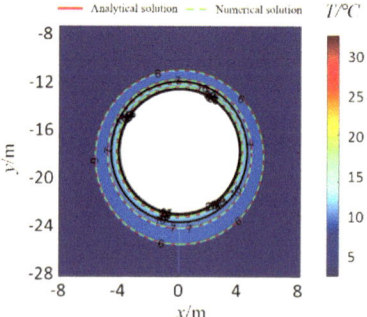

Upon observing Fig. 5, it can be noted that the presence of the insulation layer results in a significant temperature gradient near the tunnel lining, with higher numerical values of the isotherms located close to the insulation layer. Between the ground surface and the tunnel crown, the local temperature gradient in this specific area is greater compared to other regions due to the influence of the ground surface temperature. In the lower region of the tunnel, farther away from the ground surface, the local temperature gradient is relatively smaller compared to other areas. Additionally, the analytical solution calculations near the insulation layer align closely with the numerical solution calculations. This confirms the correctness and effectiveness of the analytical solution (12) proposed in this study.

5 Parameter Analysis

5.1 Influence of Tunnel Radius on Temperature

By varying the radius of the tunnel while keeping the other parameters of the tunnel model unchanged, the shallow-buried circular tunnel's steady-state temperature fields for different radii can be illustrated as depicted in Fig. 6.

Upon observing Fig. 6, it can be noticed that as the tunnel radius increases, the isotherms with equal numerical values shift outward in the direction away from the tunnel lining.

By selecting the 3 °C-isotherm line at the crown of the tunnel, the temperature gradient can be calculated and presented in Table 3.

Fig. 6 Comparison of Temperature Field in Surrounding Rock with Varying Tunnel Radius

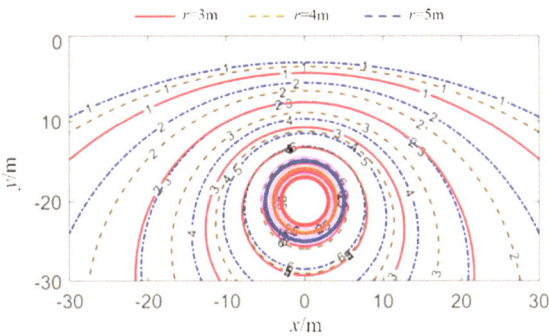

Table 3 Calculation of 3 °C temperature gradient at the crown of tunnels with different radius

Tunnel radius r (m)	3	4	5
Distance from the tunnel lining d (m)	5.48	6.13	6.87
Temperature gradient $\partial T/\partial r$ (°C/m)	25.37	18.45	11.39

From Table 3, it can be observed that as the tunnel radius increases, the 3 °C-isotherm line moves away from the tunnel lining. The farther the isotherm line is from the tunnel lining, the smaller the displacement towards the outer region, resulting in lower temperature gradients and sparser isotherm lines.

5.2 Influence of Tunnel Insulation Layer Thickness on Temperature

By altering the thickness of the insulation layer while keeping the other parameters of the tunnel model unchanged, the steady-state temperature fields of the shallow-buried circular tunnel with different insulation layer thicknesses can be depicted as shown in Fig. 7.

Upon observing Fig. 7, it can be noticed that as the thickness of the insulation layer increases, the isotherms with equal numerical values shift outward in the direction away from the tunnel lining.

By selecting the 3 °C-isotherm line at the crown of the tunnel, the temperature gradient can be calculated and presented in Table 4.

From Table 4, it can be observed that as the thickness of the insulation layer increases, the 3 °C-isotherm line moves away from the tunnel lining. Moreover, as the thickness of the insulation layer increases, the temperature gradient of the isotherm line with the same numerical value decreases at a slower rate.

Fig. 7 Comparison of steady-state temperature field in surrounding rock with varying tunnel insulation layer thickness

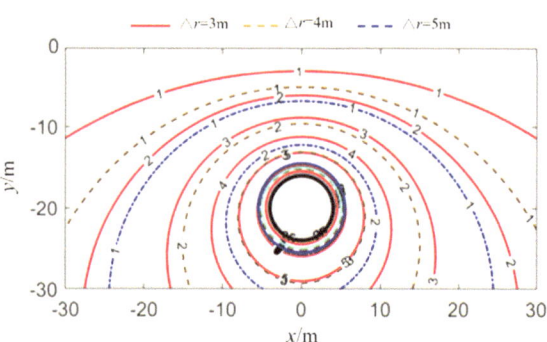

Table 4 Calculation of 3 °C temperature gradient at the crown of tunnels with different insulation layer thicknesses

The thickness of the tunnel insulation layer Δr (m)	0.5	1.0	1.5
Distance from the tunnel lining d (m)	1.23	2.98	4.67
Temperature gradient $\partial T / \partial r$ (°C/m)	34.68	23.12	10.43

5.3 Influence of Tunnel Burial Depth on Temperature

By varying the burial depth of the tunnel while keeping the other parameters of the tunnel model constant, the steady-state temperature fields of shallow-buried circular tunnels for different burial depths can be plotted. The results are presented in Fig. 8, showing the temperature distribution in the surrounding rock of the tunnel for different burial depths.

Upon observing Fig. 8, it can be noticed that as the burial depth of the tunnel increases, the isotherms with equal numerical values above the tunnel crown shift outward in the direction away from the tunnel lining. However, as the angle between the vertical line passing through the tunnel crown and the isotherm increases, the isotherms with the same numerical values gradually curve and begin to move inward toward the tunnel lining.

By selecting the 3 °C-isotherm line at the crown of the tunnel, the temperature gradient can be calculated and presented in Table 5.

From Table 5, it can be observed that as the burial depth of the tunnel increases, the 3 °C-isotherm line moves away from the tunnel lining. Additionally, as the distance between the tunnel center and the ground surface increases, the rate at which the temperature gradient of the isotherm line with the same numerical value decreases becomes smaller.

Fig. 8 Comparison of steady-state temperature field in surrounding rock with varying tunnel burial depth

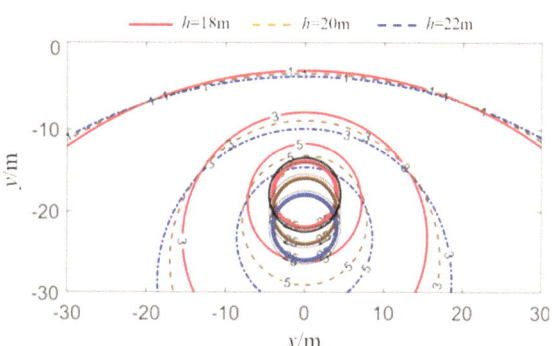

Table 5 Calculation of 3 °C temperature gradient at the crown of tunnels with different burial depths

Distance from the tunnel center to the ground surface h (m)	18	20	22
Distance from the tunnel lining d (m)	4.21	6.04	7.77
Temperature gradient $\partial T/\partial r$ (°C/m)	17.34	11.58	5.43

5.4   Influence of Thermal Conductivity of Insulation Layer on Temperature

By varying the thermal conductivity of the insulation layer while keeping the other parameters of the tunnel model constant, the steady-state temperature fields of shallow-buried circular tunnels for different thermal conductivities of the insulation layer can be plotted. The results are presented in Fig. 9, showing the temperature distribution in the surrounding rock of the tunnel for different thermal conductivities of the insulation layer.

Upon observing Fig. 9, it can be noticed that as the thermal conductivity of the insulation layer increases, the isotherms with equal numerical values shift outward in the direction away from the tunnel lining.

By selecting the 3 °C-isotherm line at the crown of the tunnel, the temperature gradient can be calculated and presented in Table 6.

From Table 6, it can be observed that as the thermal conductivity of the insulation layer increases, the rate at which the 3 °C-isotherm line moves away from the tunnel lining slows down, and the rate at which the temperature gradient decreases also decrease.

Fig. 9 Comparison of steady-state temperature field in surrounding rock with varying thermal conductivity of tunnel insulation layer

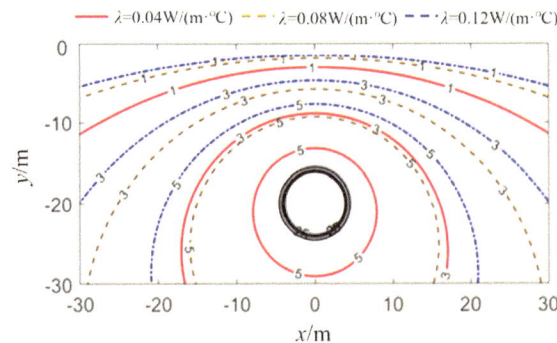

Table 6 Calculation of 3 °C temperature gradient at the crown of tunnels with different thermal conductivity conditions of insulation layer

Thermal conductivity of the insulation layer λ (W/(m·°C))	0.04	0.08	0.12
Distance from the tunnel lining d (m)	6.12	10.54	12.39
Temperature gradient $\partial T/\partial r$ (°C/m)	14.76	10.45	8.29

Fig. 10 Comparison of steady-state temperature field in surrounding rock with varying surface and tunnel lining temperature boundaries

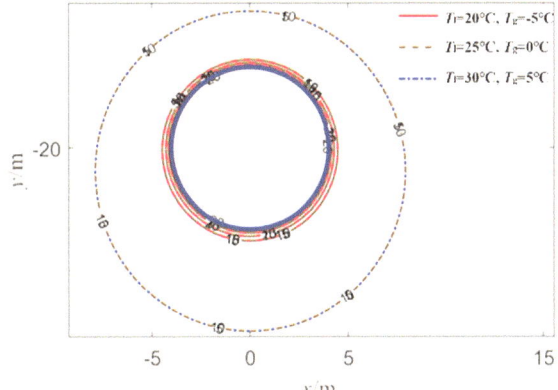

5.5 Influence of Surface and Tunnel Lining Temperature Boundaries on Temperature Distribution

By modifying the temperature boundary conditions of the ground surface and tunnel lining simultaneously, increasing them by 5 °C while keeping the other parameters of the shallow-buried tunnel model constant, three sets of isotherm plots depicting temperatures of 5, 10, 15, and 20 °C were obtained. These plots, illustrating the temperature distribution in the surrounding rock, are presented in Fig. 10.

From the zoomed-in portion of the tunnel lining in the local enlargement figure, it can be observed that when the temperature boundary conditions of both the ground surface and the tunnel lining are simultaneously increased by the same temperature increment, the isotherms corresponding to the respective boundary conditions coincide with each other for the same temperature difference. Therefore, the temperature boundaries of the ground surface and tunnel lining have an equally significant impact on the temperature distribution in the surrounding rock of the tunnel lining. This distribution pattern of the isotherms also aligns with the geometric distribution of the inner and outer boundaries of the transformed circular region after conformal mapping.

6 Conclusion

In this paper, the solution of the heat conduction equation for the surrounding rock of a shallow-buried circular tunnel was transformed into a circular annular region using the conformal mapping method. By utilizing the general solution of the Laplace equation in polar coordinates and considering the temperature boundary conditions of the tunnel lining and the ground surface, as well as the continuity of temperature and heat flux at the interface between the tunnel lining and the surrounding rock, an analytical solution for the temperature field of the surrounding rock in a shallow-buried circular tunnel was derived. Through detailed analysis and discussion, the following conclusions can be drawn.

(1) The analytical solution implemented in MATLAB software in this study shows good agreement with the results obtained from finite element software, indicating a high level of accuracy. The analytical solution satisfies the continuity conditions of temperature and heat flux functions at the interface between the insulation layer and the soil layer.

(2) Compared to finite element software, the analytical solution presented in this paper demonstrates higher computational efficiency. By inputting the corresponding model parameters, the specific values of the temperature spatial function in the surrounding rock of the tunnel can be obtained at any point. In contrast, finite element software involves a series of processes from modeling to obtaining results, which can be more complex. Specifically, the analytical solution in this paper takes only about 2 s to obtain the results using MATLAB software.

(3) Detailed parameter analysis of the tunnel radius, tunnel depth, insulation layer thickness, and properties of the insulation layer in this paper provides insights into the variation trends of the temperature spatial function under different variable conditions. This analysis has practical significance for understanding the temperature field in the surrounding rock of the tunnel under specific conditions.

(4) By modifying the soil conditions in the model, the solution of the temperature field in the surrounding rock of the tunnel can be extended to cases involving multiple soil layers, thus expanding the applicability of the model.

Acknowledgements The authors wish to acknowledge the assistance from the Dongyang Li Academician Workstation Project in Yunnan Province.

References

1. Yan, Z., & Yang, Q. (2003). Experimental study on the temperature field distribution of fire in Qinling extra long highway tunnel. *Underground Space, 2*, 191–195.
2. Lai, Y., Wu, Z., Zhu, Y., & Zhu, L. (1999). Nonlinear analysis of the coupling problem of temperature field, seepage field, and stress field in the surrounding rock of tunnels in cold regions. *Journal of Geotechnical Engineering, 5*, 529–533.
3. Zhou, N. (2022). Research on the evolution law and influencing factors of flow field temperature field in curved tunnels in cold regions under ventilation conditions. *Chongqing Jiaotong University.*
4. Xu, P., Wu, Y., Yan, X., Liang, W., & Hu, K. (2022). Analysis of temperature field and effectiveness of insulation layer in surrounding rock of high-speed railway tunnels in severe cold regions. *China Railway Science, 43*(4), 64–73.
5. Li, J. (2021). Enhanced geothermal system based on excavation - numerical simulation study on mechanics and heat transfer of tunnel surrounding rock. *Dalian University of Technology.*

6. Fan, K. (2021). The characteristics of the surrounding rock temperature field and the prediction model of rock temperature in the Nige High-Temperature Tunnel. *Chengdu University of Technology.*

7. Zhou, X., Zeng, Y., Yang, Z., & Zhou, X. (2015). Numerical solution of temperature field in surrounding rock of high-temperature tunnel. *Journal of Railway Science and Engineering, 12*(6), 1406–1411.

8. Zhang, Y., He, S., & Li, J. (2009). Analytical solution for the temperature field of surrounding rock of circular tunnel with insulation layer in cold regions. *Glacier Permafrost, 31*(1), 113–118.

9. Xia, C., Zhang, G., & Xiao, S. (2010). Analytical solution for the temperature field of tunnel surrounding rock in cold regions considering lining and insulation layer. *Journal of rock mechanics and Engineering, 29*(9), 1767–1773.

10. Lai, Y., Yu, W., Wu, Z., He, P., & Zhang, M. (2001). Analytical solution for the temperature field of the surrounding rock of circular cross-section tunnels in cold regions. *Glacier Permafrost, 2*, 126–130.

11. Hong, H., & Fang, T. (2020). Analytical solution to steady-state temperature field of Freeze-Sealing Pipe Roof applied to Gongbei tunnel considering the operation of limiting tubes. *Tunneling and Underground Space Technology, 105*, 103571.

12. Cheng, Q., Lu, A., & Yin, C. (2021). Analytical stress solutions for a deep buried circular tunnel under an unsteady temperature field. *Rock Mechanics and Rock Engineering, 54*(3), 1355–1368.

13. Liu, W., Feng, Q., Wang, C., Lu, C., Xu, Z., & Li, W. Analytical solution for three-dimensional radial heat transfer in a cold-region tunnel. *Cold Regions Science and Technology, 164*, 102787.

14. Wang, G. (2008). *Elasticity.* China Railway Publishing House.

15. Zhang, Z. (1989). *Heat transfer* (pp. 133–134). Higher Education Press.

16. Zhang, Y. (1984). Thermal stress in a semi-infinite plane with holes. *Journal of East China Institute of Water Resources, 4*, 75–83.

17. Wang, Z., Liu, M., Xie, J. (2013). Research on the problem of surface consolidation settlement caused by shield tunneling construction. *Geotechnical Mechanics, 34*(S1), 127–33.

Research on the Application of UAV Photogrammetry Technology in Highway Slope Deformation Identification

Xin Yan, Yiqaing Yu, Wei Zhan, Zhi Hu, Danqiang Xiao, and Jianjun Wu

Abstract The safety of highway slope is always one of the key works of highway management and maintenance department. However, the current highway slope maintenance inspection is mainly manual inspection, monitoring coverage is very low, only a few slopes installed automatic monitoring equipment. Due to the high cost and complex installation of automatic monitoring equipment, how to carry out efficient deformation inspection of highway slope is an important direction of highway maintenance research. Thanks to the rapid development of UAV close-range photogrammetry technology, this paper proposes to use UAV close-range photogrammetry technology to identify highway slope deformation. It mainly shoots UAV aerial survey images and obtains three-dimensional point cloud data of highway slope time series through data processing processes such as encryption and adopts ICP algorithm to register multi-phase point clouds. The change of three-dimensional point cloud is obtained by using point cloud difference calculation, so that the displacement and deformation characteristics of highway side slope are obtained. Through the application of field cases, the study verified the effectiveness of UAV close-range photogrammetry technology in highway slope deformation identification, with an accuracy of about cm level. The proposal and promotion of this technology can provide data support for subsequent highway slope protection and management and is of great significance for improving the quality and work efficiency of highway slope maintenance and management.

Keywords UAV photogrammetry technology · Highway slope · Deformation identification · ICP · Three-dimensional point cloud

X. Yan · Y. Yu (✉) · W. Zhan · Z. Hu · D. Xiao · J. Wu
Key Laboratory of Road and Bridge Detection and Maintenance Technology Research of Zhejiang Province, Zhejiang Scientific Research Institute of Transport, Hangzhou 310023, China
e-mail: yuyq@zjjtkyy.com

X. Yan
College of Geological Engineering and Geomatics, Chang'an University, Xi'an 710064, China

D. Li and Y. Zhang (eds.), *Advances in Frontier Research on Engineering Structures II*,
Lecture Notes in Civil Engineering 535, https://doi.org/10.1007/978-981-97-6238-5_43

1 Introduction

The deformation of highway slope directly affects the safety and management of highways. During the "14th Five-Year Plan" period, Zhejiang Province will build or expand thousands of kilometers of roads, followed by a large number of mountain road slopes, how to monitor the stability of mountain roads along the slope, to ensure the safe operation of mountain roads is one of the key works of the highway maintenance department [1].

Facing the increasing pressure of slope management and maintenance along the highway, the most effective means is to improve the efficiency and accuracy of highway slope safety inspection. At present, the mountain road slope safety monitoring means include manual inspection and automatic monitoring [2]. Due to high equipment cost and other problems, monitoring coverage is very low, most of the slope is mainly using manual inspection. But manual inspection work efficiency is low, subject to subjective influence, the accuracy of manual inspection is low. Therefore, the development of a new deformation identification technology applicable to mountain road slope can detect the abnormal deformation of highway slope in time, obtain its deformation characteristics in advance and carry out corresponding monitoring or protection work, which is of great significance to reduce the disaster loss of highway slope and maintain highway safety [3].

In recent years, with the rapid development of remote sensing technology, new technologies such as synthetic aperture radar interferometry (InSAR) and UAV close-range photogrammetry have been gradually applied in geological hazard identification [4]. Thanks to the advantages of wide coverage and high accuracy of satellite images, InSAR technology has played an important role in the identification of large-scale surface deformation, and has played an important role in the identification of geological hazards along the Sichuan-Tibet Railway and other national projects [5]. However, InSAR technology is limited by the return visit cycle of satellites, and the data cycle varies from several days to several months. Meanwhile, the processing process of satellite remote sensing image data is complicated, so InSAR technology cannot be widely promoted and applied in daily highway maintenance. Unmanned aerial vehicle (UAV) close-range photogrammetry technology is not limited by terrain and landform, and its operation is easy. In recent years, it has made great achievements in geological disaster investigation [6, 7].

A highway slope deformation identification technology based on 3D point cloud was proposed in this paper. The highway slope was reconstructed 3D by UAV close-range photogrammetry technology to obtain multi-phase 3D point cloud data of the highway slope. The multi-phase 3D point cloud data was registered by iterative nearest point algorithm, and finally the time series deformation data of the highway slope was obtained by difference of the point cloud data. It has been applied and verified in the actual highway slope. The proposed and popularization of this technology can effectively improve the efficiency and accuracy of highway slope inspection, and has great significance for highway slope maintenance.

2 UAV Photogrammetry Technology

UAV close-range photogrammetry technology is an emerging technology based on the development of digital imaging and photogrammetry, which integrates the application of computer technology, digital image processing, impact matching, pattern recognition and other multidisciplinary theories and methods, using the airborne camera photography to obtain high-resolution images, after processing to obtain the shape, size, position, characteristics, and mutual relations of the subject. Its essence is to construct three-dimensional space through two-dimensional images, and accurately obtain the spatial position information and surface texture information of related terrain and ground objects from the processed image results.

2.1 The Acquirement Process of UAV Image

To ensure that the results meet the accuracy requirements in the later stage, the relevant operation process must be standardized during image acquisition, and appropriate adjustments must be made according to the terrain conditions to obtain high-precision UAV remote sensing images. The acquisition process of UAV remote sensing images is shown in the Fig. 1, which generally includes data collection and arrangement, image control point layout and measurement, route planning, flight site selection, aerial photography quality inspection, etc.

Fig. 1 The acquirement process of UAV image

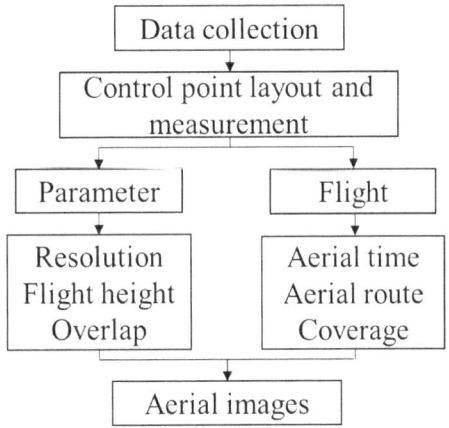

Fig. 2 The processing
procedure of UAV image

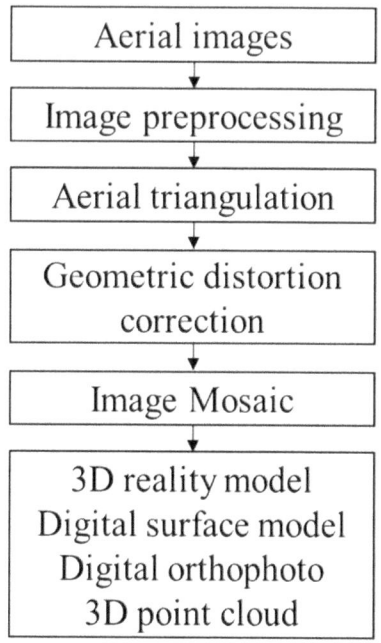

2.2   The Processing Procedure of UAV Image

After acquiring the UAV remote sensing image, a series of image processing should be carried out to obtain the required results. The UAV remote sensing image processing process is shown in the Fig. 2. Firstly, the adopted coordinate system is determined, the collected coordinate data is converted to the target coordinate system, and the aerial triangulation is carried out in combination with image POS data and image control point information. Then, the final 3D model, digital surface model DSM, digital orthophoto DOM and other results are obtained through image geometric distortion correction, image matching and fusion.

3   Case Study

3.1   General Condition

The highway slope selected in the research locates in Tonglu County. Tonglu County is mainly hilly and mountainous area with few plains, belonging to the middle and low hilly area of western Zhejiang. Surrounded by the mountains, the central is a narrow valley plain, and between the mountains and the plain are scattered hills. The Fuchun River runs through the eastern part of the county from south to north,

Fig. 3 The present state of the slope (left: the profile of the slope; right: the remains of the broken retaining wall)

and the diverting river flows into the Fuchun River from northwest to southeast. The main peak of Longmen Mountain, Guanyin, is 1246.5 m above sea level and is the highest peak in the territory. In the county's land area, mountains and hills account for 86.3%, and plains and waters account for 13.7%.

The status of the slope is shown in the Fig. 3. The lithology of the slope is strongly weathered, the rock joint is strongly developed and relatively broken, and the overlying is completely weathered residual and quaternary sedimentary soil. The slope has obvious traces of runoff and serious erosion, and a large amount of slope sediments are piled up at the foot of the slope, and the retaining wall is only a few meters long.

According to the historical disaster investigation, there was a landslide in December 2021, the retaining wall was washed out, the traffic was interrupted, the landslide volume was about 1000 m^3, and the collapsed retaining wall was broken and abandoned on the right side of the highway, as shown in the Fig. 3.

3.2 Design of Flight Parameters

The project adopts UAV close-range photogrammetry technology to investigate the development of slope disease. The UAV adopts V200 UAV of Shanghai Huace Navigation Technology Co., LTD., and the on-board camera is RIY-D2, a five-lens camera produced by Chengdu Ruibo Technology Co., LTD, each lens is 25 million pixels. To ensure flight accuracy, route planning parameters are shown in the Table 1 and Fig. 4.

The real 3D model of the slope can be obtained by 3D modeling based on the UAV aerial survey image, and the results are shown in the Fig. 5. The real 3D model can be used to better investigate the artificially difficult places such as the back edge of the slope. According to the obtained 3D model, it can be clearly found that the original cut-off ditch at the back edge of the slope broke during the historical disaster

Table 1 Design of flight parameter

Flight number	Parameter			
	Resolution (cm/pixel)	Overlapping rate (%)	Flight height (m)	Flight velocity (m/s)
1/2	1.8	90	115	8

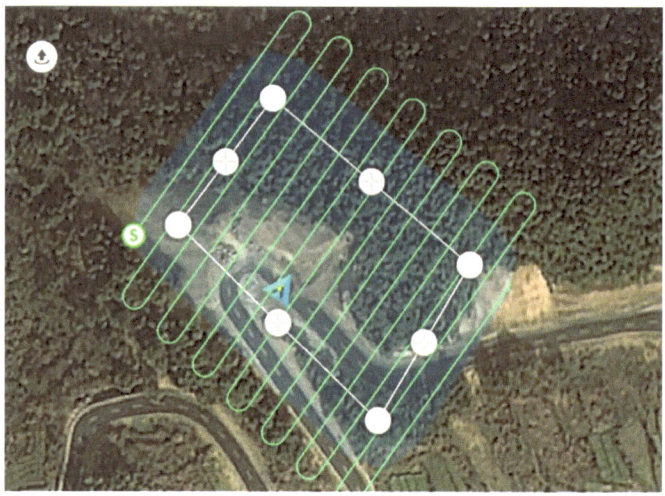

Fig. 4 Design of aerial route

process, resulting in the failure of the cut-off drainage at the back edge of the slope. The increase of rainfall infiltration and slope surface runoff further aggravated the development of slope diseases. To verify the accuracy of the 3D model, the research locates the landmarks on the spot and compares them with the coordinates in the model. The calculation results are shown in the Table 2.

3.3 3D Point Cloud

To analyze the current stability state of the slope, the study carried out two periods of UAV close-range photogrammetry on the highway slope, and established two periods of three-dimensional point cloud data, as shown in the Fig. 6. The reference point cloud data is the flight results on October 26, 2022, and the comparison point cloud data is the flight results on February 26, 2023.

Fig. 5 3D model of the slope

Table 2 Error result of ground points (unit: m)	Ground point	Horizontal error	Vertical error
	1	0.031	0.095
	2	0.053	0.102
	3	0.062	0.077
	4	0.028	0.086
	Average	0.0435	0.090

Fig. 6 Two phases of 3D point cloud data (left: 2022.10.26; right: 2023.02.26)

4 Discussion

ICP algorithm was used to perform accurate registration of two-phase point cloud data. The difference results of point cloud data are shown in Fig. 7, where red indicates settlement and blue indicates zero displacement. According to the results, it can be found that there is a large area of erosion on the side slope, the depth of slope erosion is 10–20 cm, but there is no deformation and instability overall.

Through the analysis of the historical disaster situation and the 3D reality model, it can be found that the whole slope of the highway is in a stable state. Due to the winter period from October 2022 to February 2023, there is no large-scale precipitation. Although the overall displacement and deformation have not yet occurred, serious slope erosion has still occurred on the slope. Therefore, continuous observation should be carried out on the highway slope, especially for large-scale precipitation in spring and summer, and measures should be taken to prevent soil erosion.

As shown in Fig. 7, the red area at the foot of the slope is a car, the vehicle model is Buick Regal, and the body height is 1.462 m. The difference calculation result of three-dimensional point cloud data is 1.425 m, and the calculation error is 0.037 m. Therefore, the calculation accuracy of this method is cm level.

Through the difference calculation of 3D point cloud data, the whole deformation of edge slope can be obtained well. Therefore, in daily highway slope inspection, the frequency of slope inspection varies from once a week to once a month according to the requirements of relevant regulations and the slope risk level. Maintenance personnel can use drones to efficiently complete slope inspection, obtain the deformation characteristics of highway slope through indoor three-dimensional point cloud generation—calculation, re-evaluate the slope risk level according to the deformation situation, and adjust its inspection frequency.

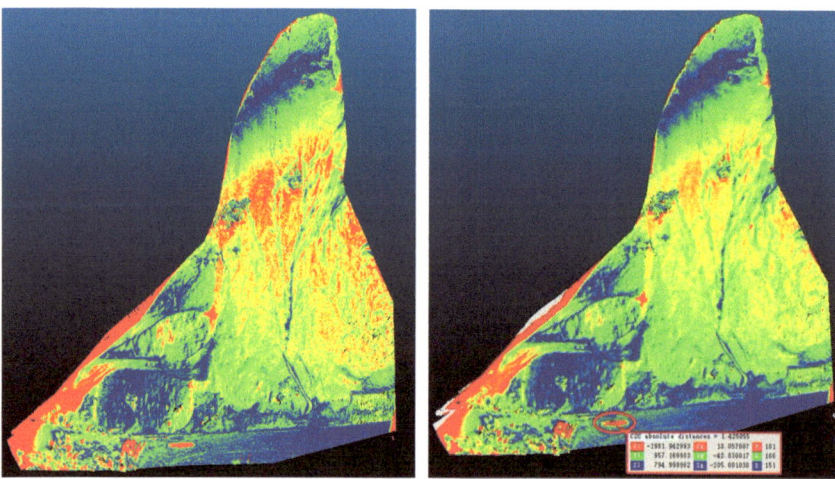

Fig. 7 Comparison of point cloud data in two phases

5 Conclusion

Through the development of UAV close-range photogrammetry technology on highway slope, the paper proposes to obtain the displacement and deformation of highway side slope by using three-dimensional point cloud data difference, and has been applied and verified on a highway slope in Tonglu County, obtaining the depth and range of highway side slope erosion successfully. It provides data basis and technical support for the subsequent management of slope soil and water conservation and overall safety protection.

Acknowledgements This research is financially supported by National Key R&D Program of China (Grant No.2020YFC1512005), the Scientific Innovation Project of Zhejiang Scientific Research Institute of Transport (Grant No. ZK202308, No.ZK202314), the Natural Science Foundation of Jiangsu Province (Grant No.BK20220421).

References

1. Yang, H. Z., Dong, J. Y., & Guo, X. L. (2023). Geohazards and risk assessment along highway in Sichuan Province, China. *Journal of Mountain Science, 20*, 1695–1711.
2. Zhang, Y., Deng, L., Han, Y., et al. (2023). Landslide hazard assessment in highway areas of guangxi using remote sensing data and a pre-trained XGBoost model. *Remote Sensing, 15*, 3350.
3. Li, Y. Z., Shen, J. H., Zhang, W. X., et al. (2023). Slope deformation partitioning and monitoring points optimization based on cluster analysis. *Journal of Mountain Science, 20*, 2405–2421.
4. Yarmohammad Touski, M., Dehghani, M., & Veiskarami, M. (2023). Monitoring and modeling of a landslide in Kahroud, Iran, by InSAR measurements and slope stability analysis. *Nature Hazards, 117*, 2249–2268.
5. Cui, P., Ge, Y. G., Li, S. J., et al. (2022). Scientific challenges in disaster risk reduction for the Sichuan-Tibet Railway. *Engineering Geology, 309*, 106837.
6. Wang, X. H., Cui, W., Zhang, G. K., et al. (2023). Identification of rocky ledge on steep, high slopes based on UAV photogrammetry. *Nature Hazards, 116*, 3201–3224.
7. Zhang, R., Shi, S., Yi, X., et al. (2023). A slope structural plane extraction method based on Geo-AINet ensemble learning with UAV images. *Remote Sensing, 15*, 1441.

Experimental Study on Defect Detection of Steel Plates Using Eddy Current Infrared Thermal Imaging Technology

Rui Ma, Ning An, and Peng Dong

Abstract In this paper, based on the principle of electromagnetic induction, eddy current infrared thermal imaging technology is used to detect steel plates with defects, and the temperature distribution diagram of steel plates is obtained. Through the experimental study, it can be seen that the temperature distribution of steel plate without defect is uniform temperature band, while the temperature distribution of steel plate with groove defect has obvious fluctuation. The location, shape, size and other information of defects can be obtained through the analysis of temperature map changes, especially the top position of defects, which is of great significance for detecting defects and their development trend.

Keywords Defects · Eddy current induction heating · Nondestructive testing

1 Introduction

In our country, long-span, extra-long-span bridges, steel box girder has been widely used. The structure of steel box girder is generally composed of steel plates of different thicknesses, connected by welding, forming orthogonal shaped plate. As the bridge deck material of orthogonal anisotropy plate, for the structure of direct bearing wheel load, under repeated load, the welding position and diaphragm has cracked, and this crack is mostly fatigue crack. For the detection of steel structure cracks, at present, the commonly used are magnetic leakage detection, eddy current detection, electro-ultrasonic detection, X-ray detection, etc., each of these NDT techniques has its advantages and disadvantages [1]. The most important disadvantage is that the detection results need to be accurately judged by personnel with a certain professional level and experience, and the display of the detection results is not intuitive enough.

Electromagnetic eddy current pulse thermal imaging detection technology is a non-contact nondestructive testing method based on the coupling of electrical, magnetic, thermal and other physical fields [2, 3]. It is a detection technology that

R. Ma · N. An · P. Dong (✉)
Research Institute of Highway Ministry of Transport, Beijing 100088, China
e-mail: 124664250@qq.com

© The Author(s) 2025
D. Li and Y. Zhang (eds.), *Advances in Frontier Research on Engineering Structures II*,
Lecture Notes in Civil Engineering 535, https://doi.org/10.1007/978-981-97-6238-5_44

combines eddy current induction heating with thermal conduction, which is suitable for the detection of various conductive materials. At present, this technology is mostly used for the detection of metal cracks [4], metal corrosion [5], composite material cracks, delamination, impact damage and other defects [6, 7]. For the pulse eddy current infrared thermal imaging detection technology, a lot of research work has been carried out at home and abroad [8, 9]. The induction heating device of eddy current infrared thermal imaging mainly consists of induction heating system, infrared thermal imager and PC [10]. As the main heat generating device, the induction heating system plays an important role in the experiment. At present, the induction heating devices used in domestic experiments are mostly imported Esahyheat series equipment. Based on the research of the principle of eddy current infrared thermal imaging detection technology, this paper adopts domestic induction heating device to conduct a comparative study on steel plate materials with different cracks, and determine the feasibility and detection accuracy of domestic equipment in this experiment.

2 Principle of Eddy Current Thermal Imaging Detection

Through the coil of alternating current, an alternating magnetic field will be generated around it. When the coil is close to the conductor under test and the conductor is a closed loop, an induced current (i.e. eddy current) will be formed in the conductor, and the eddy current will generate an induced magnetic field. As can be seen from the principle of electromagnetic induction, the induced magnetic field will always hinder the change of the original magnetic field. When there is a defect in the conductor, the distribution of the eddy current will be changed, and the distribution of the surface temperature field of the conductor caused by the thermal effect will also change.

In the process of induction heating, according to Maxwell's equations, the control equation of the eddy current field can be obtained as follows:

$$\nabla \times \left(\frac{1}{\mu} \nabla \times A \right) + \sigma \frac{\partial A}{\partial t} + \varepsilon \frac{\partial^2 A}{\partial^2 t} = J_s \tag{1}$$

In the formula (1), μ is the magnetic permeability of the material under test; A is the vector magnetic potential; σ is the conductivity of the material; ε is the dielectric constant; J_s is the external current density.

When electromagnetic induction is used as the heat source, the Joule heat generated by the eddy current first occurs on the surface of the conductor, and the internal heat is mainly transmitted by heat conduction. The temperature distribution on the surface and inside of the conductor is determined by the eddy current distribution and heat transfer process. With the increase of depth, the eddy density decreases exponentially, which is the skin effect of eddy currents. The descending speed will flatline, when the eddy density is below 1/e (equivalent to 37% of the surface eddy density). The depth at which the eddy density is 1/e of the surface eddy density is usually called the skin depth of the eddy δ.

$$\delta = \frac{1}{\sqrt{\pi \mu \sigma f}} \tag{2}$$

In the formula (2), f is the excitation frequency.

It can be seen from Eq. (2) that skin depth is inversely proportional to the excitation frequency in the induction coil. The larger the frequency, the smaller the skin depth, and the induced eddy current generated is mainly concentrated on the surface of the specimen.

According to Joule's law, eddy currents in the material will convert electrical energy into heat, resulting in a heat Q.

$$Q = \frac{1}{\sigma}|J_e|^2 = \frac{1}{\sigma}|\sigma E|^2 \tag{3}$$

In the formula (3), E stands for electric field strength. The generated Joule heat Q conducts inside the material to the surrounding space, forming a three-dimensional heat flow field and affecting the surface dimension T of the specimen. The heat conduction equation can be expressed as the following formula.

$$\frac{\partial T}{\partial t} = \frac{k}{\rho c_P}\left(\frac{\partial^2 T}{\partial x^2} + \frac{\partial^2 T}{\partial y^2} + \frac{\partial^2 T}{\partial z^2}\right) + \frac{k}{\rho c_P}Q(x, y, z) \tag{4}$$

In the formula (4), ρ stands for the density of the material; c_P stands for specific heat capacity; k stands for thermal conductivity; $Q(x, y, z)$ stands for the heat generated by the induced vortex in unit volume and unit time.

When the tested sample has defects, both the induction heating process and the heat transfer process will be affected. The temperature distribution of the specimen surface can be observed and analyzed through the images collected by the infrared thermal imager, and the surface defect distribution of the tested part can be obtained.

3 Experimental Research on Eddy Current Thermal Imaging

The experimental system used in this paper is shown in the Fig. 1. The electro-magnetic induction generator adopts the domestic electromagnetic induction heating device to study the feasibility and efficiency of high frequency (Fig. 2) and ultra-high frequency (Fig. 3) electromagnetic induction devices for steel plate crack detection.

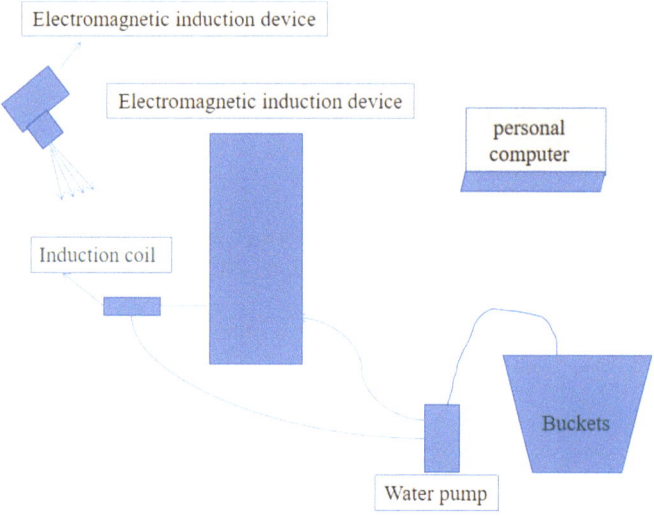

Fig. 1 Schematic diagram of the experimental system

Fig. 2 High frequency electromagnetic induction device

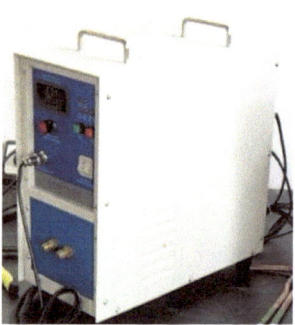

Fig. 3 Ultra-high frequency electromagnetic induction device

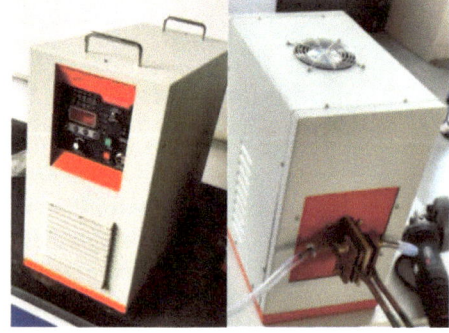

Fig. 4 The sample with
grooves

Two types of electromagnetic induction devices are selected, both of which are
produced by a company in Shenzhen. The output oscillation frequency of the ultra-
high frequency induction device is 100–200 kHz, and the excitation current is 10A-
35A. The output oscillation frequency of the high frequency induction device is
30–100 kHz, and the heating current is 100–650 A.

In the eddy current infrared thermal imaging detection experiment, another important equipment is the infrared thermal imager. The thermal imager used in this experiment is FLTR T1040 infrared thermal imager, with a temperature range of − 40–
2000 °C; the accuracy is between 5 and 150 °C, with an accuracy of ± 1 °C or ±
1%, and the temperature range is below 1200 °C at room temperature of 25 °C, with
an accuracy of ± 2° or ± 2% of the reading; the resolution is 800 × 480.

The selected sample is a 6 mm thick steel plate measuring 30 cm × 30 cm, which
has been cut using an angle grinder to create three grooves of varying lengths and
depths as defects, as illustrated in Fig. 4.

The experimental process consists of two stages: heating and cooling. The heating
stage progresses rapidly and can be divided into two processes based on the reasons
for heat generation: eddy current induction heating and heat conduction. These
processes overlap before the induction generator is turned off, resulting in a very
short heating time of 15 s. The temperature distribution during both the heating and
cooling processes are detected separately, as depicted in Figs. 5 and 6.

4 Analysis of Experimental Results

The high frequency and ultra high frequency electromagnetic induction heating
generators were used to heat the steel plate with grooves, and the temperature field
distribution of the steel plate was obtained by infrared thermal imager. Through the
observation of the temperature field, the following results can be obtained:

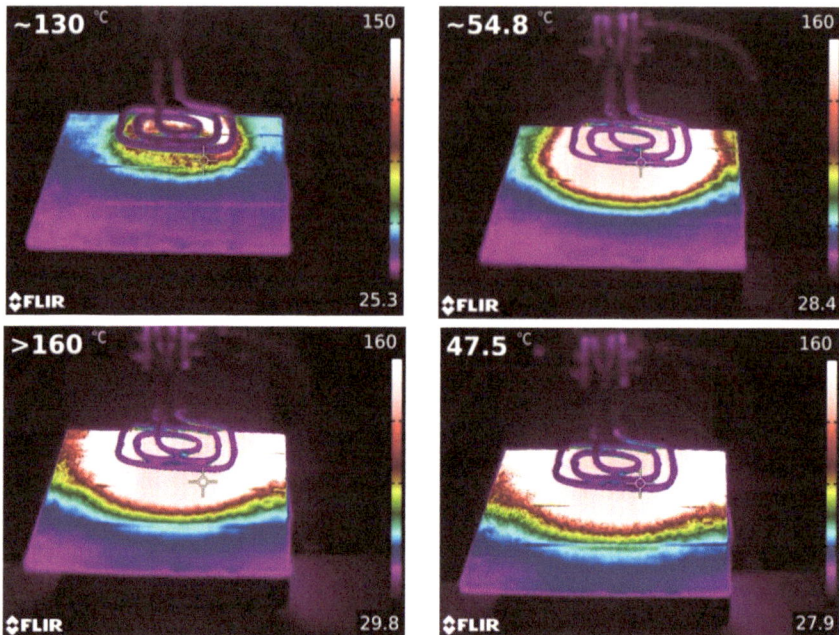

Fig. 5 Temperature distribution of the heating process

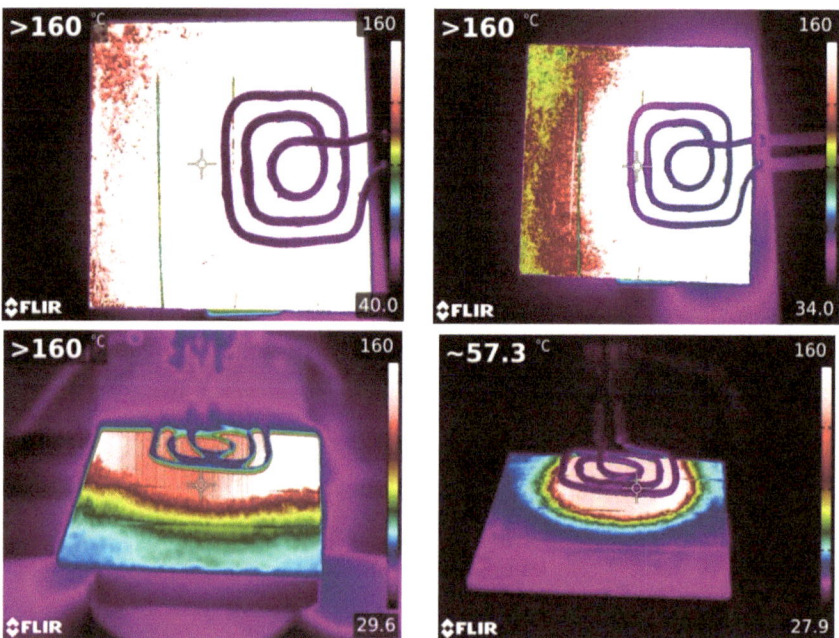

Fig. 6 Temperature distribution of the cooling process

(1) The existence of the groove will affect the temperature field distribution. In the heating process, the temperature field without the groove is uniformly distributed, and there are obvious fluctuations when there are grooves.

(2) The groove is open cracked, and the characteristics of induction heating are known. The time required for heat conduction is greater than the time for eddy induction heating. The surface temperature of the steel plate is higher than the internal temperature, and the temperature difference is large. Therefore, the temperature field distribution map obtained by the infrared instrument can be seen that the temperature at the crack is lower than the surrounding temperature.

(3) The cooling process is from high temperature to the steel plate to room temperature. The cooling process lasts for a long time, and the groove has a more obvious effect on the temperature field distribution. The groove is more obviously affected by the low temperature of the lower part, and the temperature difference is obvious with the surface of the steel plate. The location, shape, length and other information of the groove can be obtained, especially the vertex position of the groove.

5 Conclusion

Based on the principles of electromagnetic induction, eddy current, and infrared thermal imaging, this paper conducts experimental research on steel plates with grooves using a domestic electromagnetic induction emitter. Additionally, an infrared thermal imager is utilized to collect and analyze the temperature distribution map after heating. The conclusions are as follows:

(1) The eddy current induction thermal imaging detection technology effectively detects defects in steel plates. By analyzing changes in the temperature field on the surface of the steel plate, it can accurately identify the position, shape, length, and vertex position of grooves. This has significant implications for detecting crack development in steel structures.

(2) The heating effect on the steel plate includes direct heating from eddy current induction as well as heat conduction. Both effects occur simultaneously. Due to characteristics of eddy current induction, there is a noticeable temperature fluctuation with significantly lower temperatures at the bottom of grooves compared to the surface temperature of the steel plate.

(3) Eddy current induction enables rapid heating of the steel plate; within 10 s, temperatures can exceed 160 °C. Clearer defect information can be obtained during cooling stages.

Acknowledgements This work was financially supported by Central Public-Interest Scientific Institution Basic Research Fund (2021-9049a and 2021-9049b) and Open Project of National Engineering Laboratory of Bridge Safety and Technology (Beijing).

References

1. Shen, G. (2017). Development status of nondestructive testing and evaluation technology for pressure equipment. *Journal of Mechanical Engineering, 53*(12), 1–12.
2. Tong, Z. F., Xie, S. J., Liu, H. C., et al. (2020). Am efficient electromagnetic and thermal modelling of eddy of eddy current pulsed thermography for quantitative evaluation of blade fatigue cracks in heavy-duty gas turbines. *Machanical Systems and Signal Processing, 142,* 106781.
3. Jiang, L., Liu, Z. P., Liu, X. L., et al. (2014). Simulation and numerical analysis of nondestructive testing on weld cracks by eddy current thermography. *Advanced Materials Research, 1061–1062*(15), 874–880.
4. Wang, X., Hu, Y., Hou, D., et al. (2019). Eddy current thermal imaging under paint crack detection technology based on direction modulation principle. *Journal of Instrumentation, 40*(12), 56–63.
5. He, Y., Tian, G. Y., Pan, M, et al. (2014). An investigation into eddy current pulsed thermography for detection of corrosion blister. *Corrosion Science, 78,* 1–6.
6. Xu, C., Zhang, W., Wu, C., et al. (2020). An improved method of eddy current pulsed themography to detect subsurface defects in glass fiber reinforced polymer composites. *Composite Structures, 242,* 112145.
7. Wang, Q., Hu, Q., Qiu, J., et al. (2019). Differential laser infrared thermal imaging detection of internal defects in aviation composites. *Infrared and Laser Engineering, 48*(5), 127–133.
8. Yan, H., Yang, Z., Tian, G., et al. (2017). Near surface microcrack detection of ferromagnetic materials based on eddy current thermal imaging. *Infrared and Laser Engineering, 46*(3), 238–243.
9. Barakat, N., Mortadha, J., Khan, A., et al. (2020). A one-dimensional approach towards edge crack detection and mapping using eddy current thermography. *Sensors and Actuators A: Physical, 309,* 111999.
10. Wang, C., Jiang, X., Chao, Y., et al. (2023). Multi-defect detection of welding surface based on eddy current pulse thermography. *Infrared Technology, 45*(1), 84–90.

Engineering Design and Structural Simulation

Numerical Simulation Study of Deformation Characteristics of Cement Fly-Ash Gravel Pile

Jianguang Bai and Tianping Zhou

Abstract With the rapid development of high-rise buildings and infrastructure construction, higher requirements for foundations have been put forward. Cement fly-ash gravel pile is widely used in foundation treatment projects because of their high bearing capacity, low settlement, and low cost. The article adopts FLAC3D to conduct a numerical simulation study of the deformation characteristics of cement fly-ash gravel pile, the results show that cement fly-ash grave pile composite foundation produces vertical displacement with obvious zoning phenomenon under the action of upper load, horizontal displacement mainly occurs in the area of soil between piles, and the soil is extruded dense. The displacement and separation phenomenon of the interface of cushion and foundation are significantly different in the position of the pile and the position of soil between piles. The deformation of the CFG pile composite foundation is influenced seriously by the cushion as the weak zone of the CFG pile. The findings provide a reference and basis for the design and application of cement fly-ash gravel piles.

Keywords Cement fly-ash gravel pile · Numerical simulation · Deformation · FLAC3D · Lizheng geotechnical software

1 Introduction

Cement fly-ash gravel pile (CFG pile) is a kind of foundation treatment method that is made of fly-ash, cement, sand, and gravel mixed with water and has a certain bond strength pile as the reinforcement body, and the piles and the soil between piles share the load through the action of cushion to improve the foundation bearing capacity and reduce the settlement.

J. Bai (✉)
College of Energy and Transportation Engineering, Inner Mongolia Agricultural University, Hohhot, China
e-mail: b_jg@imau.edu.cn

T. Zhou
Jining Public Works Section, China Railway Hohhot Bureau Group Co., Ltd., Hohhot, China

© The Author(s) 2025
D. Li and Y. Zhang (eds.), *Advances in Frontier Research on Engineering Structures II*,
Lecture Notes in Civil Engineering 535, https://doi.org/10.1007/978-981-97-6238-5_45

With the rapid development of high-rise buildings and highway construction projects, large landmark buildings are emerging, and highways need to cross complex geological environments, and in most cases, buildings (structures) need to be built in weak soil layers with poor foundation soils, and bearing capacity and resistance to settlement deformation cannot meet the requirements, which puts forward higher requirements for foundations [1]. Cement fly-ash gavel pile can save 1/2–2/3 of the cost because it can make full use of the unique advantages of pile and soil between piles [2], and it has become one of the most commonly used foundation treatment techniques, which is suitable for foundation treatment of cohesive soil, silt, sand and other types of soil [3]. It is widely used in the foundation treatment of soft soil roadbeds of highways [4], high-rise buildings [5, 6], storage tank foundations [7], and other aspects.

However, as the applications become more and more widespread, more and more problems appear in the practical applications, which bring considerable potential safety hazards to the projects [8], CFG pile becomes an important research topic in geotechnical engineering. Yan et al. [9] analyzed the ultimate bearing capacity of CFG pile composite foundation under seismic action and the distribution character- istics of stress, strain, and plastic zone in the ultimate state, and the results showed that it is feasible to use CFG pile composite foundation scheme to treat the natural plant site of nuclear power as long as the cross-sectional size of the corner piles of the composite foundation is appropriately increased. Yin et al. [10] established a relationship model between CFG pile construction parameters and saturated fine sand liquefaction and analyzed the change law of super pore water pressure, satu- rated fine sand liquefaction phenomenon, and pile shape under different construction parameters. Xie [11] analyzed the selection of parameters and design calculation of cement fly-ash gavel piles in a granite geological area and summarized the quality problems such as necking and pile breakage during construction. Han et al. [12] used a genetic algorithm to give the optimal design calculation method for cement fly-ash gavel pile composite foundation under the lowest cost condition, which reduces the project cost under the condition of meeting the design specification of the building foundation. Based on the heat dissipation test, thermal conductivity test, and low- temperature strength test of the prepared concrete by adding fly ash, the growth law of concrete strength under negative temperature conditions was studied [13]. The pile length of the CFG pile can be designed by using a combination of long and short lengths according to the actual situation. Tang et al. [14] proposed the reinforcement ideas and design calculation methods for CFG long and short composite foundations and gave the key construction techniques. For combined cement fly-ash gravel pile composite foundations, CFG piles play a controlling role in improving the bearing capacity of the foundation and reducing the settlement [15]. Numerical simulation is one of the main tools for the study of CFG pile composite foundations. Cheng et al. [16] used numerical simulation to analyze the dynamic response of CFG pile composite foundations under dynamic load in detail. Hua et al. [17] applied a large finite element program ABAQUS with equivalent material instead of CFG composite foundation to simulate the deformation of CFG pile on surrounding soil by combining with engineering practice. Lu et al. [18] applied MIDAS to numerically simulate the longitudinal and cross sections of the road, and the results showed that the simulated settlement results were basically consistent with the field monitoring results and the settlement change trends were similar, which indicated the effectiveness of CFG pile composite foundation reinforcing soft foundation and the reliability of finite element simulation. Lai et al. [19] discussed the pile-soil stress distribution and pile-soil stress ratio of CFG composite foundation using a three-dimensional finite element model, and the deformation modulus and thickness of the cushion have important effects on the pile-soil stress ratio and overall bearing capacity of CFG composite foundation.

In this paper, we intend to use Flac3d to study the deformation characteristics of CFG pile composite foundations and provide a reference for CFG pile design and application.

2 Numerical Simulation

2.1 Modeling

The model consists of four parts: foundation model, pile model, cushion model, and Interface model, without considering the groundwater.

(1) Modeling of foundation

It is assumed that the foundation soil is homogeneous and isotropic, the type is silty clay, and the Mohr–Coulomb model is used in the structure model. The lengths of the foundation model in X, Y, and Z directions are taken as 13 m, 13 m, and 20 m, respectively, and an eight-node hexahedral element is adopted, as shown in Fig. 1a. The basic parameters are shown in Table 1, the gravitational acceleration is taken as 10 m/s^2. The elastic modulus is taken as 5 times the compression modulus, and the bulk modulus (K) and shear modulus (G) are calculated by Eqs. (1) and (2) respectively, $K = 66.7$ MPa, $G = 30.8$ MPa. The bearing capacity of the composite foundation is required to reach 480 kPa.

$$K = \frac{E}{3(1 - 2\mu)} \tag{1}$$

$$G = \frac{E}{21 + \mu} \tag{2}$$

It is assumed that the cement fly-ash gavel pile is carried out after the soil has completed consolidation, so the soil is consolidated under its weight in the model, and the displacement generated by its weight is removed. The piles are built after the soil consolidation without considering the disturbance to the soil.

(2) Modeling of the pile

Fig. 1 Model. **a** Foundation model **b** Pile model **c** Cushion model **d** Contact surface model

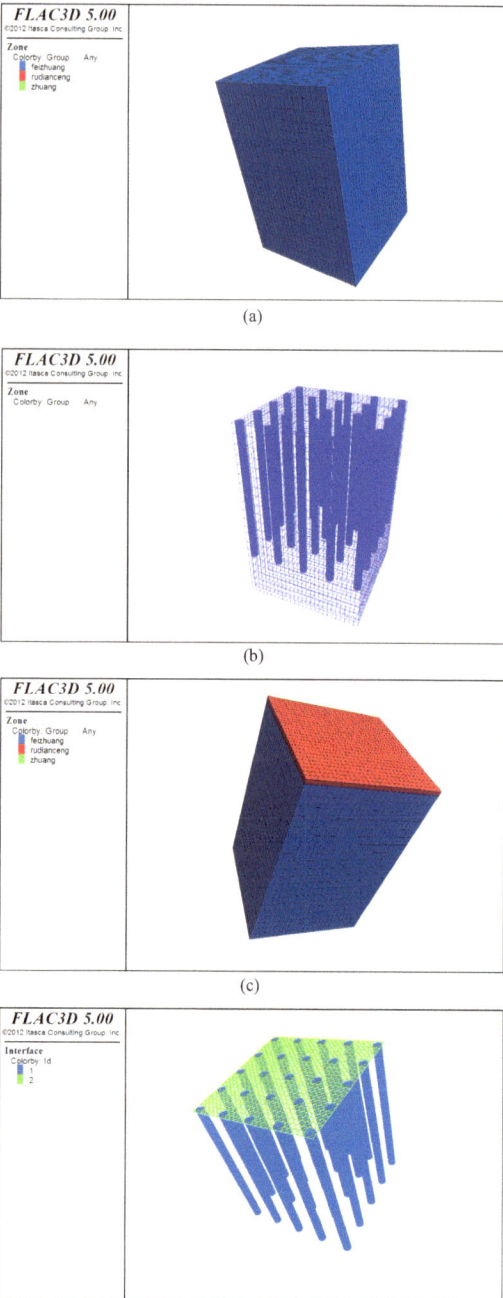

Table 1 Basic parameters

Density (kg/m^3)	1920
Cohesion (kPa)	26
Internal friction angle (°)	15
Compression modulus E_s (MPa)	16
Poisson's ratio μ	0.3
Load-bearing force (kPa)	260

The diameter of the cement fly-ash gavel pile is 0.8 m, the spacing in X and Y directions is taken as 3.0 m, and the length of the pile is 15 m (Z = 5.0 m ~ 20.0 m). The center of the side pile is 0.5 m from the edge of the foundation, and the number of piles is 25 in total. The pile material is C30 concrete, and the present structure model adopts an elastic model with eight-node hexahedral elements, as shown in Fig. 1b. The density is 2500 kg/m^3, and the modulus of elasticity is 30 GPa. The angle of internal friction and cohesion of concrete is determined according to Li et al. [20], that is

$$\tan \varphi = \frac{\sigma_c - \sigma_t}{2\sqrt{\sigma_c \cdot \sigma_t}} \tag{3}$$

σ_c-Standard value of compressive strength of concrete (MPa); σ_t-Standard value of tensile strength of concrete (MPa).

$$c = \frac{\sqrt{\sigma_c \cdot \sigma_t}}{2} \tag{4}$$

For C30 concrete, σ_c = 20.1 MPa, σ_t = 2.01 MPa, so its internal friction φ is 54.9° and its cohesion c is 3.178 MPa. The bulk modulus K is 16.7 GPa and the shear modulus G is 12.5 GPa.

(3) Modeling of cushion

The cushion material is graded medium-coarse sand. The thickness of the cushion is 0.4 m, Z = 20.0 m ~ 20.4 m, and the Mohr–Coulomb model is adopted for the present structure model with eight-node hexahedral elements, as shown in Fig. 1c. The cohesion is 1 kPa, the internal friction angle is 35°, Poisson's ratio is 0.2, and the compressive modulus is 35 MPa. The elastic modulus is 5 times the compressive modulus, bulk modulus K is 97.2 MPa and shear modulus G is 72.9 MPa.

(4) Modeling of interface

Interface 1 between the pile and the non-pile part is established by the leading in and leading out method, and the parameters kn and ks of the interface can be selected 10 times of Eq. (5), and the result is 6.67e11. The cohesion and internal friction angle is taken as 0.8 times the cohesion and internal friction angle of the soil, i.e. 20.8 kPa and 12°. The tensile stiffness of the pile-soil interface is 0, and the residual strength between the pile and the soil is 0 after the damage occurs. Interface 2 is established between the foundation and the cushion by the shifting method, and the parameters of the interface are the same as that of Interface 1, as shown in Fig. 1d.

$$\max\left[\frac{K + 4/3G}{\Delta z_{\min}}\right] \qquad (5)$$

The whole model includes 15,696 zones and 19,817 nodes. Considering the influence of the load generated at the foundation bottom, the free constraint is adopted at $z = 20.4$, lateral displacement constraint is adopted at the vertical side of the outer boundary $X = 0$, $X = 13$, $Y = 0$, and $Y = 13$, and fixed constraint condition is adopted at the bottom.

2.2 Simulation Analysis

(1) Foundation displacement

(a) Vertical displacement

Figure 2a shows the displacement curve in the Z-direction of the composite foundation surface $(Z = 0)$ from $(0,6.5,20)$ to $(13, 6.5, 20)$. As shown in Fig. 2a, the vertical displacement of the composite foundation has a clear zoning, which decreases with increasing depth and has a disc-shaped distribution [21]. The displacement in the region from 2.5 m under the pile end to the bottom of the model is 0, the displacement of the composite foundation gradually increases from 2.5 m below the pile end to 2.0 m below the bottom of the foundation, and the settlement of the pile and the soil between the piles at the same depth is basically the same. The displacement in the range of nearly 2.0 m below the bottom of the foundation reaches the maximum, and there are four peaks value and five valleys value. The valleys value is located between the piles, so the settlement of the soil between the piles is larger than that of the piles, and there is a significant difference between the settlement of the soil between piles and that of the piles at the same depth [22], in which the settlement of the foundation on both sides of the central pile is the largest, more than 1.6 cm. The deformation of each pile occurred in different degrees, the piles around produced downward displacement with displacement values of 0.4–1.28 cm, and the piles in the center were uplifted upward with the uplift displacement of 0.2 cm. Therefore, the model is limited by the constraint conditions around, the foundation will produce horizontal deformation while producing vertical displacement and squeeze toward the center, resulting in uplift.

(b) displacement

Figure 2b shows the displacement curve in the Y-direction of the composite foundation surface $(Z = 0)$ from $(6.5, 0, 20)$ to $(6.5, 13, 20)$ profile. As shown in Fig. 2b,

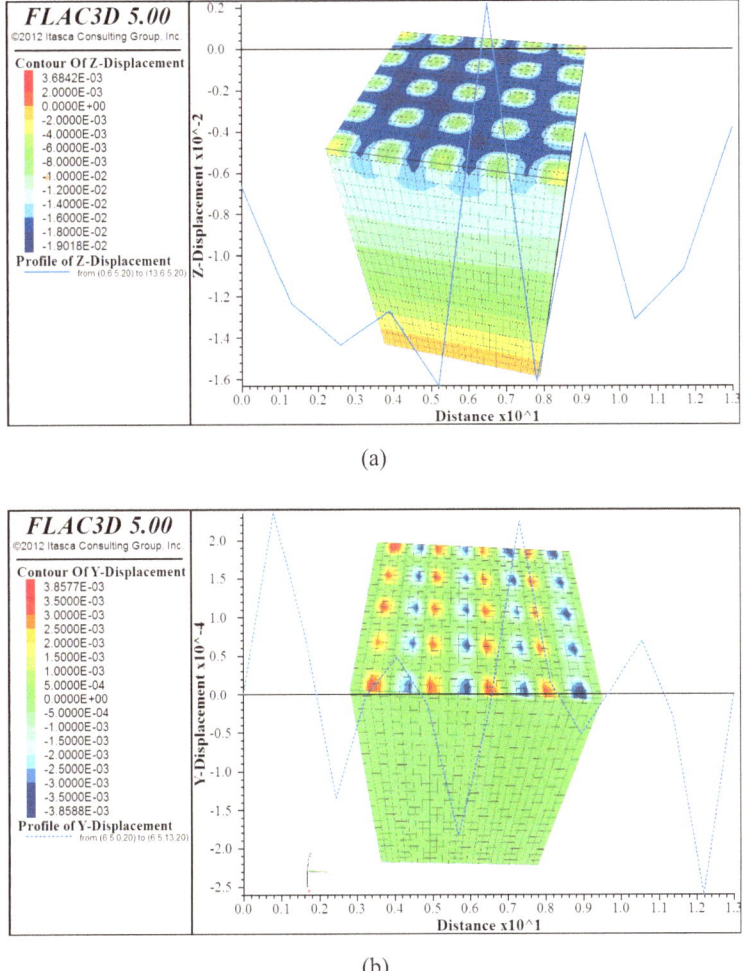

(a)

(b)

Fig. 2 Composite foundation displacement. **a** Vertical displacement **b** Horizontal displacement

the soil between piles produces horizontal displacement under the vertical load, and the displacement direction deviates from the direction of the piles, resulting in the extrusion of the soil between piles and the increase of compactness. The total displacement values between adjacent piles from left to right in the profile are 3.9, 2.4, 2.8, and 3.2 mm, so there are some differences in horizontal displacement at different positions of the foundation. The displacement around is relatively larger and the horizontal displacement in the middle part is smaller.

(2) Interface displacement

    (a) Normal displacement

Interface normal displacement between the cushion and the foundation is shown in Fig. 3a. As shown in Fig. 3a, interface normal displacement is uniformly distributed in a speckle pattern. The displacement at the pile position is approximately a concentric ring, and the displacement at the pile boundary is zero. The displacement from the pile edge to the pile center and outward gradually increases, and the displacement direction is downward. The influence range is about the circle with the pile center as the center and 2.5 times the pile diameter as the diameter. Therefore, for the contact between the rigid foundation and the cushion, the pile produces a certain penetration in the cushion [23], resulting in a local "hollow tube" effect at the pile location, while the flexible cushion is uniformly deformed at other locations and coordinated with the rigid foundation as a whole. The normal displacement of the interface between the pile and the foundation soil is 0, as shown in Fig. 3b.

The influence of the cushion on the composite foundation is great. When the thickness of the cushion is not large, the bearing effect of the soil between piles is small relative to the role of the pile, about 96% of the load will be borne by the pile [24]. When the thickness of the cushion is too large, the load-sharing effect of the pile will be reduced [25]. Therefore, the cushion is the core requirement of the CFG pile composite foundation [26]. The thickness of the cushion will have an effect on the deformation effect of the interface, and the thickness of the cushion should be considered comprehensively in practice.

(b) Shear displacement

Figure 4 shows the interface shear displacement. As shown in Fig. 4, shear displacement is generated between the cushion and the foundation, and the displacement is mainly distributed in the range of soil between piles, and the displacement values varied at different locations, with an average of about 3.5×10^{-5} mm, and the displacement values were extremely small. The interface deformation space is limited by the influence of the pile at the pile position, and the shear displacement is approximately 0, which is mainly caused by the compression deformation of the cushion and the pile piercing into the cushion. The directions of shear displacement are all pointing in the same direction, and there is no deformation to the surrounding area or the middle of the composite foundation due to the effect of uniform load.

(3) Interface separation

(a) Normal separation

Figure 5a shows the interface normal separation between the cushion and the composite foundation. As shown in Fig. 5a, the local normal separation is produced at the interface in the upper part of the pile by the influence of the pile, and the cushion in other areas outside the pile location maintains contact with the foundation. Therefore, it will produce a partial detachment at the pile position between the rigid CFG pile and the cushion, which coincides with the conclusion in Fig. 3a, indicating that the cushion becomes a weak zone of the CFG composite foundation [27] and should be paid attention to in the design. According to Fig. 5b, no normal separation

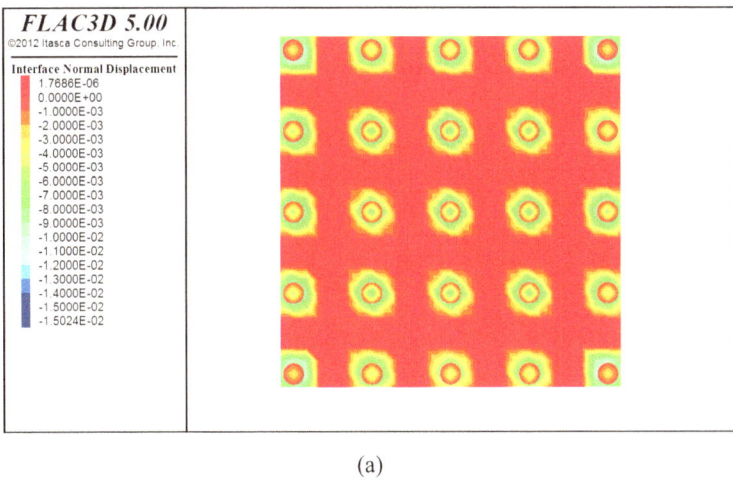

(a)

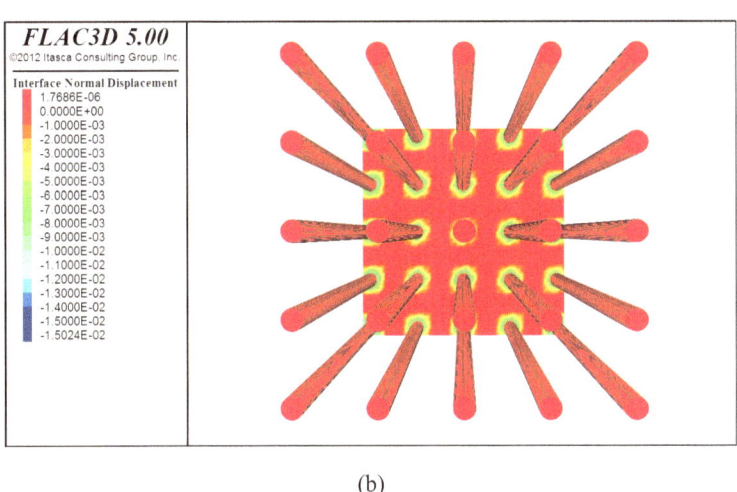

(b)

Fig. 3 Interface normal displacement. **a** Interface 2 **b** Interface 1

between the pile and the foundation is produced, and there is close contact between the pile and the soil.

(b) Shear slip

Figure 6a shows the shear slip between the cushion and the foundation. As shown in Fig. 6a, local shear slip occurs between the cushion and the foundation, which means that the horizontal deformation of the cushion occurs under the action of the upper load. Therefore, the characteristics of the cushion should be given enough consideration in practice to reduce the slip deformation between the foundation and the cushion. Figure 6b shows the shear slip between the pile and the foundation soil.

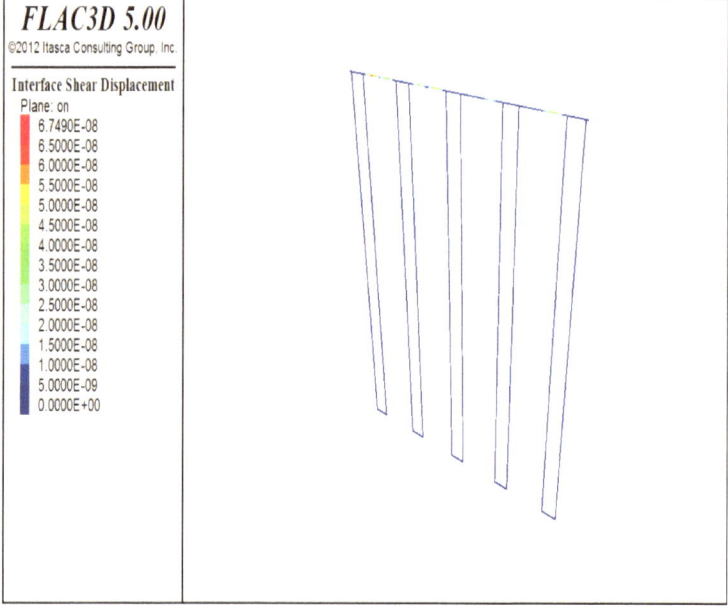

Fig. 4 Interface shear displacement

As shown in Fig. 6b, there is no shear slip between the pile and the foundation, and the deformation is coordinated.

In general, it can be seen that local normal separation and shear slip between the cushion and the foundation are produced, while no relative separation is produced between the pile and the foundation. Therefore, as a flexible material, the cushion between the foundation and the pile will be affected by the rigid material, especially the pile. There is no separation between the pile and the soil between piles. On the one hand, the distance between the pile and the pile is large, which is equivalent to a large soil thickness. On the other hand, the cushion effect of the upper cushion weakens the possibility of deformation.

3 Scheme Feasibility

Lizheng geotechnical software has powerful and excellent performance, and it is the most commonly used calculation software in engineering [28]. Lizheng geotechnical software is widely used in seepage analysis [29], slope prevention [30], slope stability analysis [31], foundation treatment calculation, and many other aspects. It is used for design calculation and FLAC3D software for simulation, and the conclusion is reliable [32]. The depth of the foundation pit adopted in the design is 10.5 m, the soil type is silty clay, and the parameters of the soil are shown in Table 1. In the

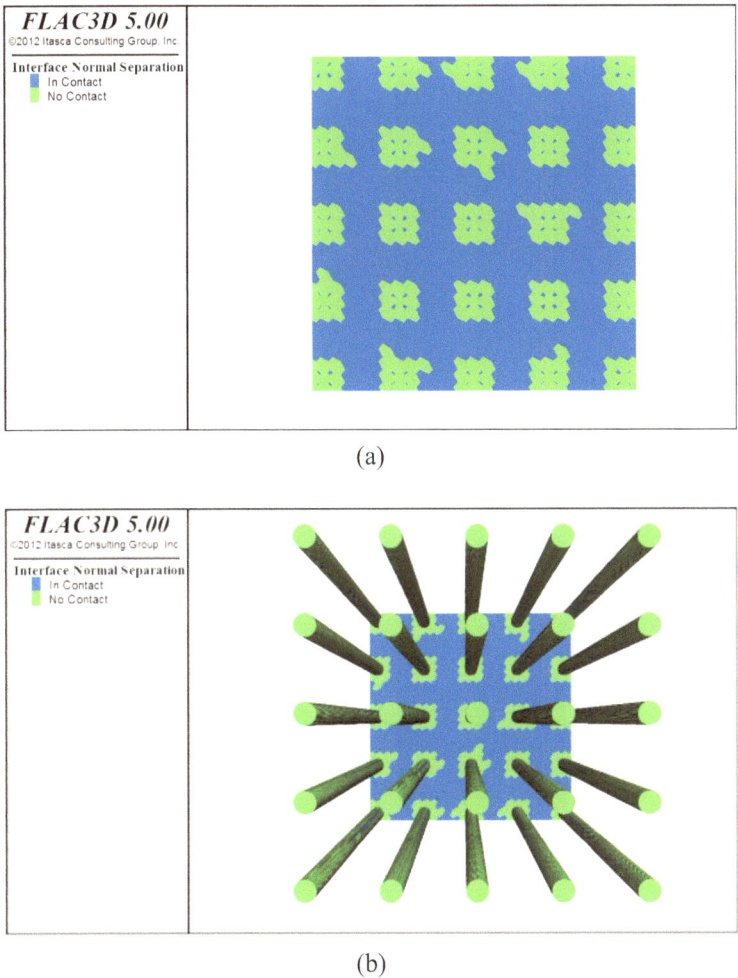

Fig. 5 Interface separation. **a** Interface 2 **b** Interface 1

design, the diameter of the cement fly-ash gravel pile is 0.8 m, the length of the pile is 15.0 m, the spacing of the pile is 3.0 m, and the square pile layout is adopted.

3.1 Bearing Capacity of Composite Foundation

The characteristic value of the bearing capacity of the composite foundation is calculated according to formula (6), and then the depth and width correction is carried out to obtain the corrected bearing capacity of the composite foundation f_z.

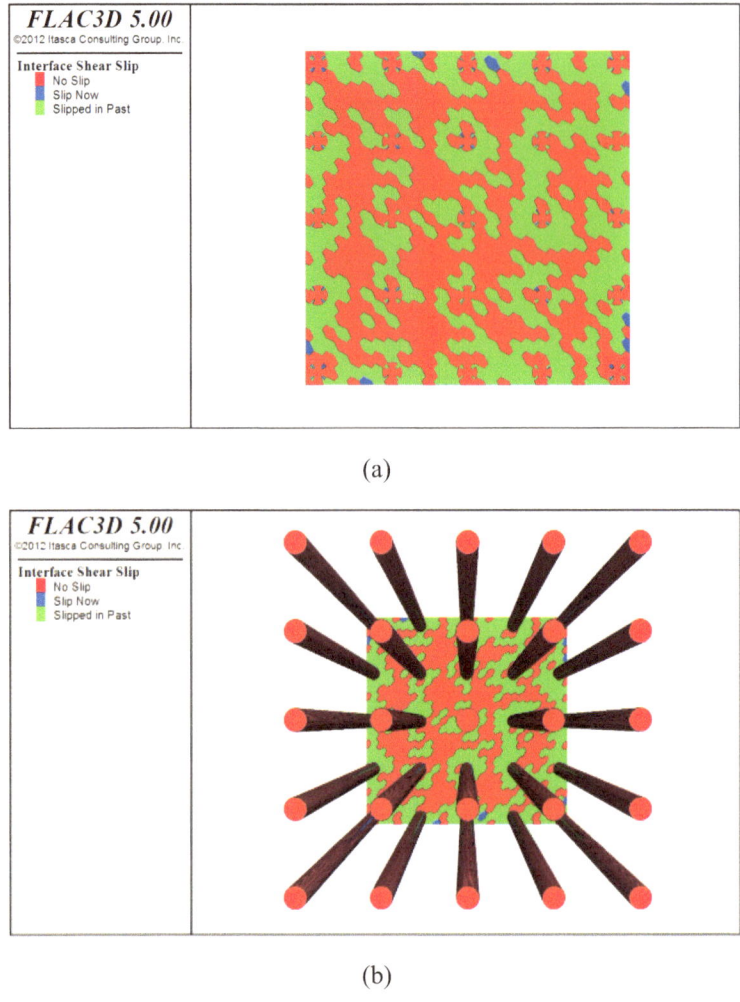

Fig. 6 Interface shear slip. **a** Interface 2 **b** Interface 1

$$f_{spk} = \lambda m \frac{R_a}{A_p} + \beta (1 - m) f_{sk} \qquad (6)$$

λ—Coefficient of bearing capacity of single pile, taken as 0.85.

R_a—Characteristic value of single pile bearing capacity, taken as 1432 kN.

A_p—Sectional area of the pile, with a diameter of 0.8 m, $A_p = 0.5024$ m^2.

β–Coefficient of bearing capacity of soil between piles, taken as 0.95.

m—Area replacement rate,m $= d^2/d_e^2$.

The d is the diameter of the pile, 0.8 m, and d_e is the equivalent circular diameter of the treated foundation area shared by one pile, $d_e = 1.13s$ while square pile layout is adopted and s is the spacing of the pile, so m is 0.056, which is greater than the critical value of the replacement rate [33].

f_{sk}—The characteristic value of the bearing capacity of the soil between piles after treatment. The improvement coefficient of bearing capacity of the natural foundation is taken as 1.1, therefore $f_{sk} = 286$ kPa.

It is calculated that $f_z = 583.8$ kPa > 480 kPa, so the bearing capacity of the composite foundation meets the requirements.

3.2 Settlement of Composite Foundation

The settlement of the composite foundation at the center point of the foundation is calculated by the stress area method, and the data of each calculation are shown in Table 2.

The equivalent value of compression modulus is calculated according to Eq. (7) [34].

$$\overline{E_s} = \frac{\sum_{i=1}^{n} A_i + \sum_{j=1}^{n} A_j}{\sum_{i=1}^{n} \frac{A_i}{E_{spi}} + \sum_{j=1}^{n} \frac{A_j}{E_{spj}}} \tag{7}$$

Z1—The distance from the bottom of the foundation to the top of this calculation layer (m).

Z2—Distance from the base of the foundation to the bottom of this calculation layer (m).

It is calculated that $\overline{E_s} = 23.05$ MPa, so the empirical coefficient of settlement calculation is taken as 0.24. The total settlement is $0.240 \times 130.07 = 31.22$ (mm), which meets the specification requirements.

The bearing capacity and settlement of the composite foundation in this project meet the actual engineering requirements, so the design scheme of the cement fly-ash gravel pile is feasible. On this basis, numerical simulation is of practical significance.

Table 2 Calculation of settlement of composite foundation

Layer number	1	2
Thickness (m)	15.4	4.60
Compression modulus E_{sp} (MPa)	24.109	16.000
Z1 (m)	0.00	15.40
Z2 (m)	15.40	20.00
Compression volume s_i (mm)	113.08	16.99
The integral value of the stress coefficient (z2a2—z1a1)	9.7924	0.9763

4   Conclusions

(1) Under the action of the upper load, the composite foundation produces vertical displacement, which decreases with the increase of depth and has an obvious zoning phenomenon. There are obvious differences between the settlement of pile and that of the soil between piles at the same depth within the range of nearly 2.0 m under the bottom of the foundation, and the settlement of pile and that of the soil between piles at the same depth from the position of 2.0 m under the bottom of the foundation to the position of 2.5 m under the pile end is basically the same. The soil between the piles produces horizontal displacement, the displacement direction deviates from the direction of the piles, and the soil is extruded and compacted.

(2) The displacement of the interface between the cushion and the foundation is approximately concentric ring at the pile position, the displacement at the pile boundary is zero, and the displacement from the pile edge to the pile center and outward gradually increases. Outside the circle with a diameter of 2.5 times of pile diameter, the displacement is approximately 0. Shear displacement is generated between the cushion and the foundation, and the shear displacement at the pile position is 0.

(3) The interface between the cushion and the pile produces local normal separation, and the cushion and the foundation in other areas outside the piles cannot be separated. The local shear slip occurs between the cushion and the foundation. There is no normal separation and shear slip between the pile and the foundation, and the deformation is coordinated.

(4) As the weak zone of the CFG pile, the cushion has a great influence on the deformation of the CFG pile composite foundation, so it should be given enough attention in the actual engineering design.

5   Expectation

The effect of cushion on the deformation of CFG pile is influenced by various factors such as the thickness of the cushion, compactness, and material type, and the mode and degree of influence need to be further studied. The properties of different types of soil vary greatly, and only one type of soil is considered in this simulation study, and relevant studies can be conducted for other types of soil. Groundwater as an important influencing factor in the stability of composite foundations is also the focus of the next research.

6 Availability of Data

All data, models, and codes generated or used during the study appear in the submitted article.

7 Competing Interests

The authors declare no competing interests.

Acknowledgements The authors would like to thank "The 13th five-year plan" for educational science research in Inner Mongolia Autonomous Region (NGJGH2019402) and Inner Mongolia Natural Science Foundation (2023LHMS05048) for this study.

References

1. Zhou, W. X. (2021). Design analysis and comparison of the pile foundation scheme and rigid pile composite foundation scheme of a high-rise residential building. *Henan Science and Technology, 40,* 63–65. https://doi.org/10.3969/j.issn.1003-5168.2021.27.023
2. Zhang, L. J., Qing, W., Sun, S. W., & Li, H. B. (2015). Researching the application of CFG pile treatment in soft foundation of expressway. *Advances in Intelligent Systems Research, 126,* 432–435.
3. Shi, J., & Wang, X. Y. (2015). Application of cement fly ash gravel (CFG pile) composite foundation in high-rise building ground work. *Geology of Anhui, 25,* 57–60. https://doi.org/10.3969/j.issn.1005-6157.2015.01.012
4. Luo, X. (2018). Treatment technology and optimization of cement fly ash macadam pile for soft soil subgrade of expressway. *Heilongjiang Jiaotong Keji, 41,* 105–106+108. https://doi.org/10.16402/j.cnki
5. Hou, H. M. (2020). Application of composite foundation of cement fly ash gravel pile (CFG Pile) in foundation treatment of high-rise building. *Adhesion., 41,* 115–119.
6. Hou, L. M. (2017). Using the concrete fly-ash gravel piles for treating the foundation of a certain construction project. *Shanxi Hydrotechnics, A01,* 21–23.
7. Qi, Y. F., & Ji, H. T. (2018). Application of cement fly-ash gravel pile composite foundation in treatment of storage tank foundation. *Oil and Gas Field Surface Engineering, 37,* 96–100. https://doi.org/10.3969/j.issn.1006-6896.2018.10.023
8. Gao, J. P. (2016). Analysis of composite foundation of cement fly ash gravel pile (CFG Pile) problems in practical application. *Fly Ash Comprehensive Utilization, A02,* 39–43.
9. Yan, Q. J., Zhong, H., & Li, H. J. (2017). Study on seismic bearing capacity of CFG piled composite foundation for nuclear power station based on quasi-static method. *Industrial Construction, 47,* 40–43+78. https://doi.org/10.13204/j.gyjz201701008
10. Yin, X. K., Du, S. Y., & Wang, T. T. (2021). Experimental study of relationship between sand liquefaction and CFG pile construction parameters. *Rock and Soil Mechanics, 42,* 2518–2524. https://doi.org/10.16285/j.rsm.2020.1859
11. Xie, B. (2015). Application of cement fly ash gravel pile (CFG Pile) in highway soft foundation treatment. *Jiangxi Building Materials, A07,* 210–211.

12. Han, Y. Q., Yang, J., Wang, C., & Zhou, Y. J. (2016). Optimization design and simulation of cement fly-ash gravel pile composite foundation. *Journal of Xi'an Technological University, 36*, 120–124. https://doi.org/10.16185/j.jxatu.edu.cn.2016.02.006

13. Liu, Z. Y., Yu, T. L., Yan, N., Piao, Z. H., & Zhang, H. X. (2021). The role of double-cylinder insulation technology in ensuring the quality of bored pile concrete under negative temperature condition. *Jordan Journal of Mechanical and Industrial Engineering, 15*, 51–58.

14. Tang, Q. F., Tong, J. X., Jia, N., Yang, X. H., Sun, X. H., & Yan, M. L. (2022). Application of composite foundation with long-short CFG piles in engineering accident treatment. *Lecture Notes in Civil Engineering, 213*, 171–179. https://doi.org/10.1007/978-981-19-1260-3_16

15. Li, J. C., Cao, J., & Cong, J. (2018). Application of combined composite foundation with CFG pile in deep soft foundation treatment. *Port & Water Engineering, A11*, 156–161+198. https://doi.org/10.3969/j.issn.1002-4972.2018.11.030

16. Cheng, X. S., Chen, J. C., Cai, X. D., Zhang, X. Y., Gong, L. J., & Gu, C. Y. (2021). Dynamic response of CFG and cement-soil pile composite foundation in the operation stage. *Geomechanics and Engineering, 25*, 385–399. https://doi.org/10.12989/gae.2021.26.4.385

17. Hua, F., Tartakovsky, D., & Sun, D. W. (2016). Effect of deformation of the CFG (Cement Fly-ash Gravel) pile group on the adjacent metro tunnel based on the equivalent material. *Journal of the Balkan Tribological Association, 22*, 507–515.

18. Lu, L. P., Zhang, Z. H., & Zhang, Y. C. (2022). Reinforcement effect of cement fly ash gravel pile composite foundation for soft foundation. *Science Technology and Engineering, 22*, 8877–8883.

19. Lai, J. X., Liu, H. Q., Qiu, J. L., Fan, H. B., Zhang, Q., Hu, Z. N., Wang, J. B. Stress analysis of CFG pile composite foundation in consolidating saturated mine tailings dam, *Advances in Materials Science and Engineering*, https://doi.org/10.1155/2016/3948754

20. Li, Y. A., Ge, X. R., Mi, C. R., & Zhang, H. C. (2004). Failure criteria of rock-soil-concrete and estimation of their strength parameters. *Chinese Journal of Rock Mechanics and Engineering., A05*, 770–776.

21. Li, J. C., Cong, J., Cao, J., & Cao, Y. L. (2018). Experimental research on deformation and stress behaviors of CFG pile composite foundation of large storage tank. *Industrial Construction, 48*, 116–121.

22. Mai, H. (2021). Analysis of bearing capacity of composite foundation with cement fly ash gavel pile. *Fly Ash Comprehensive Utilization, 35*, 40–44+87. https://doi.org/10.19860/j.cnki.issn1005-8249

23. Weng, Z. R. (2021). Application of composite foundation construction technology of cement fly ash gravel pile (CFG pile) in soft foundation treatment. *Heilongjiang Jiaotong Keji, 44*, 268–269.

24. Zhi, B., Wu, L. H. L., Li, G., Wang, Y. X., & Wang, F. (2018). Experimental analysis of efficiency factor of bearing capacity for soil between piles in CFG pile composite foundation under high stress. *Industrial Construction, 48*, 106–109. https://doi.org/10.13204/j.gyjz

25. Lan, S. H. (2017). Experimental study on the influence of cushion layer and pile parameters on cement fly ash gravel pile. *Gansu Water Resources and Hydropower Technology, 53*, 28–30.

26. Ma, J. X. (2016). Design method of cement fly ash gravel pile composite foundation. *Shanxi Architecture, 42*, 83–86. https://doi.org/10.3969/j.issn.1009-6825.2016.33.045

27. Tong, W. W., & Li, C. X. (2018). Analysis on seismic performance of cement fly ash gravel pile composite foundation. *Sichuan Building Materials, 44*, 83–84. https://doi.org/10.3969/j.issn.1672-4011

28. Zhou, Y. (2021). Results comparisons in a deep excavation case histories analysis with plaxis and lizheng software. *Soil Engineering and Foundation, 35*, 254–256.

29. Yin, L. L., Zhang, W. Y., Zhu, Q. B., Fan, C. R., & Li, J. D. (2022). Application of beijing lizheng seepage analysis software in reservoir dam seepage analysis. *Inner Mongolia Water Conservancy, A05*, 55–56.

30. Li, D. X., Deng, Y., & Cheng, J. (2018). Application of lizheng geotechnical and Flac3d in slope treatment design calculation. *Western China Communications Science & Technology, A06*, 54–575.

31. Yang, X. J., & Zhang, B. H. (2021). Application of lizheng software in slope stability calculation of asphalt core rockfill dam. *Inner Mongolia Water Conservancy, A08*, 31–32.
32. Li, D. X., Deng, Y., & Cheng, J. (2018). Application of Lizheng Geotechnical and Flac3d in Slope treatment design calculation. *Western China Communications Science & Technology, A06*, 54–57.
33. Dong, J. Y. (2021). Analysis of the influence of cement fly ash gravel pile supporting approach embankment on abutment pile on soft soil foundation. *Western China Communications Science & Technology, A01*, 162–165+178. https://doi.org/10.13282/j.cnki.wccst.2021. 01.044
34. Ministry of Housing and Urban Rural Development of the People's Republic of China. "Technical Code for Ground Treatment of Buildings (JGJ79–2012)", Peking: China Architecture & Building Press, 2013.

Synthesis and Property Analysis of a Novel Epoxy Composite Curing Agent Based on Soft Soil Curing

Shengchao Cui, Guangzheng Wang, Shu Liu, Xintong Yan, and Ming Fan

Abstract To meet the demand for shore-based engineering, engineering rescue and traffic emergency support in different complex conditions, a new low-temperature fast solidifier for epoxy resin was synthesized by Mannich reaction using thiourea, polyamine, formaldehyde, phenol and benzyl alcohol. The performance of the curing agent was investigated based on raw material ratio, reaction temperature and reaction time. Moreover, the curing time, amine value, viscosity and infrared spectrum were tested. It was compounded with other highly active curing agents and resins. The compressive strength of the cured sea sand was used as the index to determine the formula of a new epoxy composite used for the rapid curing of sand under low temperature or saturated water conditions. The results showed that the optimal synthesis conditions of the curing agent were as follows: the molar ratio of thiourea, diethylenetriamine, formaldehyde, phenol and benzyl alcohol is 1.2: 1.3: 1: 0.4: 1. The addition of a small amount of accelerator DMP-30 could effectively improve the curing reaction speed and reduce the curing time. These epoxy composites could effectively solidify the sea sand and improve its mechanical properties, which provides important utilization value for coastal engineering construction in the cold season or wet rainy season.

Keywords Epoxy resin · Curing agents · Mannich · Low temperature · Fast curing

1 Introduction

Beaches in coastal areas are naturally soft with large sand porosity, high moisture content, low cohesion, low bearing capacity and poor stability. Under certain conditions, they present great challenges to shore-related engineering construction,

S. Cui · G. Wang (✉) · S. Liu · X. Yan
College of Field Engineering, Army Engineering University of PLA, Nanjing 210007, China
e-mail: wanggzmail@aeu.edu.cn

M. Fan
Changsha Puzhao Biochemical Technology Co., Ltd., Changsha 410009, China

D. Li and Y. Zhang (eds.), *Advances in Frontier Research on Engineering Structures II*,
Lecture Notes in Civil Engineering 535, https://doi.org/10.1007/978-981-97-6238-5_46

561

engineering rescue, and emergency transportation support. If the related problem is not properly handled, it can cause various geological disasters and lead to serious engineering accidents [1]. In recent years, many studies have used polymer curing materials to solidify sand. These materials have improved the mechanical properties of sand at variable degrees, reduced soil liquefaction, and improved its weathering resistance [1–3].

Compared to other polymer curing materials, epoxy (EP) has good adhesive properties as well as excellent physical, mechanical, and electrical insulation properties [4–7]. Because of its various advantages, such as high structural strength, strong adhesion, minimal contraction, high hardening speed, and high chemical stability, it has been widely applied in hydraulic construction, aerospace and military engineering as well as in the construction of roads and bridges [8–11]. In fact, the EP does not harden separately, it can reveal its advantages only after mixing with a curing agent. In this case, the epoxy groups present in its molecular structure react with the curing agent to crosslink and polymerize into a water-insoluble polymer with a three-dimensional network structure. Therefore, EP curing materials generally consist of two components which are the curing agent component and the EP resin component. After curing, the ordinary EP resins may still incorporate some defects, such as large internal stress, poor crack resistance, brittle texture, poor impact resistance, and poor heat and humidity resistance [8, 12]. The curing effect of EP resin can be affected by temperature and moisture. Specifically, it can be easily weakened or even deactivated at low temperatures or high moisture contents, resulting in effective failing strength [13]. It is, therefore, necessary to modify the EP resin molecular structure by adding tougheners and softeners to improve its performance. Research today is greatly focused on the development and modification of curing agents.

The EP curing agents currently available on the market are mainly based on addition polymerization, which often has flexible designs and various functions. It can be classified into polyamines, acid anhydrides, polythiols, and polyphenols. Specifically, polyamines are found in most applications which have the advantages of fast reaction and room-temperature curing. However, ordinary amine-based EP curing agents are easily volatile and they can emit toxic and irritating odors, which are harmful to human health. In this case, dosage requirements are strict. Moreover, it is easy to absorb CO_2 in the air, resulting in impurities and affecting its performance. Researchers have attempted to solve these problems through chemical modifications, such as the addition of epoxy compounds, reaction with carbonyl compounds, reaction with organic acids, Michael's addition with an olefinic double bond, and Mannich's reaction with phenolic compounds. The phenylamine EP curing agent prepared by Mannich's reaction can effectively reduce the volatility and toxicity of lower aliphatic amines. Its liquid state ensures good miscibility with EP. In addition, the phenolic skeleton of the curing agent can effectively increase the heat distortion temperature of the system. The presence of active groups, such as phenolic hydroxyl, primary amino, and secondary amino groups, in its molecular structure, can accelerate the curing rate, enabling curing at low temperatures and in humid environments. This reaction was adopted in several studies to develop low-temperature EP curing agents, which achieved complete curing with water below zero degrees in half an hour to a dozen hours. Nevertheless, the previous materials are mostly used as adhesives and grouting materials, and there are few studies on their application in the rapid curing of soft marine sand. Besides, the long curing time issues and the impacts of specific conditions such as low temperature and saturated water have yet to be resolved.

In this study, a novel low-temperature fast-curing EP agent was synthesized based on Mannich's reaction with thiourea, polyamine, formaldehyde, phenol and benzyl alcohol. The influences of materials ratio, synthesis reaction temperature, and synthesis reaction time on the properties of the curing agent were investigated by infrared spectroscopy and measurements of amine value, viscosity, and curing time. Then, the curing agent was compounded with other highly active curing agents and resins. Taking the compressive strength of the cured marine sand as the evaluation index, the formula of the EP curing agent, which can be used for rapid sand curing under low temperature and saturated water conditions, was determined. The results of this study provide an important baseline for applications of EP materials in marine sand reinforcement, shore-based engineering construction, and emergency transportation support.

2 Synthesis of a Low-Temperature Fast Curing Agent

2.1 Materials and Instruments

Table 1 shows the main materials used in this study. In addition, the required instruments to carry out different experiments include a stainless-steel reaction kettle, a glass reaction kettle, a high-speed frequency conversion mixing machine, the synthesis reaction control platform, an oil bath heater, a constant temperature and humidity chamber, an electronic meter, a viscosity tester, and a Fourier transform infrared spectrometer (FTIR).

Table 1 Main materials used in the experiments

No.	Name	Specification	Manufacturer
1	Thiourea	Analytical reagent	Qingzhou Guangda
2	Paraformaldehyde	Analytical reagent	Jinsui Huagong
3	Phenol	Analytical reagent	Sinopec Yanshan
4	Diethylenetriamine	Industrial product	Jinan MingWeen
5	Benzyl alcohol	Analytical reagent	Greenhome Materials
6	DMP-30	Analytical reagent	Yueyang Zhongzhan

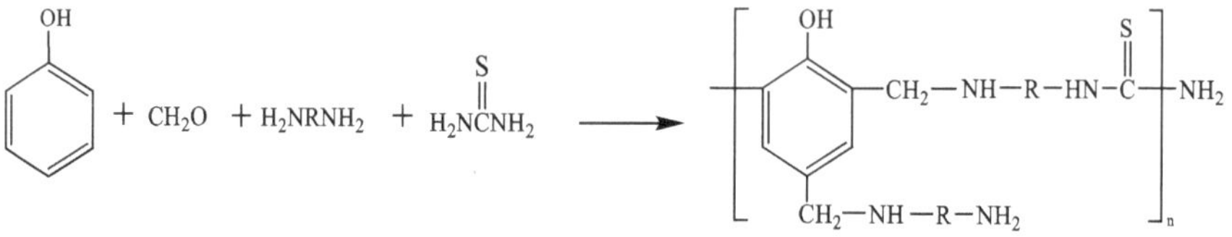

Fig. 1 Main synthesis reaction of the low-temperature curing agent

2.2 Synthesis

Raw materials were prepared according to the preset ratios, and then polyamine, thiourea, phenol, and paraformaldehyde were sequentially added to the reaction kettle. Stirring was maintained at a constant speed and low temperature until the materials were completely dissolved. Subsequently, the mixture was heated up to carry out the reflux reaction. While stirring for a certain period, DMP-30 was added to the solution. Finally, the resulting solution was cooled and discharged. The main synthesis reaction is shown in Fig. 1.

2.3 Performance Indicators and Characterization

Performance indicators of the curing agent include viscosity, amine value, and gel time. The viscosity was tested by using a rotational viscometer (GB2794-1981). The amine value was measured by using the hydrochloric acid–ethanol titration method (ZBG71005-89). The gel time was determined based on the Chinese national standard (GB12007.7-1989). In addition, the samples were prepared by the KBr tablet method, and the infrared spectrum was characterized by the FTIR technique (Thermo Fisher Scientific, USA).

3 Results and Analysis

3.1 Influence of Raw Materials Ratio on Curing Time

(1) Raw materials selection

Based on previous studies, the synthetic route of phenalkamine modification by thiourea was adopted. There are different types of amines, such as aliphatic amines, aromatic amines, alicyclic amines, and polyamines, while phenols include phenol and cardanol. To form the synthetic material, one type of amines and one type of phenols were selected and combined with thiourea and paraformaldehyde. By assessing the performances of different thiourea-modified amine curing agents, the most reactive one was selected as the base material for the low-temperature fast-curing EP agent.

In the specific synthesis process, the amount of each reactant was determined based on its molecular weight and the equivalent number of moles in the reaction, i.e., $n = m/M$, where n is the number of moles, which can be determined according to the stoichiometric coefficient in the reaction, and M is the molar mass of the corresponding reactant. Therefore, the above equation can be applied to calculate the reactant mass, m. By using different amines and phenols, paraformaldehyde, thiourea, and benzyl alcohol, five types of thiourea-modified phenalkamine curing agents were synthesized according to the reactant fractions in the basic reaction, and their performances were tested. The material ratios of each curing agent are shown in Table 2.

Table 3 shows the performance test data of the five types of curing agents. It can be noted that curing agent Type 2 has the highest active hydrogen equivalent, yet its viscosity is very high, which is not appropriate for mixing. Curing agent Type 4 has the lowest viscosity and good penetration performance, but its total amine value is low and its gel time is long. Modified by the aliphatic amine, curing agent Type 3 has a moderate viscosity, a high total amine value and the shortest gel time, i.e., it is the most reactive agent. Thus, diethylenetriamine, thiourea, phenol, paraformaldehyde, and benzyl alcohol were selected as the basic raw materials for the low-temperature fast-curing agent.

(2) Ratio of raw materials

Orthogonal tests were carried out to evaluate the influence of different ratios on the curing agent performance and, subsequently, to determine the optimal ratio of the selected basic raw materials. The experimental design is shown in Table 4.

The orthogonal test results are shown in Table 5. It can be deduced that the order of the range of agents' curing time at 20 °C is A > D > C > B. Thus, the order of impacts of the raw materials on the curing time is Thiourea > Benzyl alcohol > Paraformaldehyde > Diethylenetriamine.

It can be revealed from Fig. 2 that by increasing the amounts of thiourea and diethylenetriamine, the curing time initially decreased and then increased. The largest decrease in curing time (27%) was reached when the amounts of thiourea were

Table 2 Material ratios of each curing agent

No.	Composition (%)
1	Diethylenetriamine 17 Thiourea 18.7 Phenol 17 Paraformaldehyde 21.2 Benzyl alcohol 14.8 DMP-30 11.3
2	Diethylenetriamine 18 Thiourea 15.2 Cardanol 44.3 Paraformaldehyde 17.3 DMP-30 5.2
3	Diethylenetriamine 32 Thiourea 20.6 Phenol 22.2 Paraformaldehyde 9 Benzyl alcohol 7.8 DMP-30 8.4
4	Isophorone diamine 22 Thiourea 12.7 Cardanol 51.6 Paraformaldehyde 5.5 DMP-30 8.2
5	Meta-xylylenediamine 19 Thiourea 15 Cardanol 46.8 Paraformaldehyde 10 DMP-30 9.2

Table 3 Performance of the five types of curing agents

No.	Curing agent	Total amine value (KOH)/ (mg g^{-1})	Active hydrogen equivalent	Viscosity (20 °C)/ (Pa s)	Gel time (20 °C)/ min	Recommended dosage
1	Type 1	380–420	95–105	30,300	21 min	100: 52
2	Type 2	240–260	120–130	29,800	18 min	100: 60
3	Type 3	510–520	78–80	2500	8 min	100: 38
4	Type 4	280–300	110–112	750	20 min	100: 55
5	Type 5	340–360	95–100	1400	15 min	100: 50

Note The recommended dosage is determined by the material ratio test based on the bisphenol A epoxy resin 0164 and the active hydrogen equivalent of the curing agent

Table 4 Factors and levels of the orthogonal tests

Level	Factor			
	A (n (Thiourea): n (Phenol))	B (n (Polyamine): n (Phenol))	C (n (Formaldehyde): n (Phenol))	D (n (Benzyl alcohol): n (Phenol))
1	1.0	1.2	1.0	0.2
2	0.8	1.3	1.2	0.3
3	1.2	1.4	1.4	0.4

changed. This can be explained by the fact that thiourea sulfur atoms have high nucleophilicity at low temperatures. Consequently, the amount of thiourea significantly affected the low-temperature curing reaction. On the premise of synthesizing the desired product, the remaining small amount of thiourea can promote the activity of the amino groups of the curing agent so that the epoxy-curing reaction is fast. However, when the amino groups in the curing agent have all participated in the epoxy-curing reaction, the thiourea will lose its promoting effect even with increasing amounts. Instead, it will undergo a polycondensation reaction with formaldehyde during synthesis, increasing the curing agent viscosity, which is not beneficial for the curing reaction. The amount of benzyl alcohol has also a significant influence on the curing time but with the opposite effect. With an increasing amount of benzyl alcohol, the curing time initially increased and then decreased. When the paraformaldehyde/ phenol molar ratio was between 1 and 1.2, the curing time remained almost the same. With the continuous increase of paraformaldehyde, the curing time increased significantly. In addition, diethylenetriamine is the key component of modification and its content should be more than 30% to avoid wasting other substances and increasing viscosity and by-products. As to phenol, it provided not only the benzene ring as the basic structure for polymerization but also hydroxyl groups required to increase the grafting point. Therefore, it has improved the curing agent reactivity to a certain extent.

Overall, the obtained results suggest that 1.2: 1.3: 1: 0.4: 1 is the optimal ratio for thiourea, diethylenetriamine, formaldehyde, benzyl alcohol, and phenol.

Table 5 Orthogonal test results

Test		Factors				Amine value (KOH)/ (mg g^{-1})	Viscosity (20 °C)/ (Pa s)	Curing time (20 °C)/min
		A	B	C	D			
1		1.0	1.2	1.0	0.2	505.0	2800	10.5
2		1.0	1.3	1.2	0.3	493.0	3200	11
3		1.0	1.4	1.4	0.4	496.0	3700	11
4		0.8	1.2	1.2	0.4	495.0	3300	13
5		0.8	1.3	1.4	0.2	485.0	3900	15.5
6		0.8	1.4	1.0	0.3	504.0	2900	16
7		1.2	1.2	1.4	0.3	515.0	3300	17
8		1.2	1.3	1.0	0.4	495.0	2900	12
9		1.2	1.4	1.2	0.2	491.0	3300	15
Curing time	Average at level 1	10.8	13.5	12.8	13.7			
	Average at level 2	14.8	12.8	13.0	14.7			
	Average at level 3	14.7	14.0	14.5	12.0			
	Range	4.0	1.2	1.7	2.7			

Note A fixed amount of DMP-30 is added, the synthesis reaction time is 2.5 h, the reaction temperature is 115 °C, and the curing time is measured by using the curing agent and the bisphenol A resin 0164

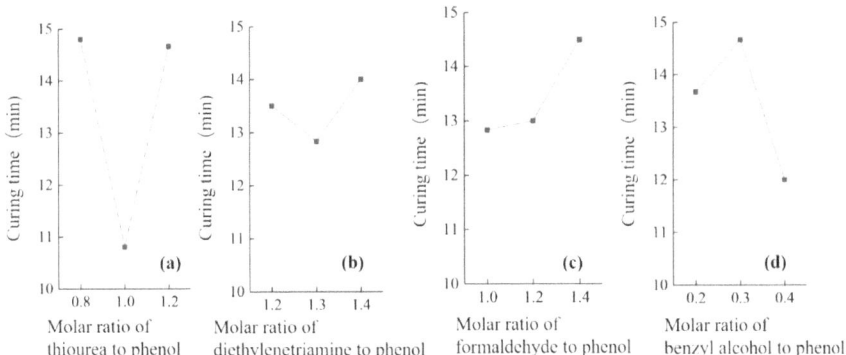

Fig. 2 Curing time variation at different ratios

3.2 Effect of DMP-30 on Curing Time

DMP-30 can improve the curing agent reactivity and accelerate the curing reaction. This was due to the presence of both phenolic hydroxyl and tertiary amines in its

Table 6 Effect of DMP-30 on curing time

Test No.	DMP-30 molar ratio	Curing time (20 °C)/min
1	0	20
2	0.03	15
3	0.06	12
4	0.09	10.5
5	0.12	9

molecular structure. Both can interact with epoxy groups, but specifically, electrons on nitrogen atoms of tertiary amines can attack the epoxy groups and promote ring-opening and cross-linking. With the optimal ratio of raw materials, n (thiourea): n (diethylenetriamine): n (formaldehyde): n (benzyl alcohol): n (phenol) = 1.2: 1.3: 1: 0.4: 1, a synthesis reaction time of 2.5 h, and a reaction temperature of 115 °C, different amounts of DMP-30 accelerator were added at the later stage of the reaction to investigate its effect on the curing time. The experimental design and results are shown in Table 6. Figure 3 shows the relationship between the curing time and the amount of DMP-30. It can be noticed that adding a small amount of DMP-30 significantly promoted the curing reaction and reduced the curing time. The higher the amount of DMP-30 was, the shorter the curing time was, and its effect gradually weakened.

Fig. 3 Relationship between the amount of DMP-30 and the curing time

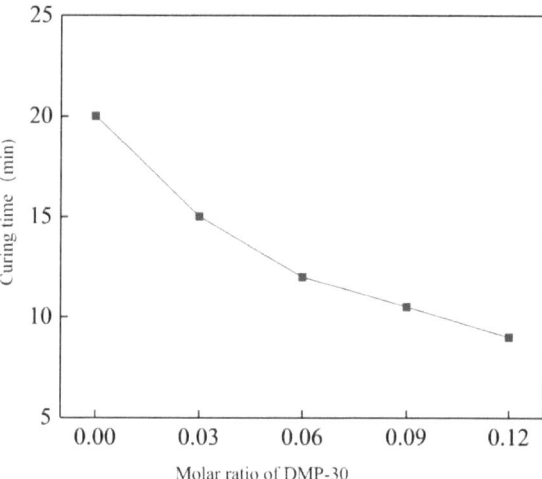

Table 7 Effect of reaction temperature on the curing performance

Test No.	Reaction temperature/°C	Amine value (KOH)/ (mg g^{-1})	Viscosity (20 °C)/ (Pa·s)	Curing time (20 °C)/min
1	105	510–520	2080	6.5
2	110	510–520	2300	7
3	115	510–520	2500	8
4	120	500–510	3400	13.5
5	125	500–510	4100	15

Note The reaction time is 2.5 h and the curing time is measured by using the curing agent and the bisphenol A resin 0164

3.3 Effect of Synthesis Reaction Temperature on the Curing Performance

During the synthesis of the curing agent, the reaction temperature affected the solubility and utilization of raw materials, the speed and efficiency of the reaction, and the proper synthesis of products. Therefore, under the optimal ratio of raw materials and a DMP-30 molar ratio of 0.12, the synthesis reaction temperature was varied to investigate its influence on the curing performance. The results are shown in Table 7 and Fig. 4. It can be noted that both the viscosity and curing time showed increasing trends with the increasing temperature. When the temperature was low, the curing time was short. The higher the reaction temperature was, the greater the viscosity was and the longer the curing time was. In contrast, the amine value slightly decreased with the increasing temperature. High temperatures triggered the polycondensation of formaldehyde with thiourea, hampered the synthesis of the target product, and increased the curing agent viscosity. In addition, if the temperature is too high, the polyamine will be decomposed, and the thiourea reactivity will be decreased. At 105 °C, the curing time was short, yet a small amount of formaldehyde and thiourea were precipitated, indicating that the reaction was not complete and there was a material loss. Therefore, the synthesis temperature should be maintained at 110–115 °C.

3.4 Effect of Synthesis Reaction Time on the Curing Performance

The synthesis reaction time has a great influence on the proper synthesis of the target product, as well as on its viscosity, activity, and molecular structure. Inadequate control of reaction time may lead to substitution reactions and the generation of impurities. Under the optimal ratio of raw materials and a DMP-30 molar ratio of 0.12, the synthesis reaction time was varied to investigate its influence on the curing

Fig. 4 Relationship between the curing time and the reaction temperature

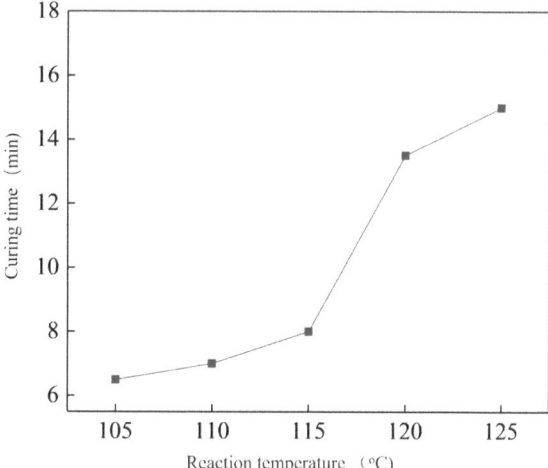

performance. The results are shown in Table 8 and Fig. 5. It can be revealed that with the increase in the reaction time, the curing time of the curing agent initially decreased and then increased. When the reaction time was within 2 h, the overall curing time remained almost steady. Above 2 h, the curing time sharply increased with the increasing reaction time. In fact, the long reaction time increased the molecular weight of the product and its viscosity. Especially at low temperatures, the mixing and penetration are incomplete, and the reactivity is low. Moreover, a short reaction time can result in an incomplete reaction, synthesis of a smaller amount of the target product, crystallization, and basic amines left over. As a result, the reactivity is low and there is a strong irritating odor. Therefore, it is appropriate to maintain the synthesis reaction time at about 2 h.

Table 8 Influence of synthesis reaction time on the curing time

Test No.	Reaction time/h	Amine value (KOH)/ (mg g^{-1})	Viscosity (20 °C)/ (Pa·s)	Curing time (20 °C)/ min
1	1	510–520	1870	7.5
2	1.5	510–520	2020	7
3	2	510–520	2300	7
4	2.5	510–520	2500	8
5	3	500–510	4000	12

Note The reaction temperature is 115 °C, and the curing time is measured by using the curing agent and the bisphenol A resin 0164

Fig. 5 Relationship between the curing time and the reaction time

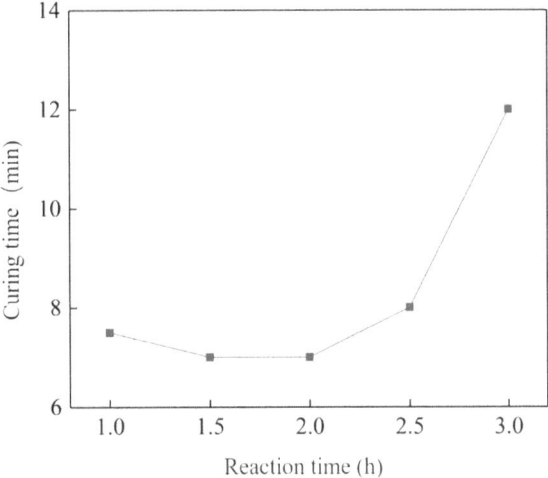

3.5 Infrared Analysis of the Low-Temperature Fast-Curing Agent

Figure 6 shows the infrared spectrum of diethylenetriamine and the low-temperature fast-curing agent. The infrared spectrum of the basic amines shows a broad peak near 3400 cm^{-1}, which is the characteristic absorption peak of the N–H stretching vibration of the amino group. The peaks at 1570 and 1483 cm^{-1} correspond to the N–H in-plane bending vibrations of primary and secondary amines. The peaks at 1108 and 1322 cm^{-1} correspond to the absorption peaks of C-N stretching vibration. In the infrared spectrum of the low-temperature fast-curing agent, the absorption peak at 3400 cm^{-1} becomes wider, which indicates the overlapping between the N–H stretching vibration peak and the introduced phenolic hydroxyl group. The absorption peak at 754 cm^{-1} is the vibration peak of the carbon chain skeleton. The peaks at 1261 cm^{-1}, 1031 cm^{-1}, and 1155 cm^{-1} correspond to the C–O vibration peaks, which are related to the addition of phenol, paraformaldehyde, and benzyl alcohol respectively. The change of the absorption peaks at 1108 and 1322 cm^{-1} of C–N stretching vibration indicates that Mannich's reaction has occurred. The C=S and N–C=S bonds correspond to the characteristic peaks at 2060 and 1406 cm^{-1} which are the new bonds generated by thiourea-modified phenalkamine. The results of infrared spectroscopy indicate that the target product was successfully synthesized.

Fig. 6 Infrared spectrum of diethylenetriamine and the modified curing agent

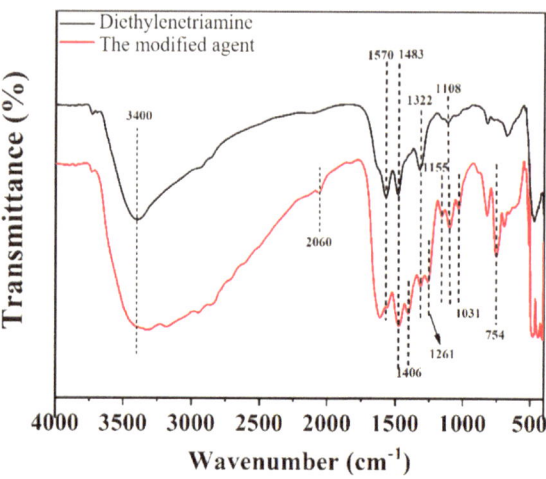

4   Material Ratio of EP Curing Agent for Sand Curing

4.1   Material Compounding and Formula Setting Methods

(1)   Compounding of curing agents

The aforementioned low-temperature fast-curing agent (denoted as XA) can effectively solve the low reactivity issues of conventional EP curing agents under low-temperature conditions. To solve the underwater curing issue, the underwater EP curing agent MS1085A produced by Changsha Puzhao Biochemical Technology Co., Ltd. was selected as the compound material. The MS1085A curing agent is synthesized by Michael's addition reaction which has hydrophobic and drainage effects under saturated water conditions. Thus, it can effectively improve the underwater curing ability.

(2)   Compounding of EP resin

To achieve high and fast reactivity of the modified low-temperature fast-curing agent, obtain a high-strength compound in a short time, and meet the requirements of low viscosity, low surface tension, and wetting and penetration properties, the phenol A resin, XY694 epoxy diluent (resorcinol glycidyl ether), and YD126 modified EP resin (acrylic monomer modified epoxy resin), which has high reactivity, were selected for compounding, as shown in Table 9. Since both XY694 and YD126 have low viscosity, large epoxy value, and much shorter gel time than bisphenol A resin, the compounded resin has a lower viscosity, better wetting permeability and higher reactivity. During compounding, several materials were added successively to the reaction kettle according to a certain ratio and then heated to the range of 45–55 °C. After stirring for 30 min at 800 r/min, the composite curing agent was obtained.

(3)   Formula of sand curing agent

Table 9 Basic performance indicators of compound epoxy resin

Resin type	Viscosity/mPa s	Epoxy value/mol/100 g	Gelation time/min
Bisphenol A resin (0164)	24,000–28,000	0.48–0.51	38
Epoxy reactive diluent (XY694)	300–500	0.75–0.82	18
YD126 modified epoxy resin	600–800	0.65–0.68	12

The low-temperature fast-curing agent XA and the underwater epoxy curing agent MS1085A were selected as the B-component of the composite curing agent, and a random combination of three resins was selected as the A-component. The proportions of A and B components were changed and then the agent was used for sand curing. By taking the compressive strength of the sand after curing for 2 h as the evaluation index, the combination with the highest strength was selected as the formula of the composite curing agent. The marine sand used in this study was taken from Ningde Beach in Fujian, and the water used for sample preparation was artificially prepared seawater with a salinity rate of about 3.6%. The samples for compressive tests were 70.7 mm × 70.7 mm × 70.7 mm. The compressive strength was tested according to "Highway Engineering Cement and Cement Concrete Testing Regulations" (JTG3420-2020). The mass ratio of the curing agent to the sand samples was 1: 7, the moisture content ranged from 0 to 45%, and the tests were conducted at 5 °C.

4.2 Test Results and the Formula

The tests were mainly conducted the experiments at 5 °C and 5% moisture content. Table 10 shows the compressive strength of the sand sample cured with different composition ratios for 2 h. Compared with the loose sand that has almost no cohesion, the cured marine sand was hardened, and its compressive strength was significantly improved. Thus, the EP composite curing agent used in this study can effectively improve the mechanical properties of marine sand in a short time.

From test 1 to test 4, it can be concluded that when the mass ratio of the A-component of both EP resin and EP curing agent was unchanged, the compressive strength of the cured sand increased with the proportion of low-temperature fast curing agent XA in the B-component. From test 5 to test 8, it can be noted that when the proportion of A and B components remained constant, as the mass ratio of the EP curing agent increased, the compressive strength of the cured sand initially increased and then decreased. The optimal mass ratio was 100: 85. From tests 7, and 9 to 14, when the mass ratio of the B-component of the curing agent and EP curing agent was fixed, the compressive strength of the cured sand did not change much with the ratio of XY694 and YD126 in the A-component. The compressive strength of the sand

Table 10 Test data (5 °C, 5% water content)

Test	Resin A	Hardener B	A: B	Compressive strength/MPa
1	0164 50% + 694 30% + 126 20%	XA 30% + 1085A 70%	A: B = 100: 50	0.1
2		XA 50% + 1085A 50%		1.3
3		XA 60% + 1085A 40%		1.5
4		XA 67% + 1085A 33%		1.8
5	694 50% + 126 50%	XA 67% + 1085A 33%	A: B = 100: 95	0.3
6			A: B = 100: 90	1.2
7			A: B = 100: 85	1.3
8			A: B = 100: 80	1.0
9	694 40% + 126 60%	XA 67% + 1085A 33%	A: B = 100: 85	1.2
10	694 35% + 126 65%			1.5
11	694 20% + 126 80%			1.1
12	694 10% + 126 90%			2
13	694 5% + 126 95%			2.1
14	694 35% + 126 60% + AGE 5%			0.9

Table 11 Formulation formula of new epoxy composite curing material

Operating temperature/°C	Resin (component A)	Curing agent (component B)	Ratio (A:B)
5–15	694 5% + 126 95%	XA 67% + MS1085A 33%	100: 85

sample in test 12 was the highest. Thus, the optimal formula for the EP composite curing agent for sand curing is determined, as shown in Table 11.

5 Conclusions

(1) By using thiourea, benzyl alcohol, phenol, diethylenetriamine, and paraformaldehyde as raw materials, a novel low-temperature fast-curing agent was synthesized based on Mannich's reaction. Through orthogonal tests, the optimal ratio of the raw materials was determined as follows: Thiourea:

Diethylenetriamine: Formaldehyde: Benzyl alcohol: Phenol = 1.2: 1.3: 1: 0.4: 1.

(2) A small amount of DMP-30 was added to greatly improve the speed of the curing reaction and reduce the curing time. The higher the amount of DMP-30 was, the shorter the curing time was, yet the promotion effect gradually weakened.

(3) The viscosity and curing time of the curing agent increased with the increasing synthesis reaction temperature, whereas the amine value slightly decreased. It is appropriate to maintain the synthesis reaction temperature at 110–115 °C. As the reaction time increased, the curing time initially decreased and then increased, and it is appropriate to control the synthesis reaction time at about 2 h.

(4) The compounding method was used to improve the reactivity of EP composite curing agents, which effectively solved the rapid curing issue of marine sand under complex conditions such as low temperature and high moisture content. The proportion of the components of the curing agent and resin and the mass ratio of the curing agent and resin were changed. Taking the compressive strength of the cured sand as the evaluation index, the EP composite curing agent formula was determined, which is suitable for different moisture content at 5–15 °C.

(5) The EP composite curing agents for curing sand can effectively improve the mechanical properties of soft sand, and therefore, it has great application value for coastal engineering constructions in cold or rainy seasons.

Acknowledgements Basic frontier project of Army Engineering University of PLA (Grant No. KYGYJKQTZQ23004).

References

1. Liu, J., Bai, Y., Song, Z., et al. (2020). Stabilization of sand using different types of short fibers and organic polymer. *Construction and Building Materials, 253*. https://doi.org/10.1016/j.con buildmat.2020.119164
2. Gao, Y. C., Gao, M., Yin, S. (2019). Experiments on static characteristics of sea sand solidified by polyurethane. *Rock and Soil Mechanics, 40*, 1–7. https://doi.org/10.16285/j.rsm.2019.0202
3. Rezaeimalek, S., Huang, J., & Bin-Shafique, S. (2017). Evaluation of curing method and mix design of a moisture activated polymer for sand stabilization. *Construction and Building Materials, 146*, 210–220. https://doi.org/10.1016/j.conbuildmat.2017.04.093
4. Zhang, G. X., Wang, W. B., Li, S. (2020). Progress in preparation and application of waterborne epoxy resin. *China Adhesives, 29*, 58–62. https://doi.org/10.13416/j.ca.2020.08.014
5. Hossain, M., & Steinmann, P. (2014). Degree of cure-dependent modeling for polymer curing processes at small-strain. *Part I: Consistent reformulation. Computational Mechanics, 53*, 777–787. https://doi.org/10.1007/s00466-013-0929-5
6. Tam, L.-H., Lau, D., & Wu, C. (2018). Understanding interaction and dynamics of water molecules in the epoxy via molecular dynamics simulation. *Molecular Simulation, 45*, 120–128. https://doi.org/10.1080/08927022.2018.1540869
7. Abali, B. E., Yardimci, M. Y., Zecchini, M., et al. (2021). Experimental investigation for modeling the hardening of thermosetting polymers during curing. *Polymer Testing, 102*, 107310. https://doi.org/10.1016/j.polymertesting.2021.107310

8. Lettieri, M., & Frigione, M. (2012). Effects of humid environment on thermal and mechanical properties of a cold-curing structural epoxy adhesive. *Construction and Building Materials, 30*, 753–760. https://doi.org/10.1016/j.conbuildmat.2011.12.077

9. Landgraf, R., Rudolph, M., Scherzer, R., et al. (2014). Modeling and simulation of adhesive curing processes in bonded piezo metal composites. *Computational Mechanics, 54*, 547–565. https://doi.org/10.1007/s00466-014-1005-5

10. Chen, Y. X., Rao, Q. H. (2014). Waterborne epoxy resin curing agent synthesis and properties. *Development and Application of Materials, 29*, 58–65. https://doi.org/10.19515/j.cnki.1003-1545.2014.01.014

11. Chen, Z. H., Ran, Y. B., Yang, J. (2022). Research progress of toughening epoxy resins. *Thermosetting Resin, 37*, 01. https://doi.org/10.13650/j.cnki.rgxsz.2022.01.012

12. Lixin, W., Suong, V. H., & Minh-Tan, T.-T. (2004). Effects of water on the curing and properties of epoxy adhesive used for bonding FRP composite sheet to concrete. *Journal of A-pplied Polymer Science, 92*, 2261–2268.

13. Zafar, A., Bertocco, F., Schjødt-Thomsen, J., et al. (2012). Investigation of the long-term effects of moisture on carbon fiber and epoxy matrix composites. *Composites Science and Technology, 72*(6), 656–666. http://doi-org-s.vpn.chd.edu.cn:8080/10.1016/j.compscitech.2012.01.010

Mechanism Model of Inherent Defects Risks for Buildings Quality Under Inherent Defects Insurance

Yingbo Ji, Wenjing Tong, Fuyi Yao, Hong Xian Li, and Xinnan Liu

Abstract Inherent defects insurance is one of the practical tools for building quality management, with the inherent defects risk assessment of buildings as its core. Unfortunately, the domestic Technical Inspection Service has failed to build a comprehensive risk system covering the project. As a result, the risk assessment mainly focuses on the construction process inspection, lacking a risk assessment model with theoretical support for the inherent defects. Therefore, the mechanism of inherent defect risks is analysed in this paper to provide a theoretical basis for the risk assessment. This paper identifies the inherent defects risks and risk factors of buildings and constructs the causal mapping and weights among building quality risks based on the Analytic Hierarchy Process of group decisions using expert interviews and questionnaires. It was found that the risk factors of survey and design, concrete pouring, roof waterproofing, and insulation engineering, and the completion review had a significant impact on the inherent defect risks.

Keywords Inherent defects risks of buildings · Buildings quality · Analytic hierarchy process of group decisions

1 Introduction

Inherent Defects Insurance (IDI) is an effective measure to improve building quality, settle owners' rights, and reduce the burden on the government. Technical Inspection Service (TIS), the core of the IDI), refers to the service entrusted by the insurance company. Using building quality insurance technology to identify and assess the

Y. Ji · F. Yao · X. Liu
School of Civil Engineering, North China University of Technology, Beijing 100144, China

W. Tong (✉) · H. X. Li
School of Architecture and Built Environment, Deakin University, Geelong 3220, Australia
e-mail: wenjing.tong@research.deakin.edu.au

inherent quality risks of the insured buildings and submit reports, providing treatments to the risk factors to improve the building quality, and reduce and avoid quality accidents.

The TIS is still in its infancy in China, resulting in a lack of mature construction risk management services, so the local provides supervision units, construction drawing review bodies, and testing units support the development of the TIS. However, the existing risk management institutes, whether construction drawing review bodies, engineering supervision units, engineering quality inspection organizations, or quality and safety supervision stations, mostly only from their perspective, fail to build a comprehensive risk system covering the project.

At present, many domestic and foreign scholars have researched the building quality risk. Ji [1] analysed the moral risk of quality inspection organizations and analysed the relationship with the existing residential quality supervision and management system using the Game Theory and put forward policy recommendations for the system design of the residential quality inspection organization in China. Su [2] established a quality evaluation system based on the IDI and proposed corresponding measures, such as establishing relevant supporting systems, improving the insurance system, broadening application project types, and diversifying the insurance models. Xie [3] established a control procedure for building quality risk based on case-based reasoning to urge risk-related parties to take the initiative to adopt targeted risk countermeasures to ensure that quality risks are in a safe state. Wang [4] put forward his opinions and suggestions on establishing the TIS and quality insurance information institutions that need IDI and look ahead to developing quality insurance in China. The literature review found that no result-oriented research on the mechanism of building quality risks has been conducted. Moreover, the mechanism research is the basis of quality risk assessment, which has important guiding significance for the risk assessment.

Due to the complex influencing factors of building quality risks [5, 6], this paper systematically explores and elaborates the mechanism of building quality risks based on the Analytic Hierarchy Process of group decisions using literature research and expert semi-structured interviews to construct the causal mapping and influencing weights between quality risk factors and risks.

2 Risk Identification of Inherent Defects in Buildings

Building quality risk is defined as the uncertainty of the degree of benefit loss caused by the deviation between the sum of each quality attribute contained in the engineering product and the expected value [7–9]. Lei Xie [3] classified the building quality risks according to the process of the whole life cycle of the project, consequences of risk occurrence, the responsibilities division of the main participants, and construction elements. The consequences of risk occurrence are divided into quality risks affecting structural safety, quality risks affecting use functions, and quality risks affecting the environment and health [7]. The building quality risks are identified in this paper, combined with the definition of building quality risk, the consequences of risk occurrence, the deviation of each quality attribute of construction products, the building quality complaint report, and the core literature.

Based on the building quality risk events summarized from the building quality complaint reports and core literature, the following risks list (Table 1) is summarized according to the basic coverage of the IDI for buildings covered in the Implementation Opinions on Promoting Inherent Defects Insurance for Commercial Residence and Guaranteed Residence in the City issued by Shanghai Housing and Urban–Rural Development Management Committee and the Interim Management Measures for Inherent Defects Insurance of Residence in Beijing issued by Beijing Housing and Urban–Rural Development Committee. However, quality risks that affect the environment and health are not in the basic coverage, so they are not discussed in this paper.

Table 1 List of building quality risk events

First-level indexes	Second-level indexes	Indexes code	Indexes interpretations
Quality Risks Affecting Structural Safety	Structural stability	Y1	Collapse risks of overall or partial building
	Structural crack	Y2	Risks of crack, deformation, breakage, and fracture affecting the structural safety in main load-bearing structural
	Settlement crack	Y3	Risks of uneven settlement of the foundation beyond the design specification
	Defects risks of cantilever components	Y4	Crack, breakage, fracture, and other risks in balconies, canopies, cornices, and other cantilever components
Quality risks affecting the use function	Exterior wall surface peeling	Z1	Risks of exterior wall surfaces collapse (including falling off)
	Defects risks of the thermal insulation	Z2	Defects risks in the insulation works of the building envelope
	Roof leakage	Z3	Roof leakage risks
	Basement leakage	Z4	Risks of basement leakage
	Kitchen and bathroom leakage	Z5	Risks of kitchen and bathroom leakage
	Leakage risks of exterior walls (including external windows)	Z6	Leakage risks of exterior wall surface (including external windows)

Table 2 Influencing factors of building quality risks

Risk factors	Interpretations of the risk factors
Survey and design	Whether the risk points of the survey and design and construction plans are not well analyzed and evaluated
Foundation engineering	Quality point situation of foundation engineering
Pile foundation engineering	Quality point situation of pile foundation engineering
Reinforcement engineering	Quality point situation of reinforcement engineering
Formwork engineering	Quality point situation of formwork engineering
Concrete pouring engineering	Quality point situation of concrete pouring engineering
Basement waterproofing engineering	Quality point situation of basement waterproofing engineering
Roof waterproofing and insulation engineering	Quality point situation of roof waterproofing and insulation engineering
Exterior wall engineering	Quality point situation of exterior wall engineering
Kitchen and bathroom waterproofing engineering	Quality point situation of kitchen and bathroom waterproofing engineering
Risk factors in the completion stage	Whether there are uncorrected and output issues
Risk factors in the review stage	Track and review of remaining quality risks in completion

3   Feed-Forward Factor Identification of Building Quality Inherent Defects Risks

The influence of different construction operations on inherent defect risks is explored based on the building quality risks list (Table 2). Therefore, this paper integrates the building quality risk factors according to the different construction operations based on the *Specification of Technical Inspection Service of Inherent Defects Insurance for Buildings.*

4   Mechanism Model Construction of Inherent Defects Risks for Buildings

To study the causal mapping between risks and risk factors, it is necessary to determine the risk factors and clarify the influencing weights of each relevant factor on the risk.

4.1 Analytic Hierarchy Process of Group Decisions

Since there has been no official data on the inherent defects and construction process defects of buildings so far, and the objective weighting method requires a large amount of accurate data, this paper determines the weights using the subjective weighting method. The analytic Hierarchy Process is a mature and easy-to-operate method in the subjective weighting method. This paper combines group decision theory with the Analytic Hierarchy Process to reveal the causal mapping between building risks and risk factors based on the knowledge background of several authoritative experts to improve the objectivity of the Analytic Hierarchy Process.

(1) Analytic Hierarchy Process

The Analytic Hierarchy Process consists of four parts: (1) establishing a system hierarchy; (2) constructing a multiple comparisons judgment matrix A; (3) calculating the weights and verifying consistency; (4) calculating the combination weights and verifying consistency.

(2) Expert Weights of Group Decisions

The expert weights of the Analytic Hierarchy Process of group decisions are: for the problem to be decided, the objective weights of decision-makers are determined based on the decision results of experts in various fields and the interrelationship of the decision results; the subjective weights of decision-makers are determined based on the professional ability of experts and their knowledge of the decision problem; the weight coefficients are introduced to construct the expert weights.

In the weight evaluation of building quality risks and risk factors index, assume that the number of experts is s. The n-order AHP judgment matrix evaluated by each expert is $A^{(k)}$ (k = 1, ω_{k2}, …, s), and the expert weights are β_k.

$$A^{(k)} = \left(a_{ij} \right)_{n \times n} = \begin{pmatrix} 1 & a_{12}^{(k)} & \cdots & a_{1n}^{(k)} \\ a_{21}^{(k)} & 1 & \cdots & a_{2n}^{(k)} \\ \cdots & \cdots & \cdots\cdots \\ a_{n1}^{(k)} & a_{n2}^{(k)} & \cdots & 1 \end{pmatrix} \tag{1}$$

The objective weights βk of the decision makers are

$$\beta^{(k)} = \frac{\gamma_k}{\sum_{k=1}^{s} \gamma_k} \tag{2}$$

The statement Analytic Hierarchy Process of decision experts is used to establish the hierarchy, construct multiple comparisons judgment matrix A, and finally derive the subjective weights $\beta_{(k)}$ using the sum method. To get the weights of decision experts that can decide the actual situation and change flexibly, the weight coefficient t ($0 < t \leq 1$) is introduced to get the final weights.

$$\omega_k = t * \beta_k^1 + (1 - t) * \beta_k^2, k = 1, 2, \ldots, s \tag{3}$$

Assuming that the nth order AHP judgment matrix given by s experts is $A^{(k)}$, (k = 1, 2, ..., s), the judgment matrix of group decisions could be constructed:

$$A^k = \left(a_{ij}\right)_{n \times n} = \begin{pmatrix} 1 & a_{12} & \cdots & a_{1n} \\ a_{21} & 1 & \cdots & a_{2n} \\ \cdots & \cdots & & \\ \cdots & & \cdots & \cdots \\ a_{n1} & a_{n2} & \cdots & 1 \end{pmatrix} \tag{4}$$

4.2 Causal Mapping Between Risks and Risk Factors and Calculation of Weights

The Questionnaire on Mapping and Weights of Building Quality Risks and Risk Factors is designed in this paper to determine the weights of risk factors. The experts for this questionnaire are engineering experts and university scholars in building management. Four questionnaires were sent out, and four valid questionnaires were returned. By combining the questionnaire results and the expert weights in the previous subsection, the judgment matrix of the criterion is constructed, and the single-ranking weights and the total weights of the risk factors are calculated after the consistency verification of the matrix.

(1) Causal Mapping between Quality Risks and Risk Factors for Buildings
(2) The Weights of Building Quality Risk Factors to Risks

For an example of the risks of a structural crack, the weights between risk factors and risks are calculated. The calculation process is as follows: (1) Constructing a judgment matrix of 4 experts and verifying consistency. From the table, the weight evaluation results of each expert have passed the consistency verification. (2) Calculating the Combined Expert Weights. (3) Calculating the Combined Expert Weights of Structural Crack Risks.

4.3 Causal Mapping Between Risks and Risk Factors and Weight Results and Discussion

The causal mapping between risks and risk factors is constructed, and the weights are calculated using the Analytic Hierarchy Process of group decisions. The weights figure is obtained as follows (Table 3).

Table 3 Weightings of risk factors associated with building quality risk events

Risk event		Survey and design in the pre-construction stage	Foundation engineering	Pile foundation engineering	Reinforcement engineering	Formwork engineering	Concrete pouring project
Structural stability risk	Fi	Survey and design in the pre-construction stage	Foundation engineering	Pile foundation engineering	Reinforcement engineering	Formwork engineering	Concrete pouring project
	Wi	0.308	0.154	0.154	0.154	0.077	0.154
Structural Crack	Fi	Reinforcement engineering		Formwork engineering		Concrete pouring project	
	Wi	0.199		0.177		0.624	
Settlement cracks	Fi	Survey and design in the pre-construction stage		Foundation engineering		Pile foundation engineering	
	Wi	0.547		0.263		0.190	
Defects in cantilevered components	Fi	Reinforcement engineering		Formwork engineering		Concrete pouring project	
	Wi	0.413		0.327		0.260	
Falling off the exterior wall	Fi	Concrete pouring project	Façade works		Risk factors in the completion stage	Review phase risk factors	
	Wi	0.140	0.542		0.159	0.159	
Roof Leakage Risk	Fi	Roof waterproofing and thermal insulation engineering			Risk factors in the completion stage	Review phase risk factors	
	Wi	0.818			0.091	0.091	
Kitchen Leakage	Fi	Kitchen and bathroom waterproof engineering			Risk factors in the completion stage	Review phase risk factors	
	Wi	0.818			0.091	0.091	
Basement leak	Fi	Basement waterproofing project			Risk factors in the completion stage	Review phase risk factors	
	Wi	0.818			0.091	0.091	
Leakage of exterior walls (including exterior windows)	Fi	Façade works			Risk factors in the completion stage	Review phase risk factors	
	Wi	0.818					

(continued)

Table 3 (continued)

Structural stability risk	Fi	Survey and design in the pre-construction stage	Foundation engineering	Pile foundation engineering	Reinforcement engineering	Formwork engineering	Concrete pouring project
	Wi	0.308	0.154	0.154	0.154	0.077	0.154
	Wi	0.778		0.111		0.111	
Thermal insulation	Fi	Roof waterproofing and thermal insulation engineering			Risk factors for façade works		
	Wi	0.167			0.833		

For the structural stability risks, foundation engineering, pile foundation engineering, and concrete engineering are critical influencing factors. For the structural crack risks, concrete engineering and formwork engineering are critical influencing factors. For the settlement crack risks, survey and design engineering are the primaries, and foundation engineering and pile foundation engineering are the secondary influencing factors. For the risks of cantilever components, reinforcement engineering, formwork engineering, and concrete pouring engineering are critical factors. The exterior wall peeling, insulation, and roof leakage are mainly affected by the exterior wall engineering. The roof leakage is mainly influenced by roof waterproofing and insulation engineering.

5 Conclusion

First, the building quality risks are identified using the literature method and report analysis method. Second, the building quality risk factors, namely the feed-forward factors of building quality risks, are summarized and explained based on the *Specification of Technical Inspection Service of Inherent Defects Insurance for Buildings*. Finally, the mechanism of building quality risks is clarified, i.e., the causal mapping between risk factors and risks and the cumulative transmission between risk factors. The causal mapping and weights between risk factors to risks are clarified using expert interviews and the Analytic Hierarchy Process of group decisions.

Acknowledgements This work was supported by the National Key R&D Program of China (2021YFF0602000) and the Research Initiation Fund of North China University of Technology (No. 110051360023XN224-50).

References

1. Ji, Y. B. (2009). Research on compulsory insurance for construction quality of commercial residential. Tianjin University.
2. Su, Y. (2020). Research on building quality evaluation for inherent defects insurance. South China University of Technology.
3. Xie, L. (2017). Research on building quality risk prediction and control. Southeast University.
4. Wang, K. (2018). Application research on building quality insurance in building risk management. Qingdao University.
5. He, S. K., & Fu, H. Y. (2006). Economic explanation and risk prevention of building quality risks. *Journal of Chongqing Jianzhu University, 6*, 106–110.
6. Xu, Y. F. (2021). Analysis of quality defects and preventions of building based on engineering management. *Ceramics, 427*(5), 126–127.
7. Li, M. (2021). Research on quality defects and treatments of building concrete. *Smart City, 7*(9), 83–84.
8. San Santoso, D., Ogunlana, S. O., & Minato, T. (2003). Assessment of risks in high-rise building construction in Jakarta. *Engineering, Construction and Architectural Management, 10*(1), 43–55.

9. Zhou, J. (2016). Research on prevention and complaint handling of residential quality defects. Shandong University.

Study on Seismic Design Method of 30-Storey Self-Centering Buckling-Restrained Brace Steel Frame Considering the Influence of High Order Vibration Modes

Binlei Wang and Xiao Tan

Abstract A design method of 30-storey self-centering buckling-restrained brace steel frame (SCBRBF) is presented considering the effect of high order vibration modes. For SCBRBF, the equal strength principle and capability design method can be used to design from BRBF. The research shows that, for the 30-storey SCBRBF, the structure is greatly affected by the high-order vibration modes, and there is a problem of conservative design by adopting the capability design method. By considering the influence of the high vibration mode of the structure in the capacity design method, the secondary design can be carried out by reasonably changing the distribution form of the corresponding brace axial force. The structure designed by this method has a small amount of steel and the seismic resistance can still meet the requirements.

Keywords Self-centering buckling-restrained steel frame, · Seismic behavior, · Design method, · Residual deformation, · High order vibration modes

1 Introduction

Disc spring self-centering buckling-restrained brace (SCBRB) takes preloaded disc-spring combination as the source of restoring force, and its restoring performance is stable. By changing the combination form of disc spring, the bearing capacity and deformation capacity of the self-centering system can be flexibly adjusted. The experimental and finite element simulation results show that the SCBRB (as shown

B. Wang (✉)
CCCC Tianjin Port Engineering Institute Co., Ltd., Tianjin 300222, China
e-mail: 651980112@qq.com

CCCC First Harbor Engineering Co., Ltd., Tianjin 300456, China

X. Tan
Tianjin Public Works Section of China Railway Beijing Bureau Group Co., Ltd., Tianjin 300010, China

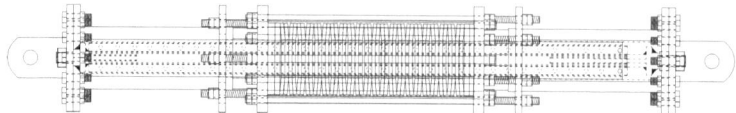

Fig. 1 Structural form of SCBRB

in Fig. 1) has stable hysteretic performance and good resetting ability and ductility [1, 2].

McCormick et al. [3] made statistics on damaged buildings after the earthquake in Japan and found that when the residual storey drift angle of the structure after the earthquake exceeded 0.5%, its maintenance cost would be higher than the reconstruction cost. SCBRB has good reset performance at the component level, but whether it can retain good reset performance after combining with steel frame is worth studying [4–6].

High-rise braced frame structures are greatly affected by high-order modes when they are earthquake-resistant. Most of the existing SCBRBF design methods do not consider the effect of high-order modes, usually use calculation software for continuous iterative optimization, which consumes more time and resources. Therefore, it is necessary to propose a design method that takes into account the effect of high-order modes of structures with fewer iterations.

2   Design Method

The structure is a four-span brace steel frame structure with 30 storey (Fig. 2). The beams and columns of supported span and non-supported span are rigid connections, and the connection form between the brace and the frame is hinged connection. The structure frame is fixed between the bottom column and the foundation. One frame is analyzed along the span direction.

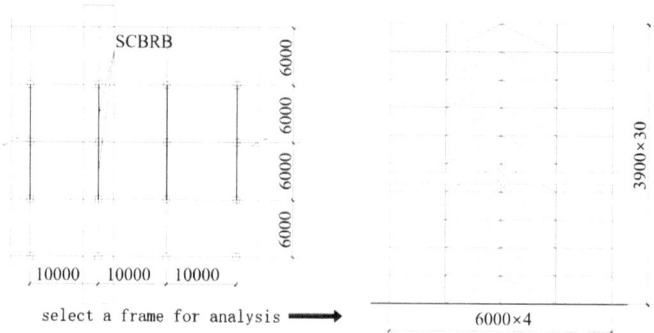

Fig. 2 Design background of SCBRBF

In the structural design, the BRBF was first designed with the help of MIDAS program, and then the buckling-restrained brace was approximately equivalent replaced with disc-spring self-centering buckling-restrained brace by the principle of equal strength [9]. During the structural design, the section size of the member changes every 3 storey.

(1) BRBF: The design process refers to BRBF's capability design method in literature [10, 11]. MIDAS program is used to adjust the section size of each component in the structure under the first-order elastic analysis, so that the brace component can meet the requirements of the design axial force ratio, and the bearing capacity of the component and the storey drift angle can meet the requirements of the code. Using the capability design method, the brace of the designed BRBF is deleted, and the maximum axial force of the BRB under the storey drift angle of 1/50 layer is decomposed and applied to the column and beam according to the form of the axial force distribution of the brace under the first-order mode shape of the structure, and whether the cross-section of the steel beam and column meets the requirements of the specification. If not, readjust the beam and column cross-section until the specification is met. For the beam and column section obtained after strengthening, the axial force ratio of the brace needs recalculated, and the brace section is adjusted accordingly, and whether the bearing capacity and deflection of each component, the storey drift and other indicators meet the requirements of the specification.

(2) SCBRBF1: Using the principle of equal strength [9], the BRB in BRBF is replaced with SCBRB, and the maximum axial force of the SCBRB under the storey shift angle of 1/50 layer is decomposed and applied to the pure frame, and the bearing capacity of the beam and column needs re-calculated. The frame section of SCBRBF1 can be seen in Table 1.

(3) SCBRBF2: Different from SCBRBF1, the influence of high-order mode shapes is considered in the process of capability design, and the structure is adjusted after reasonable adjustment of the distribution form of the maximum axial force of the brace. The brace parameter is $\alpha = 1$, and the yield strength of the brace is 235 MPa [1]. The frame section of SCBRBF1 can be seen in Table 2.

3 Calculation Result

Use the ABAQUS program to analyze the dynamic time history of the brace frame structure. B31 unit was used for beam and column, and a single steel beam was divided into 6 units and a single column into 4 units. The BRB is simplified by T3D2 unit. In order to achieve the effect of buckling- restrained, only one unit is divided into it. The self-centering system adopts nonlinear spring for simulation [7, 8].

The numerical examples of two 30-storey disc-spring self-centering buckling-restrained brace steel frames are numbered SCBRBF1 and SCBRBF2. The beam and column sections of the frame use welded H-beam and box column respectively. During the design, the steel materials of frame beam and column are the same, the

Table 1 Section of SCBRBF1

Storey	Lnner column (mm)	Outside column (mm)	Lnner beam (mm)	Outside beam (mm)	Brace section area (mm^2)	Self-centering system pre-pressure force (kN)
	$\Box C \times t$	$\Box C \times t$	H$H \times B \times t_\mathrm{w} \times t_\mathrm{F}$	H$H \times B \times t_\mathrm{w} \times t_\mathrm{F}$		
1–3	750 × 66	570 × 32	600 × 250 × 14 × 24	600 × 250 × 14 × 24	1154	439
4–6	730 × 56	560 × 26	600 × 270 × 14 × 24	600 × 250 × 14 × 24	1270	483
7–9	700 × 50	520 × 26	600 × 270 × 14 × 24	600 × 250 × 14 × 24	1078	410
10–12	660 × 48	480 × 24	600 × 250 × 14 × 24	550 × 250 × 14 × 22	968	368
13–15	620 × 40	460 × 22	600 × 250 × 14 × 24	550 × 250 × 14 × 22	895	340
16–18	570 × 32	430 × 22	550 × 250 × 14 × 22	550 × 250 × 14 × 22	812	308
19–21	490 × 28	400 × 19	550 × 250 × 14 × 22	500 × 220 × 10 × 20	776	295
22–24	460 × 18	380 × 16	530 × 220 × 10 × 20	500 × 220 × 10 × 20	731	278
25–27	420 × 16	360 × 14	530 × 220 × 10 × 20	500 × 220 × 10 × 20	512	195
28–30	350 × 14	330 × 13	530 × 220 × 10 × 20	500 × 220 × 10 × 20	283	107

yield strength is 345 MPa, the yield strength of steel brace is 235 MPa, the elastic modulus of all steels is E = 206 GPa, and the density is 7850 kg/m^3. According to the site conditions and the natural vibration period of the structure of the example, two ground motion records were selected to analyze the seismic action, namely, EI-Centro NS and Taft acceleration records [10]. For convenience of expression, the above two seismic waves are referred to as El and Tf waves respectively. The time history curve of the four ground motions under the peak acceleration of 400gal is shown in Fig. 3. Under the action of rare and frequent earthquakes, the peak acceleration of local vibrations is modulated to 400gal and 70gal respectively.

Figure 4 show the residual deformation envelope diagram of SCBRBF1 structure after earthquake under the action of Cl and Tf earthquake, which the lateral shift meets the nation standards. Figure 5 show the residual deformation envelope diagram of SCBRBF1 structure after earthquake under the action of Cl earthquake which is large, and the lateral shift angle between the residual layers of the 22nd storey is the largest, about 1/818, which meets the 0.5% interlayer lateral shift angle limit requirement [3].

Table 2 Section of SCBRBF2

Storey	Lnner column (mm)	Outside column (mm)	Lnner beam (mm)	Outside beam (mm)	Brace section area (mm^2)	Self-centering system pre-pressure force (kN)
	$\Box C \times t$	$\Box C \times t$	$HH \times B \times t_w \times t_F$	$HH \times B \times t_w \times t_F$		
1–3	660×48	570×32	$600 \times 250 \times 14 \times 24$	$600 \times 250 \times 14 \times 24$	1081	411
4–6	620×40	560×26	$600 \times 270 \times 14 \times 24$	$600 \times 250 \times 14 \times 24$	1239	471
7–9	600×36	520×26	$600 \times 270 \times 14 \times 24$	$600 \times 250 \times 14 \times 24$	1005	382
10–12	600×34	480×24	$600 \times 250 \times 14 \times 24$	$550 \times 250 \times 14 \times 22$	886	337
13–15	590×32	460×22	$600 \times 250 \times 14 \times 24$	$550 \times 250 \times 14 \times 22$	747	284
16–18	560×28	430×22	$550 \times 250 \times 14 \times 22$	$550 \times 250 \times 14 \times 22$	681	259
19–21	480×26	400×19	$550 \times 250 \times 14 \times 22$	$500 \times 220 \times 10 \times 20$	678	257
22–24	460×18	380×16	$530 \times 220 \times 10 \times 20$	$500 \times 220 \times 10 \times 20$	631	240
25–27	420×14	360×14	$530 \times 220 \times 10 \times 20$	$500 \times 220 \times 10 \times 20$	489	186
28–30	350×14	330×13	$530 \times 220 \times 10 \times 20$	$500 \times 220 \times 10 \times 20$	228	86

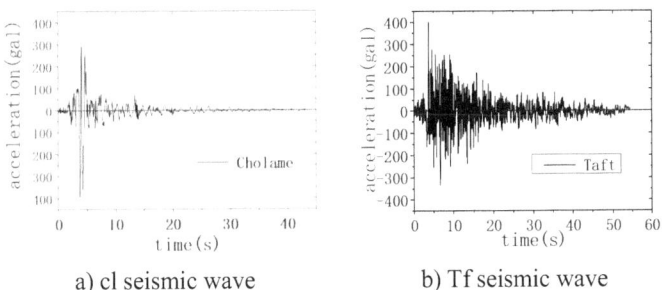

a) cl seismic wave b) Tf seismic wave

Fig. 3 Earthquake wave. **a** cl seismic wave, **b** Tf seismic wave

The brace hysteresis curve of the maximum storey drift (the 22nd storey) of SCBRBF1 structure under Cl earthquake action are shown in Fig. 6. It can be seen that the hysteresis curve of SCBRB is in a flag shape.

The brace yield state of SCBRBF1 at some time under different ground motions is extracted, and the results are shown in Fig. 7. It is known that due to the influence

Fig. 4 Storey drift under
rare earthquakes. **a** Cl, **b** Tf

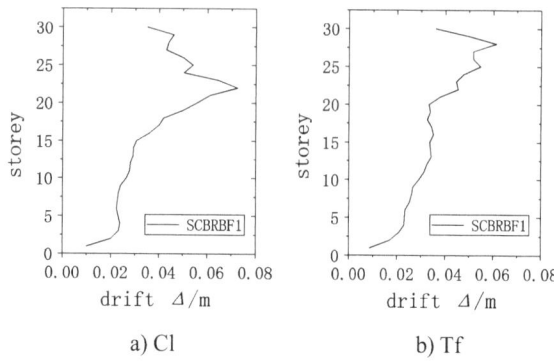

a) Cl   b) Tf

Fig. 5 Residual deformation

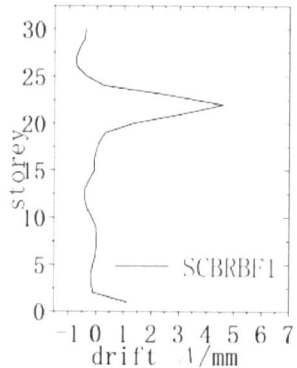

Fig. 6 Hysteresis curve

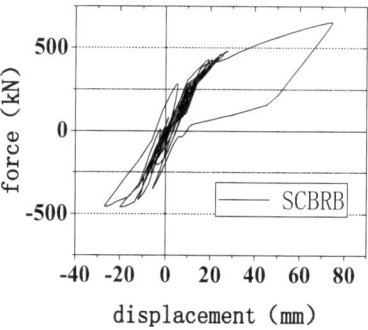

of high-order mode shapes, the same side brace of the 30-layer SCBRBF1 cannot be stretched or compressed to yield at the same time. Therefore, in the capacity design method, if the 30-storey structure is applied to the frame in the form of axial force distribution [7], which is mainly influenced by the first-order mode, the structure design will be conservative. when the capacity design method is adopted for the 30-storey SCBRBF structure, it is necessary to reasonably adjust the distribution form of the maximum axial force of the brace by considering the influence of the high vibration mode of the structure, so as to "slim down" the structure reasonably and improve its economy.

According to the brace yield situation of the example SCBRBF1, a simplified model of brace yield state (simplified stress ratio envelope diagram) was drawn, as shown in Fig. 8a. In view of the simplified model of the yield state of the brace (Fig. 8a), the distribution of the maximum brace axial force (the axial force corresponding to the brace under the storey drift angle of 1/50) of the structure in the capacity design method was applied in the form of the distribution in Fig. 8b (conservative application in the 12th storey), and then the design of the frame column was strengthened. By comparing the amount of steel used in the two structures, SCBRBF1 is 411 tons, while for example, SCBRBF2 is only 340 tons, and the amount of steel used is reduced by 16.9%.

As shown in Fig. 9, due to the low lateral stiffness of SCBRBF2, its storey drift is larger than that of SCBRBF1 in frequent earthquakes. Under the rare earthquake wave

Fig. 7 Brace stress ratio in SCBRBF1 under different seismic effects. **a** Cl, **b** Tf

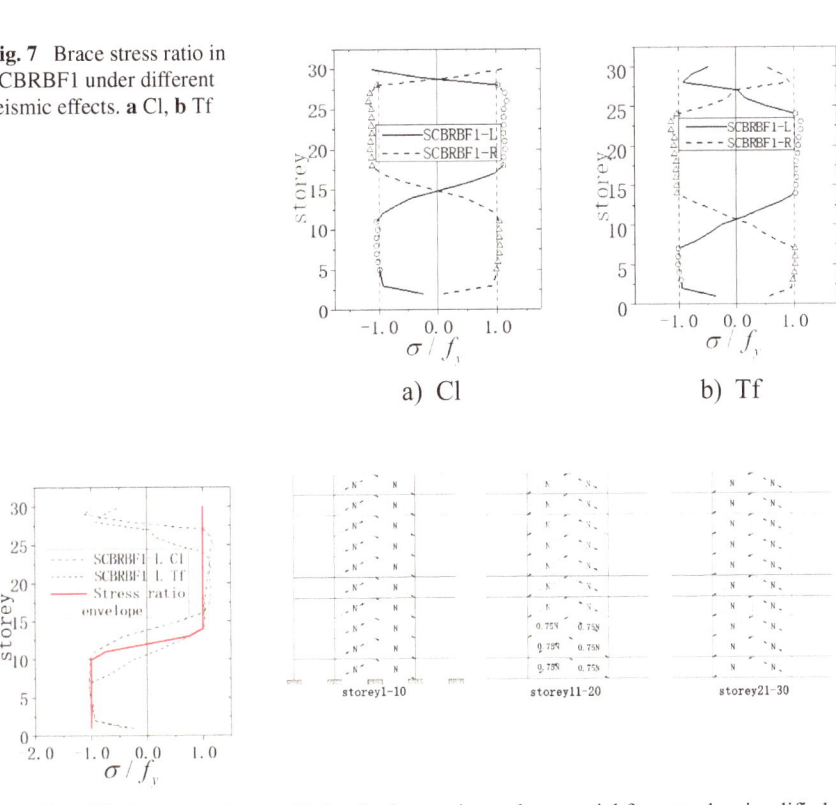

a) Simplified stress ratio envelope

b) Apply the maximum brace axial force to the simplified model

Fig. 8 Optimization design methods considering the influence of high-order mode shapes. **a** Simplified stress ratio envelope, **b** Apply the maximum brace axial force to the simplified model

Fig. 9 Storey drift under frequent earthquakes. **a** Cl, **b** Tf

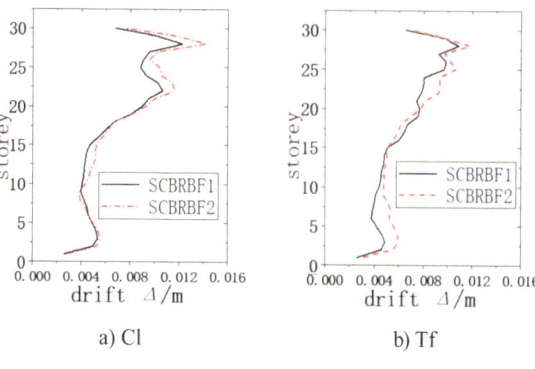

a) Cl   b) Tf

Fig. 10 Storey drift under rare earthquakes. **a** Cl, **b** Tf

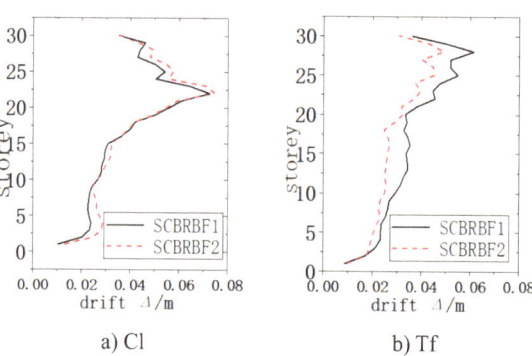

a) Cl   b) Tf

of Cl, the 28th storey shift angle of SCBRBF2 reaches 1/274, which also meets the limit requirement of 1/250 storey drift angle. Figure 10 shows the comparison of the envelope diagram of storey drift between the two structures under rare earthquakes. The storey drift of SCBRBF2 with low stiffness is larger, but not different from that of SCBRBF1, and both meet the limit of storey drift of the specification.

Figure 11 shows the statistics of brace yield state of SCBRBF2. It can be seen that its brace yield distribution is similar to that of SCBRBF1 in Fig. 8a, which is also greatly affected by high-order mode shapes. In general, the distribution envelope of brace axial force along the height direction of the structure is still consistent with the simplified envelope mode adopted in Fig. 8, indicating that the application mode of brace axial force in the aforementioned checking calculation framework as shown in Fig. 8 is feasible.

The seismic behavior of SCBRBF2 structure is similar to that of SCBRBF1. Although its displacement under some earthquakes is slightly larger than that of SCBRBF1, it still meets the requirements of the standard. Compared with SCBRBF1 structure, the amount of steel used in SCBRBF2 frame is reduced by 16.9%, which significantly improves the economy. It can be seen that, when designing 30-storey SCBRBF, the influence of high-order vibration modes can be considered in the stage

Fig. 11 Brace stress ratio distribution. **a** Cl, **b** Tf

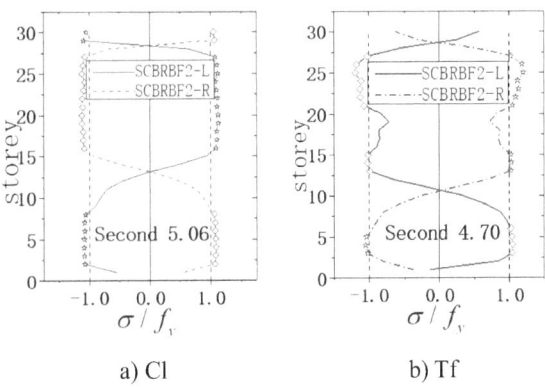

a) Cl   b) Tf

of capability design, and the maximum axial force distribution form of brace is reasonably changed. The designed SCBRBF can simultaneously have the characteristics of small amount of steel and good seismic performance. Therefore, it is worthwhile to further study the reasonable brace axial force distribution model which can be used in the capacity design of steel frames with structural height and composition changes.

4 Conclusions

The time-history analysis of two 30-storey SCBRBF shows that the 30-storey SCBRBF is greatly affected by the high-order mode. If the maximum brace axial force is applied to the steel frame according to the distribution form of first-order mode in the capacity design method to strengthen the beams and columns, the designed structure is more security and uses more steel. If the influence of the high vibration mode of the structure is considered in the capacity design method, and the distribution form of the corresponding brace axial force is changed reasonably, the designed structure uses less steel, has good economy, and the seismic resistance can still meet the requirements.

References

1. Tang, M. (2019). *Hysteretic behavior and design methods for self-centering buckling-restrained brace coaxially assembled in disc springs*. Harbin Institute of Technology.
2. Ying, Z., Jiehao, S., & Yi, X. (2021). State-of-the-art on self-centering energy dissipative braces. *Journal of Building Structures, 42*(10), 1–13.
3. McCormick, J., Aburano, H., & Ikenaga, M., et al. (2008). Permis-sible residual deformation levels for building structures considering safety and human elements. In *Proceedings of the 14th world conference on earthquake engineering* (pp. 12–17).

4. Tremblay, R., Lacerte, M., & Christopoulos, C. (2008). Seismic response of multistorey buildings with self-centering energy dissipative steel braces. *Journal of Structural Engineering, 134*(1), 108–120.

5. Miller, D. J. (2012). Development and experimental validation of self-centering buckling restrained braces with shape memory alloy. *Engineering Structures, 40*, 288–298.

6. Chou, C., Wu, T., Beato, A. R. O., et al. (2016). Seismic design and tests of a full-scale one-storey one-bay steel frame with a dual-core self-centering brace. *Engineering Structures. 111*, 435–450.

7. Li, J. H. (2014). *Seismic behavior of self-centering buckling restrained braces and SCBRB steel frames*. Harbin Institute of Technology.

8. Fan, X., Xu, L., & Lu, D. (2016). Seismic performance analysis of new self-centering energy dissipation braced frame structure. *Journal of Tianjin University* (*Science and Technology*), *49*(4), 385–391.

9. Xie, Q., Zhou, Z., Wang, W., & Meng, S. (2017). Aseismic performance analysis for braced frame systems with self-centering buckling-restrained braces with two different design criteria. *Journal of Vibration and Shock, 36*(03), 125–131.

10. Ding, Y., & Zhang, Y. (2010). Analysis of aseismic performance for dual system composed of steel frame and unbonded steel plate brace encased in reinforced concrete panel. *China Civil Engineering Journal, 43*(S1), 385–391.

11. ANSI/AISC 341–16. (2016). *Seismic provisions of structural steel buildings*. American Institute of Steel Construction.

Example of Seismic Performance-based Design of Highway Monitoring Building in High Seismic Intensity Area

Yuzheng Wang

Abstract With the technological development of resilient city and intelligent transportation, since the terminal equipment of real-time monitoring and safety monitoring system for lifeline projects is usually set up in highway monitoring halls, more and more highway monitoring centers in China's seismic high intensity areas assume the important function of emergency command centers after earthquakes. In order to ensure that the monitoring and surveillance equipment can operate normally during an earthquake, this paper puts forward a proposal to adopt seismic isolation technology for the monitoring center, and gives the performance objectives and quantitative indexes based on performance-based seismic design. Through the finite element numerical simulation and nonlinear time course analysis of engineering examples, the results show that: the inter-story shear force, inter-story displacement, floor acceleration and other dynamic response indexes are greatly reduced compared with the pre-seismic isolation, which can realize the setup goal of monitoring and controlling the building for uninterrupted use in earthquakes and assuming the function of emergency rescue and command, and effectively verifies the reasonableness of the performance goals set in this paper. At the same time, from the perspective of the whole life cycle of building and equipment maintenance, the use of seismic isolation design can also save costs. The research content of this paper provides a theoretical basis for the feasibility of promoting seismic isolation structures in the field of highway monitoring buildings.

Keywords Highway monitoring buildings · Seismic performance objectives · Finite element numerical simulation · Inter-story displacement · Floor acceleration

Y. Wang (✉)
Office of Asset Management, Tsinghua University, Beijing 100000, China
e-mail: wangyuzheng@tsinghua.edu.cn

© The Author(s) 2025
D. Li and Y. Zhang (eds.), *Advances in Frontier Research on Engineering Structures II*,
Lecture Notes in Civil Engineering 535, https://doi.org/10.1007/978-981-97-6238-5_49

597

1 Introduction

In the construction of highway in China, the traditional seismic design is usually adopted to resist seismic effects in the early stage. According to the "Standard for classification of seismic protection of building constructions", the highway monitoring room located in the area with seismic intensity of 7 degree or higher is defined as a key precautionary building, and should be strengthened according to the requirement of one degree higher than the seismic intensity in the region to strengthen its seismic measures in the seismic design. Therefore, highway monitoring buildings in high intensity areas of China used to require a significant increase in structural cost to strengthen seismic measures.

Due to the frequent occurrence of destructive earthquakes in recent years, under the premise of ensuring the safety of people's lives, maintaining the function of buildings and realizing the rapid recovery of building functions after earthquakes have become the urgent needs of seismic-resistant buildings. Therefore, relevant standards have been issued both at home and abroad, stipulating that new buildings that need to be key fortified in earthquakes need to adopt seismic isolation or damping technology. At the same time, according to the " Standard for seismic resilience assessment of buildings" [1], the seismic safety function, fundamental function and comprehensive function of buildings become the key research direction of post-disaster recovery of buildings. Therefore, for the key defense category of highway buildings, more attention should be paid to seismic performance-based design.

According to the General Specification of Highway Traffic Engineering and Roadside Facilities [2], the highway monitoring system should formulate emergency treatment plans and measures that can be collected in time, dealt with by rapid decision-making, and issue control instructions and implement relief for special traffic safety or emergency events that may occur. Therefore, the highway monitoring hall can be used as an earthquake relief command center, as shown in Fig. 1, to organize emergency relief and post-disaster reconstruction work after an earthquake [3].

The large electronic splicing screen in the post-earthquake monitoring hall is used for real-time monitoring and surveillance of the operation and safety of the lifeline project, and once damaged, it will largely hinder the disaster relief command work. If seismic isolation measures are adopted, the seismic isolation layer will be the first to enter into the energy-consuming state, thus blocking the seismic energy from being fed into the structural components and the important equipment inside, and thus guaranteeing the normal operation of the highway monitoring system during the earthquake. People can fully combine BIM and GIS technology to carry out

Fig. 1 Visualization electronics in highway monitoring center

real-time monitoring, monitoring, and evaluation of the location of the disaster and the degree of damage, and assume the important functions of emergency rescue and command and repair in the first time [4–6]. Therefore, the seismic performance design of highway monitoring halls in high intensity zones is of great significance in the field of earthquake prevention and disaster mitigation for lifeline projects.

2 Structural simulation of an engineering example

The engineering example selected in this paper is a first-class road management center general building in Kunming, which was completed by the author in 2015 according to the traditional seismic design. And the monitoring hall inside carries out systematic monitoring of the whole road. The building is a partial three-story concrete frame structure. The effect diagram and 3D model are shown in Fig. 2.

The seismic isolation design is optimized according to the Provisions on Promotion of Seismic Isolation and Damping Construction Projects in Yunnan Province and the Regulations on Seismic Management of Construction Projects [7]. In this paper, the finite element analysis software SAP2000 is used to establish the seismic isolation model and non-seismic isolation model to carry out the dynamic time course analysis under seismic action, and the main design parameters are organized as shown in Table 1.

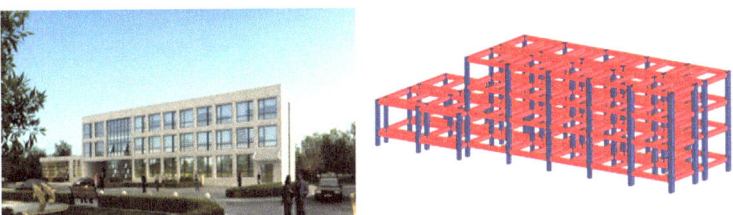

Fig. 2 Architectural rendering and 3D model drawing

Table 1 Natural conditions and parameters

Structure form	Concrete frame structure	Building area	1890 m^2
Height of building	12.55 m	Floors	3
Seismic intensity	8	Design basic seismic acceleration	0.2 g
Basic wind pressure	0.3 kn/m^2	Basic snow pressure	0.3 kn/ m^2
Site type	III	Design seismic grouping	3
Ground roughness	B	Environment category	b

Fig. 3 Diagram of Wen model and schematic of Isolator unit and Gap unit

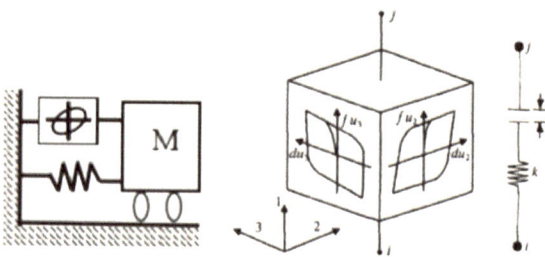

Fig. 4 Bearings plan layout (1-LNR400 2-LRB500)

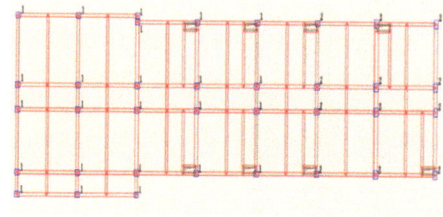

In this paper, Rubber Isolator unit is used to simulate the rubber isolation bearing and Wen unit consisting of Isolator unit and Gap unit in parallel is used to simulate the lead core rubber isolation bearing as shown in Fig. 3.

Through the trial calculation to determine the design of seismic isolation bearing for the frame structure, a seismic isolation bearing is arranged at the bottom of each frame column, and a total of 36 natural rubber seismic isolation bearings LNR400 and lead-core rubber seismic isolation bearings LRB500 are selected, and the arrangement of seismic isolation bearings is shown in Fig. 4, in which the LRB500 is arranged in the direction of Y-axis, which is used for resisting the torsional deformation of the structure.

The two natural waves are the (1) El Centro wave (Imperial Valley earthquake) EW component with original PGA = 2.14 m/s^2, and (2) Anaheim wave (Northridge earthquake) EW component with original PGA = 0.66 m/s^2. The artificial wave was fitted by SeismoArtif software according to the canonical response spectrum based on the basic parameters such as seismic intensity of 8 degrees (0.2 g), Class III site, and Group III earthquake. The waveform time course is shown in Figure 5.

Spectral analysis of the above seismic waves reveals that the waves' spectra are in accordance with the code's spectra at the moment of the characteristic period of the site $T_g = 0.65$ s, and in the range of the fundamental periods (0.58 s, 2.213 s) of the non-seismically isolated and isolated structures (see Fig. 6).

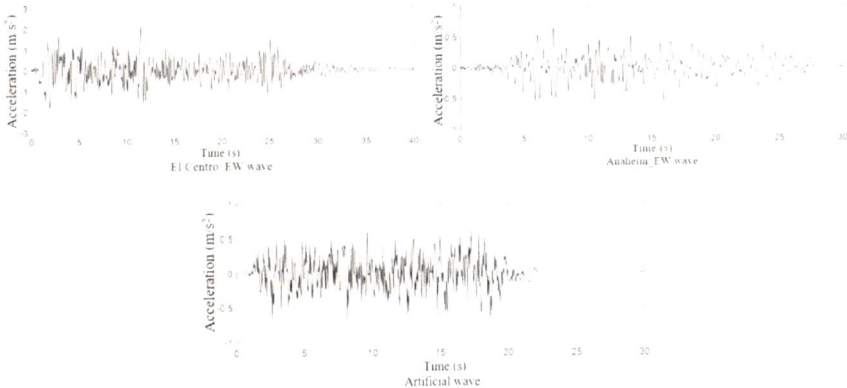

Fig. 5 Seismic wave time course curve for analysis

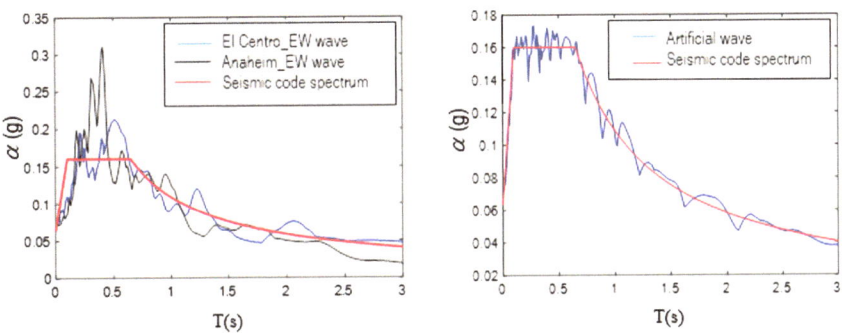

Fig. 6 Comparison of waves' spectrum and code's spectrum under frequent earthquake

3 Dynamic response analysis of the structure

In this paper, the variation rule of the seismic response of the structure is investigated by dynamic time-history analysis of the non-isolated and isolated models under the action of the selected three sets of seismic wave time course curves.

3.1 Seismic performance objectives

In order to realize the seismic performance design for highway monitoring buildings, with reference to the performance objectives of the "Guideline for seismic technology to maintain normal use of buildings in earthquakes (Exposure Draft)" (hereinafter referred to as "Guidelines") for the normal use of Class I buildings, the setup objectives of such buildings are formulated as follows: Ensure that the super-structure, non-structural components and important functional equipment (lifeline project monitoring terminals for visualization of electronic equipment) of seismic isolation buildings are basically intact under moderate earthquake, and that only mild damage occurs under rare earthquakes, and that they can be put back into use after emergency repairs. Among them, with reference [8, 9] to the study on the limit value of inter-story displacement angle in literature [10, 11] and the provisions of floor acceleration in literature [12] and "Guidelines", the proposed values of maximum inter-story displacement angle and maximum floor acceleration limit values for seis-mically isolated structures under seismic isolation and rare earthquakes, respectively, are presented, as shown in Table 2.

At the same time, it is necessary to calculate the horizontal seismic damping coefficientβ, the displacement of the seismic isolation layer, the tensile stress of the rubber bearing, wind stability and other indicators according to the relevant codes.

Among them, according to "Code for seismic design of buildings", the displace-ment value ui of each bearing under rare seismic action should be less than the maximum limit[u$_i$], which is expressed by the formula as follows:

$$u_i[u_i] \tag{1}$$

$$[u_i] = (0.55D, 3\mathrm{T}_r)_{min}$$

where, u_i is the horizontal displacement limit of the ith seismic isolation bearing; D is the effective diameter of the rubber seismic isolation bearing; T_r is the total thickness of the rubber inside the bearing.

Table 2 Seismic performance objectives for seismic isolated framed structures in key precautionary categories

Seismic level	Structural and non-structural members		Functional equipment and items	
	Performance level	Angle of inter-story displacement	Performance level	Angle of inter-story displacement
Moderate earthquake	Substantially undamaged and serviceable	1/550	Substantially undamaged and serviceable	0.2 g
Rare earthquake	Slightly damaged and can be returned to service after emergency repairs	1/250	Slightly damaged and can be returned to service after emergency repairs	0.4 g

3.2 Seismic Response Analysis of Structures under Moderate Earthquakes

Comparison of peak inter-story shear of the structure. The peak acceleration PGA of seismic waves is all normalized to 2 m/s^2 for the time-range analysis under the action of the fortification earthquake. The peak inter-story shear values of each story under the action of three seismic waves with respect to the floor are statistically obtained in Table 3. It can be found that the peak inter-story shear values of each story of the non-seismically isolated structure are much smaller than the peak inter-story shear values of the seismically isolated structure. The maximum value of the peak inter-story shear of both structures occurs in the bottom layer and then decreases uniformly with the increase of the number of layers.

Table 3 Peak inter-story shear under moderate earthquake

Floor	Seismic waves	Seismic isolated structure		Non-isolated structure		Isolated/ Non-isolated	
		X	Y	X	Y	X	Y
3	El Centro wave	634.1	586.3	4794.0	4316.0	0.132	0.136
	Anaheim wave	528.1	496.7	3961.0	3392.0	0.133	0.146
	Artificial wave	610.0	526.9	4318.0	3838.0	0.141	0.137
2	El Centro wave	1428.0	1321.0	8970.0	8833.0	0.159	0.150
	Anaheim wave	1205.0	1149.0	8314.0	7210.0	0.145	0.159
	Artificial wave	1340.0	1193.0	8501.0	8040.0	0.158	0.148
1	El Centro wave	2500.0	2363.0	12520.0	12660.0	0.200	0.187
	Anaheim wave	2044.0	2003.0	11980.0	10610.0	0.171	0.189
	Artificial wave	2197.0	2115.0	11470.0	11670.0	0.192	0.181
Isolation layer	El Centro wave	3041.0	2910.0	–	–	–	–
	Anaheim wave	2433.0	2374.0	–	–	–	–
	Artificial wave	2600.0	2533.0	–	–	–	–

Comparison of maximum inter-story displacements (angles) of the structure.
The peak inter-story displacements for each layer were tabulated and are shown in
Tables 4 and 5.

It can be seen that the average maximum inter-story displacement angle is 1/135
for the non-isolated structure and 1/858<1/550 for the seismic isolated structure,
which satisfies the seismic performance objectives set in this paper. Meanwhile, the
displacement angle of the structure in each story after seismic isolation is reduced to
1/6 of the original one.

The peak displacement of the seismic isolation layer is the horizontal displacement
of the seismic isolation bearing. The maximum horizontal displacement of the seismic
isolation bearing in this project is 104.700 mm under the action of El Centro wave,
while the maximum inter-story displacement of the superstructure is only 4.528 mm,
which is only 4.3% of the seismic isolation layer, and decreases layer by layer. It can
be seen that the deformation of the seismic isolated structure is concentrated in the
seismic isolation layer. Meanwhile, since the inter-story displacement of the same

Table 4 Peak inter-story displacements of non-isolated structure under moderate earthquake

Floor	Peak inter-story displacements/mm						Maximum value of peak inter-story displacements (angles)/mm	
	El Centro wave		Anaheim wave		Artificial wave			
	X	Y	X	Y	X	Y	X	Y
3	13.030	9.565	11.730	7.749	12.190	9.658	13.030 (1/276)	9.658 (1/373)
2	19.930	18.820	19.810	14.720	19.520	18.580	19.930 (1/181)	18.820 (1/191)
1	20.940	31.160	21.560	23.600	20.590	29.670	21.560 (1/195)	31.160 (1/135)

Table 5 Peak inter-story displacements of seismic isolated structure under moderate earthquake

Floor	Peak inter-story displacements/mm						Maximum value of peak inter-story displacements (angles)/mm	
	El Centro wave		Anaheim wave		Artificial wave			
	X	Y	X	Y	X	Y	X	Y
3	2.099	1.224	1.680	1.099	1.781	1.436	2.099 (1/1715)	1.436 (1/2507)
2	3.692	2.415	3.035	2.068	3.118	2.825	3.692 (1/975)	2.825 (1/1274)
1	4.528	4.585	3.693	3.921	3.690	4.893	4.528 (1/928)	4.893 (1/858)
Isolation layer	104.70	91.220	80.860	72.180	85.590	85.040	104.700	91.220

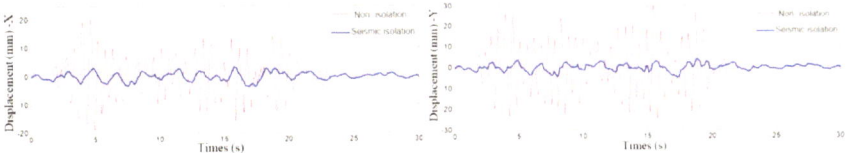

Fig. 7 Time course of layer 1 displacement before and after seismic isolation under moderate earthquake in X and Y direction

position of the non-isolated structure is 20.940 mm, the peak inter-story displacement of the seismic isolated structure is only 21.6% of the non-isolated structure. It is known that the superstructure tends to move horizontally.

Since both structures show maximum inter-story displacements at the first floor, Fig. 7 shows the time course curves of 1st floor inter-story displacements in the X- and Y-directions before and after seismic isolation of the structures under moderate earthquake of artificial wave.

As can be seen in Fig. 7, the curves of the two structures are obviously different, and the amplitude of the inter-story displacements of the seismic isolated structure is obviously reduced and the frequency is slowed down.

The variation of inter-story displacement angle in the Y-direction with respect to the floor of the two structures under artificial waves is plotted as a curve in Fig. 8. It can be seen that the reduction of inter-story displacement angle after seismic isolation is significant.

Comparison of peak structure floor acceleration. The peak floor acceleration of both structures under the moderate earthquake of the three seismic waves was counted and the data in Tables 6 and 7 were obtained. It can be seen that the peak floor acceleration of both structures reaches its maximum value at the top floor. At the same time the acceleration of the seismic isolated structure decreases significantly.

The peak top floor acceleration of the two structures under El Centro waves is plotted as a curve with respect to time, and the time-course curves of top floor acceleration in the X- and Y-directions are obtained, as shown in Fig. 9. As the

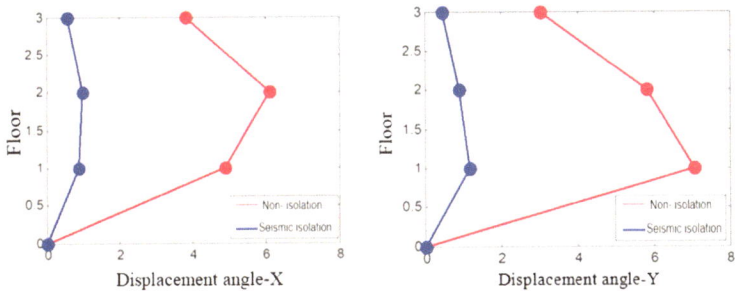

Fig. 8 Inter-story displacement angle curves in X and Y direction before and after seismic isolation under moderate earthquake

Table 6 Peak floor accelerations of non-isolated structure under moderate earthquake

Floor	Peak floor accelerations/m/s²						Maximum value of peak floor accelerations/m/s²	
	El Centro wave		Anaheim wave		Artificial wave			
	X	Y	X	Y	X	Y	X	Y
3	7.598	8.016	6.210	7.199	6.619	8.412	7.598	8.412
2	4.778	6.966	5.098	5.957	4.668	6.849	5.098	6.966
1	3.994	4.739	3.728	3.743	3.509	4.763	3.994	4.763

Table 7 Peak floor accelerations of seismic isolated structure under moderate earthquake

Floor	Peak floor accelerations/m/s²						Maximum value of peak floor accelerations/m/s²	
	El Centro wave		Anaheim wave		Artificial wave			
	X	Y	X	Y	X	Y	X	Y
3	1.130	1.113	0.878	1.064	0.932	1.321	1.130	1.321
2	1.013	0.928	0.835	0.784	0.947	1.114	1.013	1.114
1	1.021	0.860	0.777	0.751	0.928	1.093	1.021	1.093
Isolation layer	1.013	0.879	0.742	0.931	0.911	1.151	1.013	1.013

maximum value of the peak acceleration of the floor of the seismic isolated structure is $1.321 \, \text{m/s}^2 = 0.13 \, \text{g} < 0.2 \, \text{g}$, which is in line with the seismic performance objectives set in this paper, it can guarantee that the monitoring center is basically intact under moderate earthquake, and that the important instruments and equipment inside it are not subjected to functional damage and can be used uninterruptedly.

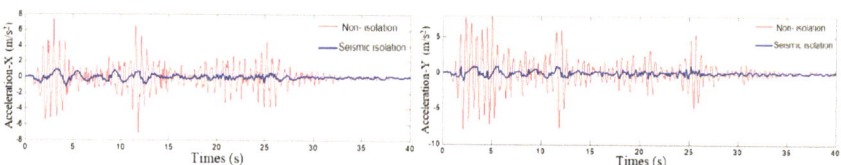

Fig. 9 Time course curve of top acceleration before and after seismic isolation under moderate earthquake in X and Y direction

3.3   Seismic Response Analysis of Structures under Rare Earthquakes

In this section, the PGA is normalized to 4 m/s^2 in the calculations for the time course analysis under rare earthquake, in which the peak inter-story displacements at each story of the two structures are counted, as shown in Tables 8 and 9.

The maximum displacement angle of the seismic isolated structure is 1/397<1/250, which meets the seismic performance objectives set in this paper. The seismic isolation displacement angle is approximately reduced to 1/6 of that before seismic isolation. The maximum horizontal displacement of the seismic isolation bearing under rare earthquake is 215.000 mm under the action of El Centro wave, while the maximum inter-story displacement of its superstructure is only 9.376 mm. At the same time, since the total thickness of the rubber layer of the seismic isolation bearing LNR400 is 82 mm, then, according to the "Code for seismic design of buildings", 215 mm<(0.55D,3T) min = 220 mm, which meets the performance target.

Figure 10 shows the time course curves of 1st floor inter-story displacements in the X- and Y-directions before and after seismic isolation of the structure, respectively,

Table 8 Peak inter-story displacements of non-isolated structure under rare earthquake

| Floor | Peak inter-story displacements/mm | | | | | | Maximum value of peak inter-story displacements (angles)/mm | |
| | El Centro wave | | Anaheim wave | | Artificial wave | | | |
	X	Y	X	Y	X	Y	X	Y
3	26.060	19.130	23.460	15.660	24.390	19.320	26.060(1/138)	19.320(1/186)
2	39.870	37.630	39.620	29.440	39.050	37.160	39.870(1/90)	37.630(1/96)
1	41.880	62.330	43.120	47.190	41.180	59.340	43.120(1/97)	62.330(1/67)

Table 9 Peak inter-story displacements of seismic isolated structure under rare earthquake

| Floor | Peak inter-story displacements/mm | | | | | | Maximum value of peak inter-story displacements (angles)/mm | |
| | El Centro wave | | Anaheim wave | | Artificial wave | | | |
	X	Y	X	Y	X	Y	X	Y
3	4.370	2.600	3.694	2.246	3.652	3.023	4.370 (1/824)	3.023 (1/1191)
2	7.778	5.366	6.710	4.613	6.560	6.068	7.778 (1/463)	6.068 (1/593)
1	9.376	9.952	8.237	8.689	8.173	10.590	9.376 (1/448)	10.590 (1/397)
Isolation layer	215.000	208.600	194.500	183.200	197.500	212.300	215.000	212.300

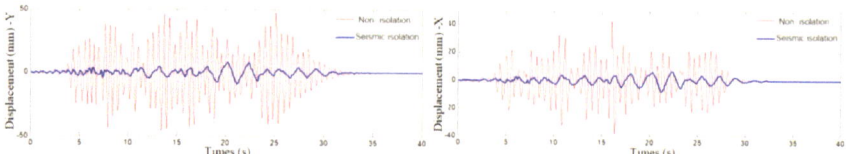

Fig. 10 Time course of layer 1 displacement before and after seismic isolation under rare earthquake in X and Y direction

Table 10 Quantity statistics

Structure type	Concrete C30 (m^3)	Reinforcement steel (kg)
Non-seismic isolation structure	557.48	80985.32
Seismic isolation structure	498.21	61365.43

under rare earthquake action of the Anaheim wave. The inter-story displacements of the structure after seismic isolation are reduced to about 1/5 of the before.

3.4 Statistics of structural works

After the dynamic time course analysis, it can be seen that after the structure is reasonably designed for seismic isolation, the seismic measures are reduced by one degree and controlled by 8 degree, and therefore the frame seismic class is reduced from Class I to Class II, which brings about a direct reduction in the cross-section dimensions of the superstructure members as well as in the amount of reinforcing bars, and the statistics of the amount of work is shown in Table 10. The total amount of concrete in the seismically isolated structure is reduced by about 11% and the total amount of reinforcing bars is reduced by about 24% as compared with before. In addition, considering the design of the piers at the seismic isolation level, it is expected that the amount of concrete in the seismic isolation structure will be reduced by about 9%, and the amount of reinforcing steel will be reduced by about 20%.

3.5 Bearing energy consumption

Figure 11 shows the X-direction restoring force curves of a lead-core rubber bearing at the mid-span location of the side span under the effect of EL-wave medium and large earthquakes. Since the hysteresis curves are fuller in shape, it can be seen that its energy dissipation is good under the action of the medium and large earthquakes.

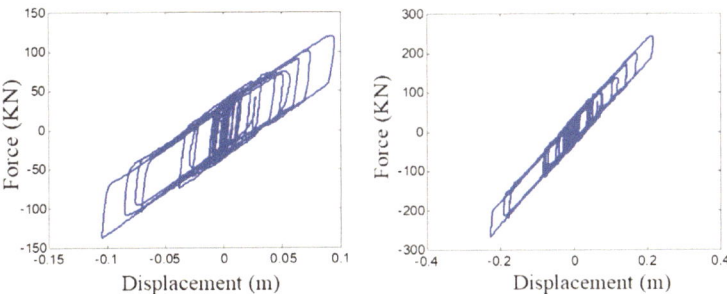

Fig. 11 Bearing restoring force-displacement curves under moderate and rare earth-quakes in El Centro wave

4 Conclusion

4.1 Results and Discussion

This paper takes the highway monitoring center located in Kunming as the research object, and analyzes the dynamic response of the non-seismic isolation model and the seismic isolation model by SAP2000 respectively, and summarizes the main conclusions drawn from the change rules of the structural inter-story shear, inter-story displacement angle, seismic isolation layer displacement, floor acceleration and other indexes as follows:

The inter-story shear force, inter-story displacement (angle), and floor acceleration of the seismic isolation structure are reduced to about 1/6–1/5 of those of the non-isolated structure under the seismic protection earthquake, which has a significant damping effect. Among them, the displacement angle and acceleration of the seismic isolation structure are 1/858 and 0.13g, respectively, which are in line with the limit values of the performance objectives set in this paper. It can be seen under moderate earthquake monitoring center structure is basically intact, the indoor power electronic splicing screen equipment can be used without interruption, directly into the emergency command.

Under rare earthquake, the inter-story shear force, inter-story displacement angle and floor acceleration of the seismic isolation structure are reduced to 1/6–1/4 of the pre-seismic isolation, and the seismic isolation story displacement, inter-story displacement angle and acceleration are 215mm, 1/397 and 0.25g, respectively, which are all in line with the limit values of the performance objectives set in this paper. It can be seen that the seismic isolation highway monitoring building, under rare earthquakes can be immediately restored to the use of the function after repair, people can be based on the real-time data and information in the equipment, tunnels, bridges and other lifeline projects for structural safety monitoring and timely implementation of rescue command.

Due to the seismic isolation structure, the amount of reinforcement and concrete work is reduced by 9% and 20% respectively. The increased cost of seismic isolation bearings and related materials and equipment is much lower than the reduced cost of concrete and steel reinforcement after seismic isolation. In addition, the whole-life maintenance cost of the seismic isolation monitoring center is significantly reduced because the structural components, non-structural components, and internal functional equipment remain intact or slightly damaged after the earthquake.

Since there is no in-depth study on the seismic performance objectives of highway monitoring buildings at home and abroad, this paper comprehensively refers to the existing codes and standards for the classification level of highway monitoring centers and the corresponding target level when setting the performance objectives. Among them, the "Regulations on Seismic Management of Construction Projects" [7] classifies the "emergency command center" as a Class I building, which is higher than the Class B building classified by the "Classification Standard for Seismic Protection".

And the monitoring center usually takes on the function of local emergency command center building according to the owner's demand, so the author thinks that when formulating the performance target, it should not be lower than the "Guidelines" for the I type of building's fortification standard, i.e., structural and non-structural components are basically intact under moderate earthquake, and the instrument and equipment work normally, and they can be continued to be used without any repairs. Under rare earthquake, the structural components and instrumentation are mildly damaged and can be used after simple repairs. This goal is basically consistent with the defense goal of high intensity area highway monitoring center. At the same time, the performance target in Appendix A of "Standard for seismic isolation design of building" [13] only specifies the macroscopic damage degree of building components, and lacks the description of the performance level of instrument and equipment, so it is not possible to accurately describe the fortification target of the highway monitoring center only by referring to this standard.

This paper refers to the "Standard for seismic resilience assessment of buildings" [1] and literature [14] for acceleration-sensitive non-structural components of the research content, the monitoring center within the large electronic splicing screen is classified as acceleration-sensitive equipment and instruments, in the setting of the performance objectives focus on acceleration limits. Finally, in combination with the actual needs of the building, in order to guarantee the normal use of the functional instruments and equipment of the monitoring center after the earthquake, with reference to the Turkish literature [12] on the floor acceleration limits of hospitals under the seismic protection, as well as the acceleration limits of the "Guidelines" on the Emergency Command Center, this paper sets the floor acceleration limits under moderate and rare earthquake as 0.2g and 0.4g.

As for the limit value of inter-story displacement angle, referring to the research of Wang Wei and others on seismic isolation and strengthening of buildings [11], and the research of Deng Xuesong and others [10] on the description of the performance state and limit value of the building in the post-earthquake "full operation" and "basic operation", taking into account that the existing highway monitoring centers in some high-severity areas do not take seismic isolation measures, and in order to easily conform to the "Standard for seismic resilience assessment of buildings" [1] for the functional recovery of the demand for the time after the earthquake, this paper sets the limit value of the displacement angle of "1/250 and 1/250" for the highway monitoring centers that have been designed and strengthened for seismic isolation under moderate and rare earthquake.

In recent years, with the in-depth research on the combination of intelligent transportation and resilient city construction technology [14, 15], the importance of highway monitoring centers, which can assist lifeline projects to enhance structural safety and quickly return to use after earthquakes, has increased unprecedentedly in the post-earthquake emergency rescue command work. However, since there is no clear index as reference for the seismic performance-based design of highway monitoring buildings in China in the past, this paper comprehensively refers to the data from domestic and international codes and literatures, and combines the functional characteristics of highway monitoring centers in high intensity zones with the needs of the owners, to supplement the study of the seismic performance objectives of this type of buildings. And through structural simulation and data analysis of engineering examples, the seismic reduction effect of the structure is verified. Meanwhile, it can be seen that the highway monitoring buildings in high intensity zones can achieve the established seismic defense objectives by adopting the performance-based seismic isolation design method. And the performance objectives set in this paper can be used as a reference index for future seismic performance-based design of buildings in this area.

4.2 Outlook

As China's "visible, measurable, controllable" requirements for traffic management continue to improve, especially in the direction of data visualization, such as the use of GIS, BIM platform, multi-source heterogeneous data integration functions, real-time view of wear and tear, and other functions [16], will become an important development direction of the future traffic monitoring system. Therefore, the vibration isolation index for monitoring, monitoring equipment and electronic equipment instruments still need to be studied in depth. For example, one can refer to other acceleration-sensitive non-structural components in the "Standard for seismic resilience assessment of buildings" of the relevant indicators, to further study the large electronic splicing screen equipment mentioned in this paper damage state rating, and improve its engineering requirements parameters.

In conclusion, the research results of this thesis provide a theoretical basis for the promotion of seismic isolation technology in the field of highway construction, and provide reference data for the improvement of seismic performance-based design methods for lifeline projects.

References

1. GB/T 38591-2020, Standard for seismic resilience assessment of buildings.
2. Ceneral Specification of Freeway Traffic Engineering and Roadside Facilities: JTG D80-2006.
3. Li, Li., Liang, Zou, & Xinghua, Luo. (2022). Planning and construction strategies for seismic resilient cities. *City and Disaster Reduction, 05*, 33–38.
4. Lei, Z., Fei, L., Guocai, J., et al. (2015). Research and design of GIS-based urban earthquake and disaster prevention planning system. *Science and Technology Outlook, 25*(28), 152.
5. Rui, C., Chunmei, H., Jian, Z., et al. (2001). Design of highway emergency rescue command software system. *Transportation and Computer, 05*, 26–29.
6. Mingzhu, Li., Yang, Li., & Liang, Zou. (2020). Research on the application of visualization of structural health monitoring of cable-stayed bridges based on BIM. *Construction Safety, 35*(12), 72–75.
7. Regulations on seismic management of construction works. Bulletin of the State Council of the People's Republic of China, 2021(23):13-19.
8. Liangkun, W., Satish, N., Weixing, S., et al. (2022). Seismic performance improvement of base-isolated structures using a semi-active tuned mass damper. *Engineering Structures, 271*, 114963.
9. TERDC. 2018 Revision of the Turkish seismic design code. The Disaster and Emergency Management Presidency of Turkey. Ankara; 2018.
10. Xuesong, D., Yongheng, G., & Yun, Z. (2008). Research on the setup level and performance target of base seismic isolation structural performance. *Journal of Guangzhou University (Natural Science Edition), 05*, 84–88.
11. Wei, W., Xia, B., & Aiqun, L. (2017). Performance-based design method for seismic isolation strengthening of existing frame structures. *Building Science, 33*(7), 99–103.
12. Erdik, M. (2018). Seismic isolation code developments and significant applications in Turkey. *Soil dynamics and earthquake engineering, 115*, 413–437.
13. GB/T 51408-2021, Standard for seismic isolation design of building.
14. Niu, Ben. (2021). Analysis of intelligent traffic platform for highway monitoring center. *Low Carbon World, 11*(08), 180–181. https://doi.org/10.16844/j.cnki.cn10-1007/tk.2021.08.086
15. Lu, X. Z., Zeng, X., Xu, Z., et al. (2017). Challenges of building earthquake resilient cities. *Cities and Disaster Reduction, 04*, 29–34.
16. Xiuwen, Chi, Zhao, Xu., Hao, Wu., et al. (2011). Research on key technology of three-dimensional GIS platform for bridge monitoring data. *Journal of Wuhan University of Technology, 33*(04), 99–103.